Planning and Optimization of 3G & 4G Wireless Networks

RIVER PUBLISHERS SERIES IN COMMUNICATIONS

Volume 8

This series focuses on communications science and technology. This includes the theory and use of systems involving all terminals, computers, and information processors; wired and wireless networks; and network layouts, procontentsols, architectures, and implementations.

Furthermore, developments toward new market demands in systems, products, and technologies such as personal communications services, multimedia systems, enterprise networks, and optical communications systems.

- Wireless Communications
- Networks
- Security
- Antennas & Propagation
- Microwaves
- Software Defined Radio

For a list of other books in this series, see final page.

Planning and Optimization of 3G & 4G Wireless Networks

J. I. Agbinya

University of Technology
Sydney
Australia

French South African Technical Institute
in Electronics
Pretoria
South Africa

River Publishers

Aalborg

Published, sold and distributed by:
River Publishers
PO box 1657
Algade 42
9000 Aalborg
Denmark
Tel.: +4536953197

ISBN: 978-87-92329-24-0

About the Editor

Johnson I Agbinya received his PhD in 1994 from La Trobe University in Melbourne in Electronic Communication engineering on ground probing radar system, MSc (Electronic Communications) from the University of Strathclyde, Glasgow, Scotland and BSc from the Obafemi Awolowo University, Nigeria. He is currently a senior lecturer at the University of Technology Sydney and an Adjunct Professor of Communications at the department of Computer Science, the University of the Western Cape and the Alcatel-Lucent Professor of Communications at French South African Institute of Technology (F'SATI) at Tshwane University of Technology, Pretoria, South Africa.

Prior to these, he was Senior Research Scientist at CSIRO Telecommunications and Industrial Physics, Sydney Australia from 1993 to 2000 in the areas of Biometrics and Signal Processing. Between 2000 and 2003 he was Principal Engineer, Vodafone Australia responsible for managing Vodafone's Industrial Research in Mobile Communications.

His current research interests are in short range wireless communications, mobile content development, vehicular networks and networks in uncovered areas, WiMAX, sensor networks, sensor web and biometric security systems. He is a rated researcher by the National Research Foundation (NRF) of South Africa. He is also a core member of Centre for Real-Time Information Networks (CRIN) at UTS and the Telkom Centre of Excellences in Tshwane University of Technology, Pretoria South Africa, French South African Institute of Technology (F'SATI) and also CoE for IP Communications at UWC South Africa.

He is the current Editor-in-Chief of the African Journal of Information and Communication Technology (AJICT) and its pioneer. He has also editor several conference proceedings. He is also the pioneer of and conference chair for AusWireless, BroadCom and ICCB. He has also been involved in the technical and organsing committees of several international conferences, such as AusWireless (2006, 2007), BroadCom 2008 and 2009, ICMB, ICT Africa, ICADST and many other conferences.

Acknowledgment

This book is the result of persistent effort of many researchers worldwide including some of my current post graduate students on topics related to their research. The review of the chapters involved key experts worldwide. Some of the chapters are derived from papers originally reviewed for BroadCom 2008 in very short formats and have been extensively extended in greater details to provide great details compared with their initial short five page limits. Hence the chapters have been reviewed again to ensure the quality of the new materials. We acknowledge the expert reviewers for their excellent work. It is impossible to mention all of them by name. I acknowledge these numerous efforts which have made this book timely and possible.

Preface

The information and communication technology is prone to rapid changes in technology and obsolescence of existing systems. Most of the current developments are in broadband and short range communications and follows Moore's law. This book covers major key new technologies which support broadband communication and specifically the beyond 3G (B3G) and fourth generation (4G) wireless networks. All the chapters have provided in depth analysis of the new technologies in a manner that makes the book very useful for undergraduate and postgraduate text material in wireless networks, network planning and optimization.

Objectives and Prerequisites

Planning of wireless networks is often not as straight forward as planning fixed networks. There are major differences and approaches that need to be adopted to design an optimal wireless network. How to design and plan such optimal networks is presented in this book. It engages the reader with ideas on how to plan different types of wireless broadband networks. The book has discussed in great details the planning of emerging wireless IP networks such as B3G and 4G technologies including

- Planning of 3G and high speed packet access networks
- Planning of WiMAX networks and the technologies which support it
- Planning of personal area networks
- Planning of wireless mesh networks
- Planning of wireless sensor networks
- Spectrum Sensing, a much needed technology for cognitive radio
- OFDM and OFDMA
- Multiple input multiple output (MIMO) technology

- Optimisation and synchronisation of 3G and 4G networks
- Cross Layer performance modelling and
- Topology control including mobility
- LTE and WiMAX link budgets are included

Modern and future wireless communication networks depend on these technologies and they are provided for the aspiring and current network architects, researchers and operators planning to deploy LTE, WiMAX and future 4G networks. The book does not assume basic understanding and overview of these technologies but rather logically takes the reader through what should be learned through algorithms, functions and examples. No assumptions are made in terms of the fundamental mathematics involved and neither is the mathematics used highly involving. The text is well suited for use as subject material in undergraduate, post graduate degree levels. It provides one of the most comprehensive source for industry-based training on these emerging networks.

Organisation of Chapters

There are 17 chapters in the book and they cover a wide range of network planning variations from 3G to next fourth generation 4G networks. The book is introduced with a chapter on high speed packet access the logical evolution of 3G networks based on wideband CDMA (WCDMA).

MIMO technology has had a tremendous impact on the evolution of 3G networks, specifically through its use in the so-called high speed packet access (HSPA) networks. It is used in HSPA to increase the access capacity beyond the raw 2Mbps in UMTS to more than 8Mbps. Chapter 1 of the book explains HSPA and how to plan a network based on the technology. The chapter gives an outline of WCDMA based UMTS (Release'99) planning process, with a comparison insight with HSPA planning procedure. Capacity and coverage limited HSPA dimensioning are discussed and compared. "*Fair Resource*" and "*Fair Throughput*" scheduling techniques are distinguished as case studies for dimensioning. The authors show also an introduced dimensioning principle based on "*Enhanced Fair Throughput*" scheduling technique. Then we will extend dimensioning model to multiple services case. Next they also present dimensioning procedure by taking into account shadowing propagation effect. In the application exercises, practical examples are given in tutorial style about how special tables can help to dimension a HSPA based UMTS network including shadowing effect and taking into account basic planning parameters such as required coverage probability, shadowing standard-deviation, service bit rate and terminal category. The chapter also includes abacuses generated for HSPA coverage limited dimensioning as well as methodological examples of their application use. The chapter ends with different HSPA planning principles and the comparison of each technique with the other through representative figures.

There are two contending technologies for 4G networks in the form of WiMAX and LTE. The IEEE 802.16 standard termed the worldwide interoperability of microwave access (WiMAX) and long term evolution (LTE) have both received mixed support worldwide. While LTE is a direct evolution of UMTS, WiMAX is a fresh air interface and thus requires completely new infrastructure. Chapters 2, 3 and 4 are related and cover the technologies used for implementing WiMAX, the planning method (Chapter 3) and the link budgets for both WiMAX and LTE in Chapter 4. They should be read together. OFDM, OFDMA and MIMO techniques are well captured in Chapter 2 and Chapter 3 discusses in great details the WiMAX PHY (fixed and mobile) and the planning techniques. Chapter 4 overviews a new propagation model the so-called Stanford University Interim (SUI) model which has been adopted by the WiMAX Forum for studying the effects of the environment on WiMAX communications.

As the industry bridges the gap between long range communication and short range communication with new air interfaces, the very short range communication, specifically personal area networks and body area networks become essential as the next frontiers of networking. Bluetooth and Zigbee based PAN are discussed in Chapters 5 and 6. The standards for PAN are compared as well.

Wireless mesh networks (WMN) is one of the most promising wireless communication designs. This design has many characteristics which will impact on the future of wireless communication; it is reliable, flexible, efficient, and cheaper to install and maintain. It is also a community-owned infrastructure which makes it possible to be extended to rural and remote areas to provide a broadband access where the cost of traditional infrastructure networks are not convenient for economically and technically reasons.

WMN consists of many nodes connected together using radio frequency (RF) in mesh topology. These nodes can be classified into two types: WMN routers and WMN clients. WMN routers form the backbone of the network. These nodes are configured automatically and reconfigured dynamically which makes the WMN to be a self-healing too. This means that the connectivity is established and maintained by the network itself. WMN routers have a conventionally gateway/bridge functionality besides additional routing supports for mesh networks, whereas the WMN client are an access point that connects the end user to the network. The planning of wireless Mesh networks is discussed in Chapter 7. This is based on both practical mesh networks and laboratory findings as the state of the art as of the time of writing. Both fully meshed and sparsely meshed networks are demonstrated and planned.

In recent years, cognitive radios have been introduced as a new paradigm for enabling much higher spectrum utilization, providing more personal and reliable radio services, reducing harmful interference, and facilitating the inter-working or convergence of different wireless networks. In Chapter 8 a study of intelligent open spectrum sharing using cognitive radios is given. This includes the discussion of the existing spectrum regulations and its challenges for spectrum scarcity. The advanced spectrum sharing and sensing is also discussed. In this context the cooperative spectrum sensing has been proposed to overcome the problem associated with the local sensing e.g. the hidden node problem due to noise uncertainty, fading, and shadowing. The chapter also presents a hard decision auto-correction reporting scheme that directly correct the error in the reported bit, and further minimizes the average number of the reporting bits by allowing only the user with a detection information-binary decision 1-to report its result. The sensing performance is investigated and the numerical result have great decrease in reporting bit without affecting the sensing performance.

From Chapter 9 and beyond, major technologies related to synchronization and optimization of emerging evolved 3G and 4G networks are explained in great details. The technologies include cooperative networking (Chapter 9), wireless sensor network planning (Chapter 10), mobility, cross-layer design, optimization of CDMA networks, interference suppression, synchronization and multimedia applications for 3G and 4G networks. The objective is the place in the hands of a telecommunication practitioner the knowledge base to make informed decisions on new networks and also for the network developer to be grounded in the required basics and principles.

Contents

Acknowledgment **vii**

Preface **ix**

1 Radio Network Capacity and Coverage Planning Methodology and Dimensioning Procedures of HSPA Based B3G Networks **1**

Anis Masmoudi and Tarek Bejaoui

1.1 Dimensioning difference between Release 99 3G UMTS and HSPA B3G 2
1.2 Coverage or Capacity Limited Dimensioning 11
1.3 HSPA Dimensioning Methodology with Shadowing Effect 17
1.4 Practical Examples and Exercises About Coverage Limited Dimensioning Methodology 19
1.5 HSPA Planning Reference Techniques 25
1.6 Summary of Chapter 29
References 31

2 Part I Overview of WiMAX Cellular Technology **33**

Mehrnoush Masihpour and Johnson I Agbinya

2.1 Overview of WiMAX 33
2.2 WiMAX Network Architecture 35
2.3 Mobile WiMAX 39
2.4 OFDMA (Orthogonal Frequency Division Multiple Access) 57

2.5 Adaptive Antenna 61
2.6 MIMO (Multiple Input-Multiple Output) Technique 63
References 64

3 Part II Planning and Dimensioning of WiMAX and LTE Networks 67

Johnson I Agbinya and Mehrnoush Masihpour

3.1 WiMAX Network Planning Parameters 67
3.2 OFDM Symbols 68
3.3 WiMAX Frame 71
3.4 Multicarrier Modulation schemes 72
3.5 General Considerations for WiMAX Network Planning 76
3.6 WiMAX Cells: Cellular Technology 77
3.7 Dimensioning for Service 87
3.8 Dimensioning for Coverage 89
3.9 Dimensioning for Capacity and Frequency Planning 95
References 102

4 Part III WiMAX and LTE Link Budget 105

Johnson I Agbinya and Mehrnoush Masihpour

4.1 Propagation Regimes 105
4.2 Transmitter 108
4.3 Signal Fading 109
4.4 Propagation Regimes and Models 116
4.5 Link Budget 119
4.6 Walfisch–Ikegami Model 123
4.7 Erceg Model 126
4.8 Path Loss Exponents 129
References 135

5 Planning of Personal Area Networks (I) 137

N. Sudha Bhuvaneswari and S. Sujatha

5.1 An Overview of PAN 137

5.2	Technology Issues Related to PAN	139
5.3	Commercialization of PAN	145
5.4	Branded PAN Products	152
5.5	WPAN Introduction	160
5.6	Future of PAN	169
	References	174

6 Planning of Personal Area Network (II) **177**

Nithya Thilak and Johnson I Agbinya

6.1	Introduction to PAN	177
6.2	Bluetooth Concept	180
6.3	Summary	198
6.4	Review Questions	198
	References	199

7 Planning of Wireless Mesh Networks **201**

Mohammad Al-Hattab and Johnson I Agbinya

7.1	Introduction	201
7.2	WMN Definition	202
7.3	Overcoming Distance Limitation	203
7.4	Advantages of Wireless Mesh Networks	204
7.5	Planning the WMN	205
7.6	Network Topology	209
7.7	Network Design	211
7.8	Routing in WMN and Cross Layers	215
	References	221

8 Spectrum Sensing and Sharing for Cognitive Radio **223**

Rania A. Mokhtar and Rashid A. Saeed

8.1	Introduction	224
8.2	Radio Spectrum Regulation	228
8.3	Spectrum Sharing and Flexible Spectrum Access	233
8.4	Conclusions	260
	References	261

9 Cooperative Networks: Optimization Through Game Theory and Mobility Prediction **265**

Dimitris E. Charilas, Theodore B. Anagnostopoulos, Athanasios D. Panagopoulos and Christos B. Anagnostopoulos

9.1	Planning Cooperative Networks	266
9.2	Incentive Mechanisms	269
9.3	Cross-Layer Optimization	271
9.4	Cooperative Relaying	274
9.5	Introduction to Game Theory	276
9.6	Optimization of Wireless Networks Through Game Theory	278
9.7	Optimization Through Mobility Prediction	282
9.8	Context Representation	285
9.9	Location Prediction Algorithms	289
9.10	Assessment of Mobility Prediction Algorithms	293
9.11	Summary of Chapter	296
	References	301

10 On the Sensor Placement Problem in Directional Wireless Sensor Networks **305**

Yahya Osais, Marc St-Hilaire and F. Richard Yu

10.1	Introduction	305
10.2	Background	308
10.3	Related Work	310
10.4	Problem Statement	314
10.5	Problem Formulations	315
10.6	Numerical Results	324
10.7	Summary and Further Research	327
10.8	Problems	328
	References	329

11 Layered Architecture for Mobility Models — Lemma **331**

Alexander Pelov and Thomas Noel

11.1	Mobility Models	331

11.2 Architecture Description 334
11.3 Existing Layers 340
11.4 Strategy Aggregation, Hybrid and Group Mobility Models 346
11.5 Summary of Chapter 349
References 350

12 Optimisation of CDMA-Based Radio Networks 353

Moses Ekpenyong and Joseph Isabona

12.1 3G Network Optimisation: Background of Issues 353
12.2 Topology Optimisation 354
12.3 Automated Optimisation 376
12.4 Network Optimisation 384
12.5 Summary of Chapter 404
References 406

13 Narrowband Interference Suppression in Wireless OFDM Systems 409

Georgi Iliev, Zlatka Nikolova, Miglen Ovtcharov and Vladimir Poulkov

13.1 OFDM Principles 409
13.2 OFDM Implementation in Multiple Access Schemes 419
13.3 NBI Suppression Methods 425
13.4 Summary of Chapter 441
References 442

14 Synchronization 445

Ramin Vali, Stevan Berber and Sing Kiong Nguang

14.1 Introduction 446
14.2 Acquisition 447
14.3 Tracking 466
14.4 Overall System Performance 477
References 478

15 Topology Control for Effective Power Efficiency in Wireless Mesh Networks **481**

Felix O. Aron, Anish Kurien and Yskandar Hamam

15.1 Introduction 481
15.2 Topology Control 483
15.3 Related Work 485
15.4 Network Model 486
15.5 Proposed Algorithm 488
15.6 Simulation Results 492
15.7 Conclusion 498
References 498

16 Cross-Layer Performance Modeling and Control of Wireless Channels **501**

Dmitri Moltchanov

16.1 Introduction 502
16.2 Adaptation Mechanisms 503
16.3 Cross-layer Modeling Principles 515
16.4 Cross-layer Design of Wireless Channels 538
16.5 The Performance Control System: An Example 543
16.6 Summary of the Chapter 549
References 552

17 Open Access to Resource Management in Multimedia Networks **555**

Evelina Pencheva and Ivaylo Atanasov

17.1 Internet Protocol Multimedia Subsystem 555
17.2 Quality of Service in IMS 559
17.3 Service Control 566
17.4 Open Access to Network Functionality 569
17.5 Parlay X Web Services 584
17.6 Summary of Chapter 590
References 597

Index **599**

1

Radio Network Capacity and Coverage Planning Methodology and Dimensioning Procedures of HSPA Based B3G Networks

Anis Masmoudi*, Tarek Bejaoui†

**University of Sfax, Tunisia*
†*University of Carthage, Tunisia*

The fast evolution of 3G networks to Beyond 3G (B3G) pushes mobile operators to dedicate a particular importance for B3G planning. Radio interface planning and dimensioning is especially crucial since it remains a bottleneck for the whole network deployment. In this chapter we investigate planning and dimensioning procedures of HSPA Based B3G networks. This includes the radio network capacity and coverage issues. First we give an outline of WCDMA based UMTS (Release'99) planning process, with a comparison insight with HSPA planning procedure. Capacity and coverage limited HSPA dimensioning are discussed and compared. "*Fair Resource*" and "*Fair Throughput*" scheduling techniques are distinguished as case studies for dimensioning. We show also an introduced dimensioning principle based on "*Enhanced Fair Throughput*" scheduling technique. Then we will extend dimensioning model to multiple services case. We will also present dimensioning procedure by taking into account shadowing propagation effect. As application exercises, we also show practical examples in tutorial style about how special tables can help to dimension a HSPA based UMTS network including shadowing effect and taking into account basic planning parameters such as required coverage probability, shadowing standard-deviation, service bit rate and terminal category. The chapter also includes abacuses generated for HSPA coverage limited dimensioning as well as methodological examples of their application use. The chapter ends with different HSPA planning principles and their comparison each technique with the other through representative figures.

1.1 Dimensioning difference between Release 99 3G UMTS and HSPA B3G

Fixed point iterative initial dimensioning method in UMTS systems (Rel'99)

Capacity and coverage in WCDMA based UMTS systems are mutually dependent due to cell breathing caused mainly by Power Control (PC) in Rel'99 basic UMTS systems. That's why radio link budget shouldn't be static but dynamic. In fact, radio dimensioning using a classical link budget doesn't provide enough accurate results since it gives either an under-dimensioning (inefficient) or an over-dimensioning (costly). So, the planner should apply a balanced result induced by cell breathing.

In this paragraph, we detail the "Fixed Point Iterative" (FPI) method to dimension each of the links (Uplink: UL and Downlink: DL) allowing determining the balanced point of the cell breathing mechanism, thus an accurate initial dimensioning. This method receives in entry a traffic density (users per service) whose prediction should be planned by phases while taking into account the required margins.

In the FPI method applied on the UL, the mobiles transmitted powers are assumed to be set to the maximum value accepted by terminal so as to determine the maximum allowed cell size, and thus minimizing the number of required sites for dimensioning. So the users' distribution is known according to services types (percentages), and the traffic density is assumed to be uniform in the studied service area (instead of giving link losses between the mobile and each node B). The iterative method below is based on cell breathing and the dependence between coverage and cell load. Since this method is based on link budget (dynamic), the soft-handover (SHO) is taken into account through gain. Moreover, an additional over-dimensioning should be performed in order to take into account the overlapping area between cells including mobiles in SHO. This over-dimensioning can be substituted by the consideration of the total number of connections $N \cdot (1 + C_{SHO})$ in the load factor, where

N is the number of mobiles without SHO, and C_{SHO} is the *SHO Overhead* (fraction of the number of mobiles with SHO versus that without SHO consideration).

In the UL initial dimensioning, the impact of traffic on the cell range is through *Noise Rise* parameter which limits Maximum Allowed Path Loss (MAPL) in the link budget. This parameter (*Noise Rise*) translates the maximum interference supported by the network (additionally to the thermal and receiver noise). The link budget takes into account mobile velocity (in the fast fading margin), service bit rate (in the processing gain), the required QoS (E_b/N_0 ratio), and macro-diversity in addition of the other classic loss and gain parameters.

The generic CDMA equation (E_b/N_0 expression) allows establishing the load factor η (for both links) expressed in (1.1) that can also be extracted from link budget:

$$\eta_{UL} = (1+f)\sum_{j=1}^{N} \frac{1}{1+\frac{W}{(E_b/N_0)_{UL,j} R_j v_j}} \tag{1.1}$$

where f is the ratio between extracellular and intracellular interference, N is the number of users in the cell, $(E_b/N_0)_{UL,j}$, R_j and v_j are respectively the required energy per bit to noise ratio, bit rate and activity factor of the user j. The *Noise Rise* isn't other than $\frac{1}{1-\eta_{UL}}$.

A CDMA access network is typically dimensioned with a maximum load factor of 50% for example (Interference level equal to that of the noise) but not exceeding 75% (*Noise Rise* of 6 dB), else there is a risk of instability.

Expression (1.1) of UL load factor can be grouped in services as follows:

$$\eta_{UL} = \sum_{j} \frac{N_j}{1+\frac{W}{R_j (E_b/N_0)_{UL,j}}}(1+f) \tag{1.2}$$

where N_j is the number of users of the service j. The users distribution according to services is assumed such that χ_j is the fraction of use of a service j defined as follows:

$$N_j = \chi_j N \tag{1.3}$$

where N is the total number of users in the service area. So (1.2) becomes:

$$\eta_{UL} = \left(\sum_j \frac{\chi_j}{1 + \frac{W}{R_j (E_b/N_0)_{UL,j}}} (1 + f) \right) N = aN \tag{1.4}$$

where a is the constant defined by:

$$a = \sum_j \frac{\chi_j}{1 + \frac{W}{R_j (E_b/N_0)_{UL,j}}} (1 + f) \tag{1.5}$$

The interference margin (*Noise Rise*) is such that:

$$Noise\_rise = -10 * \log_{10}(1 - \eta_{UL}) \tag{1.6}$$

According to link budget (depending on the interference margin), the MAPL L for each service can be written as:

$$L = b - Noise\_rise \tag{1.7}$$

where b is a constant (in dBm) grouping the sum of node B sensitivity (depending on service) and the different gains and losses implied by UL link budget.

The final cell radius value (after *cell breathing*) is the smallest one among all the services, and can be written versus the MAPL L of the most constraining service according to propagation model as follows:

$$R = 10^{\frac{L-c}{d}} \tag{1.8}$$

where c and d are constants depending on the chosen propagation model.

On the other side, since users distribution is assumed to be uniform in the area, the area density of the users is constant, and the number N of users is proportional to the cell area such that:

$$N = \alpha R^2 \tag{1.9}$$

where α is a constant (equal to 2π multiplied by the area density of users).

The system of 5 equations (1.4), (1.6), (1.7), (1.8) and (1.9) with 5 unknown quantities (η_{UL}, N, *Noise_rise*, L and R) — can be solved to obtain finally the following equation with the only unknown quantity R:

$$\beta(1 - \gamma R^2)^{10/d} = R \tag{1.10}$$

where β and γ are constants defined by:

$$\beta = 10^{\frac{b-c}{d}} \tag{1.11}$$

and

$$\gamma = a\alpha \tag{1.12}$$

The equation (1.10) cannot be solved mathematically but can be solved by the FPI method since it can be rewritten as $f(R) = R$ where $f(R)$ is the function defined by:

$$f(R) = \beta(1 - \gamma R^2)^{10/d} \tag{1.13}$$

The FPI method consists in finding the converging limit of the series x_n defined by the following recurring system:

$$\begin{cases} x_0 < 1/\sqrt{\gamma} \\ x_{n+1} = f(x_n) \end{cases} \tag{1.14}$$

The initial condition $x_0 < 1/\sqrt{\gamma}$ is required for the convergence of the iterative process. Figure 1.1 shows the system solution: intersection of the two curves $y = f(x)$ and $y = x$ with a plot materializing the convergence process by the FPI method.

We could also apply the iterative method on the initial equations according to the following process:

1) Initialize cell radius R
2) $N = \alpha R^2$
3) $\eta_{UL} = aN$

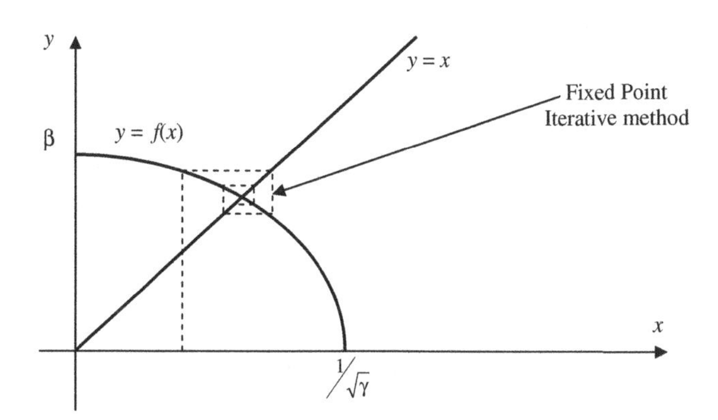

Fig. 1.1 Iterative simulation solving of UL by FPI method.

4) $Noise\_rise = -10 * \log_{10}(1 - \eta_{UL})$
5) $L = b - Noise\_rise$
6) $R = 10^{\frac{L-c}{d}}$
7) Return to 2) until convergence of the radius R (within a small error)

This iterative process is necessary instead of proceeding to a static link budget (in only one iteration). In fact, the classical method results in either a costly over-dimensioning or an under-dimensioning (or a not accurate dimensioning). Thus, it is rather more convenient to establish a prevision of traffic and services load over different phases of several years, and to apply for it the described iterative process versus the predicted area density of users and the maximum allowed load factor.

Let's move now to the DL dimensioning. By assuming one service, the total node B power can be written as follows:

$$P_T = \frac{P_N N \cdot \bar{L}}{(1 - \overline{\alpha} + \overline{f})\left(1 + \frac{W}{(1-\overline{\alpha}+\overline{f})\rho R v} - N\right)} \tag{1.15}$$

where $\overline{\alpha}$ and $\overline{f}$ are respectively the mean orthogonality factor and the mean extra to intra-cellular interference ratio, P_N is the noise power, W is the chip rate, ρ, R and v are respectively the required energy per bit to noise density ratio (E_b/N_0), the bit rate and activity factor of the service, N is the number of users served per cell in the DL, and $\overline{L}$ is the average attenuation in the cell defined by:

$$\overline{L} = \frac{1}{N}\sum_{j=1}^{N}\frac{1}{h_j} \tag{1.16}$$

where h_j is the path gain between the node B and any mobile in the cell. The total power expressed in (1.15) should not exceed the nominal maximum power P_{max} of the node B ($P_T \leq P_{\max}$). So, the capacity M required for dimensioning is obtained exactly for the maximum power $P_{\max}$. On the other side, we can establish that the ratio $r = L_{\max}/\overline{L}$ between the MAPL ($L_{\max}$) and the average attenuation ($\overline{L}$) — is constant ($r = (\gamma + 2)/2$) where γ is the distance attenuation coefficient. Thus we obtain:

$$L_{\max} = \frac{P_{\max} r\left(1 + \frac{W}{\rho R v(1-\overline{\alpha}+\overline{f})}\right)}{N P_N}(1 - \eta_{DL}) \tag{1.17}$$

where $\eta_{DL} = \frac{N}{1+\frac{W}{(1-\overline{\alpha}+\overline{f})\rho R v}}$ (one service is assumed).

Now let's apply the FPI method to the DL according to the iterative process below:

1) Initialize cell radius R
2) Calculate the number of users in this cell $N = \alpha R^2$
3) Compute the load factor η_{DL} according to the expression above.
4) Calculate the MAPL ($L_{\max}$) according to (1.17) or its approximation
5) Calculate the corresponding radius of the cell $R = 10^{\frac{L-c}{d}}$ ($d = 10\gamma$)
6) Return to 2) until convergence of the radius R (within a small error)

For the multiple services case, the required node B power can be generalized as follows [1]:

$$P = \frac{P_N \sum_{j=1}^{s} \frac{(E_b/N_0)_{DL}^{(j)} R^{(j)} v^{(j)}}{W} N^{(j)} \overline{L}^{(j)}}{1 - \sum_{j=1}^{s} \left[\frac{(E_b/N_0)_{DL}^{(j)} R^{(j)} v^{(j)}}{W} N^{(j)} ((1 - \overline{\alpha}^{(j)}) + \overline{f}^{(j)}) \right]} \tag{1.18}$$

where $\overline{\alpha}^{(j)}$ and $\overline{f}^{(j)}$ are respectively the mean orthogonality factor and the mean extra to intra-cellular interference ratio relatively to the mobiles of service j, $(E_b/N_0)_{DL}^{(j)}$, $R^{(j)}$ and $v^{(j)}$ are respectively the required E_b/N_0, the bit rate and activity factor of the service j, $N^{(j)}$ is the number of users of service j served per cell in the DL, and $\overline{L}^{(j)}$ is the average attenuation in the cell relatively to the mobiles of service j, and s is the total number of services.

Downlink dimensioning is thus limited by the most constraining service (having the lowest average MAPL). However, we cannot determine, with those only assumptions, the average attenuations of the cells relative to each of the services (Only one equation with s unknown quantities) except if we give favor explicitly to some services versus the others or if we specify rules for resource sharing according to services as an additional assumption. For example, we can set a fair distribution of the node B power among the users (*Fair Resource*) or the mobiles to have just the required E_b/N_0 or having a fair bit rate among them (*Fair Throughput*). In this last case, the problem will be assimilated to that of one service (Expression (1.15) having only one unknown average attenuation). If the traffic is assumed to be uniform (among all services), and in order to minimize the required number of node B, the sizes of cells relative to each of the services can eventually be assumed to be equal, and so we will have the same average attenuation for the cells relative to all services. This average attenuation (or path loss) of all users of the different services — is calculated from the mean attenuations of the cells of each of the services weighted by the fractions of users of the corresponding service (among all the users).

Assuming that the distribution of the number of users is according to services, and having the relations between the MAPL and the average attenuation, and between the cell radius and the total number of users in the cell, we elaborate, similarly as for the UL, an equation that cannot be solved analytically but we can solve it by the FPI method below as in the case of one service:

1) Initialize cell radius R
2) Calculate the number of users in this cell $N = \alpha R^2$
3) Compute the load factor η_{DL} according to the following expression:

$$\overline{\eta_{DL}} = \sum_j \frac{(E_b/N_0)_{DL,j}.v_j}{W/R_j}((1-\overline{\alpha}) + \overline{f}).$$

4) Calculate the MAPL ($L_{\max}$) according to its expression deduced from (1.17) or from its approximation by taking the maximum power as the node B power and in terms of average attenuation of all the services.
5) Calculate the corresponding radius of the cell $R = 10^{\frac{L-c}{d}}$ (with the same notations as the UL)
6) Return to 2) until convergence of the radius R (within a small error)

A method by dichotomy can eventually be used to solve the expression in the step 4) by acting on the cell radius, the MAPL, and the resulting number of users in the cell; and this on the basis of the computation of the required power of the node B at each iteration, then by comparing it to the maximum power $P_{\max}$ of the node B.

Reformulation of the radio coverage concept in HSPA based systems

The coverage concept in HSPA based UMTS networks is defined as the fact that the signal received by mobile guarantees a minimum required received

SINR (Signal to Interference and Noise Ratio) or power level threshold at a given coverage probability as on the expressions (1.19)–(1.22).

The area coverage probability F_u is written as follows [2, 3]:

$$F_u = \frac{1}{2}\left[1 - erf(a) + \exp\left(\frac{1 - 2.a.b}{b^2}\right).\left(1 - \frac{1 - a.b}{b}\right)\right] \quad (1.19)$$

where

$$a = \frac{x_0 - P_r}{\sigma.\sqrt{2}} \quad \text{and} \quad b = \frac{10.n.\log_{10} e}{\sigma.\sqrt{2}}, \quad (1.20)$$

σ: Standard-deviation of the shadowing effect (in dB)
x_0: Mean threshold of the power sensitivity
P_r: Average level of the power on the cell border
n: Propagation coefficient
e: Exponential constant
The difference $x_0 - P_r$ refers to the shadowing margin
erf : ⟨⟨*error function*⟩⟩" defined by :

$$erf(x) = \frac{2}{\sqrt{\pi}}\int_0^x e^{-t^2}dt \quad (1.21)$$

The coverage probability C_u on the cell edge (border) is given by:

$$C_u = \frac{1}{2}[1 - erf(a)] \quad (1.22)$$

The previous definition is equivalent to the fact that a minimal given bit rate is guaranteed (at the same probability). That has the same significance as guaranteeing a given quality indicator parameter (called in HSPA as *Channel Quality Indicator* or simply CQI [4]–[9]).

In order to validate this last new definition of radio coverage (in HSPA), we calculate, for a given shadowing standard-deviation and for different values of the distance to node B (25 m to 2 Km with a step of 25 m), the probability that CQI is above the CQI threshold value CQI_0 referring to a given service bit rate (assuming the correspondence table of the standard [10] between CQI and the *Transport Block Size* TBS, and that — instantaneous — bit rate depends on the TBS through the *Transmit Time Interval* TTI denoted as TTI_{delay} whose value is specified by the standard: Eg. TTI is equal to 2 ms in HSDPA: *High Speed Downlink Packet Access*). This calculation is repeated for four shadowing standard-deviation values (6 dB, 8 dB, 10 dB and 12 dB) and four services at different required nominal bit rates (64 Kb/s, 128 Kb/s, 384 Kb/s and 2 Mb/s). The calculation of the probability is accomplished according to the following theoretical discrete distribution model valid in the DL (HSDPA):

$$\begin{aligned} p_k &= \Pr ob(CQI = k) \\ &= \frac{1}{2}\left(erf\left[\frac{\xi \cdot Ln(d_k) - \mu}{\sqrt{2} \cdot \sigma}\right] - erf\left[\frac{\xi \cdot Ln(d_{k+1}) - \mu}{\sqrt{2} \cdot \sigma}\right]\right) \end{aligned} \quad (1.23)$$

where $d_k = 10^{-\frac{I_{inter}}{10}}[10^{\frac{P_{TX} - CQI_{ratio}(k-Offset)}{10}} - 10^{\frac{I_{intra}}{10}}]$ if $0 \le k \le E[\frac{P_{TX} - I_{intra}}{CQI_{ratio}} + Offset]$ ($I_{inter} \ll I_{intra}$) and $d_k = 1$ else; *erf* is given by (1.21), I_{inter} and I_{intra} are respectively the intercellular and intracellular interference, P_{TX} is the transmitted power in the DL per user, $\xi = \frac{10}{Ln(10)}$, $\mu = \xi \cdot Ln(\overline{L}) = 10 \cdot \log_{10}(\overline{L})$ is average logarithmic attenuation in dB referring to distance path loss, $CQI_{ratio} = 1,02$ and $Offset = 16,62$.

The probability of non-coverage (not enough quality or required quality for the $CQI = 1$ not reached) is given by p_0.

It is to be noted that $p_k = 0, \forall k \ge E\left[\frac{P_{TX} - I_{intra}}{CQI_{ratio}} + Offset\right] + 1$; or that the maximum possible value for CQI is $CQI_{\max} = E\left[\frac{P_{TX} - I_{intra}}{CQI_{ratio}} + Offset\right]$ (due to intracellular interference near node B).

The calculated probability value is compared to the coverage probability at the cell border given by (1.22) formula by taking the same values of shadowing standard-deviation, and shadowing margin calculated from the MAPL referring to the target SIR threshold and at the same distance. The MAPL (maximum path loss L_{Total}) is extracted from the SINR definition expression versus SIR threshold. This last is related to the CQI_0 value and to the BLER (BLock Error Rate) by the following expression [11]:

$$\begin{aligned} SIR_{threshold} = &\frac{\sqrt{3} - \log_{10}(\mathrm{CQI}_0)}{2} \\ &\cdot \log_{10}(\mathrm{BLER}^{-0.7} - 1) + 1.03 \cdot \mathrm{CQI}_0 - 17.3 \end{aligned} \quad (1.24)$$

The maximum relative error (among those of the different distances) is provided in Table 1.1 for each shadowing standard-deviation value and for each service.

Note that maximum relative error between both calculated coverage probabilities — is low (not exceeding 5,35% at the worst case. So, the coverage of

Table 1.1 Maximum relative error of the new definition of coverage versus the classical definition.

		Nominal service bit rate			
Maximum relative error (%)		64 Kb/s	128 Kb/s	384 Kb/s	2 Mb/s
Shadowing standard-deviation (dB)	6	5,35	1,19	0,81	2,23
	8	3,41	0,74	0,49	1,31
	10	2,46	0,52	0,34	0,87
	12	1,90	0,40	0,26	0,64

a point in the cell at a guaranteed quality (SIR threshold) is assimilated to the fact that it corresponds to equivalent bit rate and CQI referring to this target SIR (See expression (1.24)), and thus they have almost identical probabilities. Moreover, Table 1.1 validates the theoretical model of the CQI distribution according to (1.23).

Those errors are caused by those of the AMC (Adaptation in Modulation and Coding) mechanism having a discrete tracking of the propagation channel quality (different CQIs referring to determined values of instantaneous bit rate). The fact that these errors are low translates the extreme efficiency of this mechanism (AMC), which proves that it is equivalent to the power control used in basic UMTS (Rel'99).

Figure 1.2 shows the coverage probability versus the distance to node B for different services and different shadowing standard-deviation values. The plotted curves can be used as dimensioning abacuses for HSPA. Note particularly that at a coverage probability above 50% (more probable for the mobile to be covered), dimensioning is more optimistic with less shadowing (smaller standard-deviation) in contrast to a coverage probability below 50%. At an intermediate coverage probability value equal exactly to 50%, the dimensioning is independent of the shadowing standard-deviation value (corresponds to the average power received by the node B which is the same as the one obtained with a deterministic propagation model).

1.2 Coverage or Capacity Limited Dimensioning

"Fair Resource" scheduling technique study case

HSPA capacity is limited either by the number of codes HS-PSCH (*High Speed — Physical Shared Channel*) [4, 7] or by the total node B power. The cell size in the first case (R_c) is the greatest radius R verifying the equality in

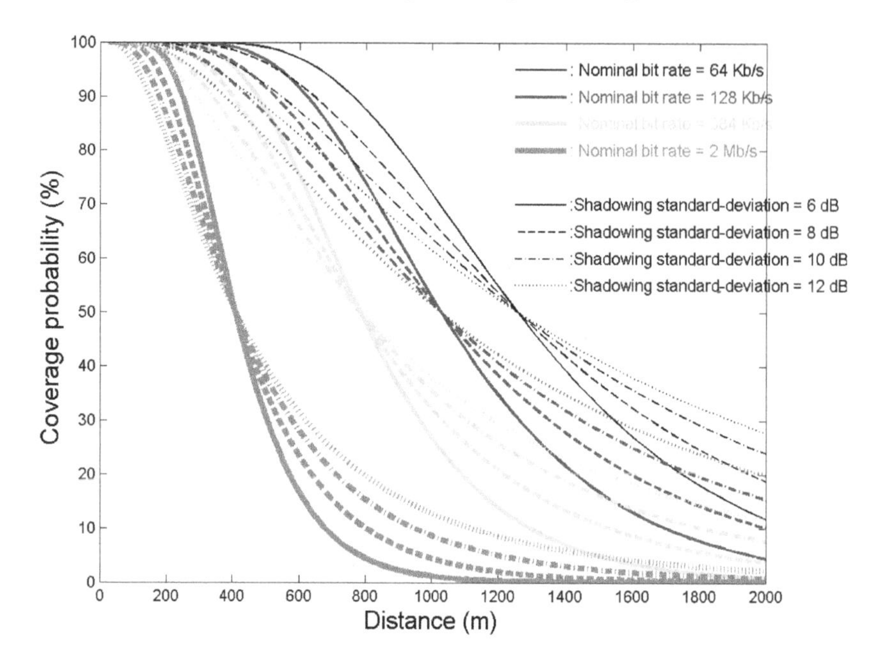

Fig. 1.2 Coverage probability versus distance to node B for different services (having different nominal bit rates) and shadowing values.

the following condition of HSPA code limitation:

$$\sum_{i \le k_0} n_i\,(r_{i+1}^2 - r_i^2) + n_{k_0+1}(R^2 - r_{k_0+1}^2) \le \frac{15}{\pi\rho} \tag{1.25}$$

where $r_i = r_{i,\min}$ and $r_{i+1} = r_{i+1,\min} = r_{i,\max}$, r_i and r_{i+1} denote respectively the lower and upper limits of the range of the sub-cell (ring) having a CQI equal to CQI_i (same modulation, coding rate and number of physical shared channels or codes n_i used thus having the same corresponding block size TBS_i according to the adequate table of [10]), and such that CQI_{k_0+1} is the CQI of the ring including the border of the cell having the size R. The number 15 in the numerator of the second term of (1.25) — refers to the number of the codes (physical shared channels) allocated to HSPA (the maximum number allocated for HSDPA: the value of 15 is taken as an example here and until the end of this chapter). In the second case (node B power limited capacity), the size R_p of the cell is such that:

$$\pi\rho R_p^2 10^{\frac{P_{TX}}{10}} = 10^{\frac{P_{Tot}}{10}} \tag{1.26}$$

where P_{Tot} is the total power of the node B. The capacity limited cell radius is: $R_{cap} = \min(R_c, R_p)$.

By assuming the dimensioning and the bit rate limited only by coverage and link quality but not by capacity, the cell radius R will depend on minimum bit rate $R_{\min}$ of the concerned service (by taking always the assumption of the "Fair Resource" as the used scheduling technique). The bit rate $R_{\min}$ refers to the "Transport Block Size" TBS_0 as follows:

$$TBS_0 = \min_i \{TBS_i / TBS_i \in \text{tables [10]} \quad \text{and} \quad TBS_i \geq R_{\min} \cdot TTI_{delay}\} \tag{1.27}$$

In this case, and by accomplishing a dimensioning without codes multiplexing, the cell radius R will be the distance r_0 referring to TBS_0 or exactly the minimum radius of the internal CQI ring (core) corresponding to the maximum CQI that radio condition and terminal capability allow. In fact, TBS_0 refers to a CQI_0 (the maximum allowed value) [10] to which refers a minimum SINR value ($SINR_{\min}$) computed as follows:

$$SINR_{\min} = CQI_{ratio} \cdot (CQI_0 - Offset) \tag{1.28}$$

From the last value, we can extract the maximum attenuation from SINR definition expression, and thus the corresponding radius r_0. If we assume the total power transmitted by the node B constant (not depending on the number of mobiles served), then the computation of the cell radius can be made immediately. Yet in reality, the power transmitted depends on the traffic, and is proportional to the number of mobiles served in the cell (since individual transmit power is constant), hence the transmitted intracellular power is proportional to the number of active mobiles of the cell, so it depends on the cell radius. Therefore, for more accuracy and precision, we must apply an iterative algorithm until its convergence to the cell radius. The last parameter shouldn't exceed, in any case and whatever the service and its required bit rate, the value of the radius referring to the minimum allowed value of CQI (equal to 1).

In contrast, by assuming the dimensioning is capacity limited (either by number of codes allocated to HSPA if $\int_0^{2\pi} d\alpha \int_0^R n(r)\,\rho\, r\, dr > 15$ where $n(r)$ is the number of codes referring to the CQI of a virtual mobile at a distance r to the node B — or by the total available power of the node B if $\pi \rho R_p^2 10^{\frac{P_{TX}}{10}} > 10^{\frac{P_{Tot}}{10}}$, or in other words assuming traffic density above some value, then the bit rate R_u guaranteed per user (assuming always a uniform traffic and the use of the "Fair Resource" scheduling technique) can be written as follows:

$$R_u = \frac{TBS_i \cdot R_{cap}^2}{TTI_{delay} \cdot r_i^2} \tag{1.29}$$

where TBS_i is the Transport Block Size referring to the ring i of the CQI $= i$ (at the border of the cell) having an outer radius $r_{i+1} > R_{cap}$ (capacity limited cell size) in case of one service. In this last case, in order to guarantee a minimum bit rate $R_{\min}$ at the border of the cell, the "Transport Block Size" TBS_0 at the cell border — should be given by:

$$TBS_0 = \min_i \left\{ TBS_i / TBS_i \in \text{ tables [1] } \quad \text{and} \quad TBS_i \geq R_{\min} \frac{TTI_{delay}}{R_{cap}^2} r_{i+1}^2 \right\} \tag{1.30}$$

In this case, the cell radius can be concluded from TBS_0 as for the previous case of coverage limited dimensioning; yet the expression (1.30) above of TBS_0 depends on the cell radius (through r_{i+1}), then the planner should apply an iterative process or by dichotomy to converge to the exact cell radius or extract it from some mathematical formula referring exactly to the required bit rate $R_{\min}$. Thus, if $TBS_{i+1} < R_{\min} \frac{TTI}{R_{cap}^2} r_{i+1}^2$ (where i is such that $TBS_i = TBS_0$), then the dimensioned cell size R_{dim} is equal to the limit size of the CQI ring with TBS_0 (i.e. r_{i+1}), else $R_{\dim} = R_{cap} \sqrt{\frac{TBS_{i+1}}{TTI \cdot R_{\min}}} (r_{i+1} \leq R_{\dim} < r_{i+2})$.

"Fair Throughput" scheduling technique study case

The "Fair Throughput" scheduling technique [8] is limited neither by coverage nor by capacity since the constant offered bit rate is below that guaranteed by the worst link (at the border of the cell), and since the total number of codes is always below that allocated for HSPA.

Expression (1.31) below gives the maximum bit rate ensured or guaranteed for all the users (with codes multiplexing) by using the "Fair Throughput" scheduling technique:

$$(R_{ens})_{FT} = \frac{15}{TTI_{delay} \cdot \sum_j \frac{n_j}{TBS_j}} \tag{1.31}$$

In the uniform traffic case study, (1.31) becomes — by reasoning on the different CQI rings of one cell with a uniform area density ρ — as follows:

$$(R_{ens})_{FT} = \frac{15}{2\pi\rho \cdot TTI_{delay} \int_0^R \frac{n(r)}{TBS(r)} r\,dr} = \frac{15}{\pi\rho \cdot TTI_{delay} \sum_i \frac{n_i}{TBS_i}(r_{i+1}^2 - r_i^2)} \quad (1.32)$$

assuming that r_i and r_{i+1} are the radii of the CQI ring borders ($r_{i+1} = r_{i+1,\min} = r_{i,\max}$).

Hence, if minimum bit rate value $R_{\min}$ to guarantee for the service is known, the maximum allowed area density of users $\rho_{\max}$ can be determined as follows:

$$\rho_{\max} = \frac{15}{TTI_{delay} \cdot R_{\min} \cdot 2\pi \int_0^R \frac{n(r)}{TBS(r)} r\,dr} = \frac{15}{TTI_{delay} \cdot R_{\min} \cdot \pi \sum_i \frac{n_i}{TBS_i}(r_{i+1}^2 - r_i^2)} \quad (1.33)$$

The planner should therefore determine the CQI ring m referring to bit rate $R_{\min}$. So, the dimensioned cell size $R_{\dim}$ guaranteeing a minimum bit rate R_{min} can be extracted from (1.32) as follows:

$$R_{\dim} = \sqrt{r_m^2 + \frac{15 \cdot TBS_m}{n_m \cdot TTI \cdot R_{\min} \cdot \pi\rho} - \frac{TBS_m}{n_m} \sum_{i<m} \frac{n_i}{TBS_i}(r_{i+1}^2 - r_i^2)} \quad (1.34)$$

"Enhanced Fair Throughput" technique study case

In order to enhance radio dimensioning performance in terms of cell size and capacity, a third dimensioning method consists in a modification of the "Fair Throughput" scheduling technique by incorporating weights to allocated resources proportionally to nominal bit rates of different services such that each mobile is satisfied by its required service bit rate in each point of the cell. This method called "Enhanced Fair Throughput" is adapted both to services and propagation conditions. We obtain a common dimensioning for all services (the same sub-cell size).

The ensured bit rate $(R_{ens})_m$ by "Enhanced Fair Throughput" at the upper bound of the ring at CQI = m is given by:

$$(R_{ens})_m = \frac{15}{2,6 \cdot \rho \cdot TTI_{delay} \sum_{j=m}^{CQI_{\max}} \frac{n_j}{TBS_j}(r_{j-1}^2 - r_j^2)} \tag{1.35}$$

(we replaced π by 2,6 in order to adapt the area to the hexagonal form of the cell) where CQI_{max} is the maximum CQI (belonging to the smaller CQI ring of the cell: It's the central sub-cell such that $r_{CQI_{\max}} = 0$) which has been established to be determined and limited by the intracellular interference level.

Multiple services study case

Let's assume s services denoted by $1, 2, \ldots, i, \ldots, s$. We call R_i the cell size referring to service i. Thus we have $R_s \leq R_{s-1} \leq \cdots \leq R_2 \leq R_1$. Let's assume that $R_{s+1} = 0$. Dimensioning should therefore be accomplished according to the most constraining service s (having the highest bit rate) since its cell radius R_s is the most limiting (the smallest radius among those of the other services).

Assuming that ρ_i is the area density of simultaneous users of the service i (with uniform distribution). The global area density of users of all the services is thus such that $\rho = \sum_{i=1}^{s} \rho_i$.

The dimensioning process is accomplished by determining cell size relative to services i one by one in the decreasing order of the required bit rates starting with the most constraining service s until the least limiting service 1. In order to guarantee a minimum bit rate $R_{\min,i}$ at the cell border relative to service i (where $R_{\min,i} \leq R_{\min,i+1}$; $\forall i = 1, 2, \ldots, s$ and assuming the "Fair Resource" scheduling technique), the Transport Block Size TBS_0 at the border of the sub-cell i without codes multiplexing is given by:

$$TBS_0 = \min \left\{ \begin{array}{l} TBS_j / TBS_j \in \text{ tables [10]such that:} \\ TBS_j \geq R_{\min,i} TTI_{delay} \\ \cdot \max\left(1, \dfrac{\sum_{m=1}^{s}\left(\left(\sum_{l=1}^{m}\rho_l\right)(R_m^2 - R_{m+1}^2)\right)}{\sum_{m=k_0+1}^{s}\left(\left(\sum_{l=1}^{m}\rho_l\right)(R_m^2 - R_{m+1}^2)\right) + \left(\sum_{l=1}^{k_0}\rho_l\right)(R_{cap}^2 - R_{k_0+1}^2)}\right) \end{array} \right. \tag{1.36}$$

where $(k_0 + 1)$ index is that of the most limiting service (coverage limited), or in other words $R_{k_0+1} \leq R_{cap} < R_{k_0}$.

Expression (1.36) is valid for both coverage and capacity limited dimensioning. For each iteration, once the radius relative to service i is determined, we can restart to find the sub-cell radius relative to service $i - 1$ by using the results relative to services i to s.

1.3 HSPA Dimensioning Methodology with Shadowing Effect

For a real time service, the minimum bit rate $R_{\min}$ should be guaranteed or ensured at least with a coverage probability of 90% of the time — for example — at the worst case (at the border of the cell). So, the HSPA network planner should determine the size TBS_0 such that:

$$TBS_0 = \min_i \{TBS_i \text{ such that } TBS_i \in \text{ tables [10] and } TBS_i \geq R_{\min} \cdot TTI_{delay}\} \quad (1.37)$$

After that, we determine the CQI_0 referring to TBS_0 [10]. Our objective is thus to obtain Prob(CQI $\geq CQI_0) = 0,90$ or:

$$F_L\left(10^{-\frac{I_{inter}}{10}}\left[10^{\frac{P_{TX}-CQI_{ratio}(CQI_0-Offset)}{10}} - 10^{\frac{I_{intra}}{10}}\right]\right) = 0,90 \quad (1.38)$$

where F_L is the Cumulative Distribution Function (CDF) of a Gaussian law having an average μ and a standard-deviation σ, thus we can deduce — from tables or abacuses giving the percentiles of a normal Gaussian law — the shadowing margin to take by the planner as follows: $10^{-\frac{I_{inter}}{10}}\left[10^{\frac{P_{TX}-CQI_{ratio}(CQI_0-Offset)}{10}} - 10^{\frac{I_{intra}}{10}}\right] - \mu$ and hence the average maximum attenuation μ, which provides automatically the maximum radius of the cell (through the deterministic part of the propagation model). In conclusion, we have established a simple dimensioning methodology of a HSPA based network (determination of the cell size), and allowing also to compute, inversely, at each point of the cell, the coverage probability guaranteeing the bit rate $R_{\min}$.

HSPA coverage limited dimensioning flow chart

Figure 1.3 shows the flow chart of the simplified HSPA dimensioning methodology:

— The block **T** relates the Transport Block Size (*TBS*) to the bit rate D by the relation $TBS = D \cdot TTI$ where TTI is the Transmit Time

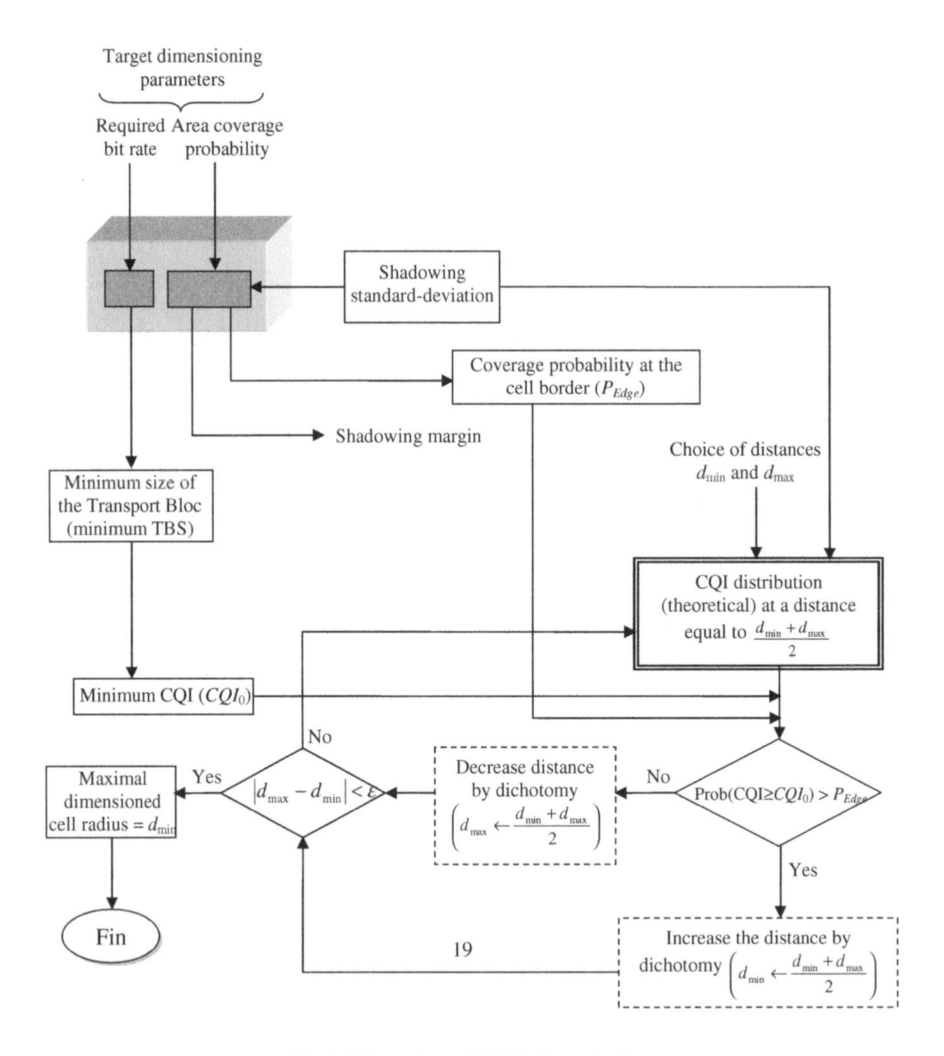

Fig. 1.3 Flow chart of HSPA dimensioning.

Interval of the Transport Blocks in the MAC-HS layer. Yet the allowed *TBS* values according to the standard [10] are already set (30 values at most for the advanced terminal categories), so we take the first value of TBS slightly above the theoretical size value referring to the minimum bit rate, hence the minimum corresponding CQI according to the right table in [10] (according to the terminal capability).

— For the block **C** of coverage, assuming that target area coverage probability and shadowing standard-deviation are known, we can conclude the coverage probability at the edge (at the border) of the cell and the shadowing margin to apply in a link budget based on a deterministic propagation model. This is accomplished iteratively and by dichotomy with coverage equations (1.19) and (1.20).

— The theoretical CQI distribution at given distance and shadowing standard-deviation is determined by the expression (1.39) below providing the discrete probabilities of the different CQI values:

$$\begin{aligned} p_k &= \mathrm{Pr}ob(CQI = k) \\ &= F_L\left(10^{-\frac{I_{inter}}{10}}\left[10^{\frac{P_{TX}-CQI_{ratio}(k-Offset)}{10}} - 10^{\frac{I_{intra}}{10}}\right]\right) \\ &\quad - F_L\left(10^{-\frac{I_{inter}}{10}}\left[10^{\frac{P_{TX}-CQI_{ratio}(k+1-Offset)}{10}} - 10^{\frac{I_{intra}}{10}}\right]\right) \end{aligned} \quad (1.39)$$

— Coverage probability C_u at the edge (at the cell border) is given by (1.22).

— $\varepsilon$is the convergence accuracy of the cell radius by the iterative algorithm (by dichotomy).

1.4 Practical Examples and Exercises About Coverage Limited Dimensioning Methodology

Tutorial example of dimensioning tables use

In this paragraph, we deal with a practical case study of the use of tables for the coverage limited dimensioning in the DL through an example of UMTS network based on the HSDPA technique.

With a shadowing standard-deviation of 6 dB, in order to have a minimum bit rate of 384 Kb/s in the cell at a probability of 95%, the maximum (allowed) cell radius is equal to 535 m according to Table 1.2 referring to the indicated shadowing standard-deviation. In other words, for a cell having a radius not exceeding 535 m, we have a probability of 95% with a bit rate at least equal to 384 Kb/s. Note that we have taken the bit rate immediately above 384 Kb/s in the list of bit rates (rows of the Table 1.2) which is, in our case, 396 Kb/s, in order to find the maximum radius guaranteeing a minimum bit rate of 384 Kb/s: Prob(bit rate $\geq$ 396 Kb/s) = 95% inside a cell having a maximum radius of 535 m.

Table 1.2 Dimensioning table (Cel size versus maximum offered bit rate and area coverage probability with a shadowing std-deviation of 6 dB).

Shadowing standard-deviation = 6 dB

Maximum distance (m)

Maximum offered bit rate at the edge of the cell (Kb/s)	Area coverage probability (%)																		
	81	**82**	**83**	**84**	**85**	**86**	**87**	**88**	**89**	**90**	**91**	**92**	**93**	**94**	**95**	**96**	**97**	**98**	**99**
68.5	1224	1203	1180	1158	1135	1112	1088	1064	1039	1012	985	956	926	894	858	818	772	715	635
86.5	1145	1125	1104	1083	1062	1040	1017	995	971	947	921	894	866	836	802	765	722	669	594
116.5	1070	1051	1032	1012	993	972	951	930	908	885	861	836	810	781	750	715	675	625	555
158.5	1001	983	965	947	928	909	890	870	849	828	805	782	757	730	701	669	631	584	519
188.5	936	919	902	885	868	850	832	813	794	773	753	731	708	683	656	625	590	547	485
230.5	875	859	844	827	811	794	777	760	742	723	704	683	661	638	612	584	551	511	453
325	817	803	788	773	758	742	726	710	694	676	658	639	618	597	573	546	515	477	423
396	764	750	736	722	708	694	679	664	648	631	614	597	578	557	535	510	481	446	396
465.5	713	700	687	675	661	648	634	620	605	590	574	557	539	520	500	476	450	417	370
631	665	654	642	630	617	605	592	578	565	550	536	520	503	486	466	445	420	389	345
741.5	621	610	599	587	576	564	552	540	527	514	500	485	470	453	435	415	392	362	322
871	579	569	558	548	537	526	515	503	491	479	466	452	438	423	406	387	365	338	300
1139.5	539	530	520	510	500	490	480	469	458	446	434	421	408	394	378	360	340	315	280
1291.5	501	493	484	475	465	456	446	436	425	415	404	392	380	366	351	335	316	293	260
1659.5	466	458	449	440	432	423	414	405	395	385	375	364	353	340	326	311	294	272	242
1782.5	432	424	416	409	400	392	384	375	366	357	347	337	326	315	303	289	272	252	224
2094.5	399	392	385	378	370	362	355	347	339	330	321	312	302	291	280	267	251	233	207
2332	367	361	354	347	340	334	326	319	311	304	295	287	278	268	257	245	231	214	190
2643.5	336	330	324	318	312	305	298	292	285	278	270	262	254	245	236	225	212	196	174
2943.5	305	300	294	289	283	277	271	265	259	252	245	238	231	222	214	204	192	178	158
3277	273	268	263	258	253	248	242	237	231	226	220	213	206	199	191	183	172	159	142
3534	239	234	230	226	222	217	212	208	203	197	192	186	181	174	167	159	150	139	123
4859.5	198	195	192	188	184	180	176	173	169	164	160	155	150	145	139	133	125	116	103
5709	138	136	133	131	128	125	123	120	117	114	111	108	105	101	97	92	87	81	72

We can eventually include other margins (fast fading, etc...) by substracting them from the MAPL equivalent to the distance 535 m (by considering a determined propagation model) then by recalculating the corresponding maximum distance (below 535 m).

Application use of generated abacuses

Figure 1.4 and 1.5 provide examples of abacuses respectively for categories 10 and 1 of terminals — allowing to dimension a HSDPA network by determining the maximum cell radius versus minimum required bit rate (guaranteed at a given cell coverage probability). These abacuses are generated on the basis of the flow chart in Figure 1.3.

For such abacuses, the specification of bit rate and required coverage probability is enough to find the maximum allowed cell radius by the help of both figures (Figure 1.4 shows the complete range of bit rates: until 5,7 Mb/s, and Figure 1.5 extracts the part of abacuses whose bit rate doesn't exceed 1 Mb/s for a better view).

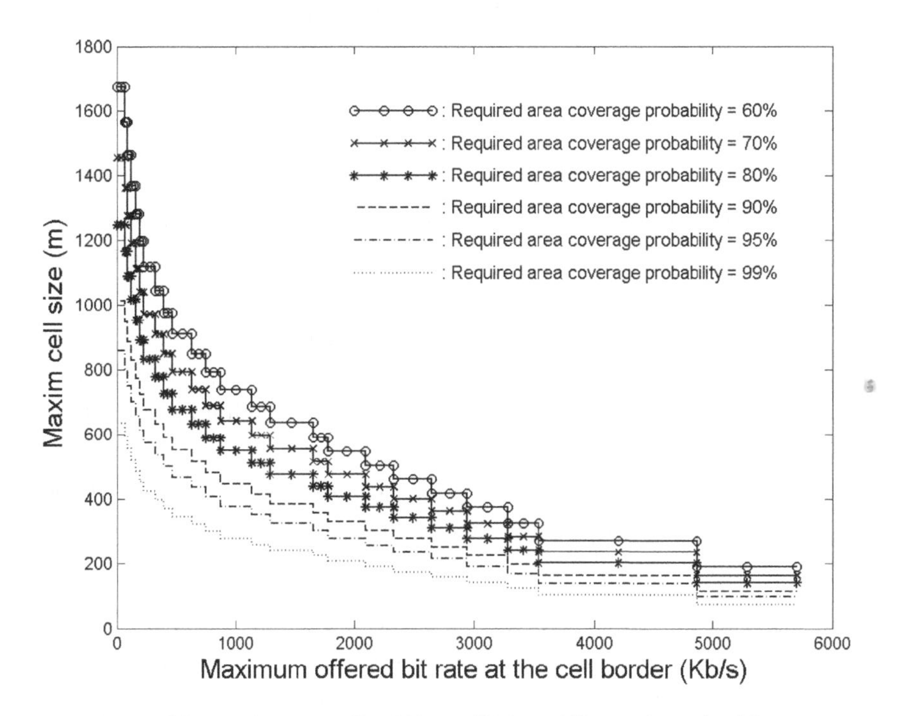

Fig. 1.4 Abacus of the cell size versus offered bit rate (High capability terminals) for different coverage probability values.

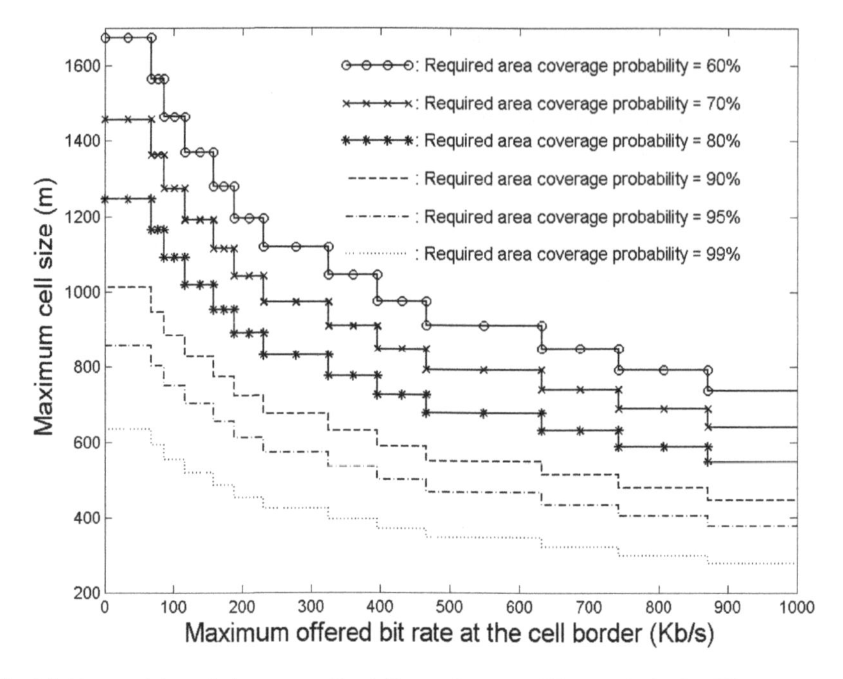

Fig. 1.5 Abacus of the cell size versus offered bit rate (Low capability terminals) for different coverage probability values.

- The discrete aspect of the abacuses translates the effect of Adaptation in Modulation and Coding affecting with a discrete manner the bit rate supported by the link (limited numbers of CQIs thus well determined Transport Block sizes).
- Note that at a given fixed coverage probability, the maximum radius doesn't exceed the maximum value corresponding to CQI = 1 (peripheral border of the cell offering the minimum offered bit rate).
- The higher the required coverage probability is, the smaller the maximum cell radius (Dimensioning with stricter conditions).

Figure 1.6 is similar to Figure 1.4 except by taking the shadowing standard-deviation as parameter instead of the required area coverage probability. It shows dimensioning abacuses of coverage limited bit rate (bit rate guaranteed in 95% of the cell). Note in particular that the dimensioned radius is lower for higher shadowing standard-deviation values. This is effectively logical since

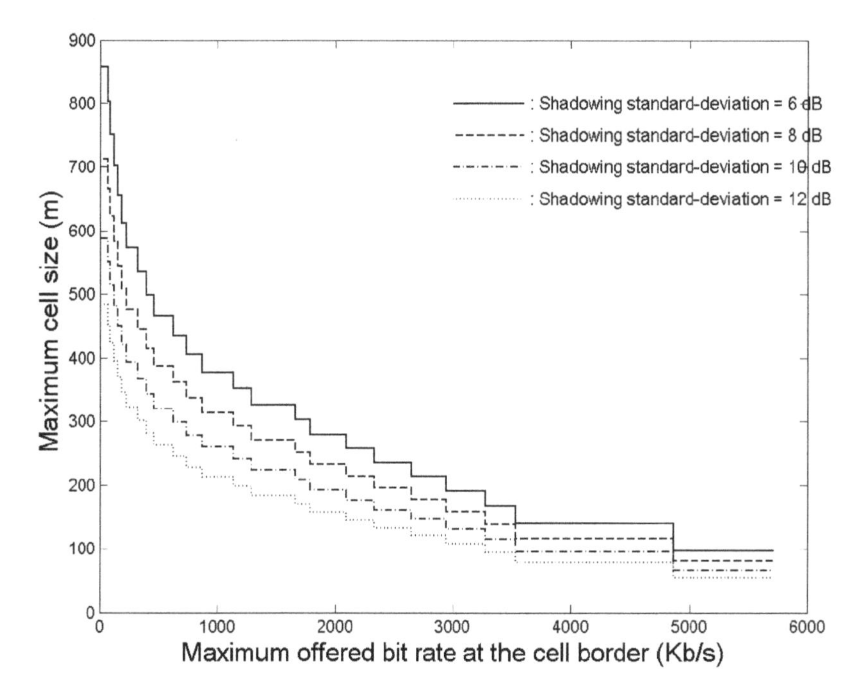

Fig. 1.6 Abacus of the cell size versus offered bit rate (High capability terminals) for different shadowing standard-deviation values.

the shadowing margin to include in the link budget — inscreases with the standard-deviation.

Figure 1.7 summarizes entirely the abacuses of the three Figure 1.4 to 1.6 while combining both the following parameters: coverage probability and shadowing standard-deviation. The same remarks can be extracted with a global vision of the impact of both parameters together (shadowing standard-deviation and area coverage probability). In particular, the smaller the coverage probability (case of 70%), the less the impact of shadowing standard-deviation is important on the dimensioned cell size (due to the impact of coverage probability on the shadowing margin).

The abacuses of Figure 1.8 allow to dimension the cell radius (coverage limited) versus required coverage probability while knowing the required bit rate and the shadowing standard-deviation.

- It is evident that the higher the required bit rate, the smaller the dimensioned cell radius (because the CQI at the edge of the cell is important).

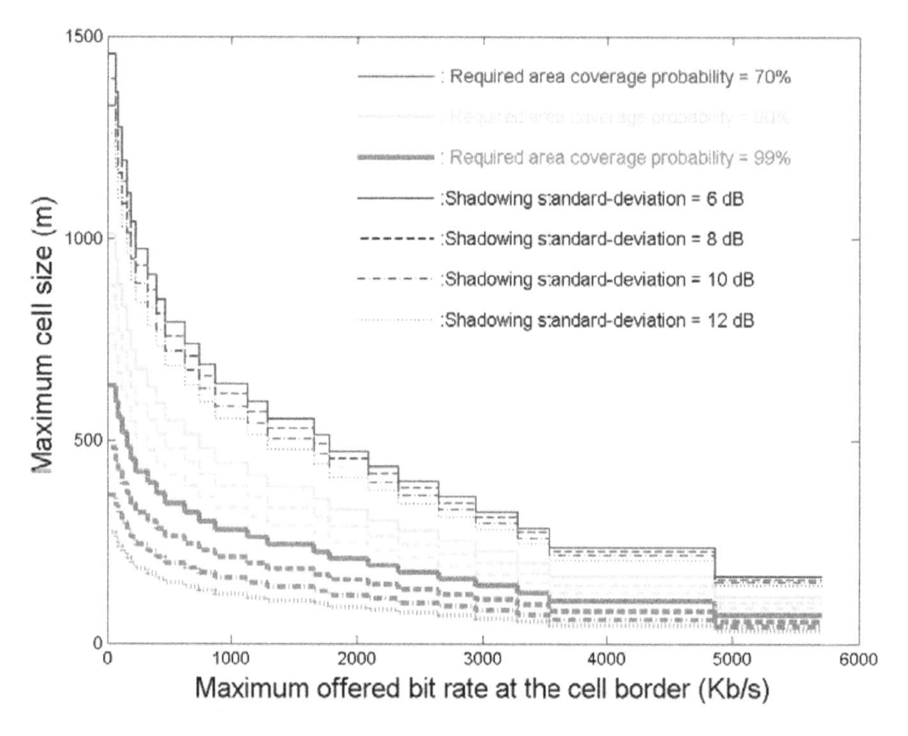

Fig. 1.7 Abacus of the cell size versus offered bit rate (High capability terminals) for different coverage probability and shadowing standard-deviation values.

- The impact of the shadowing standard-deviation on the dimensioned cell size diminishes with the increase of the maximum guaranteed service bit rate. In other words, shadowing effect is more important with a lower bit rate service. This is due to the fact that the required CIR of a low bit rate service is below that of a higher bit rate, thus more sensitive to propagation channel variations due to shadowing.
- The more the coverage probability increases and approaches to 100%, the more the dimensioned radius decreases asymptotically (near the coverage probability of 100%). This is due to the shadowing effect requiring an infinite margin to reach a coverage probability of 100% (not reached in practice).

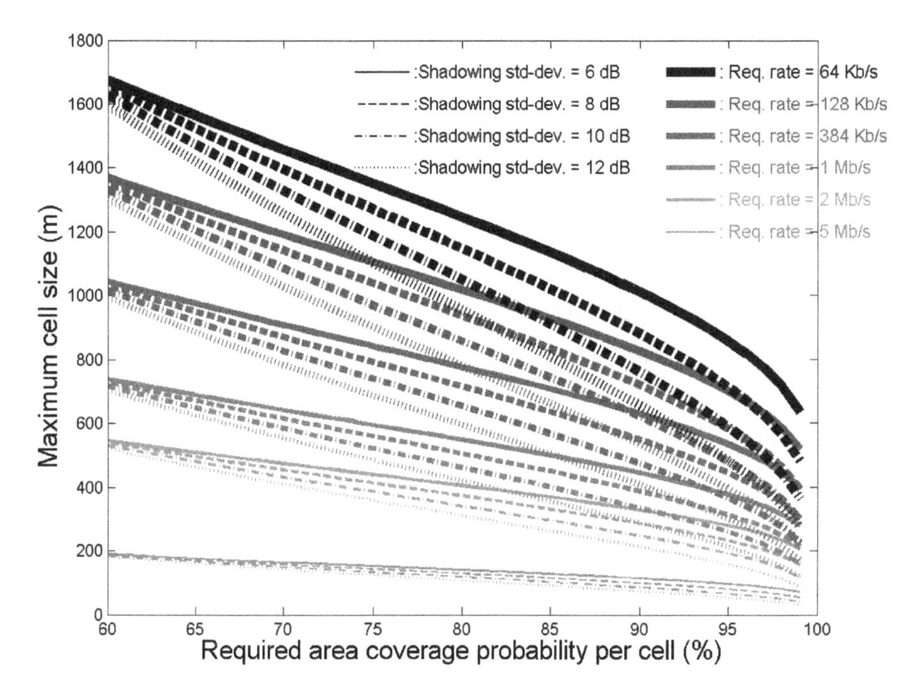

Fig. 1.8 Abacus of the cell size versus the area coverage probability for different services (different nominal bit rates) and different shadowing standard-deviation values.

1.5 HSPA Planning Reference Techniques

Different dimensioning methods description:

In this paragraph, we describe three different dimensioning methods M1, M2 and M3 based on the resource management techniques studied in paragraph 1.2. Those methods are the following:

- Dimensioning method M1: It is based on "Fair Resource" technique. It consists in dimensioning sites according to the most limiting service (in terms of coverage and capacity) in order to have access simultaneously to all services in each point of the concerned service area.
- Dimensioning method M2: It is also based on "Fair Resource" technique. It consists in dimensioning sites according the least limiting service in such a way that access to different services is accomplished through concentric sub-cells where some services

are available near the node B. If we move from one sub-cell to another away from the node B, we lose gradually access to services one by one (the guaranteed bit rate decreases by moving from one sub-cell to another away from the node B).

— Dimensioning method M2: It is based on the "Enhanced Fair Throughput" technique.

Comparison between different HSPA planning methods

Figures referring to different HSPA dimensioning techniques are presented varying the number of users per area unit (area density) in order to visualize the impact of a high or low traffic. We also visualize the impact of two different configurations of service distribution: the first is configuration A in which most mobiles use the usual services with low bit rates, and the second is configuration B in which the traffic distribution is balanced between the services (traffic load above the one of configuration A): See Table 1.3.

Dimensioning results referring to "Fair Resource" scheduling technique provide the sizes of different cells referring to different cells according to dimensioning method M2. The cell having the most limiting size (referring to service RAB 2 Mbps) shows the dimensioning result according to method M1.

In both Figures 1.9 and 1.10 referring to low spectral efficiency values, and by applying "Fair Resource" technique, we note that all the services are coverage limited if user density is below a threshold (constant MAPL), and capacity limited if the density is higher than this threshold (the MAPL decreases versus area spectral efficiency due to a scheduling among a higher number of users resulting in cell breathing). This MAPL threshold decreases when the service nominal bit rate increases due to the same reason (breathing of the coverage limited cell due to the Adaptation in Modulation and Coding procedure: AMC). The services with the lowest nominal bit rates (RAB 128 and RAB 384 in Figures 1.8 and 1.9) have their coverage decreasing (versus traffic) until

Table 1.3 Traffic distribution according to services in the configurations A and B.

Services	Configuration A: (users %)	Configuration B: (users %)
Voice (16 Kb/s): RAB 16	94%	35%
Packet (14 Kb/s): RAB 14	1,5%	25%
Packet (384 Kb/s): RAB 384	0,7%	15%
Packet (2 Mb/s): RAB 2 Mbps	1,6%	10%
Circuit (128 Kb/s): RAB 128	2,2%	15%

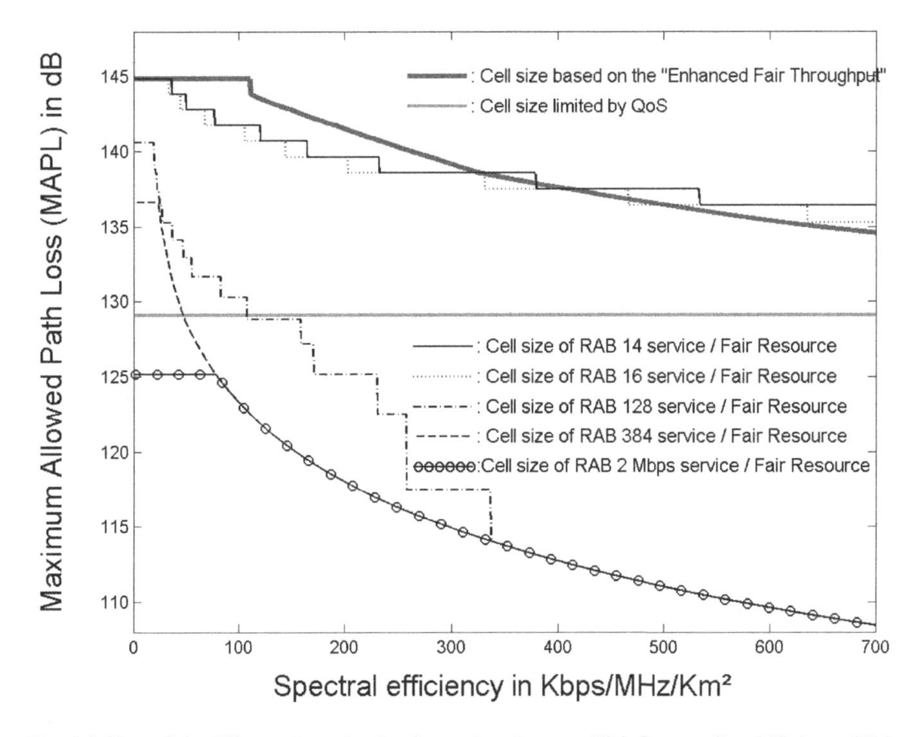

Fig. 1.9 Sizes of the different planned cells: Comparison between "Fair Resource" and "Enhanced Fair Throughput" (Low traffic case & Configuration A of services distribution).

reaching the range of the maximum bit rate service (RAB 2 Mbps according to the same figures).

The stairs form of the curves referring to services RAB 128, RAB 14 and RAB 16 is due to the granularity of instantaneous bit rates referring to Transport Block Sizes (TBS) by applying AMC.

Note that the introduced "Enhanced Fair Throughput" method improves the allowed range especially for the low traffic densities (area spectral efficiency below 300 Kb/s/MHz/Km^2 in the configuration A). Therefore, this scheduling method optimizes radio resource share and allocation. For very low users densities (below about 100 Kb/s/MHz/Km^2 for both configurations), the range of this method is constant (coverage limited range referring to the ring with the minimum CQI equal to 1).

The QoS limits low bit rate services as well as the "Enhanced Fair Throughput" method: The QoS is thus a bottleneck for dimensioning according to

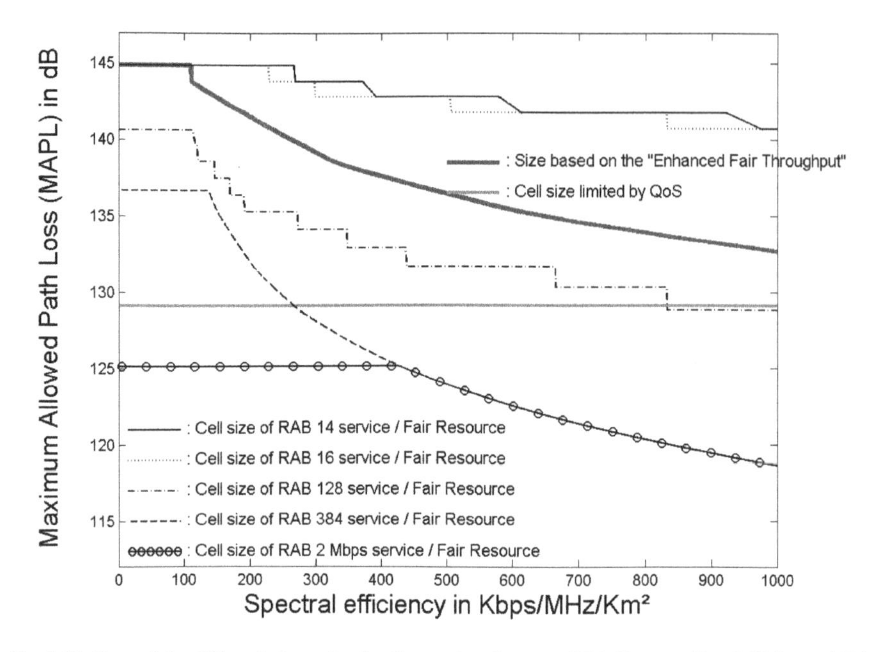

Fig. 1.10 Sizes of the different planned cells: Comparison between "Fair Resource" and "Enhanced Fair Throughput" (Low traffic case & Configuration B of services distribution).

methods M2 and M3 (for low density traffic). For higher bit rate services, it is the capacity that limits the QoS.

By moving from configuration A to B, the area spectral efficiency values have increased for the same user area density (See Figure 1.10): For example, the area spectral efficiency thresholds referring to coverage limited cells — have increased by moving from the service distribution configuration A to B for all services in "Fair Resource". It is also the reason for which the MAPL values (referring to sizes of cells) decrease less fast by moving to configuration B (i.e.: In order to have the same MAPL, area spectral efficiency in configuration B should be higher due to its higher service load).

According to Figures 1.11 and 1.12, note that for high user densities (area spectral efficiency above 1 Mb/s/MHz/Km^2), the "Enhanced Fair Throughput" technique (method M3) improves the performance in terms of coverage and capacity compared to method M1. Nevertheless, if we apply method M2 or if the dimensioning is based only on the lower bit rate services, we will have slightly less sites than with the "Enhanced Fair Throughput" technique (M3).

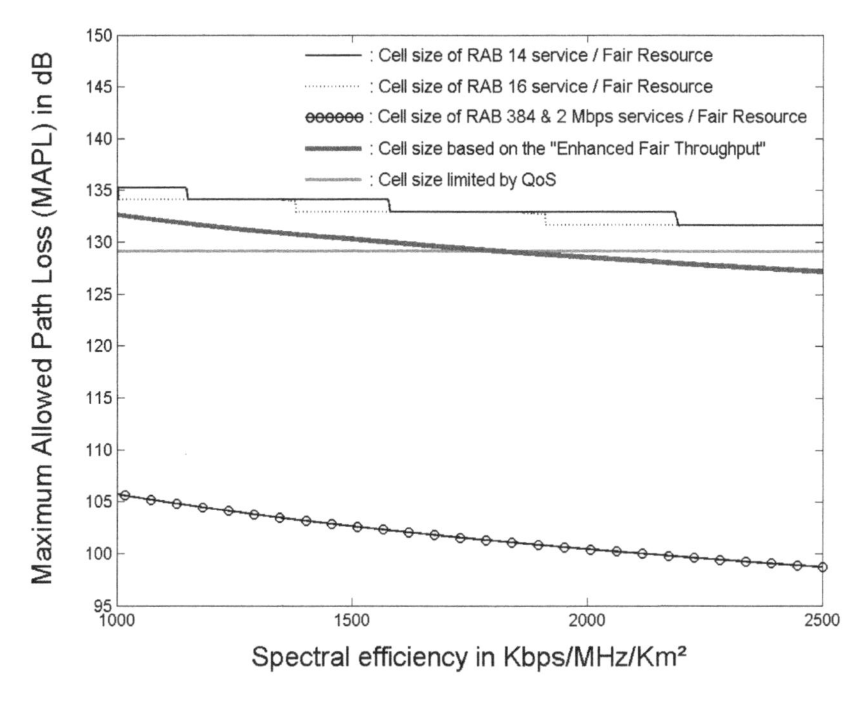

Fig. 1.11 Sizes of the different planned cells: Comparison between "Fair Resource" and "Enhanced Fair Throughput" (High traffic case & Configuration A of services distribution).

On the other hand, for a user density corresponding to a spectral efficiency above 1,8 Mb/s/MHz/Km^2, the QoS limits no more the dimensioning according M3.

The curve referring to RAB 128 service coverage appears in Figure 1.12 (Configuration B of services), whereas it is disconcerted with the most limiting service RAB 2 Mbps in Figure 1.11 (Configuration A of services). That was already explained in Figures 1.9 and 1.10 for the low traffic (In the configuration B, the MAPL decreases less fast than in the configuration A less loaded in terms of services bit rates).

1.6 Summary of Chapter

In this chapter, we have introduced HSPA dimensioning techniques with some practical examples in tutorial style using tables and abacuses to dimension a HSDPA network as a case study. We have also presented and compared

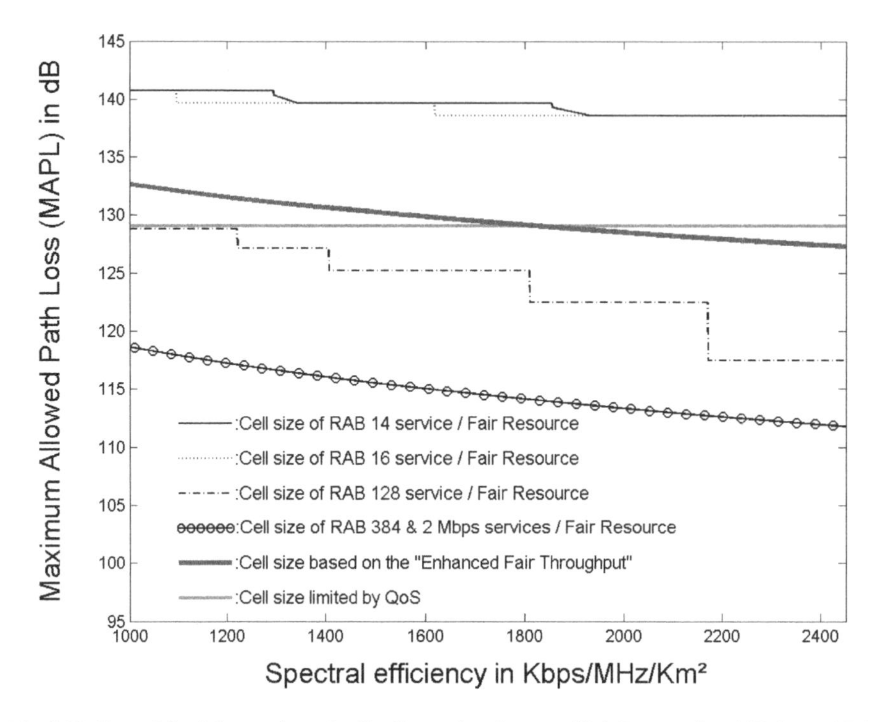

Fig. 1.12 Sizes of the different planned cells: Comparison between "Fair Resource" and "Enhanced Fair Throughput" (High traffic case & Configuration B of services distribution).

three dimensioning methods for HSPA based networks. We concluded that the "Enhanced Fair Throughput" method improves planning performance results.

Problems

Problem 1.1

Rewrite the HSPA formulae of this chapter by considering the impact of the users' traffic on intracellular interference.

Problem 1.2

Establish the fixed point equation in the DL similarly to (1.13) expression in the UL.

Problem 1.3

1) a) Determine the maximum allowed cell size offering a guaranteed bit rate of 144 Kb/s at the HSDPA cell edge at a coverage probability of 80% at the border of the cell and with a 6 dB as the shadowing standard-deviation value.

b) Is the answer of question a) refers to a coverage or capacity dimensioning?
c) By considering the HSPA codes limitation, restart the questions a) and b).

2) Restart the questions a), b) and c) in a basic UMTS network (Rel'99). What are the parameters in that should be changed for basic UMTS Rel'99 dimensioning.

N.B: To answer question 2) c), use codes with a spreading factor equal to 8.

Problem 1.4

In a HSPA based system, we assume a service area of 650 m^2 serving simultaneously 70 active mobiles using an aggregated services scheme composed of the following services:

— RAB 16: serving 60% of the total number of the users
— RAB 128: serving 30% of the total number of the users.
— RAB 2 Mbps: serving the remaining number of the users.

a) Compute the area spectral efficiency.

b) Use the appropriate abacuses in the last part of this chapter to dimension this system according to the three techniques M1, M2 and M3.

Compare the different results.

c) Assume that in addition to the previous aggregated services use, there are 25% of the RAB 16 mobile users using also RAB 128 service in two sessions simultaneously, restart the answering of questions a) and b).

References

[1] K. Sipilä, et al. "Estimation of Capacity and Required Transmission Power of WCDMA Downlink Based on a DL Pole Equation," Proc. of VTC 2000, pp. 1002–1005, May 2000.
[2] M. D. Yacoub, "Cell Design Principles," The Mobile Communications Handbook, 2nd Edition, by Jerry D Gibson, CRC Press, chapter 21, FL, USA, pp. 21-8–21-9, 1999.
[3] J. Laiho, A. Wacker, "Radio network planning process and methods for WCDMA," Annals of Telecommunications, Vol. 56, Nos. 5–6, Hermes Penton Science, pp. 317–331, May/June 2001.
[4] 3GPP TS 25.321 V6.0.0, "Technical Specification Group Radio Access Network; "MAC Protocol Specification," 3GPP Release 6, January 2004.

[5] R. Caldwell and A. Anpalagan, "HSDPA: An overview," IEEE Canadian Review, Spring 2004.
[6] T. E. Kolding, K. I Pederson, J. Wigard, F. Frederiksen, and P.E. Mogensen, "High Speed Downlink Packet Access: WCDMA Evolution," IEEE Vehicular Technology Society News, IEEE, February 2003.
[7] Holma Harri, Antti Toskala, "WCDMA for UMTS," John Wiley & Sons, Ltd, 3rd edition, 450 pages, 2004.
[8] T. E. Kolding, F. Frederiksen, and P.E. Mogensen, "Performance Aspects of WCDMA Systems with High Speed Downlink Packet Access (HSDPA)," Proceedings, VTC, Vol. 1, pp. 477–481, September 2002.
[9] 3GPP TS 25.308, "UTRA High Speed Downlink Packet Access (HSDPA); Overall description," 3GPP Release 5, Mars 2002.
[10] 3GPP TS 25.214, "Physical layer procedures (FDD)," 3GPP Release 6, December 2003
[11] F. Brouwer, et al. "Usage of Link–Level Performance Indicators for HSDPA Network–Level Simulations in E–UMTS," ISSSTA 2004, Sydney, 2004.

2

Part I
Overview of WiMAX Cellular Technology

Mehrnoush Masihpour and Johnson I Agbinya

University of Technology, Sydney Australia

2.1 Overview of WiMAX

2.1.1 Introduction to WiMAX technology

High cost of ICT (Information and Communication Technology) infrastructure leads to a gap between the developed countries and the least developed countries in terms of the access to information, known as "Digital Divides" [1]. Hence people in the countries with less financial resources lack the access to the digital information world, particularly the modern communication technologies. According to [1] the World wide interoperability for Microwave Access (WiMAX) is a potential solution to the "Digital Divides".

WiMAX is a new technology which provides broadband communication and supports user's mobility. By being cost effective and interoperable with other existing technologies, support of different user's mode such as fixed, nomadic, portable and mobile, flexible architecture and easy deployment while providing a high data rate communication in a wide coverage area; WiMAX is supposed to be the first truly global wireless broadband network. WiMAX can provide services such as Voice over IP (VOIP), IPTV, video conferencing, multiplayer interactive gaming and web browsing [2]. In addition, it is a functional alterative for the fixed networks in the case of natural disaster if the fixed networks are demolished and unable to perform.

Based on the standard IEEE802.16 known as WiMAX, it can support data rate up to 100 Mbps by using OFDM (Orthogonal Frequency Division

Multiplexing) modulation scheme, which will be discussed in detail in the Section 2.3.6. WiMAX base stations also can support a coverage range of up to 70 Km, by using smart antennas such as Multi Input Multi Output (MIMO) technique, which also results in higher throughput and performance enhancement [2]. This chapter will briefly overview the WiMAX technology first and then progress onto how to plan WiMAX networks.

2.1.2 WiMAX Physical Layer

According to the IEEE802.16, WiMAX can support four physical specifications at the physical layer as follows [3, 4]:

- Wireless MAN-SC (Signal Carrier, 10-66 GHz)
- Wireless MAN-Sca (below 11 GHz), which uses single carrier modulation
- Wireless MAN-OFDM (below 11 GHz), which uses orthogonal frequency division multiplexing and a 256-carrier OFDM scheme.
- Wireless MAN-OFDMA (below 11 GHz), which stands for Orthogonal Frequency Division Multiple Access. This air interface supports multiple user accesses, as it allocates a subset of the carriers to one receiver. However, a 2048-carrier OFDM is the scheme that is used by MAN-OFDMA [4].

Three air interfaces were defined, wireless MAN-SCa, wireless MAN-OFDM and MAN-OFDMA, which performs in the frequency band below 11GHz, can support Non-line-of-sight communication [3].

2.1.3 WiMAX MAC Layer

As IEEE802.16 standard suggests, the MAC layer of WiMAX can support:

- Reliability, as it is a connection-oriented protocol.
- Quality of service for subscriber, by adaptive allocation of uplink and downlink traffic [4].
- Compatibility with different transport protocols such as IPv4, IPv6, Ethernet, and Asynchronous Transfer Mode (ATM) [4].
- Two different connection modes, which are known as point-to-multipoint (PMP) mode and Mesh mode for multihop ad hoc networks.

- Frequency Division Duplex (FDD) and Time Division Duplex (TDD) transmission modes.

2.1.4 Quality of Service in WiMAX

WiMAX provides three different services with different required quality of services [2]. The first service is Unsolicited Grant Service (UGS), which needs guaranteed quality of service (QoS), therefore WiMAX assigns a certain amount of bandwidth to each connection statically to reduce delay and jitter. Commercial IPTV is an example which requires UGS. The second is Polling Service (PS), which requires less QoS guarantee compare to UGS; hence WiMAX allocates the bandwidth dynamically. Finally, Best Effort service (BE) requires no QoS guarantees; as a result unused bandwidth by UGS and PS traffic is assigned to BE service. For instance, web browsing and email traffic use this type of service [2].

2.2 WiMAX Network Architecture

Network architecture refers to a framework design in which the physical components of the network are connected and their functions are specified in the manner that all the components operate properly and work with each other to provide different services to the users. WiMAX architecture is based on the Internet Protocol (IP) which means the data transmission is all based on the packet switched technology, and therefore provides flexibility and modularity for the network deployment [1]. With support of IP, WiMAX can provide a range of coverage options such as small and large scale, urban and suburban also rural areas. However, it also supports fixed, nomadic, portable and mobile usage models [1]. Through its core network it can provide internetworking and interoperability with other networks.

In this section a general WiMAX network architecture and its components and the functions associated with each component is introduced.

2.2.1 Different Sub-networks

As it can be seen from the Figure 2.1, the WiMAX network is divided into two main subnets, which work seamlessly with each other to provide network access for the different types of users. The network includes Access

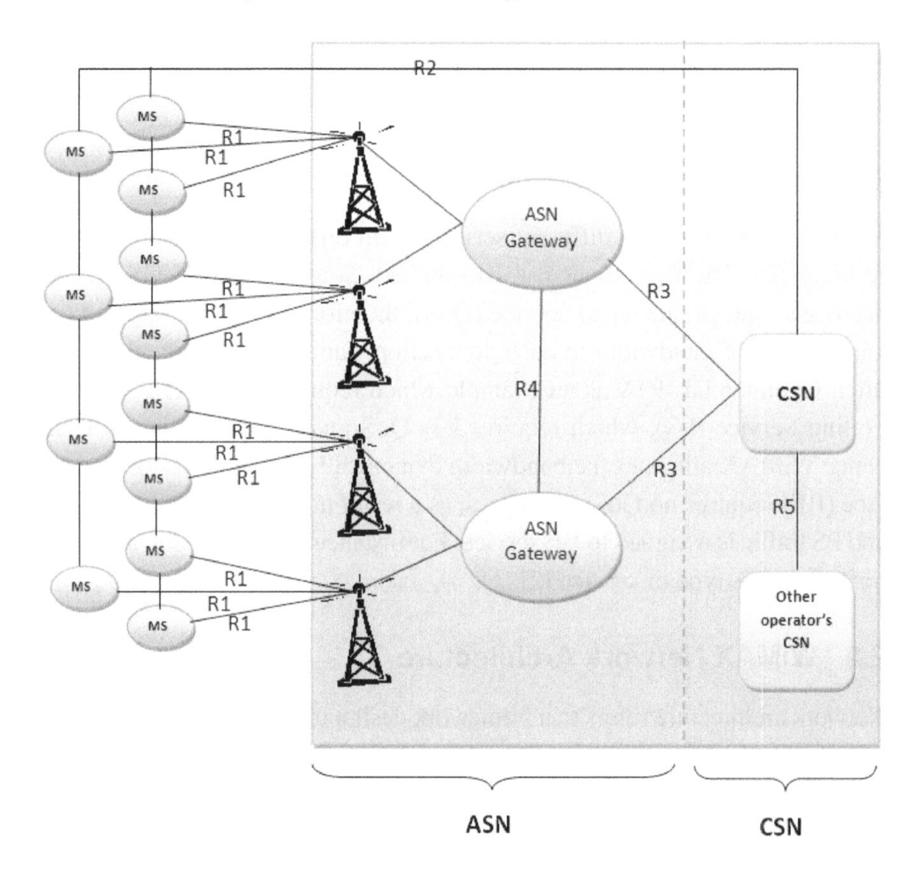

Fig. 2.1 WiMAX network architecture (Adapted from [1]).

Service Network (ASN) and Core Service Network (CSN) [1]. Each of these networks has their own components and functions, which are explored in the following sections.

To plan a logical WiMAX network, the architecture needs to address some important requirements. As cited in [1], the architecture shall provide support for interoperability of different vendors by dividing the network in terms of functionality. Moreover it should support divergent deployment schemes such as centralized, fully and semi distributed implementations and variety of usage models such as fixed, nomadic, portable and mobile. Internetworking with the other communication technologies such as Wi-Fi, 3GPP/3GPP2, also wired networks is essential. However, WiMAX architecture needs to underpin different business modes such as NAP (Network Access Provider), NSP (Network Service Provider) and ASP (Application Service Provider) [1].

2.2.1.1 Access Service Network

The ASN consists of one or a number of WiMAX Base Stations (BS) and one or more WiMAX Gateways (GW) [1]. It provides the interface between the user and CSN. Base station is responsible for provision of an interface between the user and the WiMAX network. A base station is typically located in the center or close to the borders of a WiMAX cell and have different radius range based on the elements such as the coverage area, environmental effects and the number of the users, also mobility of the subscribers. As well as the base station, there are radio equipment and the base station link to the backbone network in each cell.

Gateway is another component of ASN, which provides the connectivity between the ASN and CSN [1]. It is responsible for mobility management and connection management. By processing the user's control and bearer data traffic it performs as a provider of the inter-service network boundaries.

As WiMAX forum suggests, ASN performs a number of functions which are outlined as follows [1, 5]:

- *Radio resource management* [5]: ASN enhance the efficiency of the radio resources employment by radio resource management. This function is divided into two functional entities, which are radio resource agent and radio resource controller. Radio resource agent gathers and measures radio resource indicators to communicate with the radio resource controller if necessary also to control the local radio resource. Radio resource controller collects the radio resource indicators from the radio resource agents associated with it also to communicate with other radio resource controllers. A radio resource agent is embedded in the base station while a radio resource controller is located in each base station and gateway.
- *Mobility management* [5]: In case of deployment of a mobile WiMAX network, it is crucial to have mobility management to minimize the undesirable effects of mobility such as handoff delay or packet loss.

- *IP address assignment [5]*: Based on the usage model, ASN uses static or dynamic IP assignment. To perform that, ASN uses DHCP (Dynamic Host Configuration Protocol) for dynamic IP allocation or either via DHCP or manually in mobile Station in static IP allocation.
- *Authentication through proxy Authentication, Authorization and Accounting (AAA) Service [1]*: ASN provides authenticating and security service by an AAA server. It is also responsible for collecting the information about the usage of the resources by the subscribers and facilitates the charging process.

Handoff, interoperability with the other ASNs and relay of functionality between CSN and mobile station are other functions of the ASN [1].

NAP (Network Access Provider) is the organization that provides access to the network service provider (NSP) and benefits through the marketing of applications and services that WiMAX can offer to the users [1, 5]. Some of the services that WiMAX provides to the subscribers are VoIP, video conferencing, streaming video, interactive gaming, mobile instant message, IPTV and basic broadband wireless Internet.

2.2.1.2 Core Service Network

Core Service Network (CSN) is the other part of the WiMAX network, which consists of the Home Agent (HA), which provides roaming through the network, AAA system to ensure the unique identification of the customers, IP server and gateway to the other networks such as 3G for internetworking. CSN is mostly responsible for authentication, switching and transport. Some of the main functions of the WiMAX CSN are briefly mentioned below [1, 5]:

- Connectivity with other networks also the Internet and ASPs
- Management of the IP addressing
- Providing AA services to the users
- Tunneling support
- Roaming between NSPs also between ASNs
- Mobility and location management
- Quality and policy control according to the contracts with the users

A network service provider (NSP), which is an organization that owns the CSN, provides core network services to the WiMAX network [5].

2.2.2 WiMAX Air Interfaces

Figure 2.1 presents the main open interfaces used by WiMAX. As can be seen, interface R1 provides a connection between the subscribers and the base station within the ASN. A logical interface between the mobile station and the core service network, which corresponds to authentication, IP configuration, mobility and service management, is defined as R2. R3 is the air interface between the ASN and CSN, which is associated with the AAA, policy enforcement and mobility management. Different ASNs are connected together through the R4 interface and mobility of the mobile stations between two ASNs is the concepts that R4 is dealing with. Finally R5, which deals with internetworking between different CSNs, such as roaming, is the interface between the CSNs.

2.3 Mobile WiMAX

2.3.1 Mobile WiMAX PHY

The mobile WiMAX PHY uses a combination of TDD and OFDMA for downlink and uplink signaling and multiple user access. The unique features within the TDD/OFDMA frame provide frequency diversity, frequency reuse, and cell segmentation which improve the performance against fading and inter-cell interference.

✓ *TDD*
The WiMAX OFDMA frame is configured to support a point-to-multipoint network. The 802.16e PHY supports TDD, FDD, and half-duplex FDD operation. The initial release of the Mobile WiMAX profile will only include TDD as shown in Table 2.1. Future releases may include FDD variants to match spectrum regulatory requirements in specific countries. For interference mitigation, system-wide synchronization is required when using TDD. Synchronization is typically achieved using a global positioning system (GPS) reference at the BS. In the event that network synchronization is lost, the BS will continue to operate until synchronization is recovered, using a local frequency reference. TDD, as specified in the WiMAX profile,

enables asymmetric DL and UL traffic. Asymmetric traffic using TDD may improve the spectrum utilization and system efficiency as compared to FDD operation which typically requires equal UL and DL bandwidths. TDD uses a common channel for both UL and DL transmission allowing for a lower cost and less complex transceiver design. TDD also assures channel reciprocity which may benefit applications such as MIMO and other advanced antenna technologies. The TDD form of WiMAX uses a frame that is divided into a DL and an UL section.

✓ *Time and frequency parameters*
The IEEE 802.16e air interface as adopted by the WiMAX Forum specifies channel bandwidths ranging from 1.25 to 20 MHz. The first release of the mobile WiMAX system profile incorporated 5, 7, 8.75, and 10 MHz bandwidths as shown in Table 2.1. The bandwidth scalability in Mobile WiMAX OFDMA is achieved by adjusting the FFT size and the subcarrier spacing. For a given channel bandwidth, the subcarrier spacing is inversely proportional to the number of subcarriers and, therefore, the FFT size. The time duration of the OFDMA symbol is set by the inverse of the subcarrier spacing. Therefore by fixing the subcarrier spacing, the symbol time is automatically specified. The inverse relationship between subcarrier spacing and symbol duration is a necessary and sufficient condition to ensure that the subcarriers are orthogonal. Table 2.1 shows the subcarrier spacing and symbol time for the Mobile WiMAX 10 and 8.75 MHz (WiBRO) profiles using nominal bandwidths of 10 and 8.75 MHz respectively.

The Mobile WiMAX frame contains 48 symbols. The symbol time contains the actual user data and a small extension called the guard time. The guard time is a small copy from the end of the symbol that is inserted before the start of the symbol. This guard time is also called the cyclic prefix (CP) and its length is chosen based on certain assumptions about the wireless channel. As long as the CP interval is longer than the channel delay spread, inter-symbol interference (ISI) introduced by the multi-path components can be eliminated.

The 802.16 standard specifies a set of CP values but the initial profile specifies a CP value of 1/8, meaning that the guard time is 1/8 the length of

Table 2.1 WiMAX and WiBRO time and frequency parameters, using a1024-point FFT.

Parameter	Mobile WiMAX		WiBRO
Nominal bandwidth	10 MHz	7 MHz	8.75 MHz
Subcarrier spacing	10.9375 kHz	7.8125 kHz	9.7656 kHz
Useful symbol time (Ts = 1/Subcarrier spacing)	91.4 μs	128 μs	102.4 μs
Guard time (Tg = Ts/8)	11.4 μs	16 μs	12.8 μs
OFDMA symbol duration (Ts + Tg)	102.9 μs	144 μs	115.2 μs
Number of symbols in frame	47	33	42
TTG+RTG	464 PS	496 PS	404 PS
Frame length	5 ms	5 ms	5 ms
Sampling frequency (Fs = FFT points × subcarriers spacing)	11.2MHz	8 MHz	10 MHz
Physical slot (PS) ($/Fs)	357.14 ns	500 ns	400 ns

the symbol time. Table 2.1 shows the guard time and symbol duration for the Mobile WiMAX and WiBRO using the nominal bandwidth of 10 and 8.75 MHz respectively.

There are a number of significant differences in the UL signal compared to the DL. They reflect the different tasks performed by the BS and MS, along with the power consumption constraints at the MS. Differences include:

- No preamble, but there are an increased number of pilots. Pilots in the UL are never transmitted without data subcarriers
- The use of special CDMA ranging bursts during the network entry process
- Data is transmitted in bursts that are as long as the uplink sub-frame zone allows, and wrapped to further sub-channels as required

✓ *Preamble*

The DL sub-frame always begins with one symbol used for BS identification, timing synchronization, and channel estimation at the MS. This symbol is generated using a set of 114 binary pseudo random number (PN) sequences, called the preamble ID, of 568 length. The data in the preamble is mapped to every third subcarrier, using BPSK, giving a modest peak-to-average power ratio (compared to the data sub-channels). The preamble subcarriers are boosted by a factor of eight over the nominal data subcarrier level. There are no preambles in the UL except for systems using adaptive antenna systems (AAS). For the case when there is no UL preamble, the BS will derive the required channel information based on numerous pilot subcarriers embedded in the UL sub-channels.

✓ *FCH*

The FCH follows the DL preamble with a fixed location and duration. The FCH contains the downlink frame prefix (DLFP). The DLFP specifies the sub-channelization, and the length and coding of the DL-MAP. The DLFP also holds updates to the ranging allocations that may occur in subsequent UL sub-frames. In order that the MS can accurately demodulate the FCH under various channel conditions, a robust QPSK rate 1/2 modulation with four data repetitions is used.

✓ *DL-MAP and UL-MAP*

The DL-MAP and UL-MAP provide sub-channel allocations and control information for the DL and UL sub-frames. The MAP will contain the frame number, number of zones, and the location and content of all bursts. Each burst is allocated by its symbol offset, sub-channel offset, number of sub-channels, number of symbols, power level, and repetition coding.

✓ *Channel coding*

There are various combinations of modulations and code rates available in the OFDMA burst. Channel coding includes the randomization of data, forward error correction (FEC) encoding, interleaving, and modulation. In some cases, transmitted data may also be repeated on an adjacent subcarrier.

✓ *Randomization*

Randomization of the data sequence is typically implemented to avoid the peak-to average power ratio (PAPR) increasing beyond that of Gaussian noise, thus putting a boundary on the nonlinear distortion created in the transmitter's power amplifiers. It can also help minimize peaks in the spectral response.

✓ *FEC*
The mobile WiMAX OFDMA PHY specifies convolutional coding (CC), convolutional turbo coding (CTC), and repetition coding schemes. When repetition coding is used, additional blocks of data are transmitted on an adjacent sub-channel. CTC can give about a 1 dB improvement in the link performance over CC.

✓ *Interleaving*
Interleaving is a well known technique for increasing the reliability of a channel that exhibits burst error characteristics. Interleaving involves reordering the coded data, which spreads any errors from burst of interference over time, increasing the probability of successful data recovery.

✓ *Modulation*
There are three modulation types available for modulating the data onto the subcarriers: QPSK, 16QAM, and 64QAM. In the UL, the transmit power is automatically adjusted when the modulation coding sequence (MCS) changes to maintain the required nominal carrier-to-noise ratio at the BS receiver. 64QAM is not mandatory for the UL. Binary phase shift keying (BPSK) modulation is used during the preamble, on the pilots, and when modulating subcarriers in the ranging channel.

The BS scheduler determines the appropriate data rates and channel coding for each burst based on the channel conditions and required carrier-to-interference plus noise ratio (CINR) at the receiver. Table 2.2 shows the achievable data rates using a 5 and 10 MHz channel for both DL and UL transmissions.

Matrix A space time coding, Matrix B spatial division multiplexing (also known as multi input multi output (MIMO))

Zones can be configured to make use of multi-antenna technology, including phased array beamforming, STC, and MIMO techniques. Matrix A is an Alamouti-based transmit diversity technique, which involves taking pairs of symbols and time-reversing each pair for transmission on a second antenna.

Table 2.2 Mobile WiMAX PHY data rates.

		5 MHz Channel		10 MHz Channel	
Modulation	Code rate	DL Data rate (Mbps)	UL Data rate (Mbps)	DL Data rate (Mbps)	DL Data rate (Mbps)
QPSK	1/2CTC, 6x	0.53	0.38	1.06	0.78
	1/2 CTC, 4x	0.79	0.57	1.58	1.18
	1/2CTC, 2x	1.58	1.14	3.17	2.35
	1/2 CTC, 1x	3.17	2.28	6.34	4.70
	1/2CTC	4.75	3.43	9.50	7.06
16QAM	1/2CTC	6.34	4.57	12.07	9.41
	3/4CTC	9.50	6.85	19.01	14.11
64QAM	1/2CTC	9.50	6.85	19.01	14.11
	2/3CTC	12.67	9.14	26.34	18.82
	3/4CTC	14.26	10.28	28.51	21.17
	5/6CTC	15.84	11.42	31.68	23.52

Matrix B uses MIMO spatial division multiplexing to increase the channel capacity. For downlink MIMO, user data entering the BS is split into parallel streams before being modulated onto the OFDMA subcarriers. As with the single channel case, channel estimation pilots are interleaved with the data subcarriers. For MIMO operation, the pilots are made unique to each transmit antenna to allow a dual receiver to recover four sets of channel coefficients. This is what is needed to remove the effect of the signal coupling that inevitably occurs between transmission and reception.

The MS is initially only required to have one transmit antenna and support open loop MIMO. More advanced, closed loop MIMO operates by the MS transmitting regular encoded messages back to the BS, which provide the closest approximation to the channel seen by the MS. The BS then pre-codes the MIMO signal before transmission, according to the channel state information (CSI) provided by the MS. The BS may also have the facility to control the single transmitters from two MSs to act together to create a collaborative MIMO signal in the UL.

Matrix A and Matrix B techniques can be applied to PUSC and AMC zones, to be described next.

The IEEE 802.16e-2005 standard provides the air interface for WiMAX but does not define the full end-to-end WiMAX network. The WiMAX Forum's Network Working Group (NWG), is responsible for developing the end-to-end network requirements, architecture, and protocols for WiMAX, using IEEE 802.16e-2005 as the air interface.

The WiMAX NWG has developed a network reference model to serve as an architecture framework for WiMAX deployments and to ensure interoperability among various WiMAX equipment and operators.

The network reference model envisions unified network architecture for supporting fixed, nomadic, and mobile deployments and is based on an IP service model. Below is simplified illustration of an IP-based WiMAX network architecture. The overall network may be logically divided into three parts:

1. Mobile Stations (MS) used by the end user to access the network.
2. The access service network (ASN), which comprises one or more base stations and one or more ASN gateways that form the radio access network at the edge.
3. Connectivity service network (CSN), which provides IP connectivity and all the IP core network functions.

The network reference model developed by the WiMAX Forum NWG defines a number of functional entities and interfaces between those entities. Figure below shows some of the more important functional entities.

- **Base station (BS):** The BS is responsible for providing the air interface to the MS. Additional functions that may be part of the BS are micromobility management functions, such as handoff triggering and tunnel establishment, radio resource management, QoS policy enforcement, traffic classification, DHCP (Dynamic Host Control Protocol) proxy, key management, session management, and multicast group management.
- **Access service network gateway (ASN-GW):** The ASN gateway typically acts as a layer 2 traffic aggregation point within an ASN. Additional functions that may be part of the ASN gateway include intra-ASN location management and paging, radio resource management and admission control, caching of subscriber profiles and encryption keys, AAA client functionality, establishment and management of mobility tunnel with base stations, QoS and policy enforcement, foreign agent functionality for mobile IP, and routing to the selected CSN.

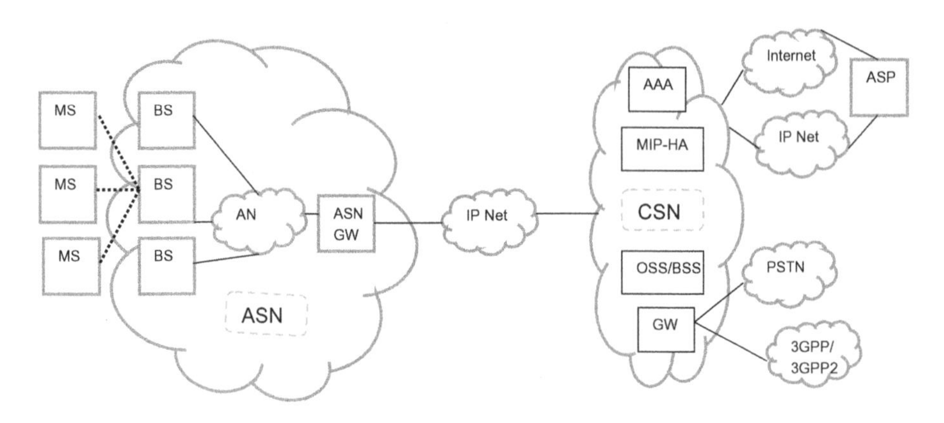

Fig. 2.2 IP-based WiMAX network architecture.

- **Connectivity service network (CSN) or Core Service Network:** The CSN provides connectivity to the Internet, ASP, other public networks, and corporate networks. The CSN is owned by the NSP and includes AAA servers that support authentication for the devices, users, and specific services. The CSN also provides per user policy management of QoS and security. The CSN is also responsible for IP address management, support for roaming between different NSPs, location management between ASNs, and mobility and roaming between ASNs.

The WiMAX architecture framework allows for the flexible decomposition and/or combination of functional entities when building the physical entities. For example, the ASN may be decomposed into base station transceivers (BST), base station controllers (BSC), and an ASNGW analogous to the GSM model of BTS, BSC, and Serving GPRS Support Node (SGSN).

The WiMAX physical layer is based on orthogonal frequency division multiplexing. OFDM is the transmission scheme of choice to enable high-speed data, video, and multimedia communications and is used by a variety of commercial broadband systems, including DSL, Wi-Fi, Digital Video Broadcast-Handheld (DVB-H), and MediaFLO, besides WiMAX.

OFDM is an elegant and efficient scheme for high data rate transmission in a non-line-of-sight or multipath radio environment

2.3.2 Sub-carrier Grouping and Sub-channels

Grouping of multiple subcarriers into sub-channels is used to improve system performance. There are two types of subcarrier allocations to form sub-channels. These are distributed and adjacent subcarrier allocations.

In *distributed allocations* the subcarriers are pseudo-randomly distributed over the available bandwidth. Distributed allocation provides inter-cell interference averaging and frequency diversity in frequency-selective fading channels. Distributed allocation of subcarriers is preferred for mobile applications.

In *adjacent allocations* subcarriers adjacent to each other in the frequency domain are grouped to form sub-channels. Adjacent allocation has advantages in slowly fading channels, frequency nonselective channels and for implementing adaptive modulation and coding (AMC). Adjacent allocation is used in low mobility and fixed applications. For this the subscriber may be assigned the sub-channel with the best frequency response.

2.3.3 Slot Allocation and Data Regions

The WiMAX PHY layer allocates slots and framing over the air. A slot is defined as the minimum time-frequency resource that can be allocated by a WiMAX system to a given link. Each slot consists of one sub-channel over one, two, or three OFDM symbols. This depends on the sub-channelization scheme used. A user data region is a contiguous series of slots assigned to the given user. This is done by the scheduling algorithms. Data regions are allocated to users based on quality of service requirements, demand and the condition of the channel. Data is mapped to physical subcarriers in two steps.

a) The first step is controlled by the scheduler. Data is mapped to one or more logical sub-channels called slots. Slots may be grouped and assigned to segments based on applications. This can be used by a BS for different sectors in a cellular network.
b) In the second step, logical sub-channels are mapped to physical subcarriers. During this process, pilot subcarriers are also assigned. Data and pilot subcarriers are uniquely assigned based on the type of sub-channelization.

To understand the sub-channel planning, consider the following table with the parameters shown. The size of the FFT varies with the throughput sought and the available frequency bandwidth.

In Mobile WiMAX the size of the FFT varies from 128, 256, 512, 1024 to 2048 subcarriers (tones). These sizes of FFT correspond to the bandwidths of 1.25 MHz, 3.5, 5, 10 and 20MHz respectively. These choices allow the subcarrier spacing to be constant at 15.625 kHz and 10.94 kHz for fixed and Mobile WiMAX respectively.

The basic resource unit in WiMAX is the symbol duration. The subcarrier value defines the symbol duration. In the fixed case the useful symbol duration is

$$T_{sf} = \frac{1}{f_{subcarrier}} = \frac{1}{15625\,Hz} = 64\mu s \tag{2.1}$$

This duration does not account yet for the pilot and null subcarrier times. For the Mobile WiMAX case, the useful symbol duration is

$$T_{sm} = \frac{1}{f_{subcarrier}} = \frac{1}{10940\,Hz} = 91.4\,\mu s \tag{2.2}$$

Mobility has been considered in the definition of the carrier spacing for Mobile WiMAX. The value of 10.94kHz was chosen as a balance between fulfilling the requirements for Doppler spread and delay spread. In the Table below, the mobile WiMAX sub-carriers listed are for downlink PUSC.

When a terminal moves its carrier frequency changes in proportion to its speed. The chosen subcarrier spacing can support delay spreads of 20 microseconds for vehicular speeds of up to 125 km/hr when operating around the 3.5 GHz range. In addition to the bandwidths specified for Mobile WiMAX additional bandwidth profiles are permitted (Table 2.3). As an example, WiBro uses 8.75 MHz with a 1024 FFT size.

In a WiMAX transmission therefore there is a DC subcarrier at the the transmission frequency and it is not used for carrying data, the pilot subcarriers that are used for synchronization, the guard subcarriers and the data carrying subcarriers (Figure 2.3). The guard subcarriers are the outer subcarriers. Thus the total number of subcarriers used for a WiMAX system is: $N_{FFT} = N_{Guard\ left} + N_{used(\max)} + N_{DC} + N_{Guard\ right}$.

Table 2.3 OFDM parameters used in fixed and mobile WiMAX.

Parameter	Fixed WiMAX OFDM-PHY	Mobile WiMAX Scalable OFDMA-PHY			
Channel bandwidth (MHz)	3.5	1.25	**5**	10	20
FFT size	256	128	**512**	1,024	2,048
Number of used data sub-carriers	192	72	**360**	720	1,440
Number of pilot sub-carriers	8	12	**60**	120	240
Number of null/guard band sub-carriers	56	44	**92**	184	368
Cyclic prefix or guard time (Tg/Tb)	1/32, 1/16, **1/8**, 1/4	None	None	None	None
Oversampling rate (Fs/BW)	Depends on bandwidth: 7/6 for 256 OFDM, 8/7 for multiples of 1.75 MHz, and 28/25 for multiples of 1.25 MHz, 1.5 MHz, 2 MHz, or 2.75 MHz.				
Sub-carrier frequency spacing (kHz)	15.625		**10.94**		
Useful symbol time (μs)	64		**91.4**		
Guard time assuming 12.5% (μs)	8		**11.4**		
Duration of OFDM symbol (μs)	72		**102.86**		
Number of OFDM symbols in a 5 ms frame	69		**48.0**		

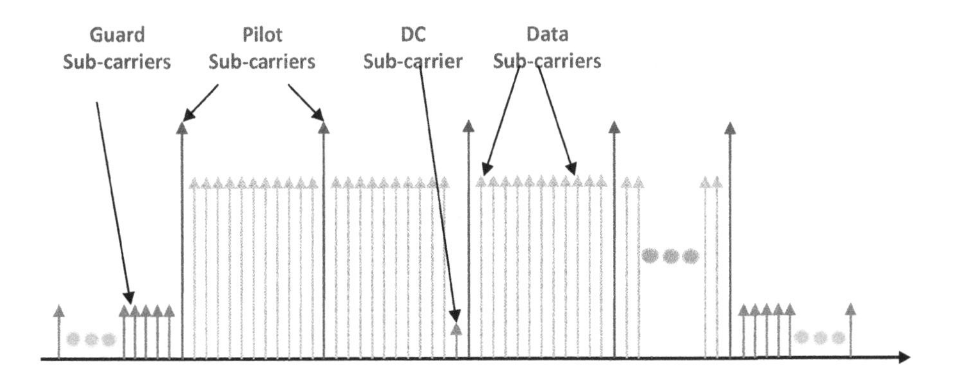

Fig. 2.3 Sub-carrier structure in WiMAX channels.

2.3.4 Permutation Zones

Permutation zones, or zones, are groups of contiguous symbols that use a specific type of sub-channel assignment. The OFDMA PHY specifies seven subcarrier permutation zone types: FUSC, OFUSC, PUSC, OPUSC, AMC, TUSC1, and TUSC2 [6]. PUSC, FUSC, and AMC are widely used in practical systems. "Except for AMC, the other zones use distributed allocation of subcarriers for sub-channelization. A frame may contain one or more zones. The DL sub-frame requires at least a zone and always starts with PUSC" [7].

Sub-channel allocation need not be contiguous. Subcarriers may be common to all sub-channels (in FUSC) or each sub-channel may allocate its own pilot carriers (PUSC).

***DL FUSC*:** In the DL Fully Used Sub-channelization (FUSC) the pilot subcarriers are common to all sub-channels and are first allocated. After this the remaining subcarriers are divided into sub-channels. In the DL and UL Partially used sub-channelization (PUSC), the set of subcarriers that are available for use data transmission and pilot are first divided into sub-channels. Then within each sub-channel, pilot subcarriers are allocated. It is reasonable for the DL FUSC to clearly specify the pilot tones because in a DL, sub-channels may be intended by a base station for different or groups of receivers while in UL, Subscriber Stations (SS) may be assigned one or more unique sub-channels by a base station and several transmitters may transmit simultaneously with the base station. This requires the specification of also known pilot subcarriers for each SS within its set of subcarriers.

***DL PUSC [7]*:** Clusters of 14 contiguous subcarriers per symbol are grouped together and two clusters form a sub-channel. The slot in this case is one sub-channel over two OFDM symbols. This default zone is required at the start of all DL sub-frames following the preamble. In this zone, on alternate symbols, pairs of pilots swap positions, averaging one in seven of the subcarriers. For dedicated pilots they are only transmitted for corresponding data. In a DL PUSC zone the sub-channels can also be mapped into larger groups called segments. The first PUSC zone is always a single input single output (SISO). Further PUSC zones can be specified for other forms of multi-antenna MIMO systems.

***UL PUSC [7]*:** "For this zone type, four contiguous subcarriers are grouped over three symbols. This grouping is called *a tile*. Six tiles make a sub-channel. For the UL PUSC, the slot is defined as one sub-channel that occurs over the three symbols. Pilots are incorporated within the slot, their position changing with each symbol. Over the course of one tile, one in three subcarriers is a pilot"[7].

***AMC*:** The AMC zone occupies a wider bandwidth compared with PUSC and FUSC [719. The sub-channel is in this case a contiguous block of subcarriers. The zone structure is the same for both the DL and the UL. The slot is defined as a collection of bins using the N $\times$ M formula, where M is the number of OFDM symbols and N is the number of bins. A bin or symbol consists of nine contiguous subcarriers. A slot is one sub-channel wide and the length changes according to the zone. The pilots in the DL change positions periodically in a rotating pattern every fourth symbol.

***DL FUSC*:** The DL FUSC zone uses all subcarriers to provide a high degree of frequency diversity. The subcarriers are divided into 48 groups of 16 subcarriers and a sub-channel is formed by taking one subcarrier from each group. For this zone a slot is defined as one sub-channel over one OFDMA symbol. "The pseudo-random distribution of data changes with each OFDMA symbol over the length of the zone, which can be useful when attempting to mitigate interference through the use of what is effectively a type of frequency hopping. The pilots are regularly distributed. Their position alternates with each symbol" [7].

In the *DL optional FUSC (DL OFUSC)* pilot subcarriers are evenly spaced by eight data subcarriers from each other. In the UL OPUSC the zone is identical with the UL PUSC "except that it uses a tile size that is three subcarriers wide by three symbols long" [7].

***TUSC1 and TUSC2*:** The total usage of sub-channels (*TUSC1 and TUSC2*) zones is only available in the DL using AAS. "They are both optional and similar to DL PUSC and OPUSC but use a different equation for assigning the subcarriers within the sub-channel" [7].

With the exception of the DL PUSC, which is assigned after the DL preamble, all of the zones can be assigned in any order within the frame. The switching points between zone types are listed in the DL MAP. Figure 2.4 shows an example of an OFDMA frame with several different types of zones [7]. From Figure 2.4 the mandatory DL PUSC zone follows the preamble in the frame. "The DL sub-frame also shows a second PUSC zone, a FUSC zone, and an AMC zone. The UL sub-frame follows the TTG and, in this example, contains a PUSC, OPUSC and AMC" [7].

2.3.5 Frame Structure

The OFDMA frame consists of a DL sub-frame and an UL sub-frame. The flexible frame structure of the TDD signal consists of a movable boundary between the DL and UL sub-frames. A short transition gap is placed between the DL and UL sub-frames and is called transmit-receive transition gap (TTG). After the completion of the UL sub-frame, another short gap is added between this

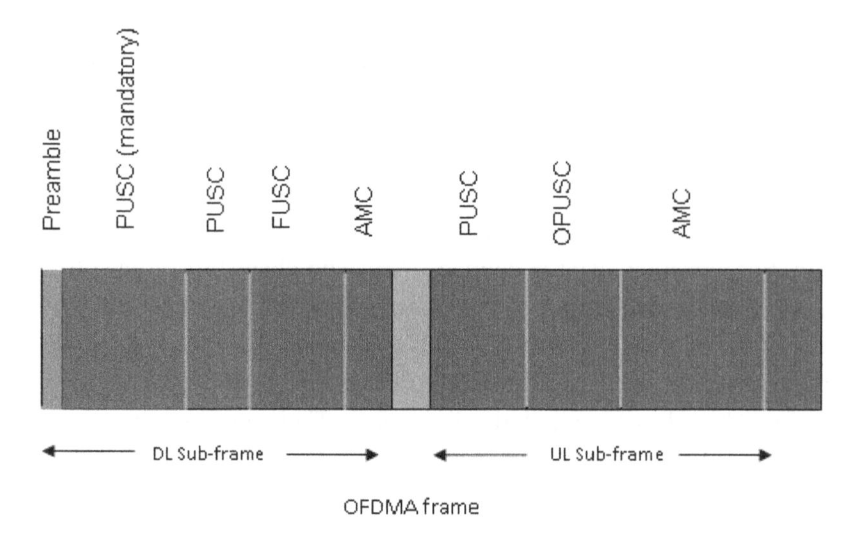

Fig. 2.4 Example of an OFDMA frame with multiple zones.

sub-frame and the next DL sub-frame. This gap is called the receiver-transmit transition gap (RTG). The minimum time durations for these transition gaps are called out in the 802.16 standard and are a function of the channel bandwidth and the OFDM symbol time. It is typical to define these transition gaps in terms of physical slot (PS) units. A PS is a unit of time defined as 4/(sampling frequency).

An example of a mobile WiMAX frame is shown in Figure 2.5. This figure shows the time-frequency relationship where the symbol time is shown along the x-axis and the logical sub-channels along the y-axis. Logical sub-channels are groupings of frequency subcarriers assigned to individual users. The concept of sub-channels and zones will be covered later in this application note. Figure 2.5 shows the DL and UL sub-frames separated by the TTG and ending with the RTG. The figure also shows the relative position of the preamble, frame control header (FCH), downlink media access protocol (DL-MAP), and uplink media access protocol (UL-MAP) whose functions will be discussed in the next section.

2.3.6 How OFDM Operates

To allocate a number of sub-channels to the sub-streams, without the need for independent radio frequencies, OFDM uses the DFT (Discrete Fourier

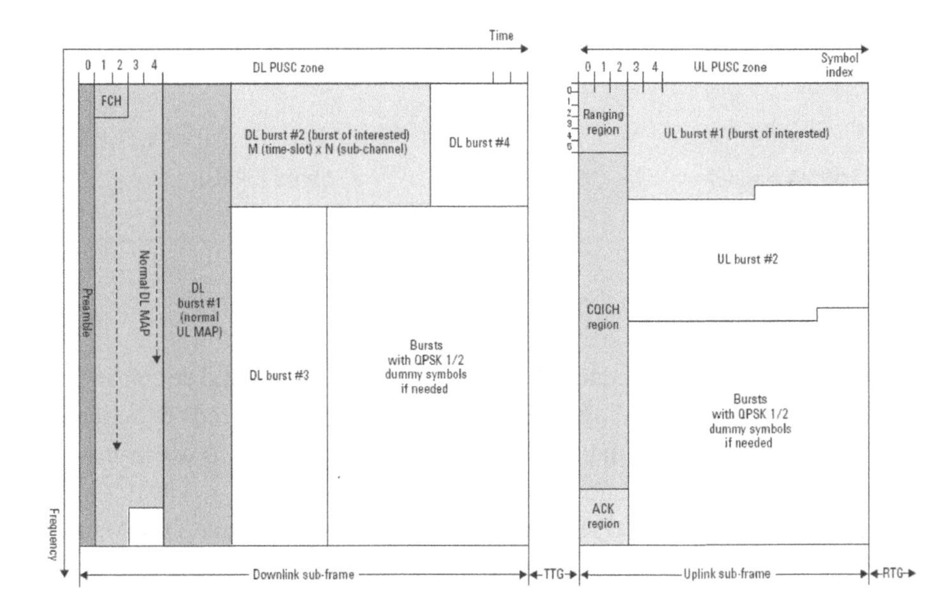

Fig. 2.5 OFDMA frame structure for TDD.

Transform) technique and specifically an efficient form of DFT known as FFT (Fast Fourier Transform) and the inverse of the technique, IFFT. Using these two techniques OFDM can divide a single radio channel into L sub-channels.

In the physical layer the WiMAX signal to be transmitted can be represented by the expression:

$$y(t) = Re\left\{ e^{j2\pi f_c t} \times \sum_{\substack{k=-N_{used}/2 \\ k\neq 0}}^{N_{used}/2} C_k e^{j2\pi k \Delta f (t-T_g)} \right\} \tag{2.3}$$

Where in this expression the variables are
Δf is the subcarrier frequency spacing
f_c is the central carrier frequency and
T_g is the guard time
N_{used} is the number of used subcarriers
C_k is a complex number representing the data to be transmitted
k is the frequency offset or subcarrier index

An OFDM symbol is formed from L data symbols and lasts for T seconds, which is equal to LT_s. To prevent the interference between the OFDM symbols

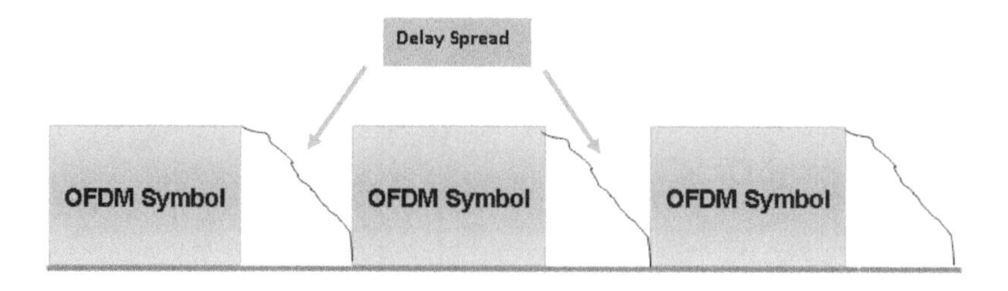

Fig. 2.6 Guard time should be larger than delay spread (Adapted from[8]).

transmitting through a channel, a guard time (Figure 2.6) should be considered between each two OFDM symbols. The guard time (T_g), needs to be larger than the delay spread of the channel and also large enough to overcome the interference between the subsequent OFDM symbols.

However, the intra OFDM-symbols interference still remains. OFDM technique solves this problem by using circular convolution. Assuming x[n] as the channel input, y[n], the channel output and h[n] as the Finite Impulse Response of the channel and that the channel is linear time-invariant, then:

$$y[n] = x[n] * h[n] \tag{2.4}$$

Re-writing equation (2.4) in terms of circular convolution, we obtain

$$y[n] = x[n] \otimes h[n] = h[n] \otimes x[n] \tag{2.5}$$

and

$$y[n] = \sum_{k=0}^{L-1} h[k]x[n-k]_L \tag{2.6}$$

Where,

$$x[n]_L = x[n \mod L] \tag{2.7}$$

Thus $x[n]$ is periodic with the period of L *samples.*

Taking the DFT of the channel output results in the output in the frequency domain:

$$DFT\{x[n]\} = X[m] = \frac{1}{\sqrt{L}} \sum_{n=0}^{L-1} x[n] e^{-j\frac{2\pi n}{L}} \tag{2.8}$$

The inverse discrete Fourier transform is given by the expression:

$$IDFT\{X[m]\} = x[n] = \frac{1}{\sqrt{L}} \sum_{n=0}^{L-1} X[m] e^{j\frac{2\pi n}{L}} \tag{2.9}$$

Equation (2.9) shows that each input $X[m]$in the frequency domain is extended by $H[m]$. It can be demonstrated that the channel is ISI free. Although a circular convolution is deployed in this technique, the natural linear channel produces a linear convolution. However the circular convolution is forged by adding a cyclic prefix to the transmitted data. A cyclic prefix technique is when bits at the end of the symbol are included at the beginning of an OFDM symbol where ISI is expected. The technique is used to form a circular convolution from a linear convolution to achieve more robustness in the system. The length of the cyclic prefix is usually equal to the guard interval or the length of the anticipated ISI. The guard interval should have at least v samples, when the delay spread of the channel has the duration of $v + 1$ samples. After adding a cyclic prefix to the original samples with the length of L, the input of the channel is $x_{cp} = [x_{L-V} x_{L-V+1...} x_{L-1} x_0 x_1, \ldots, x_{L-1}]$, where x_0 to x_{L-1} stands for the original data while x_{L-v} to x_{L-1} at the beginning of the vector shows the cyclic prefix symbols.

Based on the input of the channel the output would be $y_{cp} = h * x_{cp}$, which has the length $(V + 1) + (V + L) - 1 = L + 2V$. However, $2v$ samples or bits will be discarded at the receiver in order to eliminate interference with the preceding and subsequent OFDM symbols. Therefore L samples remain at the end. In [8], it is proved that the L sample corresponds to $y = h \otimes x$, therefore the output y is disintegrated into a channel frequency response of $H = DFT\{h\}$ and channel frequency domain input of $X = DFT\{x\}$. Although the cyclic prefix provides a circular convolution and assists to provide an ISI free channel, it is associated with some drawbacks. To implement that, more bandwidth and transmission power are needed, as the v additive samples burdens the bandwidth. The required bandwidth is extended by $(L + V/L)$, in other words the bandwidth will be $(L + V/L)$ times more. Similarly the transmission power required for transmission of the cyclic prefix is $10 \log_{10}(L + V/L) dB$. In conclusion, using the cyclic prefix is equivalent to a loss of power and data rate of $L/L + v$. To overcome the problem of decreasing the required power, some recent works have been done; however more contributions and improvement are required.

In this section, a step by step operation of the OFDM is described [8], based on the assumption that the synchronization between the receiver and the transmitter is perfect and receiver faultlessly knows the channel. The steps are

- Dividing the wideband channel into L narrowband channels and L subcarriers for each channel, to achieve the ISI free transmission for each sub stream, with using a proper cyclic prefix.
- Modulation of L subcarriers using IFFT technique for overcoming the need for L independent radio frequencies
- After IFFT operation, a cyclic prefix with length of v has to be attached to the OFDM symbols and sent to the wideband channel serially to provide orthogonality.
- At the receiver, using FFT demodulation technique, the original data is identified by dispensing of the cyclic prefix.
- Each subcarrier is equalized via a frequency domain equalizer to estimate the data symbols.

2.3.7 Synchronization in OFDM Systems

Two types of synchronization must be considered at the receiver for successfully demodulating the OFDM data symbols which are timing and frequency synchronizations [8]. Determination of the OFDM symbol's offset and optimal timing instant is known as the timing synchronization and frequency synchronization is the arrangement of the subcarriers at the receiver in the manner that is as close as possible to the transmitter's subcarriers. Thanks to the cyclic prefix the timing synchronization is relatively acquitted, however, the frequency synchronization is very crucial as the individual orthogonal data symbols need to be implemented at the frequency domain. Development of the frequency synchronization algorithm is one of the most important issues in this technique and still stimulates the scholars in this field of study.

2.3.7.1 Timing Synchronization

Although timing errors can occur due to improper timing synchronization, the system can tolerate some imperfection in timing synchronization, as long as the timing offset of τ is larger than zero and smaller than the difference in time between the maximum channel delay spread (T_m) and the guard time (T_g).

If $\tau > 0$, sampling occurs at the later time compared to the ideal instant but if the sampling is occurred at the earlier time, it is assumed that $\tau < 0$. However, ISI happens if $\tau < 0$ or $\tau > T_m - T_g$. According to the amount of τ, the receiver will lose some energy and that energy degradation results in $\Delta SNR(\tau) \approx -2(\tau/LT_s)^2$, which is the signal to noise ratio loss.

It is derived from the equation above that to minimize the timing error τ should be as close as possible to zero or as small as possible in comparison with the guard time.

2.3.7.2 Frequency Synchronization

As mentioned earlier frequency synchronization is very serious in an OFDM system and lack of an efficient algorithm for the synchronization may result in undesirable overlapping of the subcarriers instead of being isolated from each other. Based on the theory, the frequency offset (δ) should be equal to zero to provide an interference free channel, however, the frequency offset is unequal to zero in the practical world, which result from inconsistencies in the oscillators at the transmitter and receiver and also Doppler frequency shift due to mobility. Using crystal oscillator is very costly, therefore systems must tolerate to some extent frequency offsets. The formula presented below expresses that as the frequency offset increases, the SNR decreases:

$$\Delta SNR = \frac{\varepsilon_x/N_0}{\varepsilon_x/(N_0 + C_0(L\,T_S\delta)^2)} = 1 + C_0(L\,T_S\delta)^2 SNR \qquad (2.10)$$

- ε_x: Average synchronization energy
- C_0: Constant depends on the assumption

2.4 OFDMA (Orthogonal Frequency Division Multiple Access)

Whereas OFDM is a modulation technique, OFDMA is a multiple access technique used primarily in mobile WiMAX and allows different users to have access to the available channel at the same time [8]. In OFDM based system, entire subcarriers are allocated to only one user, while in OFDMA technique subcarriers are divided into different groups of subcarriers and each group has a number of subcarriers, known as sub-channels [9]. To understand the OFDMA technique, a brief overview of different multiple access techniques is provided in this section.

2.4.1 Multiple Access Techniques

To provide multiple access to the channel for the users, three different strategies [8] have been used known as FDMA (Frequency Multiple Access), TDMA (Time Division Multiple Access), and CDMA (Code Division Multiple Access). A subset of carrier frequency is allocated to each user in FDMA technique, which can be performed statically and dynamically. To provide the sub-channels to the users statically, channel assignment is conducted by a multiplexer in the digital environment and before the IFFT accomplishment. However, in dynamic channel allocation, each user is assigned a specific channel, which is more appropriate for the particular user and it can use the channel more efficient. The second form of multiple access is TDMA. In this technique, each user can access entire bandwidth for a specific amount of time called time slot. Although TDMA suits the circuit switched-based data transmission the most, it can perform in the packet switched networks using more complex and improved algorithms. CDMA is another method for multiple access which provides each user a unique code. Therefore each user can use the entire bandwidth as long as the data transmission is occurring. However it is not suitable for high data rate transmission because it is sensitive to interference.

OFDMA is a hybrid form of the FDMA and TDMA, which allocates a user a subset of subcarriers in a proportion of time dynamically and provides robustness, scalability, spectrum efficiency and better frequency reuse [8]. It assigns a channel to the user according to the conditions and suitability of the channels for particular users; hence it provides more efficient resource allocation. Based on the important features of OFDMA, such as down link and uplink sub-channelization, better frequency reuse and also scalability, it is more suitable for the mobile broadband wireless networks such as mobile WiMAX [9]. However, OFDM performs better in the fixed network. Moreover, OFDMA offers two important principles known as multiuser diversity and adaptive modulation, which result in high performance of the OFDMA.

2.4.2 Multiuser Diversity

Multiple user diversity defines the availability of the gain by choosing a user of a group of users with a 'good' channel condition [8]. It enhances the capacity also in some cases, link reliability and coverage area. Assuming k number of users, each one has the channel gain (h_k)independent from others; the probability density function (PDF) of *the* k_{th} user's channel gain is defined as:

$$p(h_k) = \begin{cases} 2h_k\, e^{-h_k^2} & if\ h_k \geq 0 \\ 0 & if\ h_k < 0 \end{cases} \tag{2.11}$$

Based on the assumption that the base station transmits to the user with the highest channel gain ($h_{\max}$):

$$p(h_{\max}) = 2kh_{\max}(1 - e^{-h_{\max}^2})^{k-1} e^{-h_{\max}^2} \tag{2.12}$$

The equation exhibits that increasing the number of users results in more probability of getting a large channel and consequently enhancing the capacity also BER (Bit Error Rate).

2.4.3 Adaptive Modulation

Adaptive modulation refers to the selection of different modulation and coding schemes by the transmitter and allows for the scheme to change on a burst-by-burst basis per link, based on the state of the channel [8]. In other words, a transmitter has to transmit the data in as high as possible data rate if the channel has a good and proper condition. However, if the condition of the channel is weak, the data rate must be as low as possible. The condition of the channel is mostly determined through the SINR measurement. To perform that, the transmitter has to know the channel information through the receiver feedback. The SINR at the receiver is attained by multiplying the transmitter power and SINR at the transmitter. According to the SINR, the best configuration is selected. For example if the SINR is high, larger constellation and higher error correcting rate should be deployed for achieving the higher data rate; Whereas, if the SINR is low, to maintain a lower data rate, lower constellation and error correcting rate has to be chosen.

In the OFDMA systems, based on the amount of the SINR and its variety among users, different blocks of subcarriers are allocated to the users. Each block provides the best configuration for a specific user or group of users. Different configurations are called *burst profiles [8]*. Table 2.4 displays different burst profiles in WiMAX.

The different modulation schemes employed in WiMAX are shown in Table 2.5.

Table 2.4 Different burst profiles in WiMAX [10].

Channel Bandwidth	3.5 MHz		1.25 MHz		5 MHz		10 MHz	
PHY Mode	256 OFDM		128 OFDM		512 OFDM		1024 OFDM	
Oversampling	8/7		28/25		28/25		28/25	
Modulation & Code Rate	PHY–Layer Data Rate (Kbps)							
	DL	UL	DL	UL	DL	UL	DL	UL
BPSK, 1/2	946	326	Not applicable					
QPSK, 1/2	1,882	653	504	154	2,520	653	5,040	1,344
QPSK, 3/4	2,822	979	756	230	3,780	979	7,560	2,016
16 QAM, 1/2	3,763	1,306	1,008	307	5,040	1,306	10,080	2,688
16 QAM, 3/4	5,645	1,958	1,512	461	7,560	1,958	15,120	4,032
64 QAM, 1/2	5,645	1,958	1,512	461	7,560	1,958	15,120	4,032
64 QAM, 2/3	7,526	2,611	2,016	614	10,080	2,611	20,160	5,376
64 QAM, 3/4	8,467	2,938	2,268	691	11,340	2,938	22,680	6,048
64 QAM, 5/6	9,408	3,264	2,520	768	12,600	3,264	25,200	6,720

Table 2.5 Different modulation and coding schemes supported by WiMAX.

	Downlink	Uplink
Modulation	BPSK, QPSK, 16 QAM, 64 QAM; BPSK optional for OFDMA-PHY	BPSK, QPSK, 16 QAM; 64 QAM optional
Coding	Mandatory: convolutional codes at rate 1/2, 2/3, 3/4, 5/6 Optional: convolutional turbo codes at rate 1/2, 2/3, 3/4, 5/6; repetition codes at rate 1/2, 1/3, 1/6, LDPC, RS-Codes for OFDM-PHY	Mandatory: convolutional codes at rate 1/2, 2/3, 3/4, 5/6 Optional: convolutional turbo codes at rate 1/2, 2/3, 3/4, 5/6; repetition codes at rate 1/2, 1/3, 1/6, LDPC

2.4.4 Scalable OFDMA (SOFDMA)

SOFDMA is the scalable form of the OFDMA, which performs well in the mobile WiMAX networks. In this scheme, the number of the subcarriers which is equal to the size of FFT, scales with the bandwidth, while maintaining each sub-channel's bandwidth constant [11]; which means the subcarrier spacing and the number of the subcarriers are independent of the bandwidth. This leads to less complexity for smaller channels while enhancing the performance of wider channels. SOFDMA can provide the bandwidth of 1.25 Mbps to 20 Mbps for each sub-channel [11].

It is mentioned in [12] that *11 kHz* is the optimum sub-carrier spacing for the wireless mobile networks, which is the tradeoff between some parameters such as the level of protection to multipath effects, Doppler shift, design cost or complexity. SOFDMA assists to maintain the optimum subcarrier spacing, by changing the FFT size based on the bandwidth, which can underpin the NLOS operations. OSFDMA also supports for MIMO technique, Advanced Modulation and Coding and some other important features.

2.5 Adaptive Antenna

In practice, a transmitted signal may be received at the receiver from different paths due to phenomena such as reflection, refraction, shadowing and scattering, which may result in two scenarios; a stronger signal may be received at the receiver, if different signal components enhance each other in phase, but if they change each other's phase the received signal will be a weak signal which leads to the reduction of the link quality by degradation of the SNR and consequently decreasing the total capacity of the channel, according to the Shannon's formula$C = B \log_2(1 + S/N)$. To overcome the problem of signal degradation arising from the multipath effects and utilize this effect, a technology known as Smart Antenna or Adaptive Antenna is used in the transmitter, receiver or both.

Using arrays of antennas as well as sophisticated smart signal processing algorithm at the transmitter or receiver is referred to as adaptive antenna. Based on the use of the antenna array at the receiver, transmitter or both, this technology is divided into three different types [13]:

- SIMO (Single Input, Multiple Output): using the array of antennas at the receiver
- MISO (Multiple Input, Single Output): using the array of antennas at the transmitter
- MIMO (Multiple Input, Multiple Output): using arrays of antennas at receiver and transmitter (Figure 2.7). In nature, many insects communicate using some form of MIMO technique (for example the two butterflies in Figure 2.7).

In this technology, the signal processor unit integrates all the signal components received at different antennas, resulting from the multipath effects and makes use of the effect and derive benefit from it to produce a strong signal which increases the channel capacity and link quality. Employing adaptive antenna leads to less multipath fading and co-channel interference [13].

Fig. 2.7 Butterfly communication system (MIMO in Nature).

Adaptive antenna gain is n times more than the gain of individual element of the antenna array, if each array consists of n individual antennas, with the equal antenna gains [14].

$$G_{(\mathrm{AdaptiveAntenna})} = nG_{(\mathrm{AntennaElements})} \tag{2.13}$$

According to the equation, the signal at the receiver increases through the combination of the signals from different antennas. This also results in the extension in the coverage range and strengthening of the SNIR (Signal per Noise Interference Ratio) result in more capacity. Adaptive antenna also magnifies the signal strength and the antenna gain at a specific desired direction to promote the link quality and reliability at that particular side and decrease the interference by minimizing the signal propagation at the directions which the interference is more likely to occur [13].

Spectral efficiency is another important advantage of adaptive antenna technology, which is accomplished through reducing the number of the required cells per area unit also increasing the amount of billable services per spectrum that profits the operators economically. Less handoff is also needed, since cell splitting is not necessary thanks to the smart signal propagation toward the desired directions [13].

Although adaptive antenna technology offers many advantages, it has some disadvantages such as complexity of the transceiver, resource management and enlarging the physical size due to using a number of antennas at the transceiver.

As described in [13] a number of techniques are applied to the technology such as Conventional Beamformer, Null Steering Beamformer, Minimum Variance Distortionless Response Beamformer , Minimum Mean Square Error Beamformer and Least Square Despread Multitarget Array.

2.6 MIMO (Multiple Input-Multiple Output) Technique

MIMO, used in WiMAX technology, is a form of adaptive antenna which employs antenna arrays in both transmitter and the receiver to combine the signal components resulting from multipath effect and provide a strong signal at the receiver. To achieve that, the MIMO system needs to be implemented in a rich-scattering environment, where a signal reaches the destination from different paths [15]. MIMO technique increases the capacity, in turn the link quality, for a given capacity and transmission power compared to SISO system (Single Input- Single Output). By using a number of antennas, MIMO technique provides three different diversities [16]. Through antenna diversity, the link quality increases by exploiting the multipath effect [16].

The first type of diversity is known as spatial diversity. Antennas located in different positions far enough from each other so that they experience different signal strengths to promote the C/I (Carrier to Interference Ratio) by integrating different signal components. The optimum distance between the antennas is suggested to be about 38% of the transmission wavelength [16]. Beam diversity is another type of diversity, where the transceiver uses multiple directional antennas. Each antenna experiences different signal strength level of the original signal based on the angle of the received signal, then a combination of the all received signal components is used to provide more C/I. However the distance between the antennas is not as serious as spatial diversity. Finally, polarization diversity uses an orthogonal pattern for locating the antenna elements to provide different channel gains in the incoming signal direction and increases the C/I. Figure 2.8 illustrates different diversity patterns.

Typically, combinations of different diversity techniques are used to achieve the best implementation of the antenna depending on the propagation

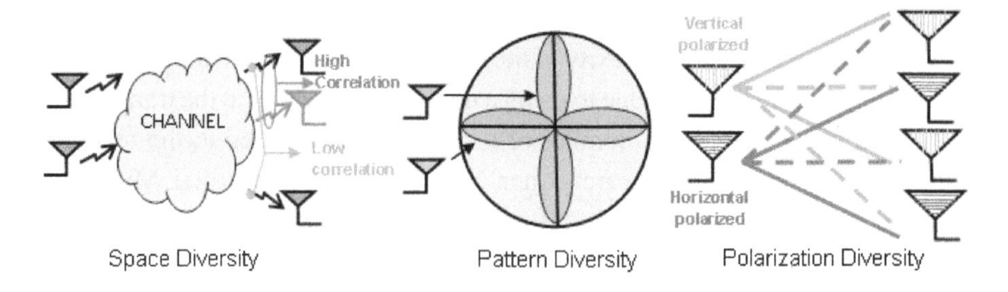

Fig. 2.8 Different diversities [15].

path between the transmitter and receiver [15,16]. However, achieving a perfect diversity is impossible in practical environments. To achieve a desirable diversity, isolation of the antenna elements is crucial. Since the mutual coupling intensifies the correlation between antenna ports, the efficiency of the diversity technique drops because for achieving a high diversity gain, correlation between the ports needs to be minimized. Maintaining the orthogonality of the antenna elements as much as possible assists in the achievement of more isolation. However, according to Chu-Harrington [16], to maintain the orthogonality, the size of the antenna needs to be enlarged. Therefore, in the concept of the MIMO antenna technology, one of the most important challenges is to implement the multiple antennas in an ever-decreasing space [16].

References

[1] Yan Zhang. *WiMAX Network Planning and Optimization.* USA: CRC Press, 2009.

[2] Johnson I Agbinya. *IP Communications and Services for NGN.* New York: Taylor and Francis, 2009.

[3] Kejie Lu, Yi Qian, Hsiao-Hwa Chen, Shengli Fu. "WiMAx Networks: From Access to Service Platform". *IEEE Computer Society* vol 22, pp. 38–45, May/June 2008.

[4] Zakhia Abichar, Yanlin Peng, J. Morris Chang. "WiMAX: The Emergence of Wireless Broadband". *IEEE Computer Society,* vol. 8., pp. 44–48, July/August 2006.

[5] Kosta Tsagkaris, Panagiotis Demestichas. " WiMAX Network". *Vehicular Technology Magazine, IEEE.* Vol. 4, pp: 24–35, June 2009.

[6] Eklund, Carl, et.al. WirelessMAN, Inside the IEEE802.16 Standard for Wireless Metropolitan Networks, IEEE Press, 2006.

[7] Peter Cain, "Mobile WiMAX PHY (RF) operation and measurement, Part 1".

[8] Jeffrey G. Andrews, Arunabha Ghosh, Rias Muhamed. *Fundamentals of WiMAX,* USA: Prentice Hall, 2007.

[9] Hujun Yin, Siavash Alamouti. "OFDMA: A Broadband Wireless Accss Technology." *Sarnoff Symposim.* pp: 1–4, 2006.

[10] "WiMAX-Physical Layer." Internet: http://digitaxis.com/web/sb-files/1201343773 phy.JPG, Jan. 26, 1008 [Sep. 18, 2009].
[11] "Introduction to FDM, OFDM, OFDMA, SOFDMA". Internet: http://WWW.conniq.com/wimax/fdm-fdma-ofdma-sofdma-01.htm, 2009 [Sep.16,2009].
[12] Vlodimir Bykovnikov, "the advantages of SOFDMA for WiMAX". Internet: http://my.com.nthu.edu.tw/~jmwu/LAB/sofdma-for-wimax.pdf, 2005 [Sep.16, 2009]
[13] Hafeth Hourani. "An Overview of Adaptive Antenna System" . *Postgraduate course in Radio Communication,* 2004/2005.
[14] M. Viberg, T. Boman, U. Carlberg, L. Pettersson, S. Ali, E. Arabi, M. Bilal, O. Moussa . "Simulation of MIMO Antenna System in Sinulink and Embeded Matlab", Internet: http://www.mathworks.se/company/events/conferences/mlnordic_conf08/proceedings/papers/mimo_antenna_systems.pdf, [Sep. 17, 2009].
[15] Aakanksha Pandey , " Researching FPGA Implementations of Baseband MIMO Algorithms Using Acceldsp" Internet: http://www.rimtengg.com/coit2008/proceedings/MS32.pdf [Sep. 18, 2009].
[16] Crown, "Antenna Design for MIMO System" Internet: http://info.awmn.net/users/images/stories/Library/Antenna%20Theory/Feeder/antenna_designs_mimo.pdf, [Sep. 18. 2009].
[17] Rhode and Schwarz, WiMAX General Information about the standard 802.16 (application note), 2006, pp. 1–34.
[18] WiMAX Capacity White Paper, SR Telecim, 2006.

3

Part II
Planning and Dimensioning of WiMAX and LTE Networks

Johnson I Agbinya and Mehrnoush Masihpour

University of Technology, Sydney Australia

3.1 WiMAX Network Planning Parameters

The key air interface technology used in WiMAX is OFDM. OFDM is an efficient form of multi-carrier modulation schemes, which is used in most of the modern communication technologies, such as 3G-LTE, DSL, 4G Cellular Systems and WiMAX.

To understand OFDM we introduce a few pertinent terminologies that are specific to it. OFDM system bandwidths are normally purchased by telecommunication operators through spectrum bids. The main system bandwidths range from 1.25 MHz, 3.5 MHz, 5 MHz, 8.75 MHz (WiBRO), 10 MHz and 20 MHz. Except for a small region of the spectrum (guard band), the rest is used for carrying data signals and for synchronization of the receiver to the transmitter. The FFT therefore occupies the spectrum in the frequency domain. The bandwidth is given by the expression

$$B = N_{c(\max)}.\Delta f \tag{3.1}$$

Where B is the OFDM signal bandwidth and Δf is the subcarrier spacing in Hz. This bandwidth does not include the guard band area. OFDM analog to digital converters sample the incoming signal at a rate which permits for adding guard bands (cyclic prefix).

OFDM is sampled using a so-called sampling factor. The sampling factor is the ratio of the sampling frequency to the OFDM bandwidth. Usually OFDM is sampled at a frequency a bit larger than the critical Bandwidth to make provision of guard bands. The sampling factor is:

$$n = \frac{F_s}{OFDM\ bandwidth} \tag{3.2}$$

Where Fs is the sampling frequency. Some of the common sampling factors are $8/7, 28/25, 78/75, \cdots$

The available spectrum is sampled using the equation (3.3):

$$F_S = Floor\left(\frac{n.BW}{8000}\right).8000 \tag{3.3}$$

Where Floor implies rounding down the result of the equation and n is the sampling factor and depends on the available spectrum (bandwidth). For WiMAX the channel spacing is

$$\Delta f = F_S/N_{FFT},$$

where N_{FFT} is the size of the FFT used to implement WiMAX. For example in fixed WiMAX $N_{FFT} = 256$. The symbol time is given by the relationship

$$T_S = \frac{(1+G)}{\Delta f} \quad \text{and} \quad G = \frac{1}{2^m}; \quad m = \{2, 3, 4, 5\} \tag{3.4}$$

3.2 OFDM Symbols

To understand how OFDM operates, in [8] some of the important terms are defined [8] and their relationships.

- *Channel delay spread* (τ): the time difference between the arrival of the first multipath component and the last one.
- Useful Symbol Time (T_b): The information bearing signal samples occupy the useful symbol time given by the relationship $T_b = \frac{1}{\Delta f}$. This time does not include the guard band time.
- Guard Period Interval/Ratio: To provide for multipath effect, a guard time proportional to the anticipated multipath is added to the useful symbol time. The guard time is related to the useful symbol time by the relation: $T_g = G \times T_b$, where $G = \frac{1}{2^m}$; $m = \{2, 3, 4, 5\}$. The cyclic prefix samples occupy this time period.

- *Overall OFDM Symbol time* (T_s): This is the time over which an OFDM symbol is valid. The symbol time is given by two times, the guard band time and the useful symbol time over which the actual information bearing signal samples are present. Therefore the total time over which the OFDM FFT is performed is given by the sum of the guard band time and the useful symbol time:

$$T_S = T_b + T_g = (1 + G)T_b \quad (3.5)$$

or

$$T_S = \frac{(1 + G)}{\Delta f}$$

- Coherence bandwidth (B_c): the bandwidth in which the channel is assumed to experience a flat fading propagation and it is approximately equal to $2\pi/\tau$

If $\tau \gg T_s$, number of symbols per second is high, which means high data rate transmission. In this case, the system is very fragile to ISI and it is desirable to minimize the ISI. Therefore, to address this problem, multicarrier modulation schemes reduce the T_s for the individual sub-bit streams to have $T_s \gg \tau$ [8].

The raw capacity of the channel per symbol can therefore be estimated and is:

$$C_{raw} = \frac{k.N_{subcarriers}}{T_S} \quad (3.6)$$

Where k is the modulation order and different modulation schemes are used to cover different areas of a cell as explained in Figure 3.1. For a 3.5 MHz bandwidth, the symbol time neglecting the guard band is $64\mu s$ and if G = 1/32, the total symbol time is $66\mu s$. With 192 subcarriers and using 64QAM modulation with 6 bits per symbol, the capacity is:

$$C_{raw} = \frac{6x192}{66\mu s} = 17.45\ Mbps$$

This is a theoretical result which assumes that the transmission channel does not introduce errors. In practice transmission channel errors are introduced and forward error correction (FEC) is required to safeguard against errors. FEC is implemented by using redundant bits in each symbol. This is

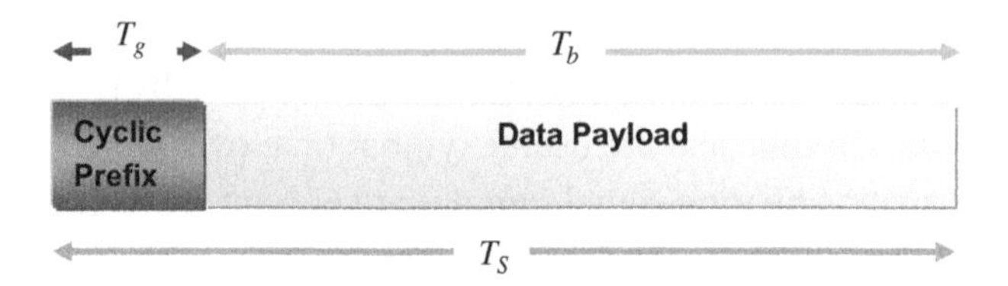

Fig. 3.1 OFDM symbol time.

Table 3.1 Mandatory FEC coding rates per modulation.

Modulation	Uncoded block size (bytes)	Coded block size (bytes)	Overall coding rate	Rs code	CC code rate
BPSK	12	24	1/2	(12, 12, 0)	1/2
QPSK	24	48	1/2	(32, 24, 4)	2/3
QPSK	36	48	3/4	(40, 36, 2)	5/6
16-QAM	48	96	1/2	(64, 48, 8)	2/3
16-QAM	72	96	3/4	(80, 82, 4)	5/6
64-QAM	96	144	2/3	(108, 96, 6)	3/4
64-QAM	108	144	3/4	(120, 108, 6)	5/6

an increase in the bits used to represent the symbol. The coding rate for the FEC is the ratio of the number of information bearing bits to the total number of bits including the redundant bits. The mandatory FEC rates for different modulation schemes are given in Table 3.1.

The effect of FEC is to reduce the raw capacity by the coding rate. Thus we have

$$C = C_{raw} \times CR = \frac{CR \times k \times N_{subcarriers}}{T_S} \tag{3.7}$$

Thus in the previous example the useful capacity when $CR = 3/4$ becomes $C = (17.45 \times 3)/4 = 13.1 Mbps$. As a measure of how effective the 3.5 MHz bandwidth has been used, the spectral efficiency is given:

$$\eta_S = \frac{C}{B} = \frac{13.1\ Mbps}{3.5\ MHz} = 3.74\ b/\sec/Hz \tag{3.8}$$

The spectral efficiency improves with higher coding rates. In other words, the less the number of redundant bits used for forward error correction, the better the spectral efficiency.

The theoretic WiMAX capacity was estimated based on this understanding and with a 20 MHz bandwidth, symbol time of $11.3\mu s$, 192 used subcarriers and (k) 6 bits per symbol at $CR = 3/4$ as:

$$C = \frac{CR \times k \times N_{subcarriers}}{T_S} = \frac{0.75 \times 192 \times 6}{11.3\mu s} = 76.46\ Mbps \qquad (3.9)$$

This theoretical WiMAX capacity estimate overlooks the overheads from the MAC layer and PHY. It also fails to highlight the fact that the range of a base station with such a capacity will be a lot shorter than often advertised for WiMAX by many authors. The range 70 km or more is normally for BPSK modulation and the large capacity is for 64QAM modulation and not distinguishing the two often leads to wrong conclusions about WiMAX capacity versus range. This confusion is created not by the standards body but by various writers on WiMAX.

3.3 WiMAX Frame

WiMAX symbols are normally sent in groups of symbols called a frame. Rather than sending the symbols as a stream, they are formatted into a time domain multiple access (TDM) frame. In practice the symbol time varies with the width of the channel. Therefore whole numbers of symbols do not necessarily fit into a frame snugly. There are often small gaps at the ends of frames that occur and not used. This overhead is often less than a symbol period in a frame and wastes capacity. The impact of this unused gap is more prominent in small frames.

As specified in the standard frames may be of lengths $T_F = [2.5,\ 4, 5,\ 8,\ 10,\ 12.5,\ 20\ ms]$. The number of symbols per frame N is:

$$N = FLOOR\left(\frac{T_F}{T_S}\right) \qquad (3.10)$$

For example in a 3.5 MHz channel with a cyclic prefix (CP) of 1/8, the symbol time is $72\mu s$ and the number of symbols in 8 ms frame is

$$N = FLOOR\left(\frac{8 \times 10^{-3}}{72 \times 10^{-6}}\right) = FLOOR\left(\frac{1000}{9}\right) = 111 \text{ symbols}$$

The unused gap at the end of the symbol is $T_F - (111 \times 72\mu s) = 8\ ms - 7.992\ ms = 8\mu s$, in this case a small reduction in the capacity (0.1% reduction).

3.4 Multicarrier Modulation schemes

In modern wideband high data rate transmission systems such as WiMAX, the effect of the ISI is very crucial, because in this technologies the delay spread $\tau \gg T_s$. In WiMAX this can easily happen when the range of the cell is large enough to cause significant multipath delay spread. The result is more ISI and consequently more errors to occur. Therefore modern communication technologies tend to use multicarrier modulation schemes in the physical layer to reduce the ISI. In the modulation schemes, a high data rate bit stream is divided into L lower rate sub-streams and transmitted on L parallel sub-channels or sub-carriers, which are orthogonal to each other in an ideal channel. The symbol time for each sub-stream is $T_s/L \gg \tau$, thus the ISI will be decreased. Although the data rate for each sub-channel is equal to R_{Total}/L, the total required data rate will be retained as all sub-channels are parallel to each other. Similarly the bandwidth of each sub-channel is derived by B_{total}/L.

To overcome the ISI in each sub-channel with a flat fading propagation, the bandwidth of each must be far less than the coherence bandwidth ($B_{total}/L \ll B_c$). It can be derived from this relationship that the more the number of sub-carriers, the less bandwidth and data rate for each sub-stream but the more the sub-streams symbol time and resulting in less ISI.

Although, there are some limitations associated with the multicarrier modulation such as high cost of low pass filters to achieve orthogonality and the requiring of a number of independent frequency channels for each sub-streams, OFDM technique copes with those problems and efficiently uses the available spectrum.

3.4.1 Guarding Against Inter-Symbol Interference

OFDM is well known for creating Inter Symbol Interference (ISI) free channels [8]. The interference between a symbol and the subsequent symbols is referred to as ISI, which is an undesirable effect of the environment, caused by either multipath propagation or Doppler effect. When a signal reaches the destination from different paths due to reflection, refraction and scattering, the propagation process is called multipath propagation.

Dividing a high data rate stream of bits into several lower data rate streams and transmitting each bit stream via parallel individual subcarriers is the technique that multicarrier modulation scheme uses for data transmission [8].

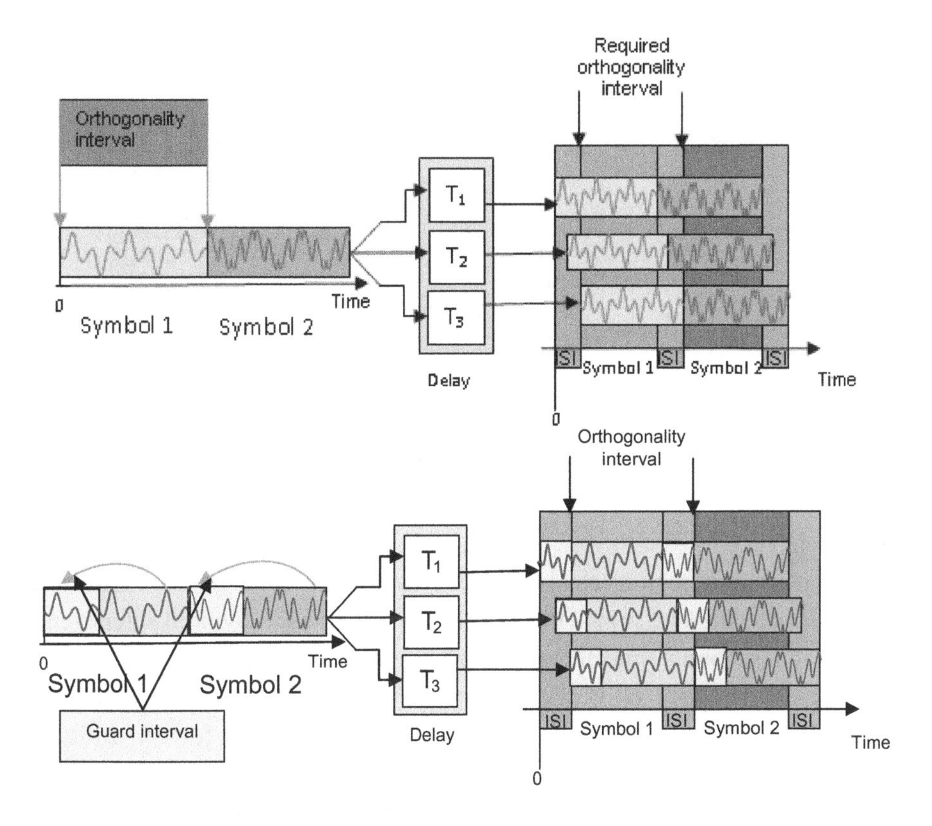

Fig. 3.2 Explanation of OFDM guard intervals and prefix (Adapted form [9]).

Radio signals often propagate from the transmitter to the receiver through many paths due to obstructions from objects in the terrain (the space separating the transmitter and receiver). Hence several copies of the same transmitted signal arrive at the receiver time shifted from each other. The time shifts are proportional to the paths taken by the signal and the effects of the channel (see Figure 3.3) on the signal. Assume that the direct path signal arrives at the receiver first. Thus delays encountered by signals arriving later can be estimated relative to the time of arrival of the direct path signal. The effect of adding the different copies of the received signal at the receiver causes signal power degradation (known as fading) due to intersymbol interference as shown in Figure 3.3. In this Figure three components of the same signal are received with the channel delaying them by T_1, T_2 and T_3 seconds respectively.

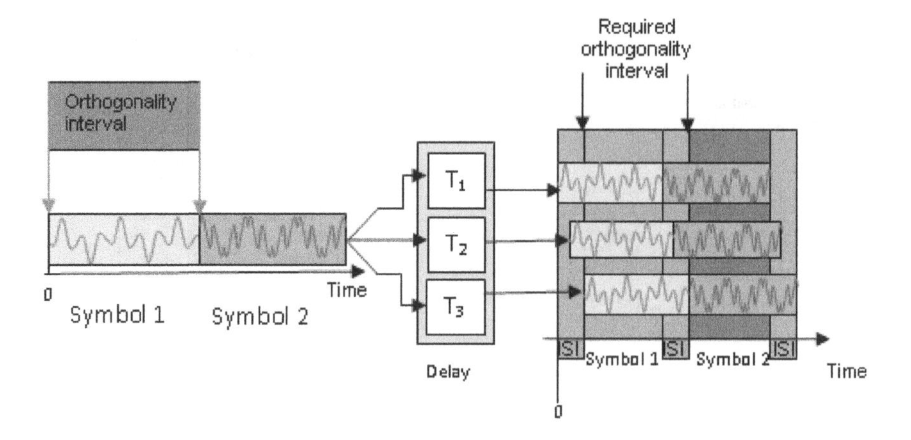

Fig. 3.3 Intersymbol interference at receiver [9].

Two separate regions can be identified in the received signals. Firstly, one due to time shifted version of the same symbol, called self-symbol-interference (SSI) and secondly interference from neighbouring symbols called inter-symbol interference (ISI). OFDM resists SSI because it uses orthogonal frequencies in the transmitter leading to zero correlation for two different frequencies. Hence SSI can be constructive in OFDM systems as the signal in such regions produce higher signal to noise ratio. This advantage is more prominent with the use of cyclic prefixes, which ensures SSI is removed.

ISI is mitigated in OFDM systems through the use of the guard period (orthogonality interval, Figure 3.2). The guard band is proportional to the maximum delay expected for all the paths the signal could take to arrive at the receiver. The guard period is implemented by adding the last part of an OFDM symbol to its front as a guard against the delay expected on arrival. Thus the desired signal could arrive at the receiver without error and at a good time distance away from copies of the same signal that have been received.

What should the length of the guard period be? The length of the guard period is variable and can only be assumed and it is proportional to the largest delay path the signal could take from transmitter to receiver. A good estimate of this value requires a simulation of the channel (path) the signal should take from transmitter to receiver. Thus information about the nature of the terrain separating them is required. If the reflection paths are long, the guard interval

Table 3.2 Example of system parameters.

Parameter	Variable and Calculations	Value
Bandwidth	B	7 MHz
Sampling factor	n	8/7
Sampling frequency	$F_S = n.B$	8 MHz
Size of FFT	N_{FFT}	256
Subcarrier spacing	$\Delta f = F_S/N_{FFT}$	31.25 kHz
Useful symbol time	$T_b = 1/\Delta f$	$32\mu s$
The guard intervals	G	1/4, or 1/32
Cyclic prefix time	$T_g = GxT_b = 32x\frac{1}{4}$	$8\mu s$ or $1.0\mu s$
delay path (given by the speed of light)	$d_{delay} = cxT_g = 3x10^8x8\mu s$ $d_{delay} = cxT_g = 3x10^8x1\mu s$	2.4 km or 0.3 km
Overall symbol time (G=1/4)	$T_S = T_b + T_g = 32\mu s + 8\mu s$	$40\mu s$
Number of Symbols	$N_{Symbols}$	25
Length of subframe	$T_{sym.frame} = N_{Symbols}xT_S = 25x40\mu s$	$1000\mu s$
Length of Frame	T_F	10 ms
Number of symbols per frame	$N = FLOOR(\frac{10\ ms}{40\mu s})$	250
User subcarriers	N_{user}	200
Pilot subcarriers	N_{Pilot}	8
Used data subcarriers	$N_{Used} = N_{User} - N_{Pilot}$	192
Modulation (QPSK), bits in subframe	$N_{bit} = N_{Used}x25x2$	9600 bits
Raw bit rate (no coding)	$R_{b(raw)} = \frac{9600}{1000\mu s}$	9.6 Mbps

should also be long and vice versa. There is a penalty in using large guard times as it causes a reduction in throughput due to wasted bandwidth.

Long range base stations in suburban areas should use larger guard interval. For example suppose the system parameters are as in Table 3.2.

3.4.2 Overcoming Doppler Shift

A moving radio transmitter relative to a receiver or vice versa causes the transmission frequency to change. This change in frequency and its effects on the received signal power is termed Doppler effect. Reducing the distance between the transmitter and receiver causes the carrier frequency to increase and increasing the distance between them causes the frequency to decrease. In an OFDM system when the transmitter and receiver move closer the carrier frequency increases and when they move apart it reduces. The change in carrier frequency is proportional to the relative velocity between the transmitter and receiver. This change is given by the expression:

$$\delta f = \frac{v}{c} \cdot f_C \tag{3.11}$$

Where, f_C is the carrier frequency and c is the speed of light. Consider a fast moving car at 120 km/hr carrying a terminal point transmitting at 6 GHZ. The Doppler shift is as high as 670 Hz. This change is substantial for a carrier frequency of 31.25 kHz (a 2.144% change in carrier position). Thus the demodulator at the receiver must be able to track this change and make corrections for it.

3.5 General Considerations for WiMAX Network Planning

Before the establishment of a communication network, a careful planning and dimensioning of the network is necessary. Through this process, network designers gather the analytical data and using practical design methodologies to obtain an optimum network design to increase the network performance, while reducing the time and expenses required for network implementation [1]. Network dimensioning and planning is the most significant part of the network design, which should be considered from three different perspectives of dimensioning for service, coverage and capacity. To achieve a proper planning, required information and Key Performance Indicators (KPI) need to be well defined. Some of the most important information that has to be identified is as follows [1]:

- Geographical area, which is due to service delivery, needs to be identified in terms of the size in km^2 also the profile that defines the area as either urban, suburban or rural environment. Furthermore the coverage area should be distinguished in terms of the network being fixed, nomadic, portable, and mobile or any combination of them.
- Subscriber's profile such as residential, small businesses and corporation customers needs to be identified.
- How subscribers are distributed through the network, which can be based on different parameters such as the number of the users per a service area, per a particular profile or per a period of time.
- Service profile is also very important, particularly for the service dimensioning. VOIP, broadband Internet access and IPTV are some of the services that WiMAX can offer.
- Available spectrum, in which WiMAX network is allowed to communicate within.

- Cartographic data to illustrate the service area through a digital map.
- KPIs as a measure of performance
- Customer requirements such as bandwidth, number of base stations and type of site.

Using the information outlined above, a WiMAX network planning and dimensioning for service, coverage and capacity can be done.

3.6 WiMAX Cells: Cellular Technology

Cellular technology is used by most of the modern communication systems such as GSM and UMTS. WiMAX also follows a cellular pattern to address the need for increasing the capacity. To understand how WiMAX technology works, it is necessary to become familiar with the main concepts of the cellular networks. In this respect, cellular technology is discussed in some detail in this section.

The growing number of subscribers requiring access to the communication networks in one hand and limited available radio frequency spectrum on the other hand, consequently the need for more capacity, has led the ICT (Information and Telecommunication Technology) scholars to develop the cellular system to allow the allocated radio frequency to be reused. A cellular network refers to a radio network consisting of a number of cells (defined by the geographical coverage of a base station), each one serving a specific geographical area by means of a base station through the air interface. In other word, a geographical area, which requires the coverage for the data transmission, is divided into a number of cells and each cell allocated a set of radio frequencies. Although, to minimize the interference between the frequencies, none of the adjacent cells can use the same set of frequencies, the cells far enough from each other can be allocated the same frequency band [6]. This is referred to as frequency reuse, which is discussed in some detail in the Section 3.6.2.

3.6.1 Cell Shape and Size

A hexagon is usually assumed to be the shape of a cell theoretically, however, in practice it is amorphous due to the environmental impact on the signal propagation and also the technology in use such as the type of antenna applied

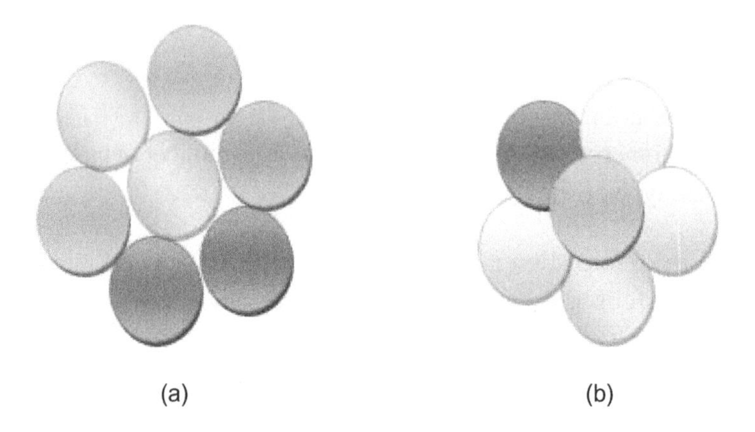

Fig. 3.4 (a) Circular cells with blind spots; (b) Overlapping circular cells.

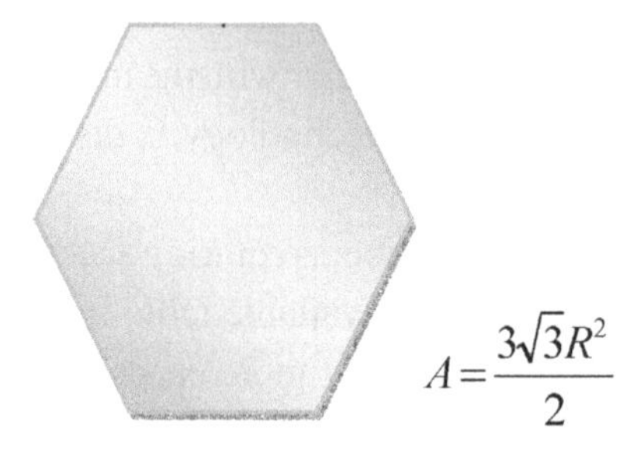

Fig. 3.5 Cell shape.

at the base station and its location. Figure 3.4(a) illustrates an ideal cell shape. Other possible shape of a cell is either a circle or square [6]. If the geographical area is divided into a number of circles, the gap between them is supposed to be a blind spot therefore it has an improper shape for a cell. In Figure 3.4(a) there are blind spots between the circles. Including the purple circle in the middle does not completely remove the blind spots. To remove the blind spots, the cell footprints must overlap as in Figure 3.4(b) which results in the waste of resources and coverage.

A square is also inappropriate because a square cell has unequal distance with its neighbors [6]. If we assume a square width of d it has four neighbors (Figure 3.6) at distance d and four at distance $\sqrt{2}d$ therefore it requires complicated computations for some required algorithms also antennas using in a cell should be none equidistant.

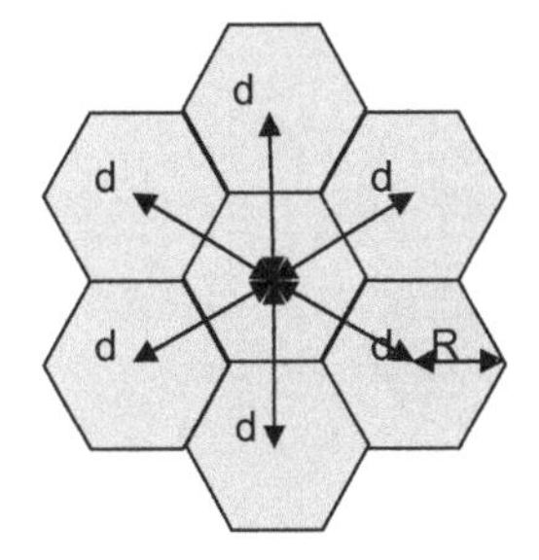

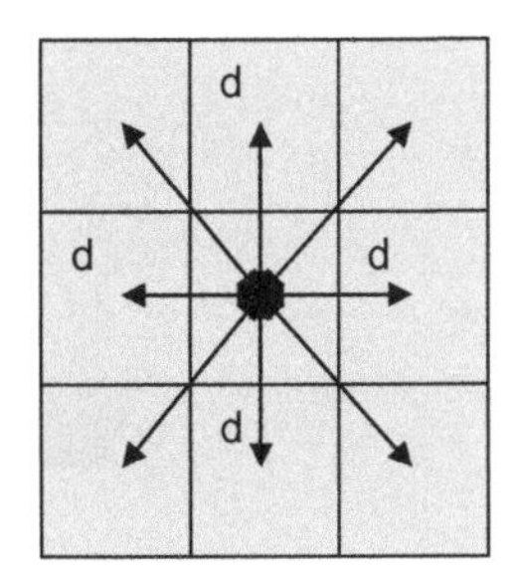

Fig. 3.6 Comparison between the cell shape and computations/square and hexagon.

A hexagonal cell, having equal distance with all adjacent cells provides the best coverage model with the least number of required base stations to serve a particular area. Assuming R as the cell radius, the distance between two adjacent cells is $d = \sqrt{3}R$ and the coverage area of a cell is approximately $2.598R^2$. However, we also the define the size of a cell in terms of its capacity and referred to as the number of users which can use a cell simultaneously.

According to the cell size, there are three different cells. Macro cell, which is large and has radius between 100 m to 10 Km or more, is suitable for the coverage in rural area with the low number if users. Micro cells are the medium size cells and have radius of about several 100 m, this type of cells are proper for crowded urban areas such as shopping malls. The smallest cells are referred to as Pico cells, which have size about a few dozen meters, can be usually implemented in networks in areas such as offices or in lifts.

As the size of the cell decreases, the required transmission power also decreases and users need less transmission power. The height of the base station antenna is lower for the smaller cells.

3.6.2 Cluster and Frequency Reuse

To increase the capacity of the network, while using a limited number of radio channels, cellular systems use the frequency reuse technology. It means that the same number of frequencies is used as many times as required in a network by imposing a careful frequency plan.

A cellular network consists of at least one cluster. A number of cells, each one using different group of frequencies from the other cells, form a cluster. The coverage range of a cluster is known as a footprint. Different cells

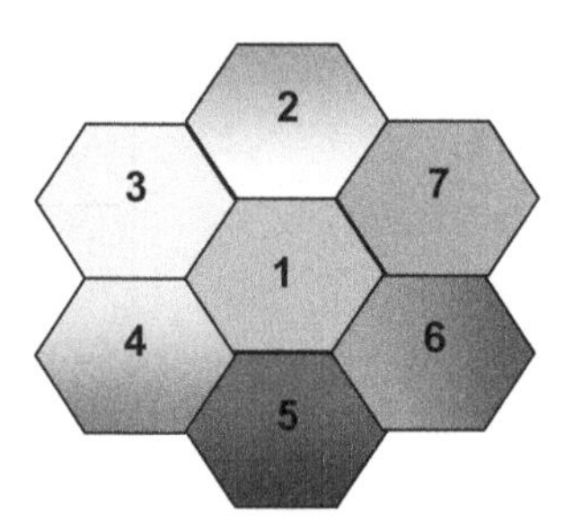

Fig. 3.7 Cluster with $K = 1/7$.

associated with a cluster must be allocated a specific and unique subset of all the available duplex channels to prevent interference between the same frequencies. Assuming S as all the available pair of channels and N as the number of channels used by each cell, the number of cells in a cluster is deduced by $K = S/N$. Inverse of this number of the cells is referred to as frequency reuse factor $(1/K)$. The possible values for $K$are 1, 3, 4, 7, 9, 12, 13, 16, 19, 21, etc [6]. Figure 3.7 shows a cluster with $K = 1/7$.

A cluster can be repeated in the network as many times as needed. If we assume the cluster to be repeated M times, the total capacity of the network will be $C = M^* S$, which is the total number of the radio channels available to be assigned to the subscribers at the same time.

To prevent the interference between the co-channel cells, which are the cells using the same group of radio frequencies, they need to be far enough from each other. The minimum required distance between the co-channel cells is calculated by:

- Assuming the base stations in the co-channel cells are located in $C(u_1, v_1)$ and $F(u_2, v_2)$ and distance between them is D derived by:

$$D = \{(u_2 - u_1)^2(\cos 30°)^2 + [(v_2 - v_1) + (u_2 - u_1)\sin 30°]^2\}^{1/2}$$
$$D = \{(u_2 - u_1)^2 + (v_2 - v_1)^2 + (v_2 - v_1)(u_2 - u_1)\}^{1/2} \quad (3.12)$$

We restrict (u_1,v_1) to integer values (i,j) and set $(u_1, v_1) = (0, 0)$ to give the expression

$$D^2 = i^2 + ij + j^2 \quad (3.13)$$

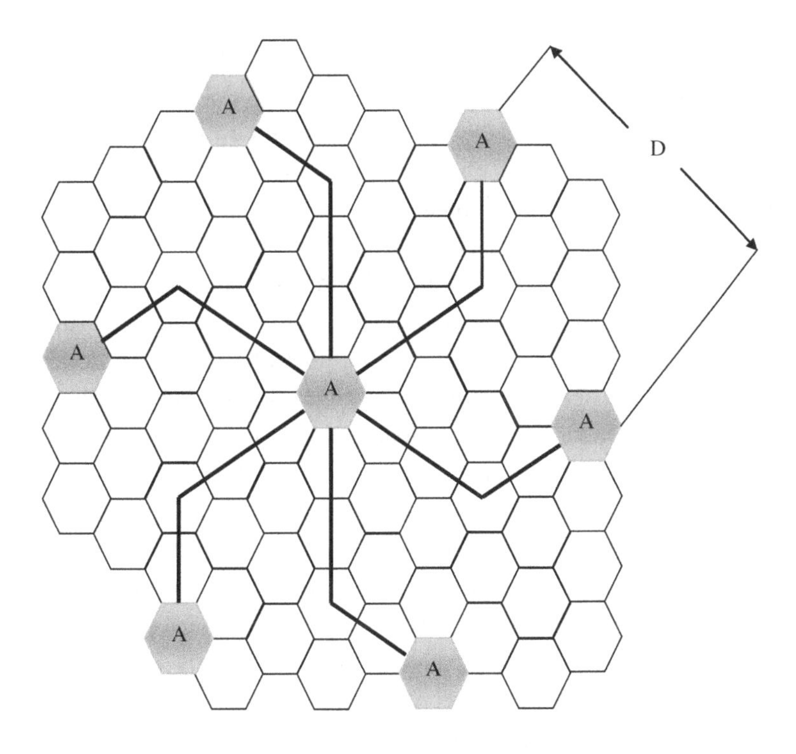

Fig. 3.8 Distance between co-channel cells.

In order therefore to connect without gaps between adjacent cells, the number of cells per cluster (cluster size), D, must satisfy:

$$D^2 = i^2 + ij + j^2 \tag{3.14}$$

Where i, j are non-negative integers. To find the nearest co-channel neighbours of a particular cell:

(1) Movie i cells along any chain of hexagons (Figure 3.8) and
(2) Turn 60 degrees counter-clockwise and move j cells.

3.6.3 Channel Assignment in Cellular Systems

Each cell can use a specific number of channels for allocation to its users. Channel assignment, which means designation of the available channels to the base stations or to the user's calls, requires a precise planning to minimize the possible interference to enhance the capacity of the network. To achieve that there are three main strategies, static or fixed, dynamic and a hybrid form of both. In the static channel assignment, which suites the network with the uniform traffic and high total traffic load in all cells, all the available channels are allocated to the cells and each base station has its own number of channels [7]. In this pattern, if all the channels are in use, new call request will be blocked. However, in dynamic channel assignment, the channel will be allocated to a call at the time that the call is made. In other word, no base station has its own specific and constant number of channels. A combination of two is used in the network with multiform traffic load but slightly the same traffic ratio, which means the traffic load have not many changes over time.

To implement a hybrid channel assignment, all the available channels are divided into two sets of static and dynamic channels. In the case that the cells have different traffic load, three main strategies can be done to prevent the calls to be blocked. Borrowing the available channels from adjacent cells is a solution for serving a user, while all channels in a cell are occupied by other users. In this case the borrowing channels are released to the original cell when the call is terminated. Another way to overcome this problem is to allocate the channels to the cells based on their needs. In this case, all the available channels are maintained in a pool and are assigned to the cells when needed. Channels can also be allocated by a central decision element, which decides allocation of the channels; however it can slow down the network.

As the number of users grows the capacity available to a user decreases. Alongside, frequency borrowing from adjacent cells, as a method of increasing the network's capacity, cell splitting (Figure 3.9(b)) also can be done to achieve this goal. Cell splitting refers to minimizing the cell size. In this scheme, a cell is divided into smaller cells, therefore the number of cell increases consequently the number of required base stations is also boosted and the achieved capacity is much higher. This scheme operates properly in crowded areas such as big cities, particularly in busy spots such as shopping malls.

Another method of increasing the capacity is cell sectoring. Using directional antennas at the base station within a cell, instead of using Omnidirectional antenna, is referred to as cell sectoring (Figure 3.9(a)). Three 120° or six 180° directional antennas are usually deployed in a base station to sector a cell. This leads to less co-channel interference but more intra cell hand over.

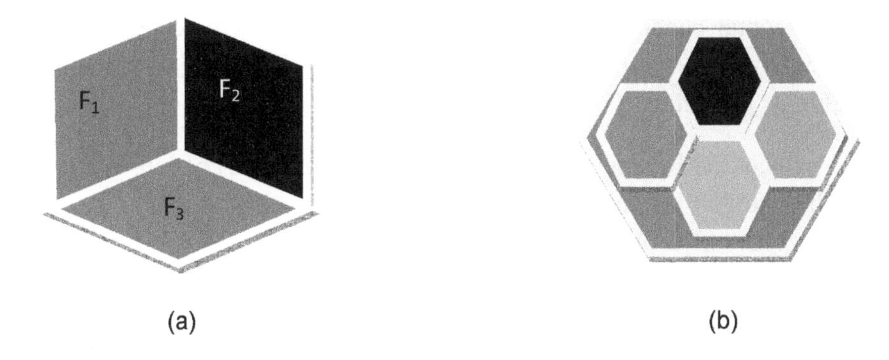

Fig. 3.9 Cells sectoring (a) and splitting (b).

These schemes are used in WiMAX networks with some unique variations which are discussed in latter sections of the chapter.

3.6.4 Handoff (Handover)

One of the most important operations in the cellular system is the concept of handover or handoff. When a mobile user changes the base station in use, which means moving from one cell to another, handover occurs. As it is described in the previous sections, different cells use different frequency channels, therefore when a user changes the cell and consequently is assigned channel the call would be dropped. Dropping a call is undesirable for a communication; therefore handoff solves the problem and maintains the connectivity while changing the cell. However, the handoff is unrecognizable to the mobile phone user.

The reason for handoff is achieving a stronger signal for a desirable communication quality. When the receiving signal is not strong enough, the mobile station requests a handoff and the call will be switched to the new base station with higher signal level.

There are two groups of handoff processes called horizontal and vertical handoff. Vertical handoff occurs when a mobile devices changes its point of attachment from a network of one type to another (eg. From a GPRS network to a UMTS or from a WLAN to UMTS etc). Handoff can occur between networks owned by different operators if commercial agreements exist between them to permit it.

Horizontal handoff occurs within the same network type (eg. within a UMTS network). There are several forms of handoff depending on the network conditions including

a) Handoff between neighboring base stations (BTS) or cells
b) Handoff between two RNCs
c) Handoff between SGSNs

This is important to perform the handover within less time as possible. To do that, three important factors should be identified, which are as follows:

- Minimum usable signal level for required quality.
- A threshold signal level
- An optimum signal level for a handoff to be performed.

Having the above factors handover margin can be defined as:

$$\Delta = \boldsymbol{P}_{r(handover)} - \boldsymbol{P}_{r(minimum\ usable)} \tag{3.15}$$

The handoff margin should be small enough to obtain optimum the handoff. Because unnecessary handovers occur if the Δ is too large, whereas too small handover margin leads to the call disconnection, as there is no enough time for switching between the cells and weak signal can drop the call.

To decide on handoff, different performance metrics can be used. Some of the metrics are defined below [6]:

- *Cell blocking probability*: the probability of blocking a new call, when the traffic load is very high and the base station cannot serve the call.
- *Cell dropping probability*: the probability of disconnecting a call in order to perform the handover.
- *Cell completion probability*: the probability of being an active call terminated successfully.
- *Probability of a successful handover*: the probability of performing a handover successfully
- *Handover blocking probability*: the probability of an unsuccessful handover
- *Handover probability*: the probability of a handover to occur before the call is terminated

- *Rate of handover*: the number of handovers occurring per a period of time

There are two types of handovers, soft handover and hard handover. Hard handoffs are operated when the connection is broken with the current base station before connecting to the target base station. However, if the connection to the new base station is performed before breaking with the current base station, soft handoffs occurs [6].

3.6.5 Traffic Engineering

Traffic engineering refers to optimizing the network performance by determination of the traffic patterns, predicting the growth rate and data analysis. In this section, traffic engineering is discussed based on the network capacity such as predicting the required capacity and data analysis on capacity issues.

As explained earlier, the capacity of a cell is determined by the number of available channels within a cell which allows the same number of simultaneous users per cell. However, it is unrealistic to have the capacity to serve all the subscribers at the same time. In practice, an expected number of simultaneous users will be identified for the frequency planning and channel allocation purposes, because not all subscribers request for calls at the same time [6]. Assuming the call capacity to be N with S subscribers, the network is described as a blocking system if $S > N$ and a non-blocking system if $S < N$. In practice systems are usually blocking systems. In case of a call being blocked two scenarios may happen based on the network configuration [6]. The call might be waiting in a queue to be served and the call might be dropped and rejected. In the second scenario, a cellular system assumes that the user hangs up and replicates the call after a certain amount of time (LLC model); however, another assumption is that user requests to call repeatedly (LCH model) [6].

For a blocking system to perform efficiently, there are some factors to be considered such as the call blocking probability and average delay, if the blocked calls are waiting to be served by the system. Traffic is determined as traffic density (A), which is stated as a dimensionless unit in *Erlang*. The traffic density is the result of an average holding time per successful calls (h) multiplied by the average rate of call request per unit of time (λ), or it can be defined as the average number of call requests during an average holding time.

$$A = \lambda * h \tag{3.16}$$

The network should be deployed in the way that can support the highest possible traffic. The highest possible traffic occurs during the busy hour of the day. To estimate the highest possible load, according to the International Telecommunication Unit-Telecommunication (ITU-T), the average traffic load of transmitted during the busy hour of 30 busiest days of the year should be identified.

In traffic engineering, it is important to consider the number of users as finite or infinite. Assuming a finite number of users is acceptable if the number of users is at least, 5 to 10 times of the capacity of the network. In this case the arrival rate is constant. However, based on the assumption of infinite number of users, the arrival rate will be dependent on the number of users, which are using the channels at the time. The arrival rate at time t is expressed as:

$$ArrivalRate = \lambda(S - K)/S \tag{3.17}$$

Where:

- S is the total number of users
- Each user average rate of call is λ/S
- K is the simultaneous number of user at time t

However, based on the assumption that the number of users is infinite and the network follows a LLC pattern, the important parameter, the Grade of Service (GoS) Probability or call blocking probability (P) should be considered. GoS determines the probability of a call to be blocked during the busy hour. The grade of service values between 0.01 and 0.001 are good values. Assuming traffic density of A and N number of channels, the grade of service is derived by the expression:

$$P = \frac{\frac{A^N}{N!}}{\sum_{x=0}^{N} \frac{A^X}{N!}} \tag{3.18}$$

From the equation, given the capacity of the system, the number of traffic can be determined for a given value of grade of service. Also given the amount of traffic, the capacity is identified for a certain grade of service.

3.6.6 Benefits and Drawbacks of Cellular Technology

To conclude the overview of cellular system, some advantages and disadvantages of this technology are summarized in this section:

Advantages:

- It is easily possible to increase the capacity and number of users through creating new cells, sectoring, cell splitting and or by simply scaling the network up progressively
- Lower transmission power
- Increasing the coverage area
- Robustness

Disadvantages:

The major disadvantages include

- Handoff within (and between) the networks is necessary
- Inter cell interference occurs due to co-channel interference and adjacent channel interference
- Fixed base stations are required. Newer versions of networks on platforms or vehicular networks seek to solve this limitation.

3.7 Dimensioning for Service

In the process of service dimensioning, different types of services offered to different customer profiles is determined and also how it affects the air interface is analyzed. Typically, there are three important services offered by WiMAX which are VoIP, broadband data and guaranteed bandwidth [1]. According to the type of the service, a portion of the total available bandwidth in each sector is allocated to that particular type of service to deliver the QoS required by WiMAX. As described in Section 2.1.4, there are three different categories of applications required divergent QoS known as Unsolicited Grant Service (UGS), Polling Service (PS), Best Effort service (BE) [2]. VoIP is categorized as a PS while broadband data which corresponds to the Internet access is an application and requires BE service and guaranteed bandwidth is equivalent of UGS.

Usually a fixed portion of the bandwidth identified as 20% [1] of the total bandwidth is assigned to the UGS during the primary configuration of the network. However, the remaining channel bandwidth is allocated dynamically to the BE and PS, each one a specific portion according to their needs. The impact of three different services on the WiMAX network is analyzed in the following sections.

3.7.1 Voice Over IP Service

VoIP refers to the vocal communication delivered in a packet-switched network. In WiMAX technology, the voice stream is broken up into a number of data packets and transmitted via air interface. One significant task in WiMAX planning is to estimate the capacity needed for delivering the VoIP service to subscribers, since any defeat to a desirable resource allocation results in degeneracy of the grade of service and subsequently degradation of several calls in the sector. To achieve that, some important characteristics of the voice service needs to be identified within a sector, such as total lines, active lines, grade of service, traffic activity and required data rate per call [1].

Mostly the required capacity per sector is needed to be calculated for a given number of simultaneous calls during the busy hour. Using Erlang B equation or table the number of active call will be identified for a given GoS:

$$\boldsymbol{P}_b = \frac{\frac{A^N}{N!}}{\sum_{x=0}^{N} \frac{A^X}{X!}} \tag{3.19}$$

Where

- A: Sector traffic activity
- N: number of active calls
- P_b: blocking probability or grade of service (GoS)

3.7.1.1 Broadband data service

This service offers access to data at high rates. It is evaluated by the peak information ratio *(PIR)* in Mbps, which is defined as the maximum data rate that a network can offer to the subscribers if the network is not congested [1]. However, since the broadband data service is a BE service, data rate will be lower when the number of simultaneous subscribers increases during the busy hour known as data over-subscription rate (O). The ratio of peak information ratio P_{pir} to data over-subscription ratio is a performance indicator that expresses the committed information ratio, P_{cir}, which is the required data rate during the busy hour.

$$P_{cir} = \frac{P_{pir}}{O} \tag{3.20}$$

$$P_{total} = \sum (P_{cir,i} \times N_{s,i}) \tag{3.21}$$

- $N_{s,i}$: corresponding number of subscribers
- P_{total}: Summation of the per profile rates
- P_{pir}: profile's CIR

The rate for FDD is derived from the *Max (DL, UL)* and for TDD as the sum of DL and UL rates [1].

3.8 Dimensioning for Coverage

Through the coverage dimensioning process, the number of base stations required for coverage of a given area in km^2 is identified in such a way that meets the satisfaction of KPIs [1]. The footprint of each cell needs to be calculated for the estimation of the number of base stations,

$$Number\ of\ required\ base\ stations = \frac{Service\ Area(km^2)}{Cell\ Footprint(km^2)} \tag{3.22}$$

Cell footprint varies from different scenarios and equipment configuration. It refers to the maximum range that a base station supports to radiate the electromagnetic waves to achieve a performance threshold based on the signal power at the receiver. The received signal strength can be calculated using the following formula [1]:

$$S(dBm) = P + G + G_{sp} - L_d - L_p - M \tag{3.23}$$

Where:

- S: Signal strength at receiver
- P: power of transmission
- G: antenna gain
- G_{sp}: signal processing gain
- L_d: system loss (shadowing, path, fading)
- L_p: penetration loss

- M: design margin (to address mobility, interference, reliability and implementation)

For different terminal profiles, the gains and losses and also design margins are different. Different terminal profiles [1] are defined as follows.

a. Fixed outdoor
    - Located on the rooftop or outside the building walls.
    - Usually for delay sensitive services such as VoIP also Ethernet
    - Connection to the indoor terminal via cable
    - High data rate is possible at a large range with low impact on the channel resources
    - This type of terminals are mostly suited to small-to-medium enterprises (SMEs)
    - Requires high cost of equipment and implementation
b. Fixed portable indoor unit
    - Located indoor close to the window or outer wall
    - Unit is portable within the indoor space and has smaller coverage range
    - Requires power supply
    - Suitable for residential access
    - Self-installation
    - Lower cost compared with the fixed outdoor terminals
c. Nomadic mobile unit
    - Mobile units can be used within the outdoor and indoor space
    - Suitable for individual customers requiring particular services

However, customers usually prefer to have a combination of different networks.

As mentioned earlier, received signal strength needs to meet at least a signal strength threshold which itself depends on SNR threshold and noise floor (N_{th}). Noise floor refers to the sum of the all noise sources and SNR threshold has to have a particular level for the system to work acceptably. The SNR threshold for a received signal is typically defined to have less than 10^{-6} bit error rate when decoding at receiver [1]. However, SNR threshold varies with the modulation and coding scheme.

To estimate the cell footprint we consider factors such as cell shape, network layout, number of sectors per cell to identify the operating system range. Operating system range depends on the deployment of different scenarios. If the network is deployed in a rural outdoor area the base station should be one that has footprint at its maximum range in use while if the same base station is used in urban area or for mobile users, the footprint is a percentage of the maximum footprint as overlapping is essential for handoff and a margin of about 10% is considered for this reason.

In terms of cell shape, although square cells are more suitable for the fixed networks, hexagonal cells are employed for implementation of the mobile network [1]. Assuming r as the operating range and F the footprint of a cell, the relationship between them is defined as:

$$F_{sq} = 2r^2 \tag{3.24}$$

$$F_{hex} = \frac{3}{2}\sqrt{3}r^2 \tag{3.25}$$

Using hexagonal cells, larger footprint are supported compared to the square cells. Therefore the number of required base stations is less than deployment by square cells. However, for achieving high capacity, using square cells is more suitable because they provide more capacity per unit area unit. For mobile network deployment hexagonal cells result in lower cost according to the number of required base stations. In some cases hybrid scenarios are in use, and deployment should be based on the worst scenario [1]. For instance if indoor and outdoor deployments are required, the planning should be based on indoor terminals because they operate in the shorter range and if the planning only meet the outdoor requirement, indoor coverage face the problem of blind spots and out of coverage areas, but if plan for the indoor terminals with shorter range the required range for the outdoor terminals has already been met.

The number of cells required is a function of the modulation scheme used. The more agile the modulation scheme is the higher the bit rate and the smaller the cell coverage. Hence, more base stations are required for modulation schemes which use more bits per subcarrier. BPSK uses 1 bit per subcarrier, QPSK 2 bits, 16 QAM 4 bits and 64 QAM uses 8 bits per carrier.

The size of the cell is a function of the modulation type and the scheme is shown in Figure 3.10. Different modulation types can reach different cell range. From Figure 3.10, the largest cell range is achieved using BPSK 1/2. The smallest cell range is provided by 64 QAM 3/4. The cell area of coverage is inversely proportional to the capacity obtainable with a given modulation type. Thus 64 QAM 3/4 provides the largest cell capacity. The area of coverage A for each modulation type is given by the expression:

$$A = \frac{3\sqrt{3}}{2}(d_k^2 - d_{k-1}^2) \tag{3.26}$$

Each modulation covers a percentage of the total area of the cell and provides some known capacity. Therefore the capacity of the cell can be estimated to be

$$C = \sum_{k=1}^{M} C_{modulation\ k} \cdot \left(\frac{A_{modulation\ k}}{A_{Total}}\right) \tag{3.27}$$

In Figure 3.10, $M = 7$.

3.8.1 WiMAX Cell Capacity

The use of over subscription ratio (OSR) as a criterion for design is a popular method when starting to develop the characteristics and parameters of a WiMAX cell (base station). OSR is the ratio of the total subscribers traffic demand over the reference capacity of the base station when taking into account adaptive modulation. The reference capacity of the base station is the available bit rate obtainable with the lowest modulation scheme. In the WiMAX case, this reference is provided by BPSK 1/2 which is the lowest modulation scheme served by the base station as shown in Figure 3.10. This reference capacity is given by the expression

$$C_{ref} = \frac{FFT_{used}}{2T_S} \tag{3.28}$$

From Table 3.2 with $T_S = 40\mu s$ and 256 FFT (7 MHz bandwidth and subcarrier spacing of 31.25 kHz), the reference capacity is 3.2 Mbps and this value

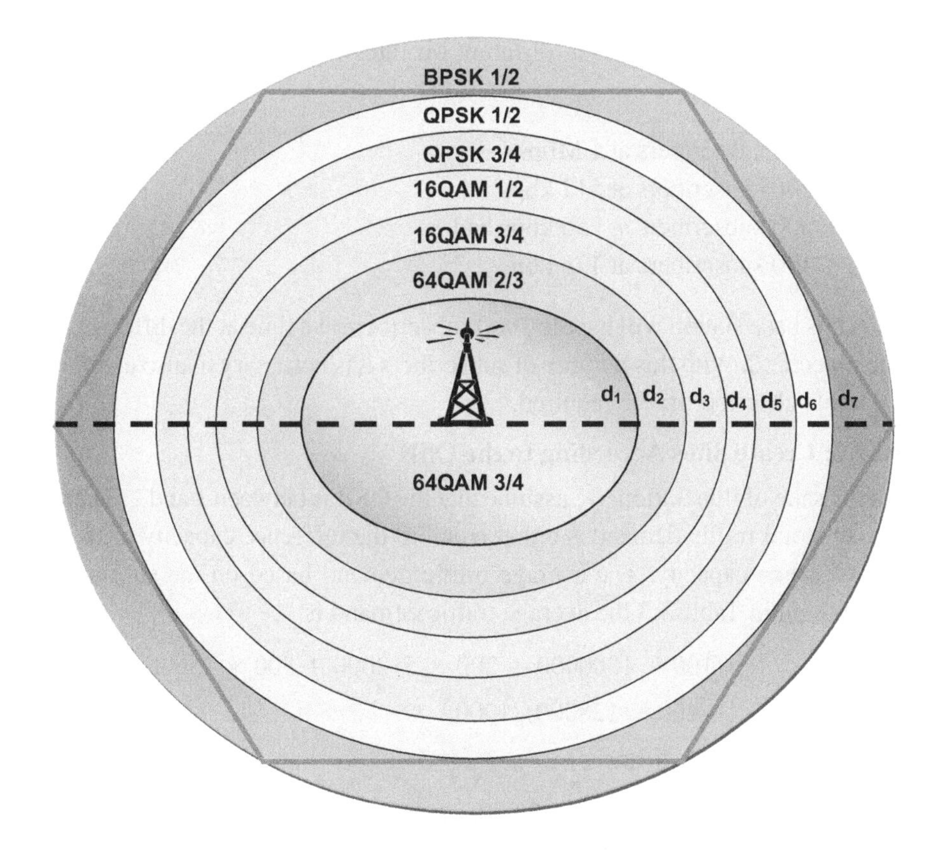

Fig. 3.10 Size of WiMAX cell as a function of modulation type.

depends on the cyclic prefix value and T_S. The subcarrier spacing used here is fairly large and better reference capacities can be obtained with more fine grain subcarrier spacing. The following steps could be followed in the design specifications for the base station.

Step 1

Define the service area of interest. This is the area where the different types of services desired by the subscribers are provided. This means for example that the city of Sydney is taken first and divided down into urban, high density urban, suburban, and rural areas. Taking for example the high density urban (CBD) area, repartition the area according to the types of subscribers who are likely going to use services at different bit rates from this base station. For example

- 100 subscribers at 1 Mbps
- 200 subscribers at 512 kbps
- 300 subscribers at 256 kbps and
- 400 subscribers at 128 kbps

Thus this base station will handle 1000 subscribers at a time at the different bit rates specified. With this number of subscribers it is necessary to also estimate the total subscriber traffic required.

Step 2: Create Sites According to the OSR

For the sake of illustration, we assume that the OSR is between 1 and 2. Therefore the total traffic demand is either equal to the reference capacity or twice the reference capacity. The average traffic demand based on the subscriber distribution in Table 3.3 the average traffic demand is:

$$\begin{aligned} C_{avg} &= (100 \times 1000000 + 200 \times 512000 + 300 \times 256000 \\ &\quad +400 \times 128000)/1000 \\ &= 330.4kbps \end{aligned}$$

Thus at a minimum bit rate of 2,820, each base station need to be able to connect 2820/330.4 = 8 (8.54) subscribers. At the high OSR of 2, the base station can handle at most 17 subscribers. At the modulation rate of 64QAM 3/4, the base station can connect 38 subscribers.

Consider a practical example where a WiMAX operator has the following service classes to attend to in the form of

- Platinum subscribers at a VBR of 1 Mbps minimum reserved (OSR 10), 3 Mbps maximum sustained (OSR 20), SME users, 5% of total number of subscribers

Table 3.3 Subscriber distribution.

Modulation scheme	Threshold (dBm)	Bit rate	C/(N + I) dB
QPSK 1/2	−97	2, 820	3
16QAM 1/2	−91	5, 640	9
16QAM 3/4	−88	8, 545	12
64QAM 2/3	−83	11, 280	17
64QAM 3/4	−82	12, 818	18

- Gold subscribers at a VBR, 540 kbps minimum reserved (OSR 10), 1 Mbps maximum sustained (OSR 20), SOHO users, 15% of total number of subscribers
- Silver subscribers at BE, 1 Mbps maximum sustained (OSR 20), residential users, 80% of total number of subscribers

The average capacity per subscriber for this case follows the same method used earlier and is

$$\begin{aligned} C_{avg} &= \left(\frac{5}{100} \times \left[\frac{1000}{10} + \frac{(3000-1000)}{20}\right] + \frac{15}{100}\right. \\ &\quad \left. \times \left[\frac{540}{15} + \frac{(1000-540)}{20}\right] + \frac{85}{100} \times \frac{1000}{20}\right) \\ &= 61.35\ kbps/subscriber \end{aligned}$$

At a raw bit rate of 9.6 Mbps for a base station in Table 3.2, the base station can therefore service (9.6 Mbps/61.35 kbps) 156 subscribers using QPSK with two bits per symbol. This number of subscribers would be reduced to a half (78 subscribers) if a FEC of 1/2 to three quarters (117 subscribers) if the FEC is 3/4. The range of this base station is 2.4 km as given in Table 3.2. Overheads in the PHY and MAC layers will reduce this subscriber base.

Step 3: Determine the number of base stations

If the total geographical area to be covered is A km square, what is the number of base stations to deploy? We assume hexagonal cells.

$$N_{BaseStations} = \frac{Total\ Area}{2.598\, R^2}$$

For a 30, 000 km square total area with base stations of radius 2.4 km, 2000 base stations are required.

3.9 Dimensioning for Capacity and Frequency Planning

After accomplishing the coverage planning, capacity dimensioning and planning needs to be completed. Capacity dimensioning is performed to ensure that the capacity offered to the customers satisfies the required capacity according to the number of subscribers, type of service using by them and data traffic transmitted throughout the network [1]. The outcome of this process is determination of the number of the sectors per cell to address the need of required

Table 3.4 Ethernet rate per modulation scheme base on 802 16e (Adapted from [1]).

PHY Mode (CTC)	SINR(dB)	DL(Mbps) 5 MHz	UL(Mbps) 5 MHz	DL(Mbps) 10MHz	UL(Mbps) 10 MHz
QPSK1/2	2.9	1.4	0.9	2.7	1.7
QPSK3/4	6.3	2.2	1.3	4.3	2.7
16QAM1/2	8.6	2.8	1.7	5.8	3.6
16QAM3/4	12.7	4.3	2.6	8.6	5.4
64QAM1/2	16.9	5.8	3.5	11.6	7.1
64QAM3/4	18.0	6.5	3.9	12.9	8.2

data rate for the subscribers using different services [1]. By estimation of the number of the subscribers per sector, total capacity of a cell can be identified, subsequently the capacity of whole system. To calculate the number of the subscriber per sector, the capacity of the sector needs to be identified.

Average throughput of a sector is prescribes by the vendor, but it mainly depends on the deployment scenario. Different modulation schemes are used by IEEE802.16e standard. Based on the modulation scheme using in each sector, the throughput varies. Table 3.3 suggests the different modulation technique and coding and corresponding throughput in both UL and DL according to the bandwidth.

According to the Table 3.3, the throughput will be maximized using 64QAM-1/2 in the 10 MHz bandwidth in the interference free environment. The equipment specification as usually explained in a table shows the modulation schemes and required SINR and Ethernet throughput for UL and DL using different bandwidth [1]. The data rate achieved through a sector is calculated using different parameters according to the modulation scheme and coding [3]:

$$R_b = R_s \frac{MC}{R_r} \tag{3.29}$$

Where:

- M: modulation gain (2 for QPSK, 4 for 16-QAM, 6 for 64-QAM)
- C: coding rate (1/2, 3/4, ...)
- R_r: repetition rate of 1,2,4,6
- R_b: bit rate
- R_s: symbol rate

Although different modulation schemes are used in different sectors, because different terminals experience divergent interference due to frequency reuse, IEEE802.16, recommends a default assumption for the average sector throughput. Based on this assumption, the average sector throughput is defined as the throughput corresponding to the16-QAM modulation scheme and 1/2 coding rate; however, the throughput will be higher if deployment conditions are desirable.

Since the average sector throughput is determined, customer data should be examined in terms of type of services, average data rate and VoIP Committed Information Ratio. After that a graph is required to show the subscribers per sector versus throughput.

The next step is to determine the number of the required sectors identified by dividing the number of the customers per area with the number of customers per sector. Assume the number of customers per unit area is S, the uplink bit rate to be $R_{b,UL}$ and the number of customers per sector is S_S, then the number of sectors is

$$N_S = \frac{S}{S_S} \tag{3.30}$$

The bit rate per sector can be determined based on system parameters and is given by the expression:

$$R_{b,DL} = B.n.\frac{N_{dataDL}}{N_{FFT}} \times bits\ per\ symbol\ x \frac{1 - overhead}{1 + T_g} xTDD_{down/upratio} \tag{3.31}$$

The TDD down/up ratio between the UL and DL times and may be 3:1 and the overhead time includes times used for synchronisation, initialisation and headers.

However, adaptive modulation is one of the important factors that affect the capacity of the network [1]. As described earlier , WiMAX is able to adopt different modulation schemes for data transmission based on the link quality. For example if the SINR is high, transmission can be done using higher order of the modulation schemes such as 64-QAM to achieve higher throughput while in the case of weak SINR lower orders of modulation schemes such as QPSK can be employed.

Alongside the adoption of different modulation schemes, to increase the capacity of the network and minimize the interference between the co-channel cells, WiMAX uses powerful sub-channelization modes which are PUSC (Partial Usage of Sub-Channels) and FUSC (Full Usage of Sub-Channels) [1, 4, 5]. FUSC refers to the usage of all sub-carriers within a cell, while PUSC is defined as allocation of specific group of sub-carriers to each sector within a cell, which there are typically 3 sectors per cell. Using PUSC mode allows the network to use the frequency reuse factor of one when the spectrum is limited without interference between adjacent sectors, because they use divergent sub-carriers. This scheme is very efficient as there is no need for guard band also provides more reliability of links in both uplink and downlink by using different sub-carriers in the neighbouring cells subsequently reducing the chance of interference [1, 4]. Whereas FUSC that is used only in downlink PUSC can be used in both direction of uplink and downlink. The reason for that FUSC can not be used in uplink is because of the nature of uplink which is more unpredictable than downlink [5]. To plan the frequency channels to achieve higher capacity and optimum use of the channels, there are some factors that need to be considered in either of uplink and downlink.

Planning for downlink is easier in comparison with uplink because the major source of interference in downlink is the neighbouring base station using the same frequency channel; therefore it is possible to calculate the SINR in each spot and prepare an interference matrix for each base station in the service area and the proper modulation scheme for each spot [5]. For example if a location has suitable condition higher modulation order can be used to enhance the capacity of the network. However, most of the downlink interference is expected at the borders of cells where the SINR is low and signal is more likely to collide with other signals on the same carrier frequency [5]. To overcome this problem, WiMAX divides the TDD frame, which consists of two portions of uplink and downlink, each includes some slots and each slot in turn consists of 48 sub-carriers, to two parts in downlink [5]. At the first part of the downlink frame, base station adopt PUSC mode to transmit data to the subscribers located at the edge of the cell to reduce the possible interference while using FUSC mode in the second part of the downlink frame, to transmit data to the subscribers near to the base station with higher SINR for increasing the capacity and finally efficient use of the spectrum [5].

In spite of downlink, uplink planning is more complicated as it is less predictable. The subscriber station can be any where within the service area also any where within a particular cell therefore it is difficult to determine the receiving level of SINR consequently the level of interference. Interference can vary not only by changing the location of the user but also by the number of the slot that it uses to transmit the data on. Using more slots by a subscriber leads to more number of sub-carriers and subscribers it collides with [5]. According to all those reasons mentioned above, the uplink can use only PUSC scheme for data transmission to achieve a robust network. To allocate an optimum modulation scheme to the link, base station estimates the SINR according to the feedback that it receives from the subscriber station through the previous received frame. Since in mobile WiMAX system, the user changes the channel frequently because of mobility, the base station needs to be aware of the current channel that user is using; therefore to allocate the best possible modulation scheme to the connection, the base station has to use any of the average SINR received previously, the best SINR or the worst SINR. The average SINR is more suitable since the selection of the worst SINR results in capacity reduction due to data transmission on lower that optimal modulation order, while assigning the modulation scheme based on the best amount of SINR causes the less robust links due to high error occurrence and data retransmission [1, 5].

What is done on capacity dimensioning within the access network will assist to determine the required capacity for the backhaul network. It is discussed in some detail in the next section.

3.9.1 Backhaul Dimensioning

The backhaul network links a number of base stations and ASN-gateways, via a radio interface or fibre rings [1]. Base stations can be connected together through a point to point, if the number of the required base stations are high, or point to multipoint radio interface, if the base stations are close to each other. Also the base stations need to be connected to a gateway within the backhaul network using an IP transport or IP link, which carries the IP packet between the base station and the gateway. To evaluate or plan a backhaul network, the information rate of the backhaul, the number of the base station and gateways also the position of them needs to be determined.

The backhaul information rate increases with the number of the sectors per a cell, also the sector's throughput in a linear manner. However, the sector's throughput varies with the deployment scenarios. Total TDD rate that can be delivered by the backhaul is the product of the number of the sectors and each sector's throughput. To calculate the FDD rate, the total throughput needs to be multiply by DL/(DL+UL) rate [1].

In accord with WiMAX forum, using MIMO system and a channel bandwidth of 10 MHz, the uplink air interface can support 28 Mbps per sector peak data rate and 63 Mbps at downlink [1]. Employing three sectors per a base station, total peak data rate that a base station can deliver in uplink and downlink is 84Mbps and 189 Mbps respectively [1]. To calculate the total data rate supported by a base station, two other factors need to be determined, which are the multiplexing technique in used and the uplink and downlink rate. Assuming FDD technique and 3:1 downlink and uplink rate, the peak downlink data rate is 46 Mbps/sector and 8 Mbps/sector for uplink, which leads to total peak data rate of 46 Mbps per sector for each base station, as the total data rate in a FDD system is equivalent to the maximum rate between uplink and downlink [1]. Therefore by multiplying the total throughput for each sector by three, total peak data rate per a base station is calculated, which is approximately 150 Mbps in this case. It means that the link between the base station and gateway needs to be able to have the capacity equal to 150 Mbps [1].

In contrast, if TDD technique is in used in the same scenario, the peak data rate per each sector in uplink and downlink is 1.83 Mbps and 13.60 Mbps respectively which results in total data rate of about 16 Mbps per sector, as the sum of uplink and downlink data rate in TDD systems result in total data rate [4]. Triple the throughput of each sector result in peak data rate of 48 Mbps per a base station.

Choosing either FDD or TDD, also the downlink and uplink rate depends on the deployment scenario and the type of application is due to be delivered in the network. However a comparison between TDD and FDD assist to select the best technique that suites the providers. Different metrics can be considered for the evaluation of the backhaul planning and the techniques that need to be deployed for achieving the better performance, such as different service type to be offered, the available bandwidth, data rate, complexity and cost also the number of subscribers that can be served by the base station.

In terms of bandwidth availability, if the frequency band is limited, TDD is more likely to have the better performance than FDD; since FDD needs different frequencies for downlink and uplink, while TDD send the uplink and downlink data traffic on the same frequency band but different time slots. However, planing for different service types, FDD is more suitable for the applications which require more bandwidth such as video, whereas TDD which is proper for application such as voice that requires lower bandwidth. Furthermore, FDD provides higher data rate than TDD but TDD is far less complex therefore lower deployment cost [4]. However, operators can benefit the high number of subscribers supported by FDD technique when they demand applications with lower bandwidth requirements.

To achieve an efficient backhaul dimensioning, it is recommended by [1] to dimension the backhaul network and anticipate the total required capacity of the WiMAX backhaul network based on the predicted demand for service by dividing the total available spectrum into three different parts, each part serving a portion of the users with the same required quality of service applications. It is suggested in [1] that this method of dimensioning requires less backhaul deployment cost initially and scale well if the number of subscribers increases by adding more backhaul. To deploy that, assuming three different types of applications are in demand defined as video calls with required high QoS guaranteed, voice calls which requires less QoS guaranteed compare to video calls and finally best effort service, which provides services such as the Internet access with no guarantee of QoS requirement. In this case, the aim is to identify the total capacity needed for the backhaul link between two base stations or between the base station and gateway to support all different service applications efficiently.

The first step is to divide the applications into three groups known as EF class (Expedited Forwarding), AF class (Assured Forwarding) and BE class (Best Effort) [1]. The applications such as video calls are placed in the EF category, which require absolute amount of delay and very sensitive to delay. AF class is include the applications with average delay and less sensitive to delay compare to the EF class applications, such as voice calls. BE class consists of the applications with no sensitivity to delay. The next step is to determine how many subscribers are using the spectrum simultaneously at each class, which is assumed to be N voice calls, M video calls and few BE type applications to be served by a base station. Based on the assumption that a portion of the spectrum is used by n EF subclasses and other portion of the spectrum is divided to be used by m AF subclasses and few BE, also using a weighted fair queuing scheduler with h as a vector defining the weights used in the scheduling to determine the distribution of the spectrum for the AF and BE class, the capacity requires by different classes are calculated by [1]:

$$c_i^{EF} = \frac{h_i^{EF} c}{h^{BE} + h^{AF} + \sum_{i=1}^{n} h_i^{EF}} \tag{3.32}$$

$$c_i^{AF} = \frac{h_i^{AF} c}{h^{BE} + h^{AF} + \sum_{i=1}^{n} h_i^{EF}} \tag{3.33}$$

$$c^{BE} = \frac{h_i^{BF} c}{h^{BE} + h^{AF} + \sum_{i=1}^{n} h_i^{EF}} \tag{3.34}$$

Where:

- c_i^{EF}: each EF subclass needs to have at least this amount of capacity
- c_i^{AF}: total bandwidth available for AF class
- c^{BE}: total bandwidth available for BE class
- c: capacity of the backhaul link
- h^{EF}: weight for EF class
- h^{AF}: weight for AF class
- h^{BE} : weight for BE class

Now the backhaul link capacity can be determined by accumulating the data result from the calculation of capacity needed for each subclass.

References

[1] Yan Zhang. *WiMAX Network Planning and Optimization*. USA: CRC Press, 2009.
[2] Johnson I Agbinya. *IP Communications and Services for NGN*. New York: Taylor and Francis, 2009.
[3] Bharathi Upase, Mythri Hunukumbure, Sunil Vadgama. "Radio Network Dimentioning and Planning for WiMAX Networks." 43.4, pp. 435–450 Available: http://www.fujitsu.com/downloads/MAG/vol43-4/paper09.pdf.
[4] WiMAX Forum. "Mobile WiMAX-Part I: A Technical Overview and Performance Evaluation". *WiMAX Forum*, pp. 1–53, 2006.
[5] Kai Dietze Ph.D, Ted Hicks, "WiMAX Uplink and Downlink Design Consideration". Internet: http://www.edx.com/files/wimax_paper_2v4-1.pdf, [Sep.19, 2009].
[6] Steve Wisniewski. *Wireless and Cellular Networks*. USA: Pearson Prentice Hall, 2005.
[7] Katzela, I. and M. Naghshineh, "Channel Assignment Schemes for Cellular Mobile Telecommunication Systems: A Comprehensive Survey," *IEEE Personal Communications*, pp. 10–31, June 1996.

[8] Jeffrey G. Andrews, Arunabha Ghosh, Rias Muhamed. *Fundamentals of WiMAX*. USA: Prentice Hall, 2007.
[9] Rhode and Schwarz, WiMAX General Information about the standard 802.16 (application note), 2006, pp. 1–34.
[10] WiMAX Capacity White Paper, SR Telecom Inc., Canada, 2006, pp. 1-34.

4

Part III
WiMAX and LTE Link Budget

Johnson I Agbinya and Mehrnoush Masihpour

University of Technology, Sydney Australia

4.1 Propagation Regimes

Network planning provides information on the feasibility of network role out. Specifically some of the information is on the estimate of the optimum number of base stations, the location of the base stations and antennas, determining the type of the antenna and also the received power at receiver and the environment characteristics of the propagation environment. These optimum values are dependent on the services and the number of users.

In wireless systems, the transmitted signal is propagated in an open environment and losses some of its power as a result of phenomenon such as scattering, diffraction, reflection and fading when received at the receiver; therefore it is important to calculate the received signal power to determine factors such as the radius of a cell or the type of the antenna [1]. The degradation of the transmitted signal through interaction with the environment is known as path loss, which means the difference between the transmitted and received signal power. In general, path loss is calculated by [1]:

$$Path\ Loss = P_T + G_T + G_R - P_R - L_T - L_R\ [dB] \tag{4.1}$$

Where, P_T shows the power at transmitter and P_R is power at the receiver, G_T and G_R is the transmitter and receiver antenna gain respectively, L_T and L_R

P_T G_T $-L_P$ G_R $-A_M$ P_R

Fig. 4.1 Link budget.

express the feeder losses [1]. This equation describes the link budget. A link budget describes the extent to which the transmitted signal weakens in the link before it is received at the receiver. The link budget therefore accounts for all the gains and losses in the path the signal takes to the receiver.

As shown in Figure 4.1, a link is created by three related communication entities:

a) the transmitter
b) the receiver
c) the channel (medium) between them. The medium introduces losses causing a resuction in the received power.

The link budget (Figure 4.1) equation therefore can be written either as in equation (4.2) to show the path loss or in terms of the receiver power to be:

$$P_R = P_T + G_T - L_P + G_R - A_M \tag{4.2}$$

This equation assumes that all the signal gains and losses are expressed in decibels. The units for these are as follows:

$A_M(dB)$, ($L_T + L_R$ in equation (4.2)) represents all the attenuation losses such as feeder loss, link margin, diffraction losses, losses due to mobility (Doppler), and the effects of rain, trees and obstacles in the signal path.

$G_T(dBi)$ is the transmitter antenna gain

$G_R(dBi)$ is the receiver antenna gain

$P_R(dBm)$ is the received power at the receiver

$P_T(dBm)$ is the transmitted power

$L_P(dB)$ is the path loss in the physical medium between the transmitter and receiver.

Radio frequency sources are often modelled as isotropic sources. The source radiates microwave energy uniformly in all directions, the so-called isotropic source (into a spherical volume). Isotropic sources do not exist in practice. A more practical source is the half-wave dipole that is used to model an effective radiated power (ERP). Effective isotropic radiated power (EIRP) is measured in terms of a half-wave dipole model and: EIRP = ERP + 2.15 dB

Interference margin is usually around 1 dB. This is used to account for interference during the busy hour and depends on traffic load, frequency reuse plan and other factors.

Penetration of microwave into buildings varies and can be very severe. *Building penetration* is usually around 5 to 20 dB and accounts for penetration into different types of building materials, for indoor coverage. Penetration is a function of the type of building and the desired signal quality at the centre of the building. In-building losses can be quite dramatic in areas such as lifts and underground shelters.

Vehicle penetration is less severe compared to building penetration. The estimate for this is around 6 dB and accounts for the attenuation of signal by the frame of a car or truck. Other margins normally considered include:

Body loss (human body) of around 3 dB: This accounts for the absorption of the signal by a mobile user's head. It is sometimes called head loss. This varies depending on hair structure and the proximity of the mobile terminal to the head. Can you predict whether a female head will result to more body loss than male? What about African and Caucasian head loss (these racial groups carry different hair textures).

Fade margin of between 4 to 10 dB accounts for multipath fading as discussed in Chapter 4. Fading dips occur for slow moving mobiles. Fast moving mobiles tend to overcome this because they move faster out of dips before the fading affects the signal. Fade margin varies with the environmental conditions. In the next sections we describe the sequence of signal reducing events in the

i) transmitter
ii) channel (medium) and
iii) receiver

The following discussions explain what happens at the transmitter, receiver and channel respectively.

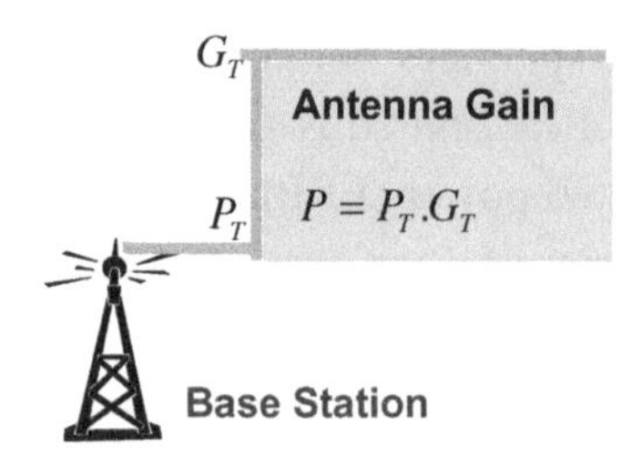

Fig. 4.2 Base station antenna power gain.

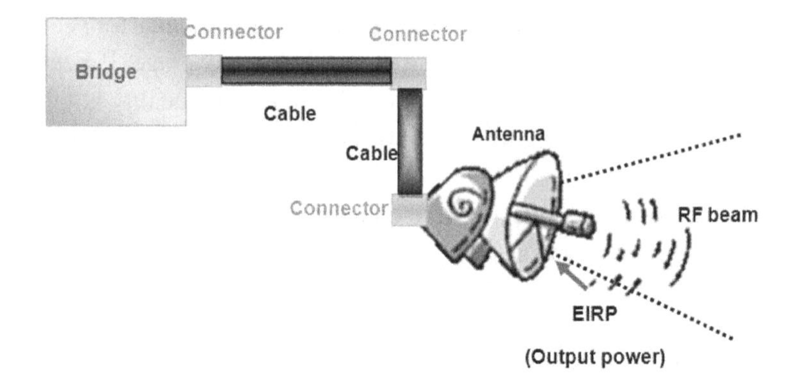

Fig. 4.3 Sources of power losses in transmitter receiver.

4.2 Transmitter

A WiMAX base station normally employs some form of MIMO system and the early implementations used 2 × 2 MIMO antennas. The base station also used adaptive antenna systems. The transmitter antenna boosts the data signal power before launching it into the channel.

The antenna creates an effective isotropically radiated power (EIRP) and outputs (radiates) it into the channel (medium), where EIRP = ERP + 2.15 dBi. Notice that the gain of the antenna is in dBi, while ERP is in dB. This conversion is essential. Due to the connectors and cables (Figure 4.3) used in the transmitter circuit, power losses are made as in Figure 4.4.

The antenna EIRP is reduced as follows

$$EIRP(dB) = P_T - L_{connector,cable} + G_T \tag{4.3}$$

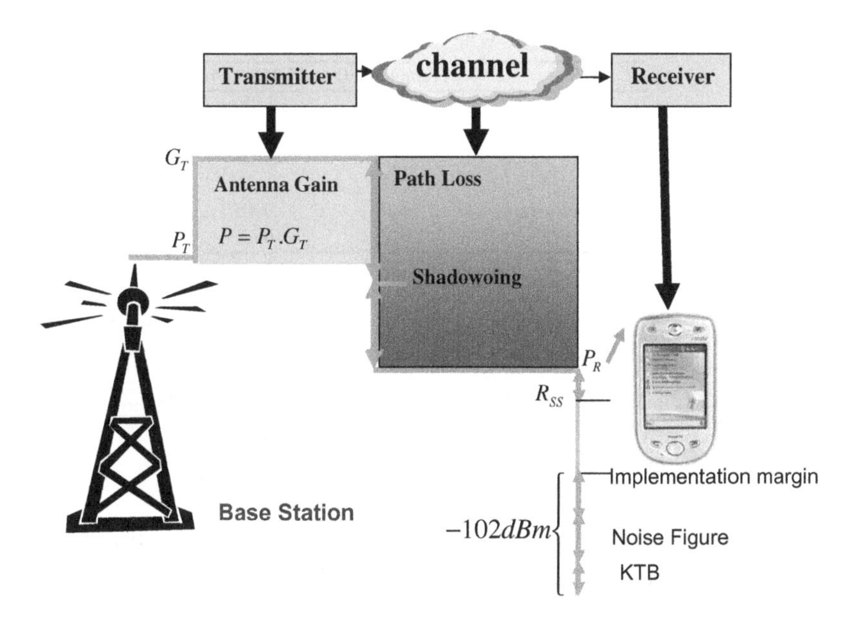

Fig. 4.4 Received signal power level after passing through the channel.

Table 4.1 parmeters of base station [2].

	Standard BS	BS with 2 × 2 MIMO	BS with 2 × 2 MIMO and 2 element AAS
DLT$_X$ power	35 dBm	35 dBm	35 dBm
DLT$_X$ antenna gain	16 dBi	16 dBi	16 dBi
Other DLT$_X$ gain	0 dB	9 dB	15 dB
UL R$_X$ antenna gain	16 dBi	16 dBi	16 dBi
Other UL R$_X$ gain	0 dB	3 dB	6 dB
UL R$_X$ noise figure	5 dB	5 dB	5 dB

For a 2 × 2 MIMO base station case, the following power losses were recorded in [2].

The parameters for a set of customer premises equipment for both fixed and mobile cases are given in Table 4.2.

4.3 Signal Fading

The WiMAX channel like all other communication channels introduces signal degradation (fading) and reduces the output signal launched to it by the transmitter (Figure 4.4).

Table 4.2 Parameters for customer premises equipment.

	Portable CPE	Mobile CPE
UL T_x power	27 dBm	27 dBm
UL T_x antenna gain	6 dBi	2 dBi
Other ULT_x gain	0 dB	0 dB
DL R_x antenna gain	6 dBi	2 dBi
Other DLR_x gain	0 dB	0 dB
DL R_x noise figure	6 dB	6 dB

The received signal power irrespective of the path loss and shadowing must be greater than the WiMAX receiver sensitivity (R_{SS}). The signal to noise ratio at the transmitter for a 2 × 2 MIMO base station is

$$SNR = P_R + 102 - 10\log_{10}\left(\frac{F_S N_{used}}{N_{FFT}}\right) \tag{4.4}$$

This expression assumes that the implementation margin is 7 dB and noise figure is 5 dB. The effects of the channel and all other degradation sources must result to the received signal power level to be at least equal to or greater than the receiver sensitivity $P_R \geq R_{SS}$.

Signal degradation has always been a disturbing factor in telecommunications, and much more so in cellular communications. There are three principal sources of degradation in a cellular environment: noise (inter-modulation noise, additive white Gaussian noise (AWGN)), multiple access interference (MAI) and fading. Noise is mainly contributed by the environment and the equipment in use. MAI is a result of sharing bandwidth and communication channel. Multiple access interference refers to inter-cell interference, intra-cell interference, co-channel interference and adjacent channel interference. Inter cell interference exists between two or more cells because of shared frequencies and frequency reuse. Intra-cell interference exists within a cell. Co-channel interference occurs between two users using the same channel. We have dealt with MAI in Chapter 3. The present chapter is dedicated to the third type of impairment in cellular systems —*fading*.

Fading is of two forms, large-scale and small-scale fading. Large-scale fading is mostly mean signal attenuation as a function of distance and signal variation around its mean value. Small-scale fading is of two forms, time spreading (which is composed of flat fading and frequency selective fading) and time variance of channel (which is composed of fast fading and slow fading). Figure 4.5 categorises these degradation sources.

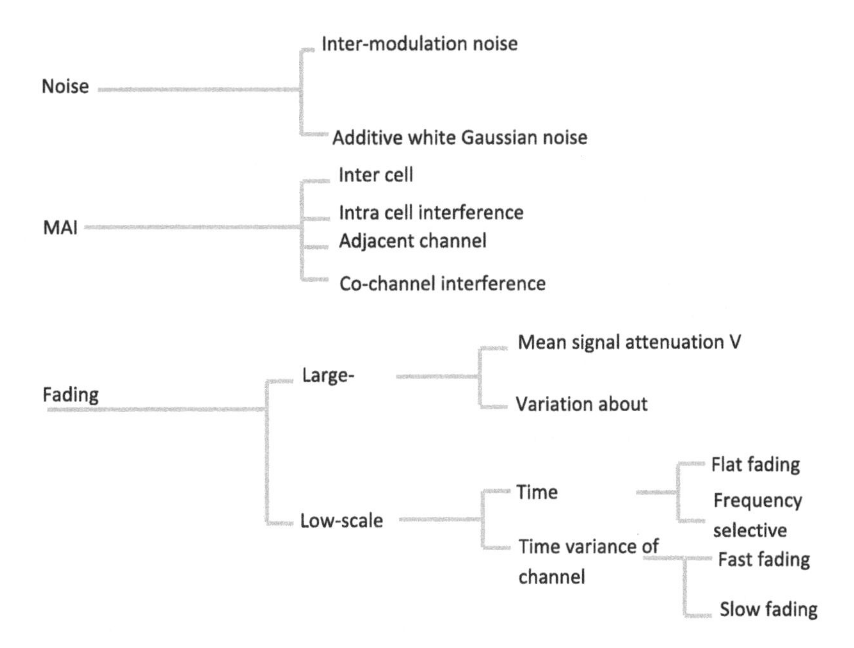

Fig. 4.5 Impairment in cellular environment.

Ideally, we would like communication signals to propagate (travel) without obstructions or disturbances. This can only happen in free space. Unfortunately, free-space is not a practical proposition in telecommunication. Scattering, reflection and diffraction of waves affect propagation of microwaves in many ways. Multipath propagation, Doppler spread and coherence time, delay spread and coherence bandwidth all conspire against the transmitted signal. These effects reduce the system signal to noise ratio, and lead to incoherent calls, dropped calls, and noisy channels.

Reflection occurs when the *path of a wave is obstructed* by a smooth surface. As the wave impinges on the surface, its direction of travel is changed. Usually the dimensions of the surface where reflection takes place are large relative to the wavelength of the wave.

When an object with large dimensions relative to the wavelength of a signal *blocks the path of a wave, diffraction* takes place. The object's sharp edges lead to diffraction or shadowing.

Scattering however occurs when a wave impinges on an object with dimensions comparable to the wavelength of the signal. The object causes the wave to spread out or scatter to different directions. In urban applications street lights, signs and foliage are the worst culprit. Scattering is a loss of useful signal strength because the receiver is unable to collect all its energy as it is tuned to a narrow view, and collects signals only in the narrow view.

What causes microwave signals to fade? The main causes of signal fading are discussed in this section. Microwave signals arrive at a receiver from many paths (multipath). The line of sight (LOS) signal is usually the preferred one in most applications as it travels the shortest path and thus arrives at a receiver with the strongest amplitude.

4.3.1 Multipath Propagation

Multipath propagation causes large variations in signal strength. The major effects are threefold:

- Time variations due to multipath delays;
- Random frequency modulation due to *Doppler shifts* from different multipath signals;
- Random changes in signal strength over short time periods;

In practice, multipath delays lead to time dispersion or 'fading effects' which is small-scale in nature.

A mobile communication multipath channel can be modelled as a linear time-varying filter with impulse response $h(t,\tau)$, where τ is the multipath delay in the channel for a fixed time t. In practice a low-pass model of the channel is easier to use than the actual complex model. The model allows a description of all other signal components relative to the signal which arrives first with delay $\tau_o = 0$. All the delayed signals arriving latter than this are discretised in terms of delays in N equally spaced time intervals of width $\Delta\tau$. All the multipath wave components in bin i are represented in terms of one component with delay $\tau_\iota = i\Delta\tau$u

Multipath fading is measured by using channel sounding through direct pulse measurements, spread spectrum sliding correlator or swept-frequency channel analyser. These techniques provide time dispersion parameters (mean excess delay, maximum excess delay at some given signal to noise ratio and rms delay spread), coherence bandwidth and Doppler spread or spectral broadening.

4.3.2 Doppler Effect

A moving object experiences Doppler frequency shifts. Mobile phone signals in fast moving cars, air plane and ships also experience Doppler shifts. The measured frequency increases as the mobile moves towards a base station. As it moves away from the base station, the frequency decreases. Doppler effects therefore leads to a variation of the signal bandwidth and is governed by the expression:

$$f_d = \frac{v}{\lambda}\cos\theta \tag{4.5}$$

where θ is the angle made by the signal path to the base station and the ground plane as shown in Figure 4.6. In this expression, velocity v is measured in meters/second, wavelength λ in meters and frequency f_d in Hertz. In the time domain, Doppler frequency shift leads to coherence time. *Coherence time* is the time duration over which two signals have strong potential for amplitude correlation. Coherence time can be approximated by the expression:

$$T_c = \sqrt{\frac{9}{16\pi f_m^2}} \tag{4.6}$$

where f_m is the maximum Doppler shift, which occurs when $\theta = 0$ degrees. To avoid distortion due to motion in the channel, the symbol rate must be greater than the inverse of coherence time ($1/T_c$).

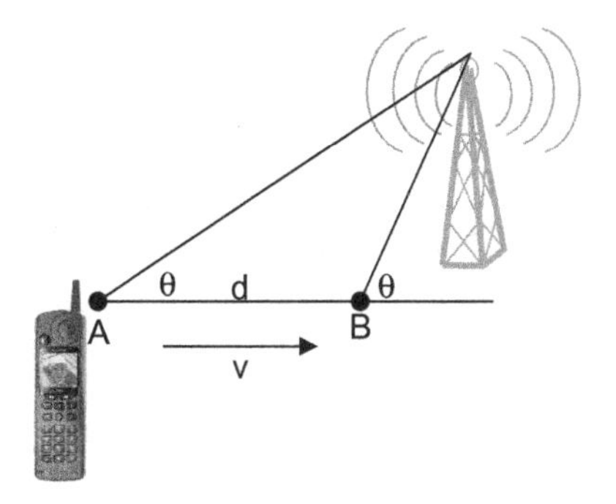

Fig. 4.6 Doppler effect in cellular communications.

4.3.3 Delay Spread and Coherence Bandwidth

Multipath delay causes the signal to appear noise-like in amplitude. We can compute its statistical averages and parameters. The standard deviation of the distribution of multipath signal amplitudes is called delay spread, σ_{τ}. Delay spread varies with the terrain with typical values for rural, urban and suburban areas: $\sigma_t \approx 0.2\mu s\ (rural)$; $\sigma_t \approx 0.5\mu s\ (suburban)$; $\sigma_t \approx 3.0\mu s\ (urban)$. Since the signal bandwidth varies due to delays, what then is the best measure of the bandwidth in practice? To answer this question, we use coherence bandwidth. It is defined to be the statistical measure of the range of frequencies over which the channel is considered constant or flat. It is the bandwidth over which two frequencies have a strong potential for amplitude correlation. Coherence bandwidth estimated for both strong and weak correlation are:

$B_c \approx \frac{0.02}{\sigma_t}$ for correlation greater than 0.9 and $B_c \approx \frac{0.2}{\sigma_t}$ for correlation greater than 0.5. Table 4.3 shows typical rms delay spreads for various types of terrain.

4.3.4 Categories of Fading

There are two major categories of fading, small-scale and large-scale fading. Large-scale fading is dependent on the distance between the transmitter and receiver. It is generally called path loss or 'large-scale path loss', '*log-normal* fading' or 'shadowing'. Small-scale fading is caused by the superposition of multipath signals, the speed of the receiver or transmitter and the bandwidth of the transmitted signal. Therefore, small-scale fading is a result of constructive and destructive interference between several versions of the same signal causing attenuation of the average signal power. This type of fading is over a

Table 4.3 Delay spread.

Delay spread figures at 900 MHz	Delay in microseconds
Urban	1.3
Urban (worst-case)	10–25
Suburban (typical)	0.2–0.31
Suburban (extreme)	1.96–2.11
Indoor (maximum)	0.27
Delay Spread at 1900 MHz	
Buildings (average)	0.07–0.094
Buildings (worst–case)	1.47

fraction of a wavelength and of the order of 20 to 30 dB. Small-scale fading is also known by other names such as 'fading', 'multipath' and '*Rayleigh*' fading. Rayleigh fading is a statistical variation of the received envelope of a flat fading signal. Multipath fading manifests as time spreading of the signal and a time variant behaviour. A time variant behaviour of the channel may be due to motion of the mobile and or changing environment (movement of foliage, reflectors and scatters).

If the impulse response of the mobile radio channel is $h(\tau, t)$ and time invariant and if it has zero mean, then the envelope of the impulse response has a Rayleigh distribution given by the expression:

$$p(r) = \frac{r}{\sigma^2} \exp\left(-\frac{r^2}{2\sigma^2}\right) \tag{4.7}$$

where σ^2 is the total power in the multipath signal. If however the impulse response has a non zero mean, then there is a component of the direct path (line of sight or specular component) signal in the channel and the magnitude of the impulse response has a Ricean distribution. (*Rice fading* is therefore the combination of Rayleigh fading with a significant non-fading (line of sight) component). Ricean distribution is given by the expression

$$p(r) = \frac{r}{\sigma^2} \exp\left(-\frac{r^2 + s^2}{2\sigma^2}\right) I_0\left(\frac{rs}{\sigma^2}\right) \tag{4.8}$$

The power of the line of sight signal is s^2 and I_0 is a Bessel function of the first kind.

The distance between either the dips or troughs in Rayleigh fading is of the order of half a wavelength. Small-scale fading occurs as either of four types:

- frequency selective fading in which the bandwidth of the signal is greater than the coherence bandwidth and the delay spread is greater than the symbol rate; Signals at some frequency components experience more fading than others
- flat fading when the bandwidth of the signal is less than the coherence bandwidth and the delay spread is less than the symbol rate
- fast fading when the Doppler spread is high and the coherence time is less than the symbol period and
- slow fading with a low Doppler spread and coherence time is greater than the symbol period

Table 4.4 Fading effects.

Type of fading	Frequency Effects	Time Effects
	Effects of Multipath Delay Spread	
Frequency Selective fading	BW of signal > coherence BW	Delay spread > symbol period
Flat fading	BW of signal < coherence BW	Delay spread < symbol period
	Effects of Doppler Spread	
Slow fading	Low Doppler spread	Coherence time > symbol period
Fast fading	High Doppler spread	Coherence time < symbol period

The first two fading are caused by multipath delay spread and the last two by Doppler spread. Table 4.4 is a summary of the conditions that exist with each type of fading.

Small-scale fading may be corrected by using adaptive equalisers or through the use of modulation techniques such as spread spectrum and error correction.

4.4 Propagation Regimes and Models

Based on different environment different path losses are experienced, consequently different radio propagation modelling is employed to estimate the electric field strength [1]. Channel models can be seen from different angles; in terms of existence of a direct line of sight between the transmitter and receiver, channel modelling methods can be categorized as free space propagation model and modelling for the land propagation when the LOS does not exist [3]. However, according to the required statistic for channel modelling, two main groups of Deterministic and Empirical modelling can be considered [1]. The deterministic model is used when the channel modelling requires the detailed information about the geometric environment such as the location and the surrounding area. In contrast, empirical models are based on the measured mean path loss for different environments, this method is less complex and cost but less accurate compared to the deterministic modelling.

One of the empirical channels modelling usually is used by WiMAX technology such as Stanford University Interim (USI) model, Cost-231 Hata model, Macro Model and Ericsson model 9999 [1]. In this section the USI and free space channel modelling are introduced in some detail.

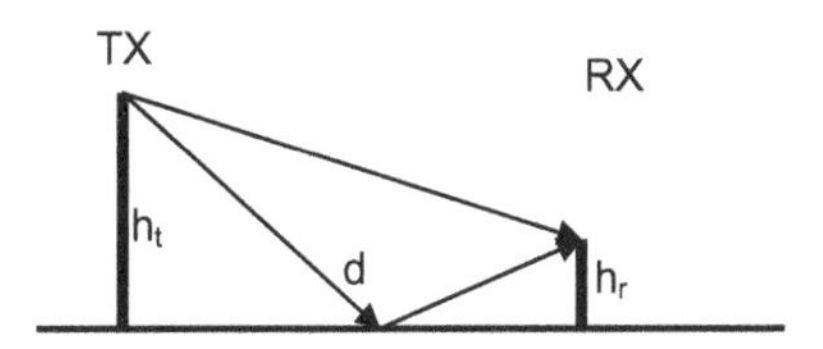

Fig. 4.7 Two-ray propagation.

4.4.1 Terrestrial Microwave Propagation

Terrestrial propagation (propagation in natural settings) is much different from free space propagation. Propagation of cellular signals in urban environment is difficult to predict because of man made structures (tall buildings, signs and other obstacles). On earth, a signal propagates through air and reflects, diffracts and scatters as discussed earlier. Theoretically, propagation can be predicted with the help of a 2-ray model and Fresnel Zones. A two-ray model is shown in Figure 4.7. The signal reaches the receiver through two paths, the direct path and a reflected ray. This model assumes a flat earth.

The path difference between the direct and reflected rays is given by the expression:

$$\Delta = \left[\sqrt{(h_r + h_t)^2 + d^2} - \sqrt{(h_t - h_r)^2 + d^2}\right] \approx \frac{2h_t h_r}{d} \tag{4.9}$$

In practice, destructive earth reflections are avoided by imposing the criteria:

$$\Delta > (2n - 1)\lambda/2, \quad n = 1, 2, 3, \ldots \tag{4.10}$$

Electromagnetic wave fronts are often divided into zones of concentric circles (separated by half a wavelength) called Fresnel zones. The zones define propagation break points. In the first Fresnel zone (no reflection) occurs and $n = 1$ so that $h_r h_t > \lambda d/4$. The first breakpoint, d_o is called the first Fresnel zone and occurs at a distance $d_o = 4h_r h_t/\lambda$. Until this point, the propagation is assumed to be free space. These two propagation models are useful for system design. The two models are used to predict microcell and indoor coverage. In many applications the distance between the transmitter and receiver is smaller than the first breakpoint, and for such cases, the Fresnel point does not help with the design.

Table 4.5 Path loss exponent.

Environment	Path loss exponent, n
Free space	2
Ideal specular reflection	4
Urban cells	2.7–3.5
Urban cells (shadowed)	3–5
In building (line of sight)	1.6–1.8
In building (obstructed path)	4–6
In factory (obstructed path)	2–3

Over ideal ground or the so-called specular ground, the received power is given by the modified free-space power:

$$P_r = P_{FS}[(1 - \exp(j2\pi\Delta/\lambda))]^2 \approx P_{FS}\left(\frac{4\pi h_t h_r}{\lambda d}\right)^2 \quad (4.11)$$

$$P_r = P_t G_t G_r \frac{h_r^2 h_t^2}{d^4} \quad (4.12)$$

As a result of the distance dependence in this expression, every time we double the distance, we lose 12 dB of signal energy. This shows that frequency reuse should be done at shorter distances. The path loss exponent varies from terrain to terrain as shown in Table 4.5.

4.4.2 Free Space Propagation

The medium separating the receiver from the transmitter plays an important role in the propagation of the signal. Free space propagation is usually modelled with the Friis formula [3]:

$$P_r = P_t \frac{G_t G_r \lambda^2}{(4\pi d)^2} \quad (4.13)$$

Where P_r, P_t, G_r and G_t are the receiver power, transmitter power, receiver antenna and transmitter antenna gains respectively. It has been assumed in this expression that radiation is into a spherical space of radius d surrounding the antenna. Thus radiation is into an area equal to the surface area of a sphere ($4\pi d^2$). The medium between the transmitter and receiver is often a dielectric, air, a piece of wire, fibre or some liquid including water. The medium varies from application to application and from terrain to terrain. The medium between the transmitter and receiver introduces a propagation loss as

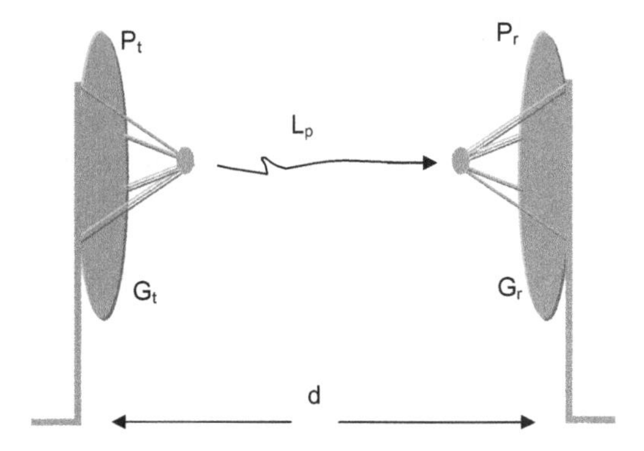

Fig. 4.8 Free space propagation.

shown in Figure 4.8. The loss is modelled with the expression:

$$L_p = 10 \times Log\left(\frac{4\pi d}{\lambda}\right)^2 \tag{4.14}$$

The propagation-exponent is a function of the terrain between the transmitter and receiver and has values in the range 2 to 5 in urban areas. A value 2 is used for free space. The propagation loss is often expressed in decibels and is:

$$L_p = 32.44 + 20Log(f) + 20Log\,(d) \tag{4.15}$$

The frequency f is measured in MHz and the distance d in kilometres. As the equation (4.14) implies, using higher carrier frequencies, the free space loss will be higher. For instance, if two receivers have the same sensitivity, using 450 MHz carrier frequency the cell radius will be 7.78 times longer than when 3.5 GHz carrier frequency is used [4].

4.5 Link Budget

Propagation losses are useful for computing link budgets. Link budgets provide the maximum allowable path losses per link and an indication of the link that will likely be a limiting factor to the system. The maximum allowable path loss also sets a limit on the maximum cell size. We will use path losses to compute link budgets in the this chapter.

The plane earth model is more appropriate for cellular channels. The model ignores the curvature of the earth's surface and considers a two-path model of direct line of sight and a ground reflected paths. In this model the heights of the transmitting and receiving antennae feature prominently in the propagation loss expression. Provided the heights of the antennas are less than the separation between the transmitter and receiver (h_b and h_r), the propagation loss expression can be shown to be

$$L_p = -10\log_{10} G_t - 10\log_{10} G_r - 20xLog\,h_b - 20xLog\,h_r + 40xLog(d) \quad dB \tag{4.16}$$

4.5.1 Practical Models

Propagation out door is difficult to predict and as such, empirical models, without real analytical basis are applied. Most of the models used are accurate to within 10 to 14 decibels in urban and suburban areas. They tend to be less accurate in rural areas because most of the data used may have been collected in the urban and suburban areas. One of the other popular models is the Okamura model. In practice there are huge variations in the types of terrain and environment to cover. The heights of antenna, clutter, tree density, beamwidth, wind speed, season (time of the year) and multipath, vary widely and affect mobile phone waves. Hence complex models are required for such situations. They are used to predict propagation loss. Beyond the Okamura model, there are the Hata and Walfisch-Ikegami models and others by Egli, Lee, Carey, Longley-Rice, Ibrahim-Parsons and many more. Although most of them are beyond the scope of this note, we will examine a few briefly.

In a study of channel models for fixed wireless application by the Institution of Electrical and Electronic Engineers (IEEE) [5], a set of "propagation models applicable to the multi-cell architecture" was presented. They assumed a cell radius of less than 10 km, variety of terrain and tree density types, directional antenna (2–10 m) installed under-the-eaves/window or rooftop, 15–40 m BTS antennas and high cell coverage requirements of between 80–90%. This section is derived from analysis.

4.5.1.1 Suburban Path Loss (Hata-Okamura Model)

The Hata-Okamura model [6] is the most widely used model for this situation. The model is valid in the 150–1500 MHz frequency range (NMT and GSM), with receiver distances greater than a kilometre from the base station and base

Table 4.6 Path loss (Terrain) correction variables.

Model Parameter	Terrain Type A	Terrain Type B	Terrain Type C
a	4.6	4	4.6
b	0.0075	0.0065	0.005
c	12.6	17.1	20

station antenna heights greater than 30m. As such, the model is applicable to mobile phone applications below 1500 MHz. The modified Hata-Okamura model extends this range to around 2 GHz [7]. This is the so-called COST 231 model. Although the model targeted 2G systems in the 900 and 1800 MHz range, it has application to systems around 2 GHz (eg., DCS1800) provided lower base station antenna heights, hilly or moderate-to-heavy wooded terrains are not involved. Corrections for these limitations were applied to cover most terrain conditions applicable to the US. The body of study for savannah, and dense forest regions in Africa and other regions need to be understood. Path loss regimes are divided into three broad categories A, B and C:

- Category A (maximum path loss): hilly terrain with moderate-to-heavy tree densities
- Category B (intermediate path loss): terrain conditions between category A and C
- Category C (minimum path loss): mostly flat terrain with light tree densities

The median path loss at 1.9 GHz for a distance d_o from a base station is given by:

$$L_p = A + 10 * n * Log_{10}(d/d_o) + s; \quad d > d_o \tag{4.17}$$

where, $A = 20xLog_{10}(4\pi d_o/\lambda)$ and λ is the wavelength in metres, n is the path loss exponent and

$$n = (a - bh_b + c/h_b) \tag{4.18}$$

The height of the base station h_b is between 10 m and 80 m, and $d_o = 100$ m, a, b, and c are constants that depend on the terrain category which is reproduced from [5] below.

The shadowing effect is represented by s and follows a log-normal distribution with typical standard deviation between 8.2 and 10.6 dB. This too depends on the terrain and tree density type. Correction terms are used to account for antenna height and frequency region. For the model to apply to frequencies outside the range of specification (2 GHz), and for receive antenna heights between 2 m and 10 m, correction terms are specified. The coarse form of the path loss model (in dB) has three correction terms as:

$$L_{pc} = L_p + \Delta L_f + \Delta L_h \tag{4.19}$$

and ΔL_f (in dB) is the frequency correction term given by the expression. The frequency correction term is given by the expression

$$\Delta L_f = 6\,Log(f/2000) \tag{4.20}$$

The frequency f is MHz, and is positive for frequencies higher than 2 GHz. The correction term for antenna height is:

$$\Delta L_h = -10.8\,Log(h/2);\ \text{for categories A and B and}$$
$$\Delta L_h = -20\,Log(h/2);\ \text{for category C}$$

The height of the receive antenna is in the range 2 m $<$ h $<$ 10 m.

4.5.1.2 Urban path loss model (Alternative Flat Suburban)

The COST 231 Hata model for propagation in an urban environment in the PCS range is calculated from the expression:

$$\begin{aligned} L_p &= 46.3 + 33.9 \times Log(f) - 13.82 \times Log(h_b) - a(h_m) \\ &\quad + [44.9 - 6.55 \times Log(h_b) * Log(d) + C_m] \end{aligned} \tag{4.21}$$

where, a(h_m) is the mobile antenna correlation factor given as:

$$a(h_m) = \begin{cases} [1.1 \times Log(f) - 0.7]h_m & \textit{for suburban urban} \\ \quad -[1.56 \times Log(f) - 0.8]dB & \\ 3.2 \times [Log(11.75h_m)]^2 - 4.97\,dB & \textit{for dense urban} \end{cases} \tag{4.22}$$

$$C_m = \begin{cases} 0\,dB & \textit{for suburban urban} \\ 3\,dB & \textit{for dense urban} \end{cases} \tag{4.23}$$

$$L_{p\,(suburban)} = L_{p(urban)} - 2 \times [Log(f/28)]^2 - 5.4 \tag{4.24}$$

and
h_b = base station height (m), 30–200 m
h_m = mobile height (m), 1–10 m
f = frequency (MHz), 1500–2000 MHz
d = distance (km), 1–20 km

It has been shown that the COST 231 Walfish–Ikegami (W-I) model provides a close match for extensive experimental data from suburban and urban areas and that the Category C in the Hata-Okamura model is in good agreement with the cost 231 W-I model. It also provides continuity between the two models. The COST 231 W-I model agrees well with measured data from urban areas, provided appropriate rooftop heights and building spaces are used.

4.6 Walfisch–Ikegami Model

The Wlfisch–Ikagami Model is valid between 800 and 200 MHz and over distances of 20 km to 5 km and was recommended by the WiMAX Forum. It is useful for dense urban and canyon-like environments where average building height is larger than the receiving antenna height. This means wireless signals undergo diffraction and are guided along the street like an urban canyon. The model applies with the following propagation parameters:
h_b = base station height (m), 3–50 m
h_m = mobile height (m), 1–3 m
f = frequency (MHz), 800–2000 MHz
d = distance (km), 0.2–5 km

This model applies therefore fairly well when WiMAX is deployed in densely populated urban areas with significant number of semi-high rise buildings (eg. Sydney CBD). The model distinguished between the LOS and the NLOS cases.

Line of Sight: For the line of sight case the loss equation is

$$\tilde{L}_P = 42.6 + 26\log(d) + 20\log(f) \tag{4.25}$$

There are no correction terms in this case and it applies directly to a LOS WiMAX situation.

For an urban canyon with line of sight the suggested propagation model is

$$\tilde{L}_P = -31.4 + 26.\log_{10}(d) + 20.\log_{10}(f) \tag{4.26}$$

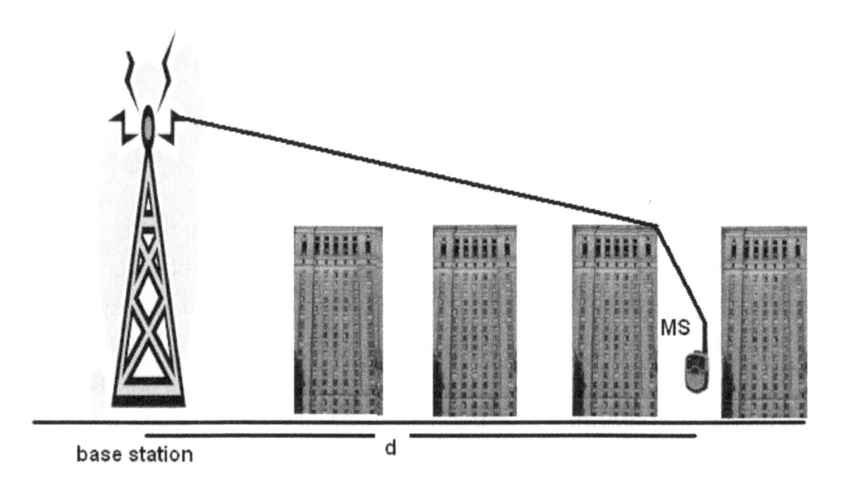

Fig. 4.9 Communication from a MS in a Canyon.

None Line of Sight: For the none line of sight case,

$$\tilde{L}_P = L_r + L_f + L_{ms} \tag{4.27}$$

The median path is given by the expression:

where Lr, Lf and Lms are the roof top, free space and multiscreening path losses.

$$\tilde{L}_P = \begin{cases} L_r + L_f + L_{ms}; & L_r + L_{ms} > 0 \\ L_f; & L_r + L_{ms} \leq 0 \end{cases} \tag{4.28}$$

The free space loss is given by the expression

$$L_f = 32.44 + 20.\log_{10}(f) + 20.\log_{10}(d) \tag{4.29}$$

The frequency f is measured in MHz and the distance in km. The rooftop diffraction model is given by the expression

$$L_{rt} = -6.9 + 10\log(w) + 10\log(f) + 20\log(d_{hm}) + L_{ori} \tag{4.30}$$

Where

$$\begin{aligned} L_{ori} &= -10 + 0.354\varphi & for\ 0 \leq \varphi < 35 \\ &= 2.5 + 0.075(\varphi - 35) & for\ 35 \leq \varphi < 55 \\ &= 4 - 0.114(\varphi - 55) & for\ 55 \leq \varphi < 90 \end{aligned} \tag{4.30a}$$

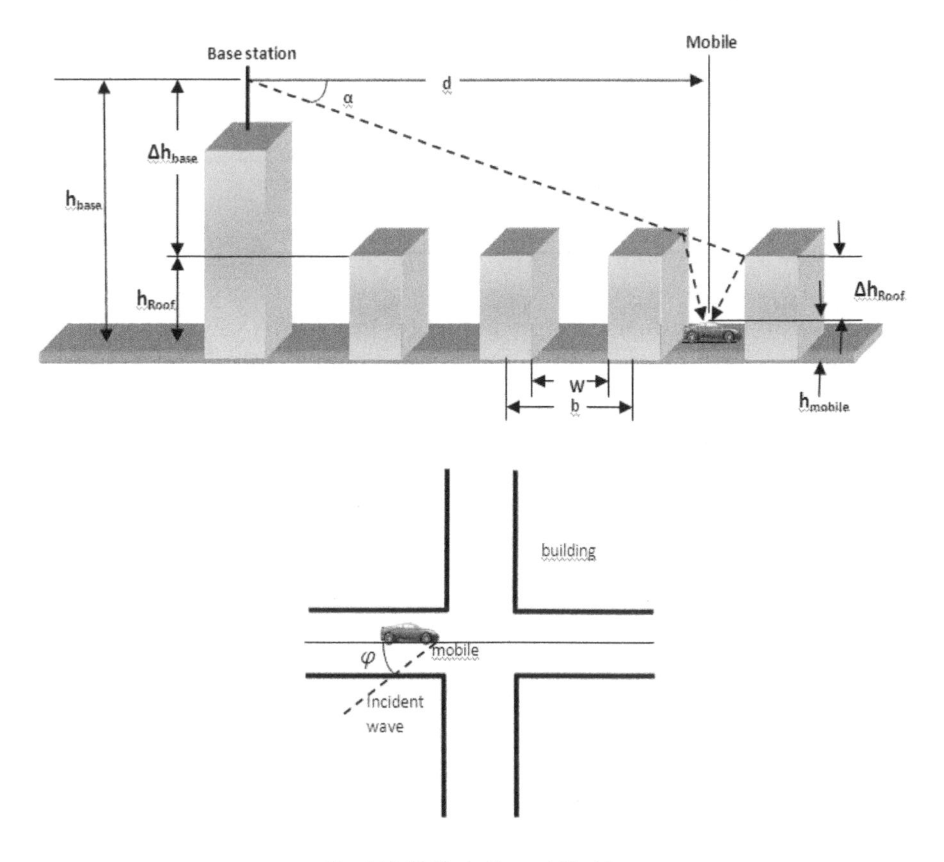

Fig. 4.10 Walfisch–Ikegami Model.

$$L_{(mult)} = k_0 + k_a + k_d \log(d) + k_f \log(f) - 9\log(W) \tag{4.30b}$$

$$\begin{aligned} k_0 &= 0, \\ k_d &= 18 - 15(d_{hb}/h_{roof}) \\ k_a &= 54 - 0.8d_{hb} \quad \text{and} \\ k_f &= -4 + 1.5\big[(f/925) - 1\big] \end{aligned} \tag{4.30c}$$

For the non line of sight case in a metropolitan area with the following parameters, the median loss model becomes

$$\tilde{L}_P = -65.9 + 38\log_{10}(d) + \left(24.5 + \frac{1.5f}{925}\right)\log_{10}(f) \tag{4.31}$$

Table 4.7

Parameter	Variable	Value
Height of base station	h_b	12.5 m
Building height	h_{bd}	12 m
Building-to-building distance	d	50 m
Building width	w	25 m
Height of mobile station	h_m	1.5 m
Orientation of all paths	φ	30°

A 10 dB fading margin is suggested by the WiMAX Forum with this formula.

4.7 Erceg Model

The Erceg model was adopted by the 802.16 group for fixed WiMAX. The model is mostly suited to fixed wireless applications. The Erceg model consists of a base model and an extended model. The data leading to the model were collected in Dallas, Chicago, New Jersey, Atlanta and Seattle in the USA, at 1.9 GHz and over 95 macrocells.

The base model has three models in one and each one accounts for specified terrain type.

i) Erceg A model is applicable to hilly terrain with moderate to heavy density of trees
ii) Erceg B model is also applicable to hilly terrain with light tree density. It applies also to flat terrain with moderate to heavy density of trees
iii) Erceg C model is applicable to flat terrain with light density of trees

The Erceg loss formula models instantaneous path loss as a sum of a median loss term and a shadow fade value given by the expression

$$L_P = \tilde{L}_P + X = A + 10\alpha \log_{10}\left(\frac{d}{d_0}\right) + X \tag{4.32}$$

Where X is the shadow fades and A is the free space path loss over a distance $d_0 = 100m$ and at a frequency f.

$$A = 20\log_{10}\left(\frac{4\pi f d_0}{C}\right) \tag{4.33}$$

Table 4.8 Parameters of the Erceg model.

Parameters	Erceg Model A	Erceg Model B	Erceg Model C
a	4.6	4	3.6
b	0.0075	0.0065	0.005
c	12.6	17.1	20
S_a	0.57	0.75	0.59
μ_s	10.6	9.6	8.2
σ_s	2.3	3	1.6

The associated path loss exponent α is modelled as a Gaussian random variable with a mean value given by the expression

$$A - Bh_b + Ch_b^{-1} \quad (4.34)$$

Hence the instantaneous value of the path loss exponent is

$$\alpha = A - Bh_b + Ch_b^{-1} + x\sigma_\alpha \quad (4.35)$$

Where x is a Gaussian random variable with zero mean and unit variance. The standard deviation of the distribution of the path loss exponent is σ_α. The parameters of the Erceg model are given below for the different terrain types.

The base model applies only at 1.9 Hz and for MS with omnidirectional antenna at a height of 2 meters and for base stations of heights 10 to 80 m. The extended model modifies the base model to enable it work over larger frequency range and with the following parameters:
h_b = base station height (m), 10–80 m
h_m = mobile height (m), 2–10 m
f = frequency (MHz), 1900–3500 MHz
d = distance (km), 0.1–8 km
The median path loss model for the extended Erceg model is

$$\tilde{L}_P = A + 10\gamma \log_{10}\left(\frac{d}{d_0}\right) + \Delta P.L_f + \Delta P.L_{hMS} + \Delta P.L_{\theta MS} \quad (4.36)$$

The correction terms in this equation are:

$$\Delta P.L_f = 6\log\left(\frac{f}{1900}\right) \quad (4.37)$$

$$\Delta P.L_{hMS} = -10.8\log\left(\frac{h_m}{2}\right); \quad \textit{for Erceg A and B} \quad (4.38)$$

$$\Delta P.L_{hMS} = -20\log\left(\frac{h_m}{2}\right); \quad for\ Erceg\ C \tag{4.39}$$

$$\Delta P.L_{\theta MS} = 0.64\ln\left(\frac{\theta}{360}\right) + 0.54\left(\ln\left(\frac{\theta}{360}\right)\right)^2 \tag{4.40}$$

The correction term $\Delta P.L_{\theta MS}$ is called the antenna gain reduction factor because it accounts for the fact that the angular scattering is reduced due to the directivity of the antenna. This correction can be significant and is about 7 dB at an antenna angle of 20°.

4.7.1 Stanford University Interim (SUI) Model

To calculate the path loss using SUI model, the environment is categorized in three different groups with different characteristics, known as A, B, and C [1]. A is referred to the hilly environment and moderate to very dense vegetation which results in highest path loss, while B refers to the hilly environment but rare vegetation or high vegetation but flat terrain. However C is referred to the flat are with rare vegetation which leads to lowest path loss.

SUI model is a suitable channel modelling for WiMAX implementation, using frequency band at 3.5 GHz, which can support for cell radius in range of 0.1 km and 8 km also the base station antenna height between 10 m and 80 m and receiver antenna height in the range of 2 m and 10 m [1]. In SUI model the path loss is calculated using formula 4.41:

$$L_p = A + 10\gamma \log_{10}\left(\frac{d}{d_0}\right) + X_f + X_h + s \quad for\ d > d_0 \tag{4.41}$$

In this formula, $d_0 = 100m$ and d is the distance between the transmitter and the receiver. s is a standard deviation which is a random variable, X_f is a correction for frequency above 2 GHz and X_h is a correction for transmitter antenna height [1].

$$X_f = 6.0\log_{10}\left(\frac{f}{2000}\right) \tag{4.42}$$

$$X_h = -10.8\log_{10}\left(\frac{h_r}{2000}\right) \quad for\ A\ and\ B\ environment \tag{4.43}$$

$$X_h = -20.0\log_{10}\left(\frac{h_r}{2000}\right) \quad for\ C\ environment \tag{4.44}$$

Where f is the frequency and h_r is the height of the antenna at the receiver. However, in formula 4.41, A is calculated by:

$$A = 20\log_{10}\left(\frac{4\pi d_0}{\lambda}\right) \tag{4.45}$$

Where λ is the wavelength of the signal in meter. Also γ is the path loss exponent which has different values between 2 and 5 for different environment and depends on the height of the base station antenna h_b and three constants of a, b and c which vary with different type of environment of A, B and C [1].

$$\gamma = a - bh_b + \frac{c}{h_b} \tag{4.46}$$

The path loss exponent in urban area when the LOS exists is 2, while it is between 3 and 5 in the urban area with absence of a LOS. However the path loss exponent will be more than 5 if the signal propagation is happening in an indoor environment [1]. SUI model is used for planning for WiMAX in rural, urban and suburban areas.

4.7.1.1 Indoor Propagation Models

In-door propagation models were originally studied for wireless LANs and cordless phones. Propagation in buildings is highly dependent on the types of building materials used, the lay out of the building and its contents and location. Log-normal shadowing applies for in-door wireless applications. In 3G networks and wireless Internet access, many picocells are deployed, hence indoor propagation modelling and estimation are of particular interest for 3G-LTE networks and WLAN.

4.8 Path Loss Exponents

Path loss exponent varies widely across propagation environments [8]. Therefore the bound on the hop distance and number is different for different types of propagation domains. For long-distance coverage the exponent for out door environments is around 4 except in none line-of-sight situations when it could be bigger than 4. The value of the path loss exponent is an indicator of how fast energy is lost between the transmitter and receiver. $\alpha < 2$ is a measure of the guiding effect of the channel and when $\alpha > 2$ the channel is considered to be scattering energy.

4.8.1 Path Loss Exponents in Different Environments

The following tables provide typical values of α and also show how different structures guide radio waves and which ones scatter them. They also provide a database of α for design of wireless networks in different environmental situations.

4.8.1.1 Out Door Environments

The value of path loss exponent out doors is a function of the terrain (free space, urban, suburban, rural and foliage type), if communication is line-of-sight (LOS) or non-LOS (NLOS), height of the antenna and the channel frequencies. Table 4.9 summarises these effects [8].

The dynamic range of α in this table is 6.3. None line-of-sight communication often means higher path loss exponents. Similarly the higher the height of the antenna the higher the expected pass loss exponent. The value of the path loss exponent after the break point is normally greater than the value

Table 4.9 Path loss exponents for out door environments.

Path Loss Exponents for Out Door Environments			
No	Location	Path loss exponent	Frequency range
1	Urban	4.2	
2	Free space	2	Micro cellular
3	Log-normally shadowing area	2 to 4	Micro cellular
4	UWB LOS up to breakpoint	2	UWB range
	UWB LOS after breakpoint	4	UWB range
	LOS urban (antenna ht = 4 m)	1.4	5.3 GHz range
	LOS urban (antenna ht = 12 m)	2.5	5.3 GHz range
	NLOS urban (antenna ht = 4 m)	2.8	5.3 GHz range
	NLOS urban (antenna ht = 12 m)	4.5	5.3 GHz range
5	LOS rural (antenna ht = 55 m)	3.3	5.3 GHz range
	NLOS rural (antenna ht = 55 m)	5.9	5.3 GHz range
	LOS suburban (antenna ht = 5 m)	2.5	5.3 GHz range
	NLOS suburban (antenna ht = 12 m)	3.4	5.3 GHz range
6	Highway micro-cells	2.3	900 MHz
	Dual Carriage Highway	7.7	1.7 GHz
7	BFWA/directional antenna (5.5–6.5 m)	1.6	3.5 GHz
	BFWA/directional antenna (6.5–7.5 m)	2.2	
	BFWA/directional antenna (7.5–8.5 m)	2.7	
	BFWA/directional antenna (8.5–9.5 m)	2.6	
	BFWA/directional antenna (9.5–10.5 m)	3.6	

before the break point. The break point distance can be approximated with the expression:

$$d_b = \frac{4\pi h_T h_R}{\lambda} \tag{4.47}$$

h_T and h_R are the heights of the transmitting and receiving antennas and λ is the wavelength of transmission. The high value of exponent for dual carriage highway is due to ground reflections from the road surface.

4.8.1.2 Indoor Environments

In Table 4.10, the path loss exponent for in door communications across a wide variation of frequencies is shown. The unpredictability of the path loss exponent is demonstrated by the range of values shown [8].

Communications indoors at various frequencies affect the path loss exponent and the predominant sources of effects are the height of the building (or height of antenna), antenna directivity, LOS or NLOS communication, the channel frequencies, the types of materials used in the construction of the buildings and the location of measurements in the building. Omni-directional antennas often result to lower path loss exponents compared to directional antennas. This is because, the omni-directional antennas collects signals from many more multipath sources. Building materials of different types lead to different path loss exponents. The dynamic range of α in this table is 8.8. Therefore the optimum hop index will vary widely in doors. Measuring α is therefore required prior to establishing the relay nodes.

4.8.1.3 Underground Environments

Communications underground such as in tunnels and mines forms a vital component of the overall wireless communication industry. In many countries, tunnels form significant sections of roads and railways. Similarly communication inside mines is also a vital support for mining and mineral exploration. Table 4.11 records typical path loss exponents reported for underground communications. Understandably, low frequency applications are prevalent.

Path loss exponent in underground communications is normally predominantly very high as seen from Table 4.11 [8]. This is due to the terrain, the materials used for construction of the tunnels and to some extent the channel

Table 4.10 Path loss exponent for indoor communications.

Path Loss Exponents for Indoor Environments			
No	Location	Path loss exponent	Frequency range
1	LOS	1.83	802.11a (5.4 GHz)
2	LOS	1.91	802.11b (2.4 GHz)
3	NLOS	4.7	802.11a
4	NLOS	3.73	802.11b
5	Omni/Omnidirectional antennas	1.55	UWB
	Omni/directional antennas	1.65	UWB
	Directional/Directional–shadow	1.72	UWB
6	Indoor CDMA	1.8 ~ 2.2	20 GHz–30 GHz
7	LOS	1.73	900 MHz
	NLOS	0.48 ~ 1.12	900 MHz
	LOS	2.23	1.89 GHz
	NLOS	−1.43 ~ 1.47	1.89 GHz
8	LOS (millimetre wave)	1.2 ~ 1.8	94 GHz
	Obstructed channel	3.6 ~ 4.1	94 GHz
	LOS	1.8 ~ 2.0	11.5 GHz
	LOS	1.2	37.2 GHz
9	Inside room of a building	0.77	900 MHz
	Inside room of a building	0.44	1.35 GHz
	LOS DECT picocells	−1.55	1.8 GHz
	NLOS DECT picocells	−3.76	1.8 GHz
10	Corridor Ground Floor	0.70	450 MHz
		0.48	900 MHz
		0.02	1.35 GHz
		−1.43	1.89 GHz
11	Corridor Floor 1 of building	1.12	450 MHz
		1.02	900 MHz
		0.07	1.35 GHz
		1.46	1.89 GHz
12	Corridor Floor 2 of building	1.79	450 MHz
		1.72	900 MHz
		0.44	1.35 GHz
		2.22	1.89 GHz
13	Indoor 3rd floor of a laboratory	1.3	2.45 GHz
		1.8	5.25 GHz
		1.7	10 GHz
		1.8	17 GHz
		1.7	24 GHz

frequencies used. The scattering properties of the terrain also affect the path loss exponent. The dynamic range of α in this table is 10.95. Path losses underground are therefore very high and hop distances must be chosen with this in mind.

Table 4.11 Path loss exponent for underground communications.

Path Loss Exponents for Underground Communications			
No	Location	Path loss exponent	Frequency range
1	Underground (train)–front	12.45	465 MHz
2	Underground (train)–rear	9.72	
3	Underground (train)–front	8.58	820 MHz
4	Underground (train)–rear	8.17	
5	Train yard (parallel to track)	2.7	
	Train yard (cross-track)	3.4	
6	Underground Mine	2.13 ~ 2.33	2.4 GHz
	Moving train–140 km track sites	1.5 ~ 7.7	320 MHz

Table 4.12 Effect of materials on path loss exponent.

Path Loss Exponents for Different Environmental Structures			
No	Location	Path loss exponent	Frequency range
1	Engineering	1.4 ~ 2.2	0.8 GHz–1.0 GHz
2	Apartment Hallway	1.9 ~ 2.2	
3	Parking structure	2.7 ~ 3.4	
4	One-sided corridor	1.4 ~ 2.4	
5	One-sided Patio	2.8 ~ 3.8	
6	Concrete Canyon	2.1 ~ 3.0	
7	Plant fence	4.6 ~ 5.1	
8	Small Boulders	3.3 ~ 3.7	
9	Sandy Flat beach	3.8 ~ 4.6	
10	Dense Bamboo	4.5 ~ 5.4	
11	Dry Tall Underbrush	3.0 ~ 3.9	

Table 4.13 SNR parameters as function of modulation schemes.

Modulation scheme	SNR CC (AWGN, BER 10–6)	SNR CTC (AWGN, BER 10-6)	Data bit per symbol
QPSK1/2	5 dB	2.5 dB	1
QPSK 3/4	8 dB	6.3 dB	1.5
16-QAM 1/2	10.5 dB	8.6 dB	2
16-QAM 3/4	14 dB	12.7 dB	3
64-QAM 1/2	16 dB	13.8 dB	3
64-QAM 2/3	18 dB	16.9 dB	4
64-QAM 3/4	20 dB	18 dB	4.5

4.8.1.4 Unspecified Environments

Path loss exponent in other terrains of interest are summarised in Table 4.12 [8].

Table 4.12 demonstrates the varying nature of path loss exponent when different types of materials and the terrain types that affect communications are considered. These tables show that there is no universally accepted path

Table 4.14 WiMAX link budget [2].

	Mobile Handheld in Outdoor Scenario		Fixed Desktop in Indoor Scenario		
Parameter	Downlink	Uplink	Downlink	Uplink	Notes
Power amplifier output power	43.0 dB	27.0 dB	43.0 dB	27.0 dB	AI
Number of tx antennas	2.0	1.0	2.0	1.0	A2
Power amplifier backoff	0 dB	0 dB	0 dB	0 dB	A3: assumes that amplifier has sufficient linearity for QPSK operation without backoff
Transmit antenna gain	18 dBi	0 dBi	18 dBi	6 dBi	A4: assumes 6 dBi antenna for desktop SS
Transmitter losses	3.0 dB	0 dB	3.0 dB	0 dB	A5
Effective isotropic radiated power	61 dBm	27 dBm	61 dBm	33 dBm	A6 = A1 + $10\log_{10}$(A2) − A3 − +A4 − A5
Channel bandwidth	10 MHz	10 MHz	10 MHz	10 MHz	A7
Number of subchannels	16	16	16	16	A8
Receiver noise level	−104 dBm	−104 dBm	−104 dBm	−104 dBm	A9 = −174 + $10\log_{10}$(A7*$le6$)
Receiver noise level	8 dB	4 dB	8 dB	4 dB	A 10
Required SNR	0.8 dB	1.8 dB	0.8 dB	1.8 dB	A11: for QPSK, R 1/2 at 10% BLER in ITU Ped. B channel
Macro diversity gain	0 dB	0 dB	0 dB	0 dB	A12; No macro diversity assumed
Subchannelization gain	0 dB	12 dB	0 dB	12 dB	A13 = $10\log_{10}$(A8)
Data, rate per subchannel (kbps)	151.2	34.6	151.2	34.6	A14; using QPSK, R 1/2 at 10% BLER
Receiver sensitivity (dBm)	−95.2	−110.2	−95.2	−110.2	A15 = A9 + A10 + A11 + A12 − A13
Receiver antenna gain	0 dBi	18 dBi	6 dBi	18 dBi	A16
System gain	156.2 dB	155.2 dBi	162.2 dB	161.2 dB	A17 = A6 − A15 + A16
Shadow-fade margin	10 dB	10 dB	10 dB	10 dB	A18
Building penetration loss	0 dB	0 dB	10 dB	10 dB	A19; assumes single wall
Link, margin	146.2 dB	145.2 dB	142.2 dB	141.2 dB	A20 = A17 − A18 − A19
Coverage range	1.06 km (0.66 miles)		0.81 km (0.51 miles)		Assuming COST-231 Hata urban model
Coverage range	1.29 km (0.80 miles)		0.99 km (0.62 miles)		Assuming the suburban model

loss model for indoor, out door or underground channels. The path loss model varies, from building to building and from terrain to terrain. All path loss models in use are approximations for only a few conditions.

4.8.1.5 Receiver Sensitivity

The receiver power is limited by its design (implementation margin), thermal noise in the receiver, its noise figure and signal-to-noise ratio. Thus the receiver sensitivity is given by the expression:

$$R_{SS} = SNR_R + NF_R + L_{implementation} + N_{Thermal} \tag{4.48}$$

These factors are given by the expressions

$$N_{Thermal} = -147 + 10\log_{10}(\Delta f) = -147 + 10\log_{10}\left(B.n.\frac{N_{Used}}{N_{FFT}}\right) \tag{4.49}$$

Where Δf is the subcarrier spacing. The thermal noise is affected by its thermal noise density which is

$$N_0 = K.T.B \approx -174\,dB \tag{4.50}$$

K is Boltzman constant. The SNR is a function of the modulation scheme used and these have been provided in the standard and are given in Table 4.13.

The receiver noise figure is caused by the electronics in its RF chain. It is a ratio of the input SNR to a device to its output SNR. This is normally measured at 290 Kelvin. It provides a measure of the performance of a device. It is given by the expression:

$$NF = \frac{SNR_{in}}{SNR_{out}} \tag{4.51}$$

Table 4.14 is a summary of a WiMAX link budget [2].

References

[1] Josip Milanovic, Snjezana Rimac-Drlje, Krunoslav Bejuk, "Comparison of Propagation Models Accuracy for WiMAX on 3.5 GHz." *IEEE International Conference*, pp. 111–114, 2007.

[2] Mobile WiMAX Group, "Coverage of mobile WiMAX", pp. 1-18.

[3] Rana Ezzine, Ala Al Fuqaha, Rafik Braham, Abdelfettah Belghith. "A New Generic Model for Signal Propagation in WiFi and WiMAX Environment". *Wireless Days*, pp. 1–5, 2008.

[4] Tamaz Javornik, Gorazd Kandus, Andrej Hrovat, Igor Ozimek. *Software in Telecommunication and Computer Networks*. pp 71–75, 2006.
[5] Yan Zhang. *WiMAX Network Planning and Optimization*. USA: CRC Press, 2009.
[6] Johnson I Agbinya. *IP Communications and Services for NGN*. New York: Taylor and Francis, 2009.
[7] Kejie Lu, Yi Qian, Hsiao-Hwa Chen, Shengli Fu. " WiMAx Networks: From Access to Service Platform". *IEEE Computer Society*, 22, pp. 38–45, May/June 2008.
[8] Johnson I Agbinya, "Design Consideration of Mohots and Wireless Chain Networks", Wireless Personal Communication", © Springer 2006, Vol. 40, pp. 91–106.

5

Planning of Personal Area Networks (I)

N. Sudha Bhuvaneswari and S. Sujatha

Dr. G. R. Damodaran College of Science, India

In this chapter we study the basic concepts on Personal Area Networks (PAN) and its related technology issues, architecture, potential security risks and mechanisms. This chapter also highlights on the commercial PAN products available in the market currently and the various brands of these products leading the technology market. This chapter further elaborates the basics of wire free Personal Area Networks (WPAN), its working, models, standards, performance, security and privacy concerns. The chapter concludes with a note on the future developments of WPAN including next generation networks.

5.1 An Overview of PAN

A PAN is a collection of fixed, portable, or moving components within or entering a Personal Area, which form a Network through local interfaces. A Personal Area is a sphere around a person (stationary or in motion) with a typical radius of about 10 meters. The definition includes components that are carried, worn, or located near the body, e.g., personal digital assistants (PDAs)/handheld personal computers (HPCs), printers, microphones, speakers, headsets, bar code readers, sensors, displays, pagers, mobile phones, and smart cards. A list of basic PAN reference model terms is given below:: -

Component/Service/Application: A PAN consists of components. Each *component* is an independent computing unit. That is, it must have processing capabilities as well as digital memory. A component must have at least one local interface that it can use to connect *directly* to at least one other component but need not be directly connected to every PAN component. A component can be both stationary and mobile.

A *service* is a communication or computing service offered by a component either locally (i.e. through a user interface of some sort), or remotely to other components. A service need not be security related. Each component keeps a list of services it offers as well as rules/policies for access and service discovery and/or advertisement. An *application* is a process running on a component. An application can be a service offered within a component or to other components. An application might try to connect to other components and utilize the services they offer.

User/Owner: The *user* of a component is the person who physically controls and operates the component in accordance with the policies configured in the component. Each component has a single *owner*. By specifying an appropriate policy, the owner of a component might allow users to temporarily use his or her device.

Local interface/Global network interface: Each component has at least *one local communication interface* suitable for direct connection to other PAN components. We consider both fixed and wireless PAN interfaces. Apart from one or several PAN interfaces a component may also have a *global network interface*.

Security policy: Each component has different *security policies*. We distinguish between two different types of security policies: local and remote. The local security policy determine which resources on the component that a user is allowed to manage and if authorization is demanded or not. It also describes how configuration and executables should be installed. The remote security policy determines the requirements on access to the component services and the communication between the service and the entity in the PAN that utilize the service. This includes authentication and encryption requirements as well as access rules.

5.2 Technology Issues Related to PAN

Frequency Bands

The **Personal Area Network (PAN)** is a package that uses the Bluetooth technology. With Bluetooth, a user can create a personal area network by physically moving near a Bluetooth product.

Bluetooth in its most basic form is cable replacement. Where cables now connect many devices, a wireless Bluetooth connection will provide low-cost wireless communications and networking between PCs, mobile phones, and other devices. This will enable untethered, wireless connectivity, to the Internet and other devices, anytime, anywhere. Bluetooth is based on a global radio-frequency (RF) standard, which operates on the 2.4 GHz ISM band, providing license-free operation in the United States, most of Europe and Japan.

Bluetooth technology supports both point-to-point and point-to-multipoint connections. Several piconets can be established and linked together ad hoc, and all devices in the same piconet are synchronized. The topology can best be described as a multiple piconet structure. The full duplex data rate within a multiple piconet structure with 10 fully loaded, independent piconets is more than 6 Mbps.

Characteristics of Frequency Bands

Normal range 10 m (0 dBm)
Optimal range 100 m (+20 Bm)
Normal transmitting power 0 dBm (1 mW)
Optional transmitting power −30 to +20 dBm (100 mW)
Receiver sensitivity −70 dBM
Frequency Band 2.4 GHz
Gross Data rate 1Mbits
Max Data Transfer 721+56 kbit/3 Voice Channels

PAN Standards

Each unique technology requires its own standard and standards making body to provide an adequate forum for discussion and debate. This forum establishes a clearinghouse for inputs from a great variety of sources. They range from semiconductor manufacturers responsible for producing chips based on the standard, to users who will employ the devices and applications made possible by the standard. This comprehensive standard addresses such priorities as

Table 5.1 Comparative table for technologies.

Specification Comparison			
Specification	Bluetooth	IrDA	Home RF
Data Rate (Kbps)	1000	4000	2000
Distance (m)	10	1	50
No. of Devices	8	2	127
Voice channels	3	1	6
Topology	Point to Multipoint	Point to Point	Network

network economy, frequency, performance, power consumption and data-rate scalability. The standard includes all the elements needed for reliable QoS.

The creation of a standard is not a dry, detached process. Instead, it is a lively, sometimes emotional, exchange between some of the world's experts on the subject. As may be expected, the experts sometimes disagree. It is a process of give and take, where everyone has to give at least a little.

Additionally, there are five imperative principles that drive the standards process: due process, openness, consensus, balance, and the right of appeal. The IEEE 802 rigorously enforces these principles.

Mobility Management in PAN

Mobility is one of the most invigorating features, having an enormous impact on how communication is evolving into the future. Mobility Management is one of the major functions of a GSM or a UMTS network that allows mobile phones to work. The aim of mobility management is to track where the subscribers are, so that calls, SMS and other mobile phone services can be delivered to them.

Few mobility management issues highlighted by researchers are: Triggering and Handover.

Triggering: Different kinds of events: traditional radio link specific conditions, context dependent, security-related, upper-layer requirements and other system, application or user-dependent events within a pervasive environment can trigger mobility management actions. These triggers are internal to the network, and the triggering functional area has to coordinate and develop mechanisms to compile triggers that could be relevant to the decision process. A general framework is required to resolve conflicting triggers generated simultaneously by different components, on the basis of predefined policies and rules.

Handover: In the emerging 4G networks which are both multi-domain and multi-technology, handover requests could be based on a number of different needs or policies such as cost reduction criteria, network resource optimization, service related requirements, etc. Various handover solutions have been devised to provide seamless transfer of services across heterogeneous boundaries. The handover techniques can be classified as IP Based, IDMP Based and Agent Based.

Handoff Technique

In cellular telecommunications, the term handoff refers to the process of transferring an ongoing call or data session from one channel connected to the core network to another. In satellite communications it is the process of transferring satellite control responsibility from one earth station to another without loss or interruption of service.

Necessity of Handoff

- when the phone is moving away from the area covered by one cell and entering the area covered by another cell the call is transferred to the second cell in order to avoid call termination when the phone gets outside the range of the first cell;
- when the capacity for connecting new calls of a given cell is used up and an existing or new call from a phone, which is located in an area overlapped by another cell, is transferred to that cell in order to free-up some capacity in the first cell for other users, who can only be connected to that cell;
- in non-CDMA networks when the channel used by the phone becomes interfered by another phone using the same channel in a different cell, the call is transferred to a different channel in the same cell or to a different channel in another cell in order to avoid the interference;
- again in non-CDMA networks when the user behavior changes, e.g. when a fast-traveling user, connected to a large, umbrella-type of cell, stops then the call may be transferred to a smaller macro cell or even to a micro cell in order to free capacity on the umbrella cell for other fast-traveling users and to reduce the potential interference to other cells or users (this works in reverse too, when a user is

detected to be moving faster than a certain threshold, the call can be transferred to a larger umbrella-type of cell in order to minimize the frequency of the handoffs due to this movement);
- in CDMA networks a soft handoff may be induced in order to reduce the interference to a smaller neighboring cell due to the "near-far" effect even when the phone still has an excellent connection to its current cell;

Categories of Handoff

- A hard handoff is one in which the channel in the source cell is released and only then the channel in the target cell is engaged. Thus the connection to the source is broken before the connection to the target is made — for this reason such handoffs are also known as *break-before-make*. Hard handoffs are intended to be instantaneous in order to minimize the disruption to the call. A hard handoff is perceived by network engineers as an event during the call.
- A soft handoff is one in which the channel in the source cell is retained and used for a while in parallel with the channel in the target cell. In this case the connection to the target is established before the connection to the source is broken, hence this handoff is called *make-before-break*. The interval, during which the two connections are used in parallel, may be brief or substantial. For this reason the soft handoff is perceived by network engineers as a state of the call, rather than a brief event. A soft handoff may involve using connections to more than two cells, e.g. connections to three, four or more cells can be maintained by one phone at the same time. When a call is in a state of soft handoff the signal of the best of all used channels can be utilized for the call at a given moment or all the signals can be combined to produce a clearer copy of the signal. The latter is more advantageous, and when such combining is performed both in the downlink (forward link) and the uplink (reverse link) the handoff is termed as *softer*. Softer handoffs are possible when the cells involved in the handoff have a single cell site.

Security Issues in PAN

The PAN security architecture defines a higher layer method for establishing security associations between first party components through imprinting. This initial security association is used as a key management facility for securing internal PAN communication. It may be foreseen that in the future personal devices are capable of running a general-purpose transport layer protocol for transporting communication between the devices. Then the natural approach would be performing authentication of the devices above or at the transport layer, and secure the communication data also at the transport layer. However, this is still not always the case, and may not be in the future, either. The new applications such as ubiquitous computing make use of PAN networks consisting of simple devices that may have only link layer capabilities. Therefore the PAN security architecture supports security services at the link layer in order to make full benefit of the existing link layer security systems.

In addition to securing the communication between different components within a PAN it is mandatory to manage the security features and policies for shared use of PAN components to protect the data and resources. Due to the limitations of the PAN components on computation power, memory space, and communication bandwidth, the security methods have to take these restrictions into account. It is also important to rely only on the features provided by the two communicating PAN components since it could be possible that no external or centralized infrastructure (e.g. CA and PKI services) is available, e.g. due to the absence of wireless network coverage.

Policy handling and authorization

The PAN device security policies are used to define the authorization to

- Access private or public data on the PAN component. This may include user data (files, directories) as well as device configuration and status information.
- Access services and applications provided by the PAN component by a user or by other components in the PAN.
- Access resources on the reference component, such as available memory space and CPU performance, consumables (e.g. paper on a printer component) or other cost incurring services (e.g. a remote network access service) available for the requesting component.

- Use the secure execution environment on the PAN component for remote program code to be downloaded and executed.
- Set up the communication between the PAN components in a secured (encrypted) or unsecured way.

The identity of a service requesting remote PAN component can be derived from the authentication procedure and might be required for checking the access control information carrying PAN certificates or PAN tickets described below.

To reduce the user interaction requirements for policy configuration to a minimum, default policies for at least the three basic categories of PAN component trust classes shall be available. These default policies may be used as a starting point to refine the authorization and access policies depending on the party requesting a certain service.

Access control mechanisms

For the access control mechanisms, distinction can be made on those methods using local storage for the access control information on the reference component, which provides certain service and the remote storage. Assuming that each manageable object on the PAN component (e.g. file, service, application,) has attached access attributes for the three basic trust classes (first party — second party — untrusted). In the case that more refined access control is required (e.g. depending on the requestor's identity), the additional attributes could be stored on the serving PAN component by means of an access control list (ACL). Such an access control list therefore belongs to the local storage category and sufficient memory space needs to be allocated for the attribute records of all possible users and/or PAN component trust levels. To limit the effect of continuous growth of memory space for the attribute records, aging mechanisms to limit the lifetime of the permissions shall be implemented. In the second storage category, the access attribute records are generated by the serving PAN component when a new requestor is registering for a service. The attribute record is secured against modification and delivered to the requestor for storage and presentation when accessing the serving PAN component. Different approaches for securing the attribute records can be chosen:

- PAN certificate: contains the attributes, which are signed by the private key of the issuer. When presented to the serving PAN component, the signature is validated and the access permissions are granted to the specified level.

- PAN ticket: due to the higher computational requirements of public key algorithms compared to secret key methods, an alternative approach would cryptographically hash the attributes with a secret key only known to the serving PAN component. It is the only one who could then also check for the validity of the access permission record.
- Encrypted PAN ticket: this extension of the previous approach keeps the attribute information secret to the requestor. If required, the PAN certificates could be generated by a different first party component e.g. in the case that the serving component has no appropriate user interface to manage the access policies. Both the PAN certificate and PAN ticket approach have the advantage of lower memory requirements for the serving component compared to the ACL approach. However, the latter approach has the advantage of simple modification of permissions, even if the device to which these permissions were granted is (temporarily) not connectable by the serving component. ACLs may still be needed for managing the local user's permissions even if the access control information for remote components is stored remotely. In all cases, local management of already consumed resources (CPU time, memory, consumables, external network access time, etc) relevant for accounting is needed to check for compliance with the resource attribute values stored remotely in the PAN certificate or ticket.

5.3 Commercialization of PAN

Packet Analyzers

A packet analyzer or a network sniffer or sniffer is a computer software or computer hardware that can intercept and log traffic passing over a digital network or part of a network. As data streams flow across the network, the sniffer captures each packet and eventually decodes and analyzes its content according to the appropriate RFC or other specifications.

A Packet Analyzer is a powerful but easy to use network monitor and analyzer designed for packet decoding and network diagnosis. With it real-time monitoring and data analysis, we can capture and decode network traffic transmitted over local host and local network.

A Packet Analyzer is designed for the following purposes:

- Network master — To monitor network traffic and bandwidth user activities and to identify unauthorized activities.
- Network administrators — To diagnose network problems and troubleshoot issues.
- IT professionals — To supervise contents inside your corporate network.
- Security managers — To complement network security monitoring.
- Network Consultants — To help you solve problems for customers.
- Network application developers — To debug your network application and examine network protocols.
- Parents — To find out what your children are doing on the Internet.

Commercial Packet Analyzers

Commercialization is the process or cycle of introducing a new product into the market. The actual launch of a new product is the final stage of new product development, and the one where the most money will have to be spent for advertising, sales promotion, commercialization and other marketing efforts.

- dSniff
- Ettercap
- Fluke Lanmeter
- Microsoft Network Monitor
- NetScout Sniffer
- Network Instruments Observer
- Network Security Toolkit
- PacketTrap pt360 Tool Suite
- snoop (part of Solaris)
- tcpdump
- WildPackets OmniPeek (old name AiroPeek, EtherPeek)
- Wireshark (formerly known as Ethereal)

Test Tools and Network Management Applications

Network management refers to the activities, methods, procedures, and tools that pertain to the operation, administration, maintenance, and provisioning of networked systems.

A Test tool for Network Management applications is designed for the following purposes:

- Operation deals with keeping the network (and the services that the network provides) up and running smoothly. It includes monitoring the network to spot problems as soon as possible, ideally before users are affected.
- Administration deals with keeping track of resources in the network and how they are assigned. It includes all the "housekeeping" that is necessary to keep the network under control.
- Maintenance is concerned with performing repairs and upgrades — for example, when equipment must be replaced, when a router needs a patch for an operating system image, when a new switch is added to a network. Maintenance also involves corrective and preventive measures to make the managed network run "better", such as adjusting device configuration parameters.
- Provisioning is concerned with configuring resources in the network to support a given service. For example, this might include setting up the network so that a new customer can receive voice service.

Commercial Network Management System

Airbee-ZNMS is a Network Management System product designed for low-rate Wireless Personal Area Networks (WPAN). This powerful management platform has been developed in line with the Open System Interconnect (OSI) specifications of FCAPS (Fault Configuration-Account Performance and Security Management). Airbee-ZNMS is a JAVA based web-enabled technology. Airbee-ZNMS uses the proprietary Airbee-ZNMP protocol, a variant of the standard SNMP protocol tailored to suit the requirements of managing WPAN networks.

The major components of Airbee-ZNMS are the Airbee-ZAgent and the Airbee-ZNMS Server. The Airbee-ZAgent, similar to the SNMP agent, resides in ZigBee devices and communicates with the server using the Airbee-ZNMP protocol.

Application for Portable Systems

Portable systems can be a wearable or handheld communication device used as a component in personal area networks. A portable phone is equipped with a personal area network (PAN) detection mechanism to detect all portable phones and electronic identifiers (e.g., key fobs) in its PAN. A buddy list is included in the portable phone. If an electronic identifier is detected, such as from a key fob, but its corresponding portable phone is not, a proxy signal mechanism signals to the wireless telephone network that the phone may receive calls for a different phone. In response, the wireless telephone network routes calls for the different phone to the proxy phone instead. In this manner a phone may change functions automatically as devices enter and leave the PAN.

In the mid-1990s, IBM's Almaden Research Center developed a method that let people transfer information by touch. It worked by sending a billionth of an amp (nanoamp) of current through the body, which is actually a thousand times less than the current generated by combing your hair.

Numerous applications were cited; for example, business card data could be exchanged by shaking hands. By touching a pager in one hand, the calling telephone number could be sent to a cellphone in the other hand. A PAN-enabled unit worn on the wrist could transmit a user's ID to an ATM or security checkpoint.

Application for Cell Phones

Personal area networks (PAN) link people to local services and the web while also filtering incoming calls, blocking ads and summarizing e-mail attachments. First-generation devices include browser-enabled cell phones, Internet-ready PDAs and services customized for the devices, such as Geo-Works' pager information service. At present a micro browser/PDA/cell-phone hybrid that uses the Wireless Application Protocol (WAP), Bluetooth and XML to provide wireless access to LAN and web servers.

One of the most exciting of the new wireless technologies is the ability of mobile phones and other wireless devices to run Web browsers and Web applications. The Wireless Application Protocol, or WAP, is a standard that is designed to allow a mobile phone or other wireless device to run Web applications, served up by a Web Server located on a private corporate network, portal site or the Internet.

Another Enhancement that promises many interesting application possibilities is the Bluetooth technology, a wireless protocol architecture designed to support a Personal Access Network (PAcNet).When two or more Bluetooth-enabled devices come within range of each other (about 10 metres), they instantly establish a wireless network between them. This technology, founded by a consortium of companies including IBM, Intel, and Nokia, is gaining significant interest from hardware and software manufacturers.

Sensor Network Monitor

During the year 2006, The Chicago Fire Department tested a wireless system that will pinpoint the location of firefighters in burning buildings. At the University of California, Berkeley, researchers developed the system in response to a request by the fire department following the events of Sept. 11, 2001, when rescue workers in the Twin Towers using incompatible two-way radios were unable to communicate with each other.

Developed by the school's mechanical engineering department, together with the Center for Information Technology in the Interests of Society (CITRIS), the Fire Information and Rescue Equipment (FIRE) system provides firefighters and command chiefs details about rescue workers' positions in a building. The two-way radios that most fire departments use have limitations in this respect, as they require firefighters to themselves provide status reports on their locations.

The FIRE system, which the Chicago Fire Department began testing in the spring, consists of two elements — SmokeNet and FireEye. The SmokeNet is a wireless network using Moteiv's Tmote Sky wireless sensing platform and sensors, as well as its Boomerang software, which enables wireless sensor devices to register and report changes in the environment to firefighters. The sensors, which use two AA batteries, can be installed in smoke detectors, on ceilings, or in door jams throughout a commercial building. The sensors use active 2.4 GHz RFID tags with a read range up to 100 feet. As part of its pilot program, the Chicago Fire Department has installed these sensors throughout one of its facilities. The FIRE system is also installed in some UC Berkeley buildings and is being examined by several other cities.

The in-building sensors send out an RF signal every two seconds, scanning for firefighters whose air tanks are equipped with wireless sensors that both receive and send transmissions to and from the building sensors installed. The sensor in each firefighter's air tank includes a unique number; as a firefighter passes an in-building sensor, the air tank sensors communicate its ID number to in-building sensor, establishing the firefighter's location. This position information is sent via a wireless ZigBee network to chiefs' or incident commanders' tablet PCs. The PCs will have access to AutoCAD drawings, provided by the city, of buildings in which the sensors have been installed, and firefighters' locations will appear on the screen as dots. Knowing the location of the firefighters helps commanders make tactical decisions, such as when to have firefighters evacuate.

The in-building sensors also can be equipped to measure smoke levels and temperature, alerting firefighters about conditions around them. There can be dozens of sensors in the building transmitting to a firefighter's corresponding sensor at any time.

UC Berkeley student Joel Wilson has developed a system known as FireEye, which includes a head-mounted display screen attached to the nose guard inside the helmet of a firefighter. The FireEye displays an interactive floor plan map featuring the firefighter's current location, or that of other company members, on a postage stamp-sized LCD screen positioned below the right eye of the fire fighter. This technology provides a breakthrough for safety, efficiency, and effectiveness of first responders. The system pilot is funded by the Chicago Fire Department as well as by Ford Motor Co., a CITRIS associate corporate member. Early prototypes of the FireEye were built at the Ford Rapid Prototyping Lab, a 2,000-square-foot design studio within the UC Berkeley's mechanical engineering department.

In 2007, Ident NA Technology was founded to market SKINPLEX in the U.S. Developed in Germany by Ident Technology AG. SKINPLEX uses the human skin as the transmission medium.

The prototype of the PAN system, shown in Figure consists of a battery-powered transmitter and receiver, and a host computer running a terminal program. The PAN prototypes measure 8 × 5 × 1 centimeters, about the size of a thick credit card. The transmitter contains a microcontroller that continuously transmits stored ASCII characters representing an electronic business card. The devices are located near the feet, simulating PAN shoe inserts.

Fig. 5.1 Sensor through Handshaking.

When the woman and man depicted in the figure are in close proximity, particularly when they shake hands, an electric circuit is completed, allowing picoamp signals to pass from the transmitter through her body, to his body, to the receiver by his foot, and back through the earth ground. ASCII characters are sent to the receiver, demodulated, and sent via serial link to the host computer where they are displayed. Thus, when they shake hands, the woman downloads her electronic business card to the man.

Location Tracking System

GPS receivers allow palmtop to plot location on a map and provide driving directions to an address, dynamically updating as we travel. This application can also list popular destinations like restaurants, hotels, banks, and businesses. This type of technology is useful not only for maintenance crews trying to locate a piece of equipment, but can also be used in the control room for tracking the location of the crews and managing their deployment. Mobile workers could also make use of traffic conditions, now being posted on Web sites in many cities. An application that modifies driving directions based on live road conditions would be a money and time saver for mobile crews and for regular commuters.

In Real-Time Control systems, data is collected from the field and brought to a central control room. Using wireless technology, information can now be automatically sent from the system directly to a specific person's cell phone or palmtop. Using a WAP-enabled mobile device, the recipient could then connect to the equipment issuing the alert, and subsequently access a Web page showing a diagnostic analysis, operating history, and possible remedies.

The user could conceivably take corrective action, or if necessary, disable the equipment until a repair crew could attend to it. This translates into improved workflow, faster response times and better focus by the operator on his or her core responsibilities.

Applications within the RealTime Control area could also include "proximity displays." In a typical situation, for example, where a maintenance person is performing a scheduled site visit to a pipeline field station, his mobile phone or PDA could activate a display depicting a map of the station, including the location of each piece of equipment. The operator would then select the equipment item he or she is interested in, after which another display would pop up showing the manufacturer's specifications for that equipment, its maintenance history, a diagnostic report of its current health and even real-time sensor readings (such as vibrations, bearing temperatures, RPM, pressures, etc.). Should the maintenance person need to send a picture of the equipment, or other data, to someone back at the office, his PDA (or even his cell phone) could handle this too. Once the maintenance had been completed on the equipment, the field personnel could then update the centralized maintenance records while still at the site, without plugging in a single wire.

5.4 Branded PAN Products

Low Cost Handheld Messaging System

Brand is a process of "name, term, sign, symbol or design, or a combination of them intended to identify the goods and services of one seller or group of sellers and to differentiate them from those of other sellers. Zipit Wireless, Inc. has come out with Zipit Wireless Messenger 2 (Z2) that provides the users with the ability to send and receive SMS text messages The Z2 brings together IM and SMS text messaging into one sleek device giving consumers the ability to communicate to cell phones, other Zipit devices and personal computers while freeing up the family computer without the risk of a surprise phone bill at the end of the month. This is the first and only purely Wi-Fi based device to offer the flexibility of both types of messaging. The new texting option is available at $4.99 per month and includes no overage penalties, no hidden fees, and no service cancellation restrictions — making it the only no-surprises, no-strings-attached texting option on the market

With more than half of the country's teens sending billions of instant messages every day, IM is the number one text-based communication choice for teens, followed by text messaging. The Z2 allows teens to have multiple, concurrent conversations with their friends for free at any Wi-Fi enabled home, at any of the more than 9,000 participating Wi-Fi enabled McDonald's®restaurants in the U.S., powered by Wayport, or at any open or free wireless "hotspot." The new text message service will be integrated within Zipit's existing IM friends list where multiple text conversations can be held simultaneously and mixed in with IM conversations. Instead of inbox and sent messages, conversations are managed in sequential order and stored in a single conversation window.

Z2's text messaging communicates with more than 20 cellular carriers and supports all major IM platforms used by teens — AOL, Yahoo and MSN. To introduce the new feature, all Z2 users will receive the full text messaging service, up to 3000 messages per month, for free through January 31, 2008. Beginning February 1, 2008, Z2 users will be able to add 3,000 messages per month for only $4.99.

The Z2 is designed to give teens and preteens a cool device that has all the features they want while giving parents the peace of mind that their kids are safe. And now the new text messaging feature allows Z2 users to stay in touch no matter where their friends are or what type of device they are using.

The web-accessed parent portal (http://parents.zipitwireless.com/) enables parents to manage the amount of time teens can use the Z2, specify the times and days of the week that the Z2 can be used and monitor incoming and outgoing text messages. This cutting-edge technology provides parents with an added layer of security and comfort that complements the safety features inherent to the Z2 design — such as protecting users from well-known Internet hazards that include spam, viruses and access to inappropriate Web sites.

The Z2 also includes a fully capable music player called MyTunez that provides instant access to music collections stored on an optional mini-SD card or streamed from the Internet. Teens can also view photos and slideshows with the built-in MyPhotoz feature. Zipit Wireless, Inc. is a Greenville, S.C.-based developer of consumer electronics. The company focuses on wireless communication and entertainment devices that leverage the Internet and the explosion of Wi-Fi networks.

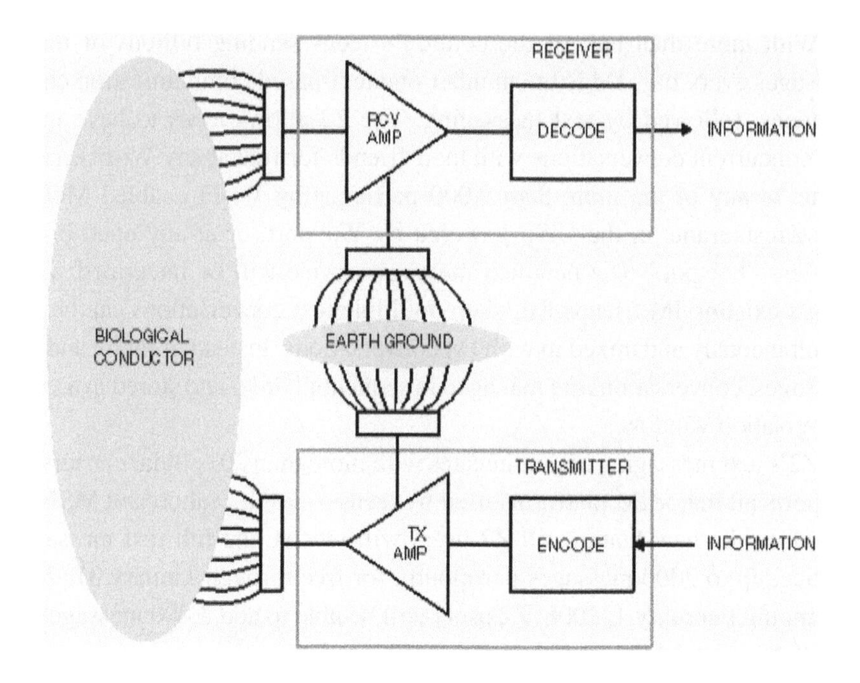

Fig. 5.2 Block Diagram of PAN.

Message Passing System for PAN

Basic concept of a PAN communication channel

Figure 5.2 shows a PAN transmitter communicating with a PAN receiver. Both devices are battery powered, electrically isolated, and have a pair of electrodes. The PAN transmitter capacitively couples a modulating picoamp displacement current through the human body to the receiver. The return path is provided by the "earth ground," which includes all conductors and dielectrics in the environment that are in close proximity to the PAN devices. The earth ground needs to be electrically isolated from the body to prevent shorting of the communication circuit.

Lumped model of communication channel: Symmetry breaking

In Figure 5.3 the PAN transmitter is modeled as an oscillator, and the receiver is modeled as a differential amplifier. The basic principle of a PAN communication channel is to break the impedance symmetry between the transmitter electrodes and receiver electrodes. The transmitter's and receiver's intraelectrode impedances are ignored since the former is a load on an ideal voltage source

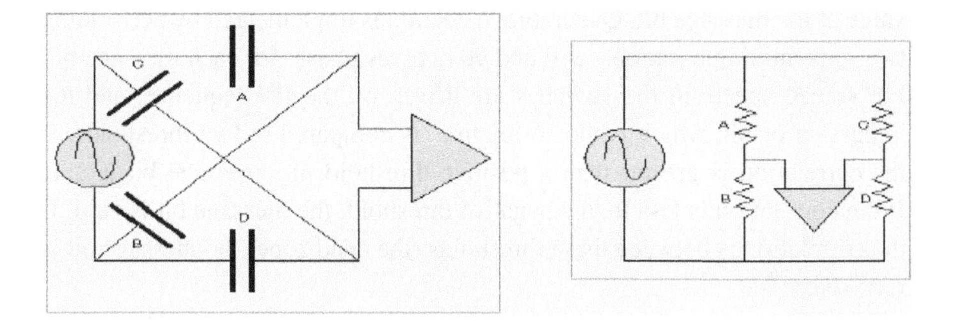

Fig. 5.3 Electrical lumped model of PAN transmitter and receiver.

and the latter is modeled as an open circuit. The four remaining impedances are labeled A, B, C, and D.

The circuit is rearranged to show how PAN device communication works by breaking the symmetry between the four electrodes. The circuit is a Wheatstone bridge where any imbalance of the relationship A/B = C/D will cause a potential across the receiver. Since the ratios must be exactly equal in order to null the circuit, and body-based PAN devices are constantly in motion, there will nearly always be an electrical communication path, as long as the receiver is sensitive enough to detect the imbalance.

Modulation strategies for Communication Channel

Two modulation strategies were examined for PAN communication: on-off keying and direct sequence spread spectrum. On-off keying turns the carrier on to represent a message bit one and turns the carrier off for a message bit zero. The signal-to-noise performance is improved by increasing the transmit voltage. Direct sequence spread spectrum modulates the carrier with a pseudonoise (PN) sequence, producing a broadband transmission much greater than the message bandwidth. Symbol-synchronous PN modulation is used where a message bit one is represented by transmitting the entire PN sequence and a message bit zero is represented by transmitting the inverted PN sequence. The signal-to-noise performance increases with the length of the PN sequence.

The prototype hardware is capable of detecting either on-off keying or direct sequence spread spectrum, determined by microcontroller coding. For on-off keying the bipolar chopper switches are driven at the carrier frequency and the integrated result is compared to a fixed threshold to determine the value of the message bit. Quadrature detection is implemented by performing two sequential integrations, at 0 and 90 degrees phase, for each message bit. For spread spectrum the switches are driven by the PN sequence, and the integrated result, which is the correlation, is compared to two thresholds. If the correlation is greater than a positive threshold, the message bit is one. If the correlation is less than a negative threshold, the message bit is zero. If the correlation is between these thresholds (the dead zone), no message bit is received.

Once the message has been successfully received and demodulated, the microcontroller transmits the message to a host computer over an optical link, which electrically isolates the transceiver allowing evaluation and debugging independent of an electrical ground reference.

Regulatory Issues

The Bluetooth Logo Certification Program requires Bluetooth products to interoperate with products manufactured by other vendors; those products that don't interoperate will not be allowed to use the Bluetooth logo.

The Federal Aviation Administration (FAA) and other aviation regulatory bodies worldwide are currently reviewing the use of Bluetooth products on private and commercial aircraft. In the U.S. the FAA is the governing body to grant approval for Bluetooth product use on aircraft.

Major Global technology leaders Ericsson, Nokia, IBM, Intel and Toshiba founded the Bluetooth SIG in 1998. These companies are now supported by over 1,000 other organizations with a wide range of expertise, including Widcomm, Inc and all these companies to interoperate with each other, they have to cope up with regulatory issues governed by authorized bodies.

Companies likely to adopt Global technology include, but are not limited to, software developers, network vendors, silicon vendors, peripheral and camera manufacturers, mobile PC and handheld device manufacturers, consumer electronics manufacturers and more.

There are several patents on different parts of the technology. Because of this, all licensees will have to sign a zero cost license agreement to cover IP and naming.

Cable Replacement

There are a number of wireless technologies which are used over a short distance (usually up to about 10 meters) to connect devices to each other. As such networks are based on the immediate area around the individual user; they are called Wireless Personal Area Networks. Examples of WPAN technologies are Bluetooth, Ultra wideband and ZigBee/802.15.4

Bluetooth is a low-cost radio solution that can provide links between devices. Originally, and more typically the range of these devices is up to 10 meters. Bluetooth has access speeds of up to 721 Kbps; considerably slower than the various 802.11 Wireless LAN standards. Bluetooth technology is embedded in a wide range of devices, e.g. mobile phones, printers, video cameras, PDAs, computer mice and keyboards etc. Bluetooth is primarily used as a wireless replacement for a cable to connect devices assuming they are configured to share data. Although Bluetooth was not originally intended to be used for 802.11 wireless networking, it is possible to buy access points for Bluetooth LAN and combined 802.11b/Bluetooth access points. The Bluetooth standard is relatively complex and it is therefore not always easy to determine if any two devices will communicate. Bluetooth operates in the 2.4 GHz band so can cause interference with Wireless LAN (802.11b and 802.11g) equipment.

Ultra-Wideband (UWB) is a wireless technology intended to provide high speed, low power wireless connections (100 Mbps–2 Ghz) over short distances (10 m). It is expected to be used for cable replacement applications and for multimedia networking in the home. Ultra-Wideband is based on pulsing a signal in very short bursts across a very wide bandwidth. Data is sent by altering the amplitude, phase or position of the pulses. OFDM and frequency hopping techniques have also been developed. The IEEE standards process for Ultra-Wideband (IEEE 802.15.3a) has now been stopped due to lack of agreement between industry groups proposing two different solutions:

1) MultiBand Orthogonal Frequency Division Multiplexing (MB-OFDM) UWB, backed by the WiMedia Alliance
2) Direct sequence-UWB (DS-UWB), backed by the UWB Forum

Companies that are members of the WiMedia Alliance and the UWB Forum are developing and launching UWB products based on the two incompatible solutions. UWB is likely to be used to provide wireless versions of existing cable technologies such as USB 2, 1394 (FireWire), Bluetooth and video connections (eg DVI).

Wireless versions of Hi-Speed USB (USB 2.0) is the first consumer application of UWB technology. CableFree USB (UWB Forum solution) products were shown at the Consumer Electronics Show in January 2006 and Wireless USB (WiMedia Alliance solution) enabled devices are likely to appear later in the year. The WiMedia Alliance USB solution is backed by the USB Implementers Forum and is called Certified Wireless USB.

ZigBee/802.15.4

ZigBee is a wireless sensor network technology specification based on the IEEE 802.15.4 standard. The ZigBee Alliance is a trade body that oversees testing and certification for ZigBee products. ZigBee is intended to be a low cost, low power, low data rate wireless networking standard for sensor and control networks. The technology will primarily be used for industrial and home sensor networks and building control systems, such as security systems, smoke detectors/alarms, and heating and lighting controls. ZigBee enabled products can create mesh networks, routing traffic via other ZigBee devices. ZigBee works in the 2.4 GHz band and provides maximum data rates of 250 Kbps.

Automated meter reading

Micrel Inc. and Cyan Holdings plc are working together on a technology partnership that involves developing subsystem level module products aimed at automated meter reading infrastructure, public lighting management, Ethernet gateways and RF sensor network markets. Micrel has provided the RF chips, while Cyan to supply the microcontrollers and system software. This partnership provides customers with production-ready modules complete with drivers, protocol stacks, RF mesh networking and even embedded Web servers. Radio Wire technology enables a family of low-power RF-enabled Ethernet/ USB gateways and nodes for a range of industrial applications. The modules are user-customizable and supplied with a free graphical development tool providing an easy to use and cost-effective solution to industrial control and communications.

Micrel and Cyan collaborated on developing subsystem level module solutions for growing and emerging global markets. The MICRF505 is a family of true single chip, FSK transceivers that are intended for use in half-duplex, bidirectional low-powered RF links. These multichanneled FSK transceivers are targeted at UHF radio equipment in compliance with the North American FCC parts 15.247 and 249, China Standard No. 423 and the European Telecommunication Standard Institute specification, EN300 220. The MICRF506 is designed to operate in the unlicensed 433MHz ISM band. These chips support data rates up to 200Kbit/s and are part of Micrel's recently launched Radio Wire family covering the UHF band. The products are aimed at the automated meter reading, alarm and security, and personal area network markets and are suited for a wide variety of applications including gas, water and energy meters; home, building and security monitors; and industrial controls and sensor networks.

Sensor based pollution monitoring system

As electronic devices become smaller, lower in power requirements, and less expensive, people have begun to adorn their bodies with personal information and communication appliances such devices include cellular phones, personal digital assistants (pdas), pocket video games, and pagers. When these devices start sharing data, it can reduce functional i/o redundancies and allow new conveniences and services. The concept of personal area networks (pans) is presented to demonstrate how electronic devices on and near the human body can exchange digital information by capacitively coupling picoamp currents through the body. A low-frequency carrier (less than 1 megahertz) is used so no energy is propagated, minimizing remote eavesdropping and interference by neighboring PANs.

We are already towards an electronic future where information will be accessible at users fingertips, whenever and wherever needed. Some of the computing and communication equipment required to provide this intimate and immediate access to information will be incorporated into people attire. Just as a glance at today's wristwatch saves a trip to the nearest clock, a glance at tomorrow's wristwatch will replace finding a terminal to check e-mail.

A person who carries a watch, pager, cellular phone, personal stereo, personal digital assistant (PDA), and notebook computer is carrying five displays, three keyboards, two speakers, two microphones, and three communication devices. The duplication of i/o components is in part a result of the inability of the devices to exchange data. With proper networking these devices can share i/o, storage, and computational resources.

The ability to share data increases the usefulness of personal information devices, providing features not possible with independent isolated devices. Imagine the following scenario: I am at home preparing for the day and want to find the time of my first meeting. I call out "when is my first meeting?" the microphone in my watch transmits my voice through a series of transponders distributed throughout my house to a voice recognition computer that searches my calendar and sends back a response to a speaker or visual display in my watch. When I leave my house the door senses my departure and sends a message to my colleagues. When I approach my office building, the door acknowledges me by opening, sends a message of my arrival to my colleagues, and uploads any new messages.

The scenario requires the user to wear a device that periodically transmits a unique user code to allow a nearby stationary transceiver to identify, locate, and exchange messages with the user's device. Clearly, privacy is a big issue in such scenarios. Privacy is both a right and a commodity. to maintain privacy control, a wearer must determine when the identification beacon is activated and what type of information can be transmitted. The capabilities of autonomous yet interconnected devices may transform the notion of ubiquitous computing to the concept of ubiquitous i/o. as wireless network ports become more common, there will be less need to carry around power-hungry processors and bulky mass storage. As the growth of wireless services (e.g., cellular phones, pagers, and radio frequency local area networks [RF LANs]) fills up the limited RF spectrum, near-field communication offers an alternative to congesting the airwaves with data.

5.5 WPAN Introduction

Wireless Personal Area Networks (WPANs) have moved quickly to the mainstream and are now found in many educational institutions, homes, businesses and public areas. Organizations and consumers have been keen to take advantage of the flexibility adding wireless networks can offer. A recent report from In-Stat predicts that the wireless market will grow from 140 million wireless chipsets a year in 2005 to 430 million in 20091. The emergence of new security standards has also increased confidence in WPANs. Users are becoming more familiar with the technology and are increasingly expecting wireless access to be available. There is a wide range of products and standards involved in WPAN technology and more continue to emerge.

WPAN definition

A WPAN (wireless personal area network) is a personal area network — a network for interconnecting devices centered around an individual person's

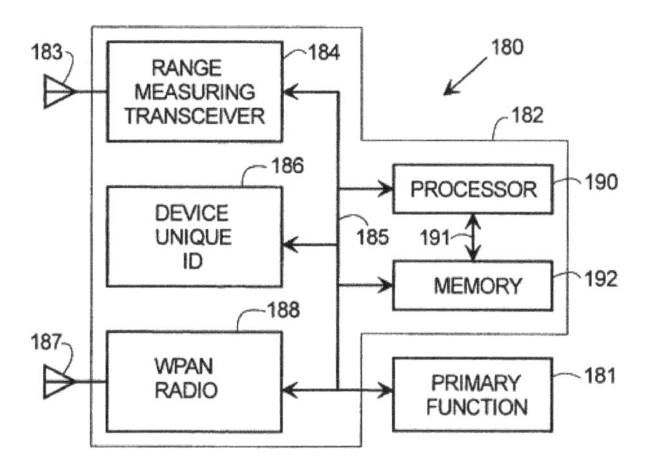

Fig. 5.4 WPAN functional subsystem.

workspace - in which the connections are wireless. Typically, a wireless personal area network uses some technology that permits communication within about 10 meters - in other words, a very short range. One such technology is Bluetooth.

WPAN operation

A wireless personal area network (WPAN) associates function subsystem coupled to the primary function subsystem, wherein the WPAN association function subsystem includes a range measuring transceiver having radio frequency identifier (RFID) tag components, wherein the range measuring transceiver is addressable by one or more other PAN elements to exchange identifiers with them. PAN element is also adapted to determine ranges and relative motions of at least some of the one or more other PAN elements using RFID measurements, and automatically to associate with a WPAN that includes a subset of the one or more other PAN elements that satisfy predetermined association criteria that include range and relative motion.

The multisphere trust model

The tremendous growth in wireless technology demands a reference model with multiple parties with human-centric applications. This has led to the development of Multisphere trust model that includes human being surrounded by six spheres of equipments, networks and functions. They are from the inner

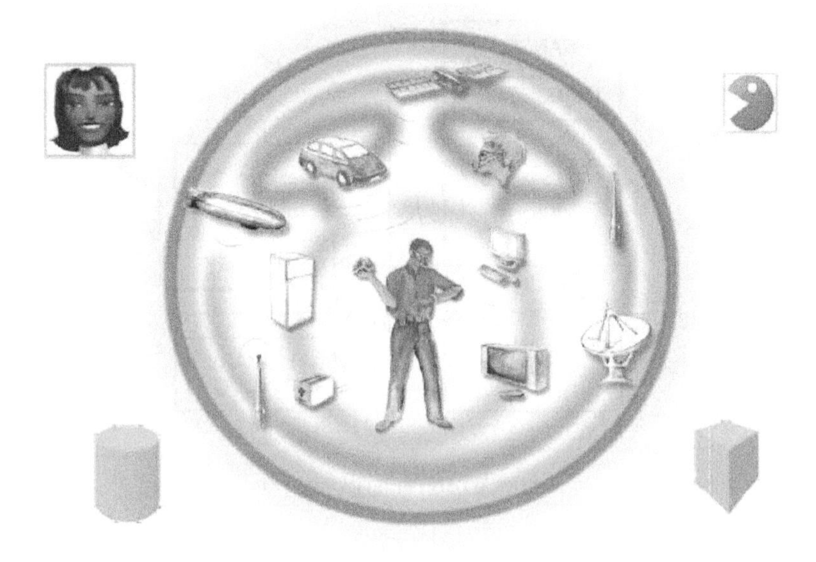

Fig. 5.5 A visualization of the multisphere reference model.

to the outmost sphere:

Different levels of multisphere model

1. *PAN (Personal Area Network)*: This sphere contains the data components that are closest to the user and that he/she carries all the time, e.g. mobile phones, watches, cameras, glasses, and other body near equipment.
2. *The immediate environment*: This is the immediate surroundings consisting of fixed or little movable equipments, e.g. TV, PC Workstation, refrigerator
3. *Instant partners*: This spherical level is reached when equipment is added, which we need to communicate and interact with people around us.
4. *Radio access*: The sphere of equipment and networks needed to obtain ubiquitous coverage for distributed systems. It is a fundamental requirement that all radio interfaces shall offer access for PAN and "Instant Partners", i.e. spheres 1 and 3.

5. *Interconnectivity*: This sphere is a functional layer that gives interconnectivity to mobile systems. The user is offered a real mobile Internet independent of radio access network and terminal.
6. *Cyberworld*: This is the outermost sphere that makes up an enhanced reality (Cyber world) created by all the applications offered in next generation mobile system. Today's inhabitants of the Cyber world, where our interests, needs and wishes are maintained by individual agents, which are autonomous data programs in the complex system represented by the Wireless World and it is yet to grow to meet the demands of users.

Security & Privacy Issues

Security is an issue of concern for wireless communications as these transmissions in an *ad hoc* network can be captured by other devices. Therefore it is important for users to understand the security issues involved in all wireless technologies. The WPAN technology is able to be made secure but is primarily intended to have high levels of compatibility with other enabled devices. The frequency hopping of IEEE 802.15 devices goes someway to preventing eavesdropping by other devices, as only those devices in the piconet know the hopping sequence. However, when a new device is added to the piconet, the users of the other devices are able to manage the security levels for those newcomers. There are three modes of security available to WPAN technology.

Mode 1 — this is a non-secure mode to be used to seek out other devices. In this mode, the device transmits its Device Access Code but does not start a security, authorisation or encryption procedure.

Mode 2 — Using a private user key for authentication, this mode will allow flexible polices for access to applications and services and is used for applications running in parallel but have differing security requirements.

Mode 3 — This mode requires the device to have authentication and encryption before any connection can be established. When devices join in a piconet, settings from previous sessions can be reestablished.

This will allow trusted devices to begin the session without reestablishment of the security settings. New devices will need to go through the security preferences with these trusted devices. A device can be set so that unknown devices are not aware of the physical presence of the device or not able to connect to the device without authorization. Before the creation of the first session, the users of the two devices need to enter a PIN of up to 16 digits. This then creates a key that can be used for future sessions.

Performance

An enormous amount of multimedia data will be transmitted by various devices connected to the wireless personal area network, and this network environment will require very high transmission capacity. To take over this problem multiple antennas to the MB-OFDM UWB system can be used for achieving high performance. This multiple antenna approach requires the antennas to remain orthogonal in the time domain, the channel estimation can be applied to the MB-OFDM specification in the case of more than 2 transmit antennas. By using the multiple-antenna scheme and proposed channel estimation technique, the reliability and performance of the MB-OFDM system can be improved.

When multiple ZigBee wireless personal area networks (WPANs) are in close proximity to each other, contentions and collisions in transmissions will lead to increased packet delays. However, there is no existing study on how delay performance would be affected in a crowded real-life environment where each person walking down a busy street would be wearing a ZigBee WPAN. A mobility pattern can be applied for analyzing a real-life video trace and estimate the packet delay by combining data collected from ZigBee experiments.

Use of silicon germanium makes the concept of fast personal networks possible. IBM has improved the performance of a chip-making material that could be used to make advanced wireless devices such as automobile radar and high-bandwidth personal area networks. Silicon germanium is the ingredient within IBM manufacturing technology that allows radio chips to run at high frequencies while taking advantage of the benefits of silicon manufacturing techniques with a speed of 60 GHz. Chips capable of 60 GHz could be used to create a wireless personal-area network that offers a high-speed Internet connection over a very short distance. The combination of fast download speeds and a short coverage area could reduce the ability of outsiders to steal wireless signals from office workers.

WPAN Standards

In March 1999 the IEEE 802.15 standards working group was created to develop a family of communications standards for WPANs. In the first meeting of the new working group in July 1999, the Bluetooth SIG submitted

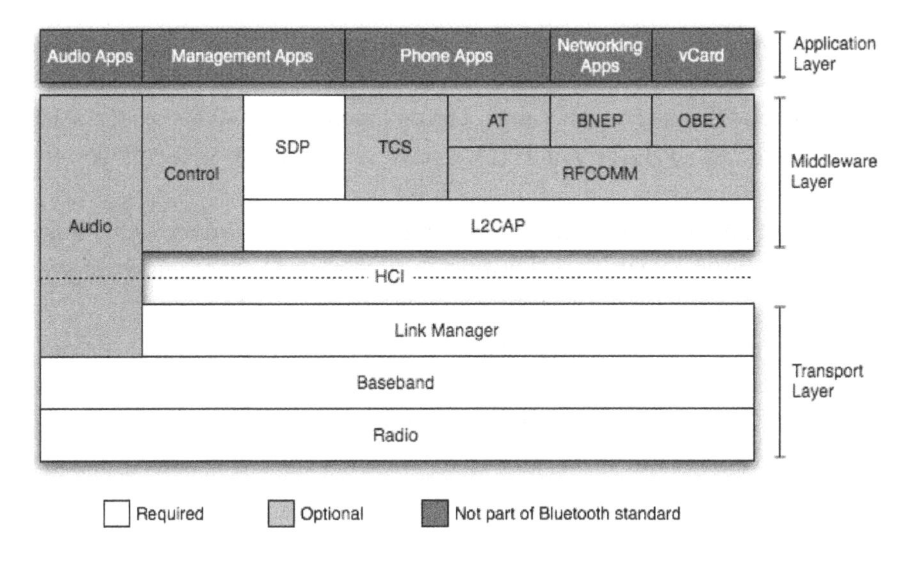

Fig. 5.6 The bluetooth protocol stack.

the just created Bluetooth specification as a candidate for an IEEE 802.15 standard. The Bluetooth proposal was chosen to serve as the baseline of the 802.15.1 standard. In addition to the IEEE 802.15.1 activity, the IEEE 802.15.2 task group studies coexistence issues between 802 wireless technologies. The 802.15.3 task group is developing standards for high-rate radios (>20 Mb/s). Finally, the 802.15.4 task group is developing standards for low-rate radios (<200 kb/s).

The top speed of wireless personal area networks (WPAN) has just jumped to 55 Mb/s from 1 Mb/s (megabits per second) under a new standard from the Institute of Electrical and Electronics Engineers (IEEE). This increase opens the door for the broad use of multimedia, digital imaging, high-quality audio and other high-bandwidth WPAN applications that need a wireless solution combining low-cost and low-power with high data rates and robust quality of service (QoS).

802.15.1, more commonly known as Bluetooth, is a low-data-rate, low-power wireless networking standard aimed at replacing cables between lightweight devices. The Bluetooth protocol stack is shown in Figure 5.6. The Bluetooth stack defines many components above the PHY and MAC layers, some of which are optional.

IEEE 802.15.2 was assigned the task of providing recommendations as to the coexistence of WPAN and WLAN. IEEE 802.15.2 is divided into 2 groups: Non-collaborative and collaborative. The collaborative solutions require a data link between the WPAN and WLAN devices, while the non-collaborative solutions assume there is no such data link.

The collaborative solutions offered in the IEEE standard are: Alternating wireless medium access, Packet traffic arbitration, and Deterministic interference suppression.

The non-collaborative solutions include: Adaptive interference suppression, Adaptive packet selection, Packet scheduling for ACL links, Packet scheduling for SCO links, and Adaptive frequency-hopping.

The new standard, IEEE 802.15.3, "Wireless Medium Access Control (MAC) and Physical Layer (PHY) Specifications for High Rate Wireless Personal Area Networks (WPAN)," allows a WPAN to link as many as 245 wireless consumer devices in a home at data rates to 55 Mbps at distances from a few centimeters to 100 meters.

IEEE 802.15.3 provides for high-rate wireless connectivity in the 2.4 GHz unlicensed frequency band among fixed and portable devices. It specifies raw data rates of 11, 22, 33, 44 and 55 Mbps, which can provide data throughputs in excess of 45 Mbps. The rate chosen affects typical transmission range, for example, as much as 50 m at 55 Mbps and 100 m at 22 Mbps. The highest rate accommodates low-latency, multimedia connections and large-file-transfer, while 11 and 22 Mbps provide long-range connectivity for audio devices.

IEEE 802.15.4 is a proposed standard addressing the needs of low-rate wireless personal area networks or LR-WPAN with a focus on enabling wireless sensor networks. The standard is characterized by maintaining a high level of simplicity, allowing for low cost and low power implementations. Its operational frequency band includes the 2.4 GHz industrial, scientific and medical band providing nearly worldwide availability; additionally, this band is also used by other IEEE 802 wireless standards. Coexistence among diverse collocated devices in the 2.4 GHz band is an important issue in order to ensure that each wireless service maintains its desired performance requirements. The ZigBee standard extends 802.15.4 to provide Bluetooth-like interoperability features. ZigBee builds on top of 802.15.4's

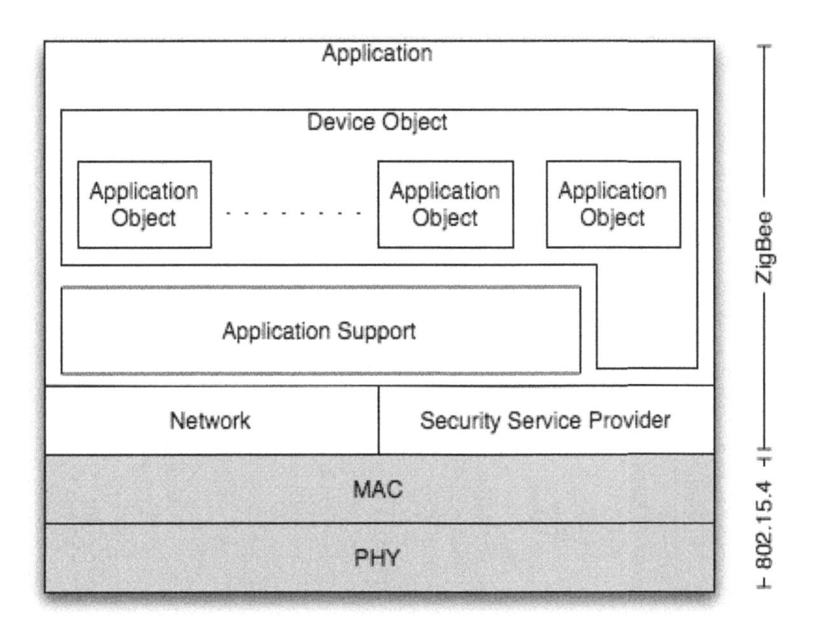

Fig. 5.7 ZigBee protocol stack.

radio layer, specifying network, security, and application layers. The resulting architecture is shown in Figure 5.7.

Ultra-wide band (UWB) radios take a drastically different approach from Bluetooth and 802.15.4. Where the latter two radios emit signals over long periods using a small part of the spectrum, UWB takes the opposite approach: UWB uses short pulses (in the ps to ns range) over a large bandwidth (often many GHz). According to Shannon's Law, the maximum data rate of a radio link can be increased much more efficiently by increasing its bandwidth than by increasing its power; hence, UWB radios offer very high data rates (hundreds of Mbps or even several Gbps) with relatively low power consumption. The use of short pulses over a wide spectrum also means that the signal is below the average power output defined as noise by the FCC (-41.3 dBm/MHz), and that UWB signals are not susceptible to noise or jamming. UWB is a much simpler technology than Bluetooth and ZigBee, since there are currently no mandatory or optional middleware layers that build on top of the basic PHY and MAC layers.

Comparison of WPAN standards

Table 5.2 Comparative table for WPAN standards.

Standard	Bluetooth	UWB	ZigBee	Wi-Fi
IEEE spec	802.151	802.15.3a*	802.15.4	802.11a/b/g
Frequency band	2 4 GHz	3.1–10.6 GHz	868/915 MHz; 2.4 GHz	2.4 GHz; 5 GHz
Max signal rate	1 Mb/s	110 Mb/s	250 Kb/s	54 Mb/s
Nominal range	10 m	10 m	10–100 m	100 m
Nominal TX power	0–10 dBm	−41.3 dBm/MHz	(-25)-0 dBm	15–20 dBm
Number of RF channels	79	(1–15)	1/10; 16	14 (2.4 GHz)
Channel bandwidth	1 MHz	500 MHz–7.5 GHz	0.3/0.6 MHz; 2 MHz	22 MHz
Modulation type	GFSK	BPSK, QPSK	BPSK(+ASK), O-QPSK	BPSK, QPSK COFDM, CCK, M-QAM
Spreading	FHSS	DS-UWB, MB-OFDM	DSSS	DSSS, CCK, OFDM
Coexistence mechanism	Adaptive freq. hopping	Adaptive freq hopping	Dynamic freq. selection	Dynamic freq. selection, transmit power control (802.11h)
Basic cell	Piconet	Piconet	Star	BSS
Extension of the basic cell	Scattemet	Peer-to-peer	Cluster tree, Mesh	ESS
Max number of cell nodes	8	8	>65000	2007
Encryption	EO stream cipher	AES block cipher (CTR, counter mode)	AES block cipher (CTR, counter mode)	RC4 stream cipher (WEP), AES block cipher
Authentication	Shared secret	CBC-MAC (CCM)	CBC-MAC (ext. of CCM)	WPA2 (802.11i)
Data protection	16-bit CRC	32-bit CRC	16-bit CRC	32-bit CRC

5.6 Future of PAN

Next Generation Networks

Mobile networks are evolving beyond traditional network intelligence to an expanded view of what it means to have an Intelligent Mobile Network. This expanded view takes into account the fact that intelligence is becoming more distributed throughout the network. In fact, the very definition of the "network" is challenged as intelligence and control is distributing to mobile communications devices as well as the Internet.

Next generation networks of the near future will leverage all of these technologies in conjunction with traditional network intelligence for more efficient and effective operations as well as the provision of advanced and value-added services. Next generation networks of the future will require advanced operational support systems (OSS) and business support systems (BSS) to support the new technologies and related needs such as more flexible customer data management and billing. It is therefore important to recognize that intelligence is not regulated to the real-time processing environment alone.

Most currently deployed mobile data services suffer from lack of personalization. One way to personalize services is to make them personal preference driven. Personal profiles enable the user to specify what type of data and when it is important to receive. In addition, intelligent agents can monitor usage to learn and/or predict what type of information and content the user will most likely desire.

Technologies such as Bluetooth allow for Personal Area Networks (PAN) and the ability to engage in wireless communications in an ad hoc basis. In addition to creating a convenient means of personal wireless communications or personalization, this technology has significant implications for various applications including mobile commerce and location based services.

Location Services

The ability to pinpoint the location of mobile users provides the ability to offer a variety of location-based services such as location-based billing, information, emergency, and tracking services. Location also enhances other value-added applications such as mobile gaming, mobile chat/messaging, and friend finder services. As next generation networks evolve, location will become a part of most every call and data session.

Presence and Availability

Closely related to location, presence provides information about the general whereabouts of a mobile user while availability indicates the user's ability to engage in communications such as mobile chat. Next generation mobile networks will be presence, device, and privilege aware systems, capable of determining when and where users wish to engage in various types of communications.

Alice has a PAN, a Personal Area Network on her body: she has a Bluetooth enabled PDA, mobile phone and laptop that she is carrying, and are all currently turned on, and forming a network. Her laptop also has the ability to connect using an available WLAN, and her mobile phone has the ability to connect through GPRS, though GPRS is slower and much more costly for Alice to use. She is now on the move, and her laptop is downloading her emails using the GPRS connection on the mobile:

Laptop → (Bluetooth) → Mobile → (GPRS) → Mobile phone network

While walking, she passes into an area covered by a free WLAN hotspot: Her PAN now immediately starts to initiate a connection with the hotspot. This is called "merging" of the networks (that of the hotspot and that of her PAN). Once this merging is complete, the downloading of her email continues totally unaffected, but instead of using the expensive and slow GPRS connection, it is now using the newly established WLAN connection. If she now wants to browse the web with her PDA, the PDA will also use the WLAN connection of the laptop:

PDA → (bluetooth) → Laptop → (WLAN) → Hotspot

Mesh Networks

A wireless mesh network is an emerging paradigm in the field of wireless networking technologies ranging from personal area networks (PANs) to wide area networks (WANs).

Mesh networking technologies for both high-rate and low-rate wireless personal area networks (WPANs) are under development by several standardization bodies. They are considering adopting *distributed* TDMA MAC protocols to provide seamless user mobility as well as a good peer-to-peer QoS in WPAN mesh. It has been, however, pointed out that the absence of a central controller in the wireless TDMA MAC may cause a severe performance degradation: e.g., fair allocation, service differentiation, and admission control may be hard to achieve or can not be provided.

Multi-Protocol Networks

Although the underlying technology is changing, a rich legacy of networked application software remains "mission-critical" for today's enterprises. Many of these application investments are tied to a particular networking protocol and remain so, despite convincing arguments advanced in the "protocol wars" waged in the trade press. It is not unusual in today's enterprise to find a variety of networking protocols such as Systems Network Architecture (SNA), Transmission Control Protocol/Internet Protocol (TCP/IP), and DECnet**. Thus, it has become part of the modern paradigm not only to seek industry-wide standards for new protocols but also to accommodate the old through convergence and coexistence within unified networks.

Networking protocols are being deployed across an ever-growing assortment of media and carrier services. On-campus (i.e., the geographic and physical facilities of an institution or business) media include coaxial cable, telephone twisted pair, fiber-optic channel, fiber-optic local area network (LAN), infrared, and radio frequencies. Each has its peculiar characteristics of speed, error rate, and distance supported. Wide area network media have also become more diverse. Carriers are augmenting traditional analog telephone facilities with many more sophisticated offerings, including X.25, Integrated Services Digital Network (ISDN), frame relay, and, most recently, asynchronous transfer mode (ATM). Cellular communications and other wireless schemes are also being offered in the wide area network context.

This diverse networking infrastructure is being used to support an increasing variety of applications, each with different requirements for security, integrity, bandwidth, response time, and dependability of service.

We can relate the Open Blueprint to the OSI model, bearing in mind that the various layers of the Open Blueprint embrace a broader set of industry standards than are included in the OSI suite. The lowest layer matches the OSI physical layer. *Subnetworking* (as the term is used in the Open Blueprint) corresponds to those parts of the lower OSI layers that contain functions dealing with specific communications facilities, as well as frame- and packet-handling formats and dialing procedures. *Transport networking* corresponds to the OSI networking and transport layers. The Signaling and Control Plane is not present explicitly in the OSI model but is derived from the signaling and control plane in the Broadband Integrated Services Digital Network (B-ISDN) model. *Distributed systems services* correspond to OSI layers 5-7.

Sensor Networks

Wireless Sensor Networks (WSN's) are an emerging technology that offers the ability for a large number of sensors to coordinate in a single network. WSN's are generally made up of multiple sensors, each of which combines sensing ability, signal processing, and wireless communications on a single chip. WSN's have been used for a variety of applications, including monitoring manufacturing processes, enemy detection in tactical environments, monitoring traffic in urban areas, deployment in forests to detect fires, and gathering information in disaster areas. Most WSN's are deployed as an ad hoc network and then cooperate in order to sense physical phenomenon. Many of these applications require the use of a large number of sensors (>1000) operating at extremely low data rates. Low power operation is a necessity for a network of sensors this large. This is because the sensors are usually spread out over a large area and because there are so many of them, replacing their batteries often is undesirable.

Cognitive Radio

Cognitive Radio is a paradigm for wireless communication in which either a network or a wireless node changes its transmission or reception parameters to communicate efficiently avoiding interference with licensed or unlicensed users. This alteration of parameters is based on the active monitoring of several factors in the external and internal radio environment, such as radio frequency spectrum, user behavior and network state. For example, cellular network bands are overloaded in most parts of the world, but amateur radio and paging frequencies are not. Independent studies performed in some countries confirmed that observation, and concluded that spectrum utilization depends strongly on time and place. Moreover, fixed spectrum allocation prevents rarely used frequencies (those assigned to specific services) from being used by unlicensed users, even when their transmissions would not interfere at all with the assigned service. This was the reason for allowing unlicensed users to utilize licensed bands whenever it would not cause any interference (by avoiding them whenever legitimate user presence is sensed). This paradigm for wireless communication is known as cognitive radio.

Although cognitive radio was initially thought of as a software-defined radio extension (Full Cognitive Radio), most of the research work is currently focusing on Spectrum Sensing Cognitive Radio, particularly in the TV bands. The essential problem of Spectrum Sensing Cognitive Radio is in designing high quality spectrum sensing devices and algorithms for exchanging spectrum sensing data between nodes. It has been shown that a simple energy detector cannot guarantee the accurate detection of signal presence, calling for more sophisticated spectrum sensing techniques and requiring information about spectrum sensing to be exchanged between nodes regularly. Increasing the number of cooperating sensing nodes decreases the probability of false detection [9].

Applications of Spectrum Sensing Cognitive Radio include emergency networks and WLAN higher throughput and transmission distance extensions.

Evolution of Cognitive Radio toward Cognitive Networks is under process, in which Cognitive Wireless Mesh Network (e.g. CogMesh) is considered as one of the enabling candidates aiming at realizing this paradigm change.

The main functions of Cognitive Radios are:

- *Spectrum Sensing*: Detecting the unused spectrum and sharing it without harmful interference with other users, it is an important requirement of the Cognitive Radio network to sense spectrum holes, detecting primary users is the most efficient way to detect spectrum holes. Spectrum sensing techniques can be classified into three categories:
    - *Transmitter detection*: cognitive radios must have the capability to determine if a signal from a primary transmitter is locally present in a certain spectrum, there are several approaches proposed:
        - matched filter detection
        - energy detection
        - cyclostationary feature detection
    - *Cooperative detection*: refers to spectrum sensing methods where information from multiple Cognitive radio users are incorporated for primary user detection.
    - *Interference based detection.*

- *Spectrum Management*: Capturing the best available spectrum to meet user communication requirements. Cognitive radios should decide on the best spectrum band to meet the Quality of service requirements over all available spectrum bands, therefore spectrum management functions are required for Cognitive radios, these management functions can be classified as:
    - o *spectrum analysis*
    - o *spectrum decision*
- *Spectrum Mobility*: is defined as the process when a cognitive radio user exchanges its frequency of operation. Cognitive radio networks target to use the spectrum in a dynamic manner by allowing the radio terminals to operate in the best available frequency band, maintaining seamless communication requirements during the transition to better spectrum.
- *Spectrum Sharing*: Providing the fair spectrum scheduling method. one of the major challenges in open spectrum usage is the spectrum sharing. It can be regarded to be similar to generic media access control MAC problems in existing systems

References

[1] O. Pummakarnchana, N.Tripathi, J.Dutta, Air Pollution Monitoring and GIS Modeling, Science and Technology of Advanced Materials, 2005.
[2] Greg Hackmann, 802.15 Personal Area Networks, 2006.
[3] IEEE, IEEE 802.15 WPAN Task Group 5(TG5), IEEE Task Group http://www.ieee802.org/15/pub/TG5.html
[4] MobileIN.com, Intelligent Mobile Networks, 2004 http://www.mobilein.com
[5] Henrik Petander, A Network Mobility Management Architecture for a Heterogeneous Network Environment, Espoo, 2007.
[6] Dr.Robert Heile, New Standard opens door for high-rate wireless personal area networks, IEEE 802.15 Working Group Chair, 2003.
[7] Isameldin M Suliman, Janne Lehtomaki, Ian Opermann, Performance evaluation of TCP in an integrated WPAN and WLAN Environment, Center for Wireless Communications.
[8] Christian Gehrmann, Thomas Kuhn, Kaisa Nyberg, Peter Windirsch, Trust Model, communication and configuration security for Personal Area Networks.
[9] Eric Lee Eckhoff, Wireless sensor networks and personal area networks for data integration in a virtual reality environment, Iowa State University, 2004.
[10] Sam Churchill, Intels My WIFI:Personal Area Networking, dailywireless.org, 2009.
[11] Stephen Lawson, IBM shows glimpses of Bluetooth future, InfoWorld, 2000.
[12] Network Magazine, Bluetooth:Connectivity without wires, 2001.

[13] Filip Louagie, Luis Munoz, Sofoklis Kyriazakos, Paving the way for the 4G: A New family of WPAN.
[14] Zipit, New Hand-Held Wireless instant messaging device first and only to offer text messaging, 2007.
[15] Bruce Schechter, The Body Electric, IBM Think Research.
[16] M. L. Hess, J. A. Lorrain, G. R. McGee, Multiprotocol Networking-a blueprint, IBM Systems Journal.
[17] Karney, James, Brave New Unwired World, All Business, 2000.
[18] T. G. Zimmerman, Personal Area Networks:Near-field intrabody communication, IBM Sytems Journal, Vol 35.
[19] Intermec, Guide to Wireless Personal Area Networks, Intermec Personal Area Network Solutions, 2000.
[20] Claire Swedberg, ZigBee-based RFID System, RFID Journal, 2006.

6

Planning of Personal Area Network (II)

Nithya Thilak and Johnson I Agbinya

University of Technology, Sydney Australia

6.1 Introduction to PAN

6.1.1 Acronyms

PAN	Personal Area Network
WPAN	Wireless Personal area network
IrDA	Infrared Data Association
UWB	Ultra-wide band
AM-ADDR	Active member address
PM-ADDR	Parked member address
LMP	Link manager protocol
HCI	Host controller interface
L2CAP	Logical link control and adaptation protocol
SDP	Service Discovery Protocol
PCM	Pulse code modulation
CVSD	Continuous variable slope delta modulation
BNEP	Bluetooth network encapsulation protocol
TCP	Transmission control protocol
UDP	User datagram protocol
SIG	Special interest group
BT	Bluetooth Technology
MTU	Maximum transfer unit.

LAN Local area network
DH Data high rate
DM Data medium rate
CRC Cyclic redundancy check.

6.1.2 Introduction

Personal area network (PAN) is a computer network used for communication among computer devices which including telephones and personal digital assistants close to one person. The devices may or may not belong to the person. Its range is typically a few meters. It is used for intrapersonal communication, connecting higher level network and the internet [1].

Personal area networks can be wired with computer buses such as USB and FireWire. A wireless personal area network (WPAN) can also be made possible with network technologies such as IrDA, Bluetooth, UWB, Z-Wave and ZigBee. The different technologies are explained below.

6.1.2.1 Bluetooth

A Bluetooth PAN is also known as *piconet*, it is composed of 8 active devices in a master-slave relationship .The first Bluetooth device in the piconet is the master, and all other devices are slaves that communicate with the master. The range used for communication using this technology is 10 to 100 meter.

6.1.2.2 IrDA

The Infrared Data Association (IrDA) is a physical specifications communications protocol standards used for short range exchange of data over infrared light in personal area networks (PANs).Communication via IrDA Requires direct line of sight.

6.1.2.3 UWB

Ultra-wideband (UWB) is a radio technology used at very low energy levels for short-range high-bandwidth communications by using a large portion of the radio spectrum. Recent applications are target sensor data collection, precision locating and tracking applications.

6.1.2.4 Z-Wave

Z-Wave is a wireless communications standard designed for home automation, specifically to remote control applications in residential and light commercial environments. It uses a low-power RF radio embedded or retrofitted into home electronics devices and systems, such as lighting, home access control, entertainment systems and household appliances.

6.1.2.5 ZigBee

ZigBee is a wireless network used for home, building and industrial control. WPAN requires ZigBee when operating at 868 MHz, 909–928 MHz and 2.4 GHz specification. It has ability to form Mesh network between nodes, which allows short range of an individual node to expand and multiples that help in covering large area. One ZigBee can contain more than 65,000 nodes. The ZigBee was designed for low power, wireless monitoring, sensors and remote control solution. Node can communicate maximum of 75 meter depending upon power output and environmental characteristics, more typical for 10–20 meter.

WPANs (Wireless Personal Area Networks) are short range communication systems (from a few centimetres to about 10 metres) that allow exchanging information among devices organised around an individual person. Nowadays Bluetooth is by far the most widely utilised technology for deploying WPANs.

The PAN (Personal Area Network) profile specifies how two or more Bluetooth devices can create an ad-hoc network, and how to access remote networks through access points. The main advantage of the PAN profile is that it enables an IP-based service. Thus Bluetooth nodes can be directly addressed in an independent and transparent manner from any IP network. For this purpose, the PAN profile employs BNEP (Bluetooth Network Encapsulation Protocol), inspired by Ethernet and specifically devised for the transport of IP data over Bluetooth. However, the joint employment of BNEP, IP and the transport protocol related to IP (UDP orTCP) introduces an overhead that can affect the performance of the Bluetooth transmissions.

In the Chapter, there are significant proposals to optimise the efficiency of Bluetooth connections [5, 6]. Most of these proposals empirically investigate the practical throughput and end-to-end delay that are achieved as a function of the distance between the origin and the destination nodes, the Bit Error Rate or the coexistence with 802.11 networks. However, these studies normally do not consider the effect of the election of a particular BT profile and the data segmentation performed at the upper layers. This Chapter proposes an analytical model to estimate the lower bound of the delay in transmissions of user data of an arbitrary size when the PAN profile is employed.

6.1.3 Purpose of the Document

In this Chapter one of the technologies (Bluetooth in PAN) are explained in detail. Bluetooth is by far the most employed technology to develop practical applications of Wireless Personal Area Networks (WPAN). This chapter gives idea about the performance of Bluetooth transmissions that make use of the Bluetooth PAN (Personal Area Network) profile. In particular, the study offers an analytical model that defines the optimal bound for the end-to-end data delay. The basic concept of Bluetooth technology must be known to understand the concept of end to end transmission delay in Bluetooth PAN.

'On Completion of this chapter you will be able to have good knowledge on PAN and bluetooth technology. This will help you in designing PAN using other wireless technology.

6.2 Bluetooth Concept

6.2.1 Introduction

Bluetooth is an open specification for a radio system that provides the network infrastructure to enable short range wireless communication of data and voice. It comprises of a hardware component and a software component.

What does Bluetooth technology do? Cable replacement

6.2.2 Bluetooth Architecture

Bluetooth devices can interact with one or more other Bluetooth devices in several different ways. The simplest scheme is where only two devices are involved. This is referred to as point-to-point. One of the devices acts as the master and the other as a slave. This ad-hoc network is referred to as a piconet. As a matter of fact, a piconet is any such Bluetooth network with one master and one or more slaves. A diagram of a piconet is provided in Figure 6.1. In the case of multiple slaves, the communication topology is referred to as

Table 6.1 Comparison of Bluetooth and wired network technology.

	Bluetooth technology	Wired network
Topology	Support up to 7 simulation links	Each link requires another cable
Flexibility	Goes through walls, bodies, cloths	Line of sight or modified environment
Data Rate	1 MSPS, 720 Kbps	Varies with use and cost
Power	0.1 watt active power	0.05 watt active power or higher
Size/weight	25mm* 13mm* 2mm, several grams	size is equal to range.Typically 1–2 m.weight varies with length(ounces to pounds)
Cost	Long-term $5 per endpoint	$3–$100/meter(end user cost)
Range	10 meter or less up to 100 meters with PA	Range equal to size. typically 1–2 meters
Universal	Intended to work anywhere in the world	Cables vary with local customs
Security	Very , link layer security, SS radio	Secure (it's a cable)

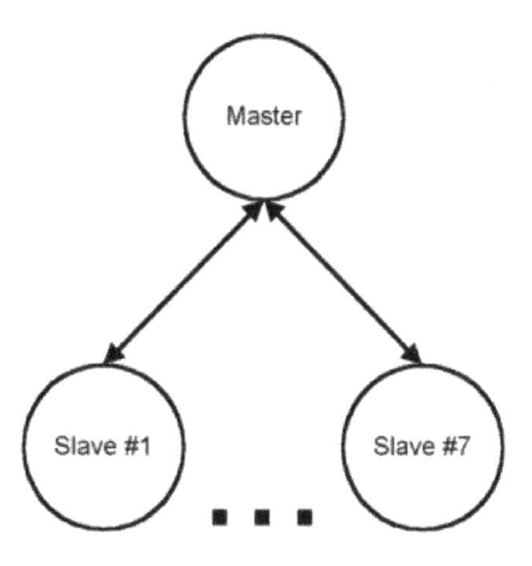

Fig. 6.1 Bluetooth piconet.

point-to-multipoint. In this case, the channel (and bandwidth) is shared among all the devices in the piconet. There can be up to seven active slaves in a piconet. Each of the active slaves has an assigned 3-bit Active Member address (AM_ADDR). There can be additional slaves which remain synchronized to the master, but do not have an Active Member address. These slaves are not active and are referred to as parked. For the case of both active and parked units, all channel access is regulated by the master. A parked device has an 8-bit Parked Member Address (PM_ADDR), thus limiting the number of parked members to 256. A parked device remains synchronized to the master clock and can very quickly become active and begin communicating in the piconet.

6.2.3 System Architecture

Bluetooth communication occurs in the unlicensed ISM band at 2.4 GHz. The transceiver utilizes frequency hopping to reduce interference and fading. A typical Bluetooth device has a range of about 10 meters [3]. The communication channel can support both data (asynchronous) and voice (synchronous) communications with a total bandwidth of 1 Mb/sec. The supported channel configurations are as follows:

Bluetooth devices are classified according to three different power classes, as shown in the following table.

6.2.4 Protocol Architecture

6.2.4.1 Radio Layer

This is the first layer of Bluetooth protocol. Everything in Bluetooth runs over the Radio Layer, This layer defines the requirements for a Bluetooth radio transceiver, which operates in the 2.4 GHz band. It also defines the sensitivity levels of the transceiver, establishes the requirements for using Spread-spectrum Frequency Hopping and classifies Bluetooth devices into three different power classes which are shown in Table 6.3.

6.2.4.2 Baseband Layer

The next level in the Bluetooth protocol stack is the Baseband Layer, which is the physical layer of the Bluetooth. It is used as a link controller, which works with the link manager to carry out routines like creating link connections with

Table 6.2 Bluetooth technology configuration.

Configuration	Max. Data Rate Upstream	Max. Data Rate Downstream
3 Simultaneous Voice Channels	64 kb/sec X 3 channels	64 kb/sec X 3 channels
Symmetric Data	433.9 kb/sec	433.9 kb/sec
Asymmetric Data	723.2 kb/sec or 57.6 kb/sec	57.6 kb/sec or 723.2 kb/sec

Table 6.3 Range and power classification of bluetooth.

Power class	Maximum output	Power	Range
1	100 mW	(20 dBm)	long rang devices (100 m)
2	2.5 mW	(4 dBm)	normal or standard range devices (10 m)
3	1mW	(0 dBm)	short (10 cm)-range operation

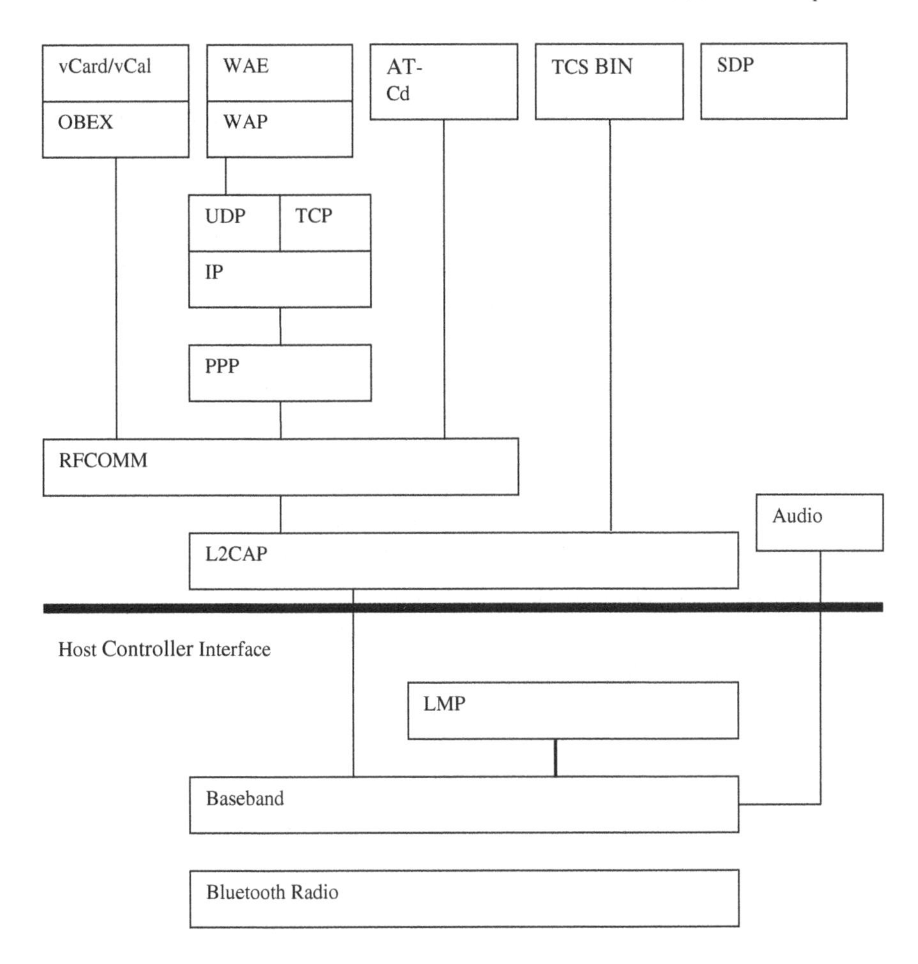

Fig. 6.2 Bluetooth protocol stack.

other devices. It controls device addressing, channel control (how devices find each other) through paging and inquiry methods, power-saving operations, and also flow control and synchronization among Bluetooth devices.

6.2.4.3 Link Manager Protocol (LMP)

A Bluetooth device's Link Manager Protocol (LM) carries out link setup, authentication, link configuration and other protocols. It discovers other LMs within the area and communicates with them via the Link Manager Protocol (LMP).

6.2.4.4 Host Controller Interface (HCI)

Next in the protocol stack, above the LMP is the Host Controller Interface (HCI), which is there to allow command line access to the Baseband Layer and LMP for control and to receive status information. It's made up of three parts: 1) The HCI firmware, which is part of the actual Bluetooth hardware, 2) The HCI driver, which is found in the software of the Bluetooth device, and 3) The Host Controller Transport Layer, which connects the firmware to the driver.

6.2.4.5 Logical Link Control and Adaptation Protocol (L2CAP)

Above the HCI level is the Logical Link Control and Adaptation Protocol (L2CAP), which provides data services to the upper level host protocols. The L2CAP plugs into the Baseband Layer and is located in the data link layer, rather than riding directly over LMP. It provides connection-oriented and connectionless data services to upper layer protocols.

Protocol types are first identified in the L2CAP. Data services are provided here using protocol multiplexing, segmentation and reassembly operation, and group abstractions occur. L2CAP allows higher-level protocols and applications to send and receive data packets up to 64 kilobytes. The L2CAP spends a lot of its time handling segmentation and reassembly tasks.

6.2.4.6 RFCOMM

Above L2CAP, the RFCOMM protocol is what actually makes upper layer protocols think they're communicating over a RS232 wired serial interface, so there's no need for applications to know anything about Bluetooth.

6.2.4.7 Service Discovery Protocol (SDP)

Also relying on L2CAP is the Service Discovery Protocol (SDP). The SDP provides a way for applications to detect which services are available and to determine the characteristics of those services.

6.2.5 Working Concept of Bluetooth

An interesting aspect of the technology is the instant formation of networks once the bluetooth devices come in range to each other. A piconet is a collection of devices connected via Bluetooth technology in an ad hoc fashion. A Piconet

Table 6.4 Brief description of function of various layers in bluetooth protocol stack.

Layers	Layer Name	Functions
First	Radio	Defines the requirements for a Bluetooth radio and sensitivity levels of the transceiver.
Next level	Baseband	Used as a link controller, controls device addressing.
Third	LMP	carries out link setup, authentication, link configuration and other protocols
Next level to LMP	HCI	Allow command line access to the Baseband Layer
Above the HCI layer	L2CAP	Provides data services to the upper level host protocols
Above L2CAP	RFCOMM	makes upper layer protocols think they're communicating over a RS232 wired serial interface
Relying on L2CAP	SDD	detect which services are available, Determine characteristics of those services

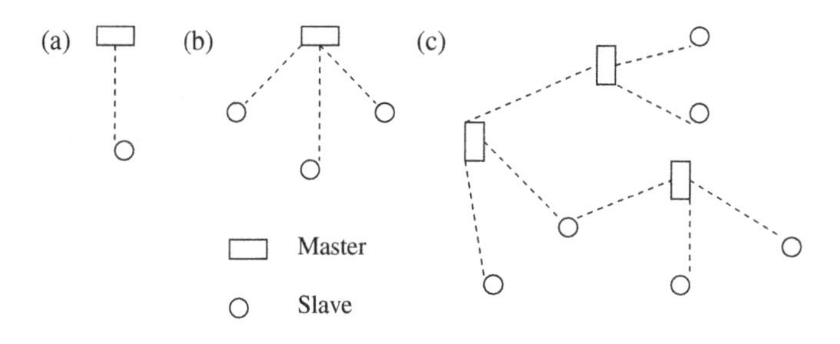

Fig. 6.3 Pico nets with single slave operation (a); Multi slave operation (b); scattered operation (c).

can be a simple connection between two devices or more than two devices. Multiple independent and non-synchronized piconets can form a scatternet [4]. Any of the devices in a piconet can also be a member of another by means of time multiplexing. i.e a device can be a part of more than one piconet by suitably sharing the time. The Bluetooth system supports both point-to-point and point-to-multi-point connections.

When a device is connected to another device it is a point to point connection. If it is connected to more that one (upto 7) it is a point to multipoint connection. Several piconets can be established and linked together ad hoc, where each piconet is identified by a different frequency hopping sequence. All users participating on the same piconet are synchronized to this hopping sequence. If a device is connected to more than one piconet it communicates in each piconet using a different hopping sequence. A piconet starts with two connected devices, such as a portable PC and cellular phone, and may grow to eight connected devices. All Bluetooth devices are peer units and have identical implementations.

However, when establishing a piconet, one unit will act as a master and the other(s) as slave(s) for the duration of the piconet connection. In a piconet there is a master unit whose clock and hopping sequence are used to synchronize all other devices in the piconet. All the other devices in a piconet that are not the master are slave units. A 3-bit MAC address is used to distinguish between units participating in the piconet. Devices synchronized to a piconet can enter power-saving modes called Sniff and hold mode, in which device activity is lowered. Also there can be parked units which are synchronized but do not have a MAC addresses. These parked units have a 8 bit address, therefore there can be a maximum of 256 parked devices.

Voice channels use either a 64 kbps log PCM or the Continuous Variable Slope Delta Modulation (CVSD) voice coding scheme, and never retransmit voice packets. The voice quality on the line interface should be better than or equal to the 64 kbps log PCM. The CVSD method was chosen for its robustness in handling dropped and damaged voice samples. Rising interference levels are experienced as increased background noise: even at bit error rates up 4%, the CVSD coded voice is quite audible.

6.2.5.1 Bluetooth Pairing

Bluetooth pairing occurs when two Bluetooth devices agree to communicate with each other and establish a connection.

In order to pair two Bluetooth wireless devices, a password (passkey) has to be exchanged between the two devices. A Passkey is a code shared by both Bluetooth devices, which proves that both users have agreed to pair with each other.

This is the normal process that occurs with Bluetooth pairing:

Bluetooth device A looks for other Bluetooth devices in the area

In order to find other Bluetooth devices, Bluetooth device A must be set to discoverable mode. When set to discoverable, Bluetooth device A will allow other Bluetooth devices to detect its presence and attempt to establish a connection.

When the discover setting is off, no other Bluetooth device will be able to find it. Undiscoverable devices can still communicate with each other but they have to initiate communication themselves.

Bluetooth device A finds Bluetooth device B

Usually the discoverable device will indicate what type of device it is (Such as a printer, cell phone, headset, etc.) and its Bluetooth device name. The Bluetooth device name is the name that you give the Bluetooth device or the factory name that originally was programmed.

Bluetooth Device A prompts you to enter a password (PassKey)

With advanced devices, both users must agree on the Passkey and enter it into their device. The code can be anything you like as long as it is the same for both Bluetooth wireless devices. On other devices, such as Bluetooth headsets, the Passkey stays the same.

Bluetooth device A sends the Passkey to Bluetooth device B

Bluetooth device B sends the Passkey back to Bluetooth device A

If both Passkeys are the same, a trusted pair is formed. This will happen automatically.

Bluetooth device A and B are now paired and able to exchange data

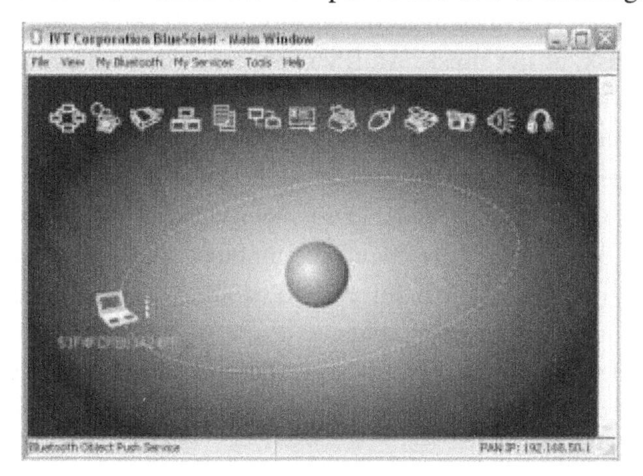

6.2.6 Planning of a Network

6.2.6.1 Addressing of Nodes

Designing of a network contains of Determining the network Hardware, obtaining network IP Number, Deciding an IP addressing format for the Network, Naming Entities of the Network.

Determining the network Hardware

When you design the network, must decide what type of network meets the needs of your organization. Some of the planning decisions you must make involve the following network hardware:

- The network topology, the layout, and connections of the network hardware
- The number of host systems the network can support
- The types of hosts that the network supports
- The types of servers that might need
- The type of network media to use: Ethernet, Token Ring, FDDI, and so on
- Whether you need bridges or routers extend this media or connect the local network to external networks
- Whether some systems need separately purchased interfaces in addition to their built in interfaces

Based on these factors, we can determine the size of your personal

Deciding an IP addressing format for the Network area network

An IPv4 network is defined by a combination of an IPv4 network number plus a network mask, or **net mask**. An IPv6 network is defined by its **site prefix**, and, if sub netted, its **subnet prefix**.

Unless your network plans to be private in perpetuity, your local users most likely need to communicate beyond the local network. Therefore, you must obtain a registered IP number for your network from the appropriate organization before your network can communicate externally. This address becomes the network number for your IPv4 addressing scheme or the site prefix for your IPv6 addressing scheme.

Internet Service Providers provide IP addresses for networks with pricing that is based on different levels of service. Investigate with various ISPs to

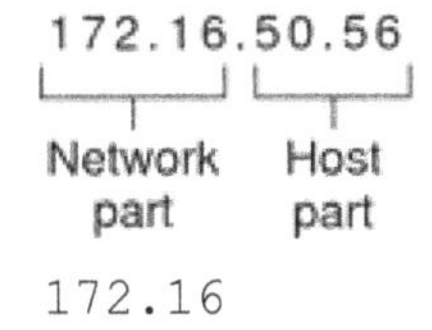

Fig. 6.4 IPv4 address format.

determine which provides the best service for your network. ISP's typically offer dynamically allocated addresses or static IP addresses to businesses. Some ISPs offer both IPv4 and IPv6 addresses.

Designing an IPv4 Addressing Scheme

This section gives an overview IPv4 addressing to aid you in designing an IPv4 addressing plan. Each IPv4-based network must have the following:

- A unique network number that is assigned by either an ISP, an IR, or, for older networks, registered by the IANA. If you plan to use private addresses, the network numbers you devise must be unique within your organization.
- Unique IPv4 addresses for the interfaces of every system on the network.
- A network mask.

The IPv4 address is a 32-bit number that uniquely identifies a network interface on a system. An IPv4 address is written in decimal digits, divided into four 8-bit fields that are separated by periods. Each 8-bit field represents a byte of the IPv4 address. This form of representing the bytes of an IPv4 address is often referred to as the **dotted-decimal format**.

The following figure shows the component parts of an IPv4 address, 172.16.50.56.t

Registered IPv4 network number: In class-based IPv4 notation, this number also defines the IP network class, Class B in this example that would have been registered by the IANA.

50.56

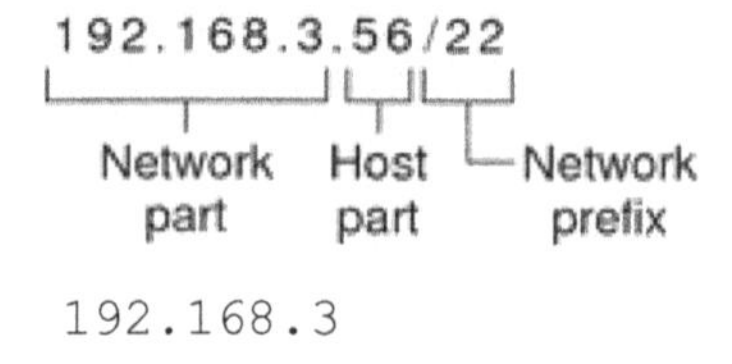

Fig. 6.5 CIDR format IPv4 address.

Host part of the IPv4 address: The host part uniquely identifies an interface on a system on a network. Note that for each interface on a local network, the network part of the address is the same, but the host part must be different.

The next example shows of the CIDR format address 192.168.3.56/22

Network part, which consists of the IPv4 network number that is received from an ISP or IR

56 Host part, which you assign to an interface on a system.

/22 Network prefix, which defines how many bits of the address comprise the network number. The network prefix also provides the subnet mask for the IP address. Network prefixes are also assigned by the ISP or IR.

A Solaris-based network can combine standard IPv4 addresses, CIDR format IPv4 addresses, DHCP addresses, IPv6 addresses, and private IPv4 addresses.

Designing Your IPv4 Addressing Scheme

This section describes the classes into which standard IPv4 address are organized. Though the IANA no longer gives out class-based network numbers, these network numbers are still in use on many networks. You might need to administer the address space for a site with class-based network numbers. For a complete discussion of IPv4 network classes.

The following table shows the division of the standard IPv4 address into network and host address spaces. For each class, "Range" specifies the range of decimal values for the first byte of the network number. "Network Address" indicates the number of bytes of the IPv4 address that are dedicated to the network part of the address. Each byte is represented by *xxx*. "Host Address" indicates the number of bytes that are dedicated to the host part of the address. For example, in a class A network address, the first byte is dedicated to the

Table 6.5 Division of the IPv4 classes.

Class	Byte range	Network number	Host address
A	0–127	xxx	xxx.xxx.xxx
B	128–191	xxx.xxx	xxx.xxx
C	192–223	xxx.xxx.xxx	xxx

Table 6.6 Range of available IPv4 classes.

Network Class	Byte 1 Range	Byte 2 Range	Byte 3 Range	Byte 4 Range
A	0–127	1–254	1–254	1–254
B	128–191	Preassigned by IANA	1–254	1–254
C	192–223	Preassigned by IANA	Preassigned by IANA	1–254

network, and the last three bytes are dedicated to the host. The opposite designation is true for a class C network.

The numbers in the first byte of the IPv4 address define whether the network is class A, B, or C. The remaining three bytes have a range from 0–255. The two numbers 0 and 255 are reserved. You can assign the numbers 1–254 to each byte, depending on the network class that was assigned to your network by the IANA [8].

The following table shows which bytes of the IPv4 address are assigned to you. The table also shows the range of numbers within each byte that are available for you to assign to your hosts.

IPv4 Subnet Number

Local networks with large numbers of hosts are sometimes divided into subnets. If you divide your IPv4 network number into subnets, you need to assign a network identifier to each subnet. You can maximize the efficiency of the IPv4 address space by using some of the bits from the host part of the IPv4 address as a network identifier. When used as a network identifier, the specified part of the address becomes the subnet number. You create a subnet number by using a netmask, which is a bitmask that selects the network and subnet parts of an IPv4 address.

Designing Your CIDR IPv4 Addressing Scheme

The network classes that originally constituted IPv4 are no longer in use on the global Internet. Today, the IANA distributes classless CIDR format addresses to its registries around the world. Any IPv4 address that you obtain from an ISP is in CIDR format, as shown in Figure 6

Table 6.7 CIDR prefixes and their decimal equivalent.

CIDR Network Prefix	Available IP Addresses	Dotted Decimal Subnet Equivalent
/19	8,192	255.255.224.0
/20	4,096	255.255.240.0
/21	2,048	255.255.248.0
/22	1024	255.255.252.0
/23	512	255.255.254.0
/24	256	255.255.255.0
/25	128	255.255.255.128
/26	64	255.255.255.192
/27	32	255.255.255.224

The network prefix of the CIDR address indicates how many IPv4 addresses are available for hosts on your network. Note that these host addresses are assigned to interfaces on a host. If a host has more than one physical interface, we need to assign a host address for every physical interface that is in use.

The network prefix of a CIDR address also defines the length of the subnet mask. Most Solaris 10 commands recognize the CIDR prefix designation of a network's subnet mask. However, the Solaris installation program and /etc/netmask file require you to set the subnet mask by using dotted decimal representation. In these two cases, use the dotted decimal representation of the CIDR network prefix, as shown in the next table.

How IP Addresses Apply to Network Interfaces

To connect to the network, a system must have at least one physical network interface. Each network interface must have its own unique IP address. During Solaris installation, you must supply the IP address for the first interface that the installation program finds. Usually that interface has the name *device-name*0, for example eri0 or hme0. This interface is considered the primary network interface.

If you add a second network interface to a host, that interface also must have its own unique IP address. When you add the second network interface, the host then becomes multihomed. By contrast, when you add a second network interface to a host and enable IP forwarding, that host becomes a router.

6.2.7 Transmission Delay

6.2.7.1 PAN Profile: - Minimum end to end delays

In this section the minimum delay for transmitting N user data bytes is estimated assuming ideal conditions, that is, the information flows from the Bluetooth master to the slave with a zero Bit Error Rate (no retransmissions occur) and a negligible storage time in the buffers. In order to incorporate the impact of all the protocols involved in the transmission under the PAN profile, the analysis must take into account the overhead of the headers added by all the layers, as well as the need for fragmentation in the $(i + 1)$-th layer to avoid exceeding the i-th layer MTU (Maximum Transfer Unit). The PAN profile allows the transport of TCP/IP or UDP/IP packets over L2CAP (Logical Link Control and Adaptation Protocol) using the BNEP protocol. BNEP replaces the typical Ethernet header of a LAN (Local Area Network) transmission with a specific header.

The header size is 15 or 3 bytes depending if a general or a compressed packet format is employed. The compressed format is utilized when both the origin and the destination of a BNEP packet correspond to a master-slave pair in a Bluetooth piconet. Every BNEP header and its payload are encapsulated in a Bluetooth L2CAP data PDU (Packet Data Unit) or frame. Since the BNEP frames encapsulate IP datagram's, carrying in turn UDP or TCP data, the transmission delay to be calculated will be equal to the transmission delay at the transport layer. If UDP is employed, the time required at the UDP layer (*tUDP*) to transmit N-byte user data can be estimated as:

$$t_{UDP}(N) = t_{IP}(N + H_{UDP}) \tag{6.1}$$

HUDP is 8 bytes (the size of the UDP header)
tIP(N) is the delay at the IP layer.

The computation of this delay, defined in equation (2), must contemplate the fragmentation that occurs at the BNEP layer when the BNEP MTU ($M'B$) is exceeded:

$$\begin{aligned} t_{ip}(N) = {} & \mathrm{N_{frag}(N).t_{ack}(M'_B + H_B + H_{L2CAP})} \\ & \mathrm{+ t_{TX}(L_{rem}(N) + H_{IP} + H_B + H_{L2CAP})} \end{aligned} \tag{6.2}$$

M'$_B$ BNEP MTU (1500 bytes, as the length of the maximum Ethernet payload). value M'_B is lower than the L2CAP MTU for BNEP, which is 1691 bytes, every BNEP packet is encapsulated in a single L2CAP frame

H$_{IP}$ the number of bytes in the standard IP header (20bytes)

N$_{frag}$ the number of non-final BNEP fragments, computable as

$$N_{frag}(N) = \left\lceil \frac{N}{M'_B - H_{IP}} \right\rceil - 1 \qquad (6.3)$$

x indicates the rounding to the lowest integer higher than x

H$_B$ the number of bytes in the BNEP header (3 bytes for the compressed format)

H$_{L2CAP}$ the size of he L2CAP protocol header (4 bytes)

L$_{rem}$(N) the numbers of bytes of the last BNEP/L2CAP frame, which is calculated as

$$L_{\mathrm{rem}}(N) = ((N - 1) \bmod (M'_B - H_{IP})) \qquad (6.4)$$

The formula in (2) also includes the segmentation that Bluetooth (BT) performs when more than one BT baseband packet is required to transport a L2CAP frame. In this sense, the formula considers two components, t_{ACK} and t_{TX}, defined as follows:

The term $t_{ACK}(N)$describes the time (estimated in terms of BT slots) that is required by Bluetooth to send an intermediate BNEP/L2CAP frame:

$$t_{ACK}(N) = \begin{cases} 2.T_S & \mathrm{N} \leq \mathrm{L}_1 \\ 4.T_S & \mathrm{L}_1 < \mathrm{N} \leq \mathrm{L}_3 \\ 6.T_S & \mathrm{L}_3 < \mathrm{N} \leq \mathrm{L}_5 \\ 6.T_s \cdot \left\lfloor \frac{N}{L_5} \right\rfloor + tACK(N \bmod L_5) & \mathrm{N} > \mathrm{L}_5 \end{cases} \qquad (6.5)$$

where $[x]$ denotes the highest integer lower than *x, TS* is the duration of a Bluetooth slot (625 μs), and L_1, L_3 and L_5 are the maximum payload sizes for a 1, 3 and 5-slot Bluetooth packet, respectively. These sizes are 27, 183 and 339 bytes for DH (Data High -Rate) packets and 17, 121 and 224 bytes for DM (Data Medium-Rate) packets [5].

As long as a BT packet will not be transmitted until the acknowledgement of the previous one is received, the recursive expression in equation (5) takes into account the time necessary to acknowledge the intermediate BT packets into which the BNEP/L2CAP frames are decomposed. Therefore, for each intermediate BT packet there is a fixed delay of 2, 4 or 6 slots, depending on whether the current segment is transmitted in a 1, 3 or 5-slot packet. The term tTX(N) defines the time required for transmitting the final BNEP/ L2CAP frame. In this case, as the transmission will be completed when the last bit of the final fragment is received in the BT slave, neither the final acknowledgement slot nor the complete final slot of the BT packet are computed for the estimation of the delay. Specifically, this time t_{TX}(N) can be calculated as a function of the number of transmitted bits in the following way.

$$t_{TX}(N) = \begin{cases} N_B(N).T_B & N \leq L_5 \\ t_{ACK}(L5) \cdot \left\lfloor \frac{N}{L_5} \right\rfloor + t_{TX}(N \bmod L5) & N > L_5 \end{cases} \tag{6.6}$$

T_B is the transmission time for 1 bit (1 μs at the peak data rate of 1 Mbps)

N_B(N) is the size (in bits) of the final BT packet. This size can be computed as:

$$N_B(N) = N_{ov} + N_{pl}(N) \tag{6.7}$$

Where:

N_{ov} represents the Bluetooth packet header of 126 bits, obtained by adding the number of bits in the packet header (54 bits) and the access code (72 bits)

N_{pl}(N) is the number of bits in the Bluetooth payload and body calculable as

$$N_{PL}(N) = \begin{cases} (N + HS + HCRC).8 & DH\,packets \\ \left\lceil \frac{(N+HS+HCRC).8}{10} \right\rceil .15 & DM\,packets \end{cases} \tag{6.8}$$

where $H_{CRC} = 2$ corresponds to the 2 bytes of the CRC (Cyclic Redundancy Check) field while H_S is a header of 1 or 2 bytes depending on the number of slots of the BT packet ($H_S = 1$ forl slot and $H_S = 2$ for 3 and 5 slot-packets, respectively). The previous equation takes into account that for DM packets, which are protected with FEC 2/3 (Forward Error Correction), for every 10 information bits 5 redundancy bits are added. Consequently, if the number of bits is not a multiple of 10, the packet must be filled with extra bits after the CRC. Finally, note that the equation in (6) also considers that if the final BNEP/L2CAP frame exceeds the size of a 5-slot BT packet, more than one BT packet will be required. Thus, it also computes the time of the acknowledgments of the corresponding intermediate 5-slot BT packets.

6.2.8 Bluetooth Security

Today, all communication technologies are facing the issue of privacy and identity theft. Bluetooth technology is no exception. The information and data we share through these communication technologies is both private and in many cases, critically important to us.

Everyone knows that email services, company networks, and home networks all require security measures. What Bluetooth users need to realize, is: Bluetooth requires similar security measures. Recent reports have surfaced describing ways for hackers to crack Bluetooth devices security codes.

Are the Threats Serious?

Most of the recent Bluetooth security scares, like most scares, are overdramatized and blown out of proportion. The truth is, these issues are easily combatable, and various measures are already in place to provide for the secure use of Bluetooth technology.

According to the Bluetooth Special Interest Group (SIG), in order to break into a Bluetooth device, a hacker must:

Force two paired Bluetooth devices to break their connection

Steal the packets used to resend the PIN, then

Decode the PIN

The hacker must of course be within range of the Bluetooth device and, according to the Bluetooth SIG, be using very expensive developers equipment. The SIG suggests users create a longer PIN (8 digit is recommended).

6.2.9 Specification

Bluetooth devices in a piconet share a common communication data channel. The channel has a total capacity of 1 megabit per second (Mbps). Headers and handshaking information consume about 20 percent of this capacity.

In the United States and Europe, the frequency range is 2,400 to 2,483.5 MHz, with 79 1-MHz radio frequency (RF) channels. In practice, the range is 2,402 MHz to 2,480 MHz. In Japan, the frequency range is 2,472 to 2,497 MHz with 23 1-MHz RF channels.

A data channel hops randomly 1,600 times per second between the 79 (or 23) RF channels.

Each channel is divided into time slots 625 microseconds long.

A piconet has a master and up to seven slaves. The master transmits in even time slots, slaves in odd time slots.

Packets can be up to five time slots wide.

Data in a packet can be up to 2,745 bits in length.

There are currently two types of data transfer between devices: SCO (synchronous connection oriented) and ACL (asynchronous connectionless).

In a piconet, there can be up to three SCO links of 64,000 bits per second each. To avoid timing and collision problems, the SCO links use reserved slots set up by the master.

Masters can support up to three SCO links with one, two or three slaves.

Slots not reserved for SCO links can be used for ACL links.

One master and slave can have a single ACL link.

ACL is either point-to-point (master to one slave) or broadcast to all the slaves.

ACL slaves can only transmit when requested by the master.

6.2.10 Advantages

Table 6.8 Advantages of bluetooth technology.

BT Devices are wireless	No need to worry about bringing along all of your connecting cables.
Inexpensive	Cheap for companies to implement, which results in lower over-all manufacturing Costs.
Automatic	Doesn't require you to think about setting up a connection or to push any buttons.
Standardized Protocol = Interoperability	A high level of compatibility among devices is guaranteed.
Low Interference	Avoids interference with other wireless devices by using spread-spectrum frequency hopping technique
Low Energy Consumption	Since it uses low power signal it requires less energy
Share Voice and Data	allows compatible devices to share both voice and data communications

6.3 Summary

Bluetooth is most common wireless technology used for the Personal Area Network compared to other technology. BT replaces cable.

When only two devices used for communication it is referred as point-to-point communication. Here one acts as master other act as slaves. This ad-hoc network is referred as piconet.

Bluetooth communication occurs in unlicensed ISM band at 2.4GHz.

In the protocol architecture everything in the bluetooth runs over radio layer. Base band layer is the physical layer of the Bluetooth.

LMP layer carries out link setup, HCI Allows command line access to base band layer.

L2CAP Provides data services to the upper level host protocols. RFCOMM makes upper layer protocol think they are communicating over a RS232 wired serial interface.

SDD detect which services are available and determine characteristics of those services.

Bluetooth system supports both point-to-point and point-to- multi-point connection.

In order to pair two bluetooth wireless devices, a password (passkey) has to be exchanged between them.

Bluetooth devices in a piconet share a common communication data channel. A piconet have a master and up to seven slave. Master transmits in even time slot, slaves in odd time slots.

Bluetooth allows compatible devices to share both voice and data communication.

6.4 Review Questions

What is the name of the first bluetooth device in Piconet?
Range used for communication using bluetooth technology?
Application of UWB?
Main advantage of PAN profile?
Why do we need Bluetooth technology?
What is the normal or standard range device for Bluetooth technology?
Function of Host control interface?
What is number of bytes in standard IP header?

6.4.1 To be solved:-

Draw the Bluetooth scattenet having 2 maser and 1 slave devices?

When UDP is employed in transmission, the delay is calculated by $t_{UDP}(N) = t_{IP}(N + H_{UDP})$. Is this same for TCP, justify it?

Try Bluetooth pairing concept for two devices, later try it for three or more devices, keeping one of the device as master and others as slave.

6.4.2 Answers to Review Questions

Master.

10–100 meter.

UWB- Ultra–Wideband, recent application are target sensor data collection, precision locating and tracking application.

It enables an IP- based service.

For Cable replacement

10 meter.

Allows command line access to base band layer.

20 bytes

References

[1] http://en.wikipedia.org/wiki/Personal_area_network
[2] http://www.tutorial-reports.com/wireless/bluetooth/architecture.php
[3] http://www.wirelessdevnet.com/channels/bluetooth/features/bluetooth.html
[4] http://www.mobileinfo.com/bluetooth/how_works.htm
[5] Bluetooth Special Interest Group (SIG), "Specification of the Bluetooth
[6] System vol. 2: Profiles", Version 1.1, February, 2001.
[7] P. Huang, and A. C. Boucouvalas, "Delay Analysis for Bluetooth Baseband ACL Packets", In Proc. Of Convergence of Telecommunications, Networking & Broadcasting Symposium (PGNET 2005), Liverpool, June 2005, pp. 396–401.
[8] http://docs.sun.com/app/docs/doc/816-4554

7

Planning of Wireless Mesh Networks

Mohammad Al-Hattab and Johnson I Agbinya

University of Technology, Sydney, Australia

7.1 Introduction

Wireless mesh network (WMN) is one of the most promising wireless communication designs. This design has many characteristics which will impact the future of wireless communication; it is reliable, flexible, efficient, and cheaper to install and maintain. It is also a community-owned infrastructure which makes it possible to be extended to rural and remote areas to provide a broadband access where the costs of traditional infrastructure networks are not convenient for economic and technical reasons.

Wireless mesh networks are a very good option when installing a temporary network because the backbone of the network is easy to install or remove when the purpose of the network is over; for example, in disaster areas like earthquakes, emergency crews can install many nodes to form a backbone of a wireless network to communicate and distribute data among the rescue crew.

Another good example is when installing a network within historical and heritage buildings where it is not possible to install cables and/or conduit because of the direct effects on the historical value. Moreover, the cost of installing cables in buildings that do not have conduits is much more expensive than the wireless routers at access points.

A WMN consists of many nodes connected together using radio frequency (RF) in mesh topology. These nodes can be classified into two types: WMN routers and WMN clients. WMN routers form the backbone of the network.

These nodes are configured automatically and reconfigured dynamically which makes the WMN a self-healing network. This means that the connectivity is established and maintained by the network itself. WMN routers have a conventionally gateway/bridge functionality besides additional routing support for mesh networks, whereas the WMN client is an access point that connects the end user to the network.

WMN nodes could use a multi-radio interface which would increase the capacity of the networks. It is also a multi-hop multi-path network which means there are many redundant paths for data between the source and the destination. On the other hand transmitting data over a multi-hop improves the coverage and reduces the required power.

WMN and Ad hoc networks are very similar. Many authors consider WMN as Ad hoc networks. The only clear difference is the type of nodes used.

In WMN, nodes can be categorised into mesh nodes or client nodes. Mesh nodes are usually used for routing data into the networks, while client nodes are the end user of the network. Different softwares are installed in each type of these nodes to perform their functions.

In Ad hoc networks, all nodes have the same routing capability. Therefore the concept of backbone and client nodes does not exist.

Nodes in WMN are almost stationary. However, this is not a concrete fact as some nodes may join the network and others may leave or move. Because nodes in WMN are more likely to be stationary the topology of the WMN is less dynamic than Ad hoc network topology where nodes are mobile.

7.2 WMN Definition

WMN is a wireless communication network that is formed by radio nodes connected together in mesh topology. These radio nodes form the backbone of the network and they have the capability of routing packets within the network. WMN are able to respond to any topology change because nodes are re-configured dynamically. This self-sufficient relationship between mesh nodes removes the need for centralised management [1].WMN uses intelligent routing protocols that allow nodes to choose one path among the available paths to the destination based on routing metrics to improve the network performance.

7.3 Overcoming Distance Limitation

One of the big advantages of WMN is that it overcomes the nodes transmitting range problem. When many nodes are deployed in the area of the network they can send data through multi-hop, because each node functions as a router and a repeater forwarding data to the next node until the data reaches the final destination. If we compare this with a traditional WLAN where nodes can only send and receive from the access point, even if they are in the proximity of each other, we can realise how the range of node increases by hopping the data through different nodes without the need of a centralised point or without increasing the physical range of the node. Figure 5.1a shows a traditional WLAN with an access point and some nodes at a certain area, nodes A, B and C are connected to the network through the access point, node D is not connected because it is out of the access point rang. Despite the fact that node D is close to node B, they are not connected to the network. Because, node B can communicate only through the access point. Figure 5.1b shows the same structure in WMN design, because node B can communicate directly with any other node within its range then node D is connected to the network through node B.

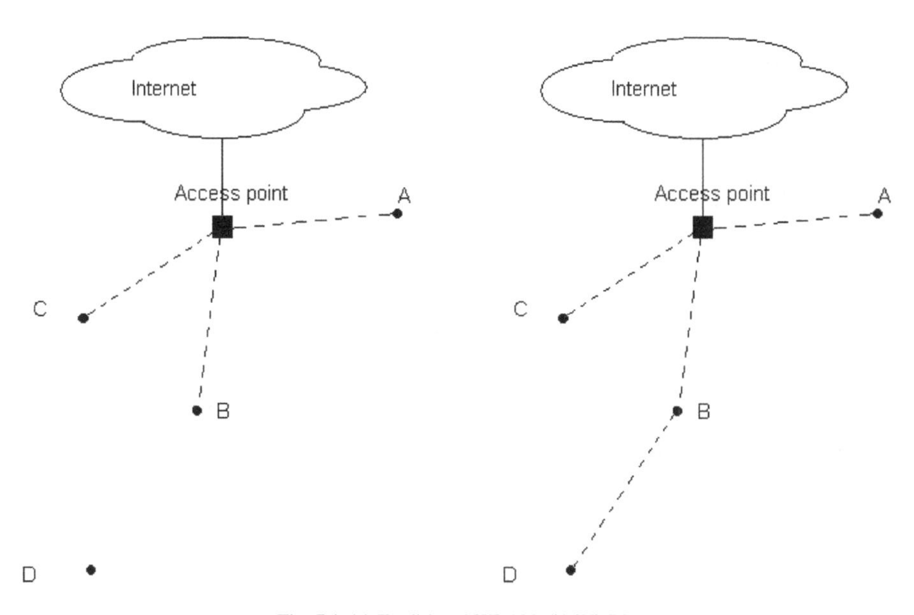

Fig. 7.1 (a) Traditional WLAN, (b) WMN.

7.4 Advantages of Wireless Mesh Networks

Wireless networks have many advantages that attract researchers and venders to consider this structure and develop it.

7.4.1 Reliability

The decentralisation in WMN gives nodes the ability to enter and leave the network at any time. Nodes can dynamically change their routing table based on their neighbours, so any changes to the node's neighbourhood should be treated quickly in order to keep the performance of the network at an acceptable level.

7.4.2 Self Healing

Nodes are able to reconfigure their own routing tables without the need for network's central administration. Any changes in the networks such as link failure, node departure or node arrival can be recognised by the node itself. As a result the network will be able to adapt to any loss of paths.

7.4.3 Self Configured

Because nodes can recognise changes in the topology of the network, they do not need central configuration.

7.4.4 Fast Operations

Nodes can cooperate with each other in the network, so they can communicate directly without any need to consult or report a central control or server. This property makes the operation and communication between local nodes faster.

7.4.5 Existing Technology

Wireless mesh networks rely on the same Wi-Fi standards (802.11a, b and g), which is already in place for most wireless networks.

7.4.6 Low Network Cost

WMN uses much less wiring than traditional LAN which lowers the cost of the network installation as fewer wires mean a lower cost, particularly for

Table 7.1. Path loss exponent for different environments [2].

Environment	Path loss exponent α
Free space	2
Urban area cellular radio	2.7–3.5
Shadowed urban cellular	3–5
In building line-of-sight	1.6–1.8
Obstructed in building	4–6
Obstructed in factories	2–3

large areas of coverage. Fewer technicians are required because there are less wiring and no conduit. WMN routers are also cheap because it is an existing technology.

7.4.7 Scalability

Wireless mesh nodes are easy to install and uninstall, making the network extremely adaptable and expandable as more or less coverage is needed. Moreover, mobile nodes can enter and leave the mesh network because they operate a software which is compatible with other nodes in the networks.

7.5 Planning the WMN

The first step in planning the network is to plot an initial diagram for the network; this plot may show the locations of mesh nodes, it can be done using any map such as Google maps or Google earth. The locations of mesh nodes are not required to be accurate because they are relative locations and can be done in later stages. These relative locations may give the designer intial measurements for the distance between nodes. The traffic demand (number of client nodes) is another important factor to consider.

The deployment of mesh nodes into the area of the network should satisfy a good connectivity and coverage. The distance between nodes can be determined by many ways:

1. Mathematical analysis: The received signal at any mesh node should be higher than a threshold value called the sensitivity. To calculate the received signal at a certain distance many parameters should be known, the transmitted power, the path loss and the propagation models.

The received signal level is proportional to the transmitted signal power and the distance between the transmitter and the receiver, it is also affected by the environment.

The received signal at distance d given signal strength at distance d0 can be expressed by:

$$P_{dBm}(d) = P_{dBm}(d0) - 10\alpha \log_{10}\left(\frac{d}{d0}\right) + \varepsilon \tag{7.1}$$

where ε is a Gaussian random variable with zero mean caused by the shadowing effect and α is the path loss exponent, the value of α depends on the environment.

Table 5.5.1 lists typical path loss exponent obtained in various mobile radio environment.

The path loss model is given by the following equation:

$$PL(d)[dB] = PL(d0) + 10\alpha \log\left(\frac{d}{d0}\right) + \varepsilon \tag{7.2}$$

The received power at distance d can be expressed by the total path loss and transmitted power (total gain):

$$\Pr(d)[dBm] = Pt[dBm] - PL(d)[dB] \tag{7.3}$$

Since ε is a random variable then PL(d) is a random variable with a normal distribution in dB about the distance dependant mean, so is Pr(d), and the Q-function may be used to determine the probability that the received signal is above a particular level (threshold):

$$\Pr o[\Pr(d) > \gamma] = Q\left(\frac{\gamma - \Pr(d)}{\sigma}\right) \tag{7.4}$$

The free space path loss

$$\text{PL(dB)} = 10\log\frac{P_t}{P_r} = -10\log\left(\frac{G_t G_r \lambda^2}{(4\pi)^2 d^2}\right) \tag{7.5}$$

For unit gains and a frequency of 2.45 GHz used in WMN

$$PL(d)[dB] = 40 + 20\log(d) \tag{7.6}$$

In general for a system of frequency 2.45 GHz and an allowed value of loss L the path loss can be expressed in:

$$PL(d)[dB] = 40 + 10\alpha \log(d) + L(allowed) \tag{7.7}$$

L is caused by different factors like attenuation and is not easy to calculate. Experience shows that trees add about 20 dB and walls add 10–20 dB, so when we design the link we should have a margin of 10 to 15 dB above the sensitivity of the receiver to ensure a good link quality.

As an example, let us estimate the feasibility of a 5 km link, with one access point and one client radio. The access point is connected to an omnidirectional antenna with 10 dBi gain, while the client is connected to a sectorial antenna with 14 dBi gain. The transmitting power of the AP is 100 mW (or 20 dBm) and its sensitivity is −89 dBm. The transmitting power of the client is 30mW (or 15 dBm) and its sensitivity is −82 dBm. The cables are short, with a loss of 2dB at each side.

Adding up all the gains and subtracting all the losses for the AP to client link gives:
20 dBm (TX Power Radio 1)
+ 10 dBi (Antenna Gain Radio 1)
− 2 dB (Cable Losses Radio 1)
+ 14 dBi (Antenna Gain Radio 2)
− 2 dB (Cable Losses Radio 2)
The Total Gain = 40 dB
The path loss for a 5 km link, considering only the free space loss is:
Path Loss = 40 + 20 log(5000) = 113 dB
Subtracting the path loss from the total gain
40 dB − 113 dB = −73 dB

Since −73 dB is greater than the minimum receive sensitivity of the client radio (−82 dBm), the signal level is just enough for the client radio to be able to hear the access point. There is only 9 dB of margin (82 dB–73 dB) which will likely work fine in fair weather, but may not be enough to protect against extreme weather conditions.

Next we calculate the link from the client back to the access point:
15 dBm (TX Power Radio 2)
+ 14 dBi (Antenna Gain Radio 2)
− 2 dB (Cable Losses Radio 2)
+ 10 dBi (Antenna Gain Radio 1)
− 2 dB (Cable Losses Radio 1)
The Total Gain = 35 dB
Obviously, the path loss is the same on the return trip. So our received signal

level on the access point side is:
35 dB − 113 dB = −78 dB

Since the receive sensitivity of the AP is −89 dBm, this leaves us 11 dB of fade margin (89 dB–78 dB). Overall, this link will probably work but could use a bit more gain. By using a 24 dBi dish on the client side rather than a 14 dBi sectorial antenna, you will get an additional 10 dBi of gain on both directions of the link (remember, antenna gain is reciprocal). A more expensive option would be to use higher power radios on both ends of the link, but note that adding an amplifier or higher powered card to one end generally does not help the overall quality of the link.

2. Link planing software: Many link planing software are available online free of charge. They can be used to calculate the level of received signal, these packages help the network designer to set up the level of transmitting power and the distance between mesh nodes. they include the digital terrain elevation model and the types of obstacles between nodes which give an accurate path loss calculation. One of the most user friendly online tools is the Bay Professional Packet Radio group (GBPRR) tool.
This software is available at:
(*http://my.athenet.net/~multiplx/cgi-bin/wireless.main.cgi*)

3. Measurements: The received signal strength from each node at the proposed mesh router location is measured. Since this requires measurements data for all nodes, the other two methods are easier to apply.

This method might be used to find the parameters for the propagation model using some measured data and then use this data to calculate the path loss exponent and the variance for the shadowing component.

For example, if we have the following received signal levels with the corresponding distances:

Distance from transmitter	Received power
100 m	0 dB
200 m	−20 dB
1000 m	−35 dB
3000 m	−70 dB

To calculate the minimum mean square error estimates for the path loss exponent $A(n) = \sum_{i=1}^{4} (p_i - \hat{p}_i)^2$ Where p_i is measured value and $\hat{p}_i$ is the estimated value of the signal at distance di

Using the equation

$$P_{dBm}(d) = P_{dBm}(d0) - 10\alpha \log_{10}\left(\frac{d}{d0}\right)$$

We can find that:

$$\begin{aligned} A(n) &= (0-0)^2 + (-20-(-3n))^2 + (-35-(-10n))^2 \\ &\quad + (-70-(-14.77n))^2 \\ &= 327.153n^2 - 2887.8n + 6525 \\ \frac{\partial}{\partial n}(A(n)) &= 654.306n - 2887.8 \end{aligned}$$

Setting this equal to zero gives us the value of n as n = 4.4
Substitute this in A(n) gives us the vale of $\sigma^2 = 38.09$

7.6 Network Topology

The simplest design for wireless mesh network is to deploy the nodes that form the network into a fully meshed topology where each single node can communicate directly with all other nodes. The network can also be connected to the Internet via a gateway [1] as shown in Figure 7.2.

This design is possible with a small number of nodes situated within a limited geographical area.

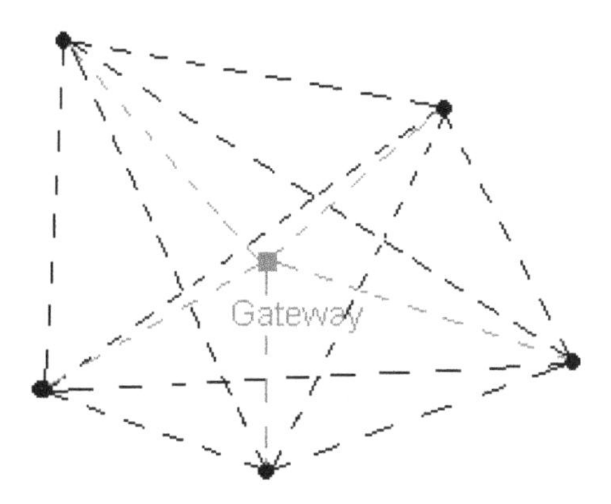

Fig. 7.2 Full mesh topology.

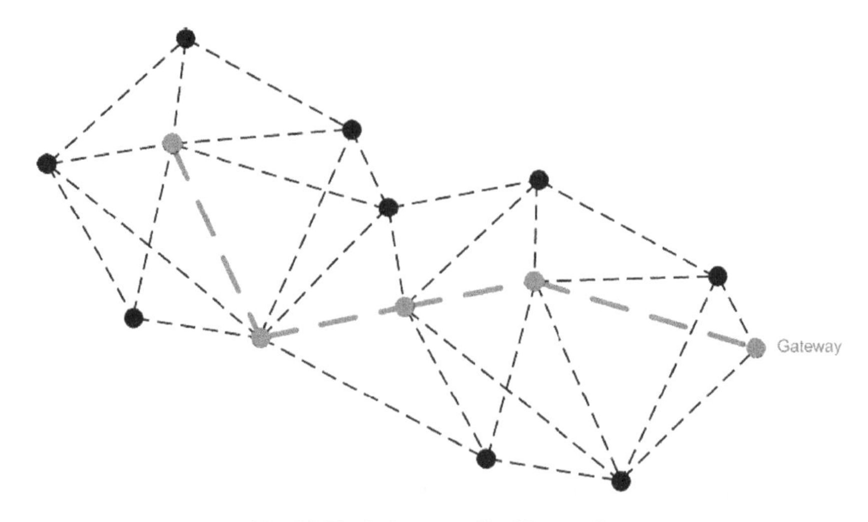

Fig. 7.3 The deployment of backbone nodes.

When the size of the network increases, the area of the network will be increased. Therefore, the nodes will not be able to communicate in peer-to-peer connections, this means many nodes will need to hop through other nodes to reach the gateway or to communicate with each other. These multi-hops will reduce the efficiency of the network and result in an unbalanced distribution in the bandwidth between nodes because the distances from the gateway to these nodes are different, so the solution of this problem is to define a group of nodes called 'backbone node' and use them for routing packets into the network. Backbone nodes must be distributed in the network such that they can be reached by a small number of hops as shown in Figure 7.3. Backbone nodes are usually separated by a longer distance than normal mesh nodes.

One could suggest the deployment of several gateways into the network such that all WMN nodes are close enough to the gateway. This, however, is a costly option compared to the backbone nodes.

When the network is not dense and the nodes are sparsely deployed into several geographical areas we can divide the network into clusters and deploy a backbone node in each cluster. In this case nodes in each cluster are connected to the backbone through one ore two hops as shown in Figure 7.4 and data can be routed efficiently through the backbone nodes.

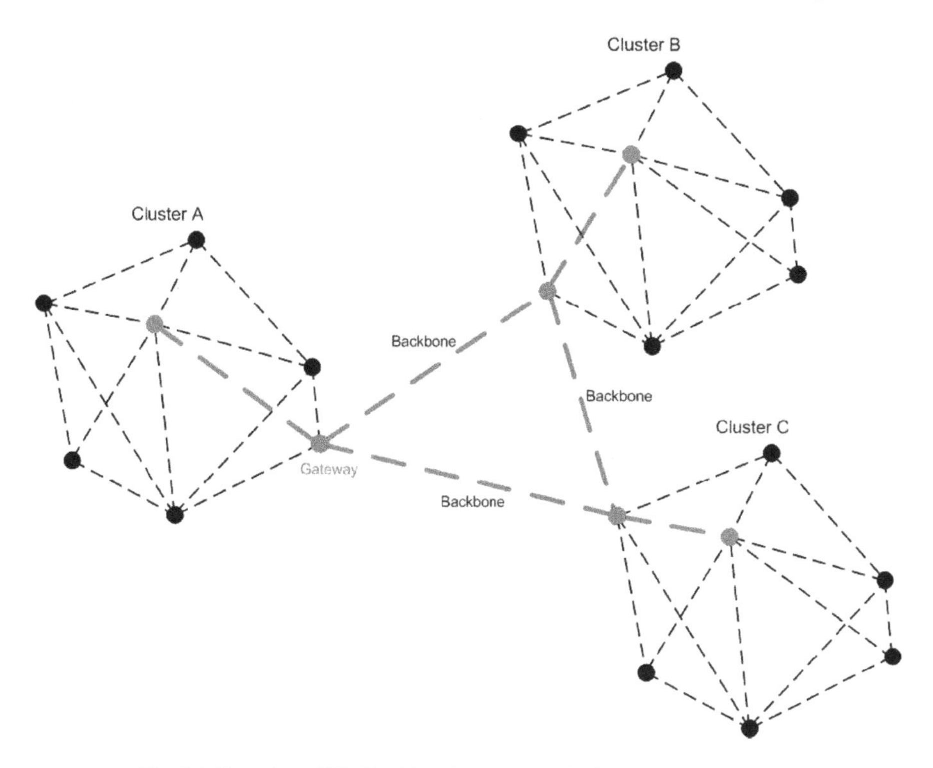

Fig. 7.4 Clustering of WMN with backbone nodes deployment at each cluster.

7.7 Network Design

7.7.1 Nodes Types

In wireless mesh networks, nodes are classified into two types: mesh nodes and client nodes. Each type of these nodes should communicate separately from the other type to gain a maximum efficiency and to extend the coverage of the network as we will explain later. The backbone nodes are the same as mesh nodes but with one difference: they have more than one interface to communicate. These interfaces enable them to communicate with each other and with other mesh nodes.

Mesh WMN nodes: WMN mesh nodes consist of static wireless routers that form the backbone of the network. These nodes form a mesh of self-configuration and self-healing links. With gateway functionality, mesh routers can be connected to the Internet [3]. Client WMN can be connected directly to

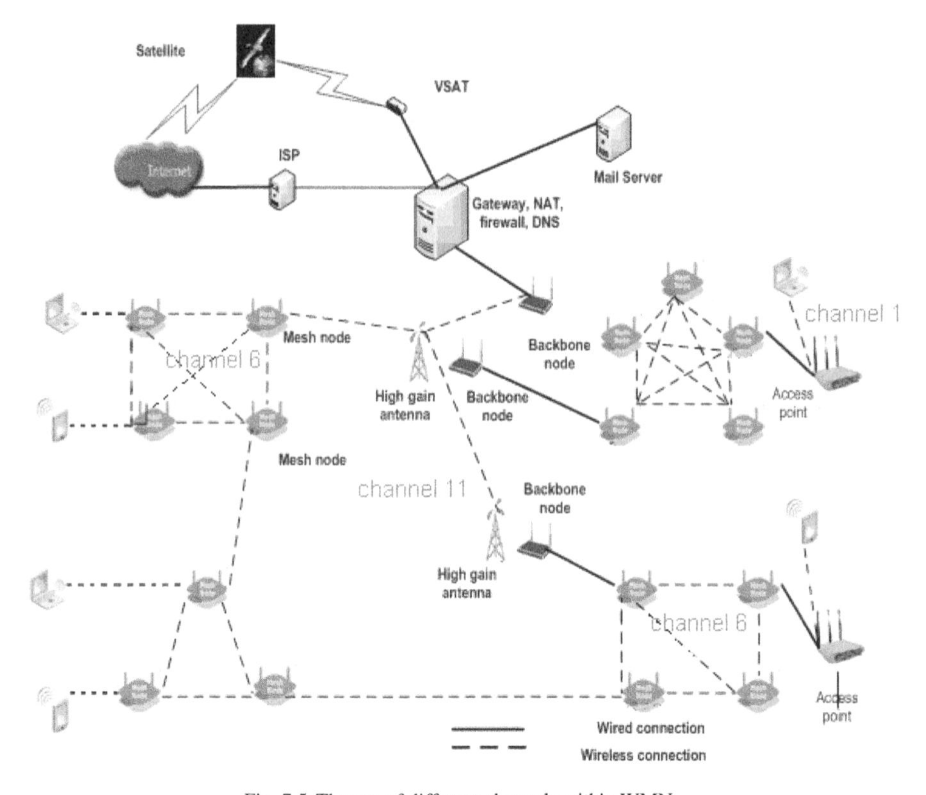

Fig. 7.5 The use of different channels within WMN.

these routers through a wired connection via Ethernet links or via a radio link in the case of multiple-interface radio. Different channels are allocated to each type of nodes, for example WMN mesh nodes use different channels than the channels used in WMN client nodes. Different wireless standards may also be used to prevent interference between client WMN nodes and mesh WMN nodes and to improve the bandwidth utilisation in the network [4]. The use of different channels and different standards makes it easer to extend the range of the WMN especially at the backbone level. Figure 7.5 shows a WMN consists of backbone/mesh nodes and client nodes with a different use of channels in each type of node.

Client WMN nodes: This type of node is the end use node. It could be a mobile or a static wireless router that creates a hot spot to connect several devices to the network, or it could be any other Wi-Fi enabling device.

A client node can communicate directly with others if they have a routing capability or through an access point as in classical WLAN.

7.7.2 Channel and Radio Allocation

To prevent interference between backbone nodes, mesh nodes, and client nodes, different radios and/or different channels should be allocated to each type of nodes. If the same radio and the same channel are used, then the efficiency of the network will be dramatically reduced because all types of nodes will be able to listen to each other therefore the concept of backbone nodes will vanish.

In general, it is recommended to design the WMN backbone nodes with two radios: one to communicate with other backbone nodes and the other to communicate with normal mesh nodes. This radio allocation will prevent interference between backbone nodes and mesh nodes within the backbone network. On the other hand client nodes and mesh nodes should operate on different channels to prevent interference within the mesh network.

If the network uses one radio between all nodes, non-overlapping channels should be used for each type of nodes. For example, in 802.11 b/c channels 1, 6 and 11 are the only non-overlapping channels. To create a node that operates in two channels, two routers may be connected back-to-back using an ethernet port.

Figure 7.5 shows how different channels are assigned to different types of nodes; it also clarifies how two routers are connected together to create one node with two channels.

Channels and radio allocation enable the creation of three possible network designs. These designs include Infrastructure/Backbone WMNs, Client WMNs, and Hybrid WMNs.

Infrastructure/Backbone WMNs: In this architecture mesh nodes form an infrastructure for the clients. They form a mesh of wireless routers using various types of technology such as 802.11/b/c, 802.16, or any other standards. Clients can be connected via a wired connection to the mesh or through an access point which has been connected to the mesh router. Figure 7.6 shows this architecture.

Client WMNs: In this architecture client nodes have a peer-to-peer connection; they form the actual network, and perform the routing functionality.

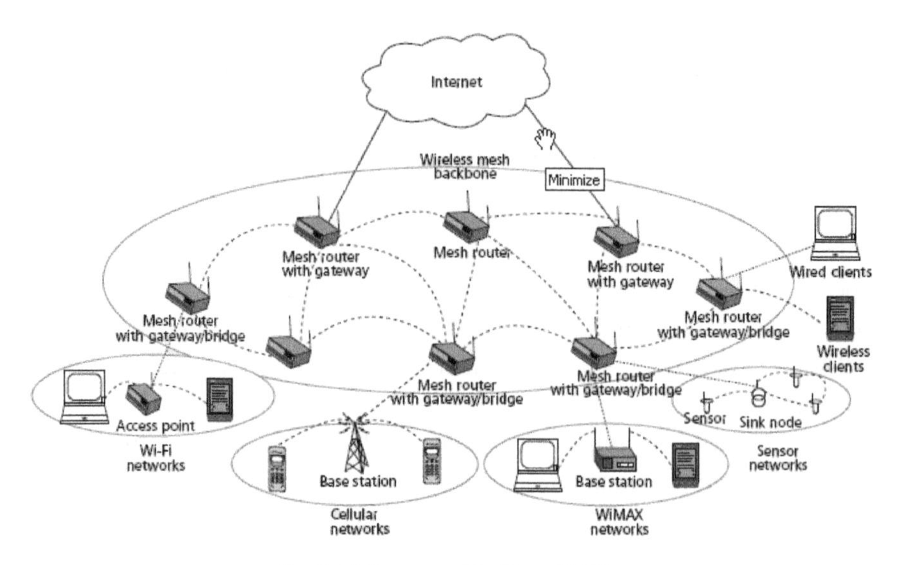

Fig. 7.6 Infrastructure/Backbone WMNs.

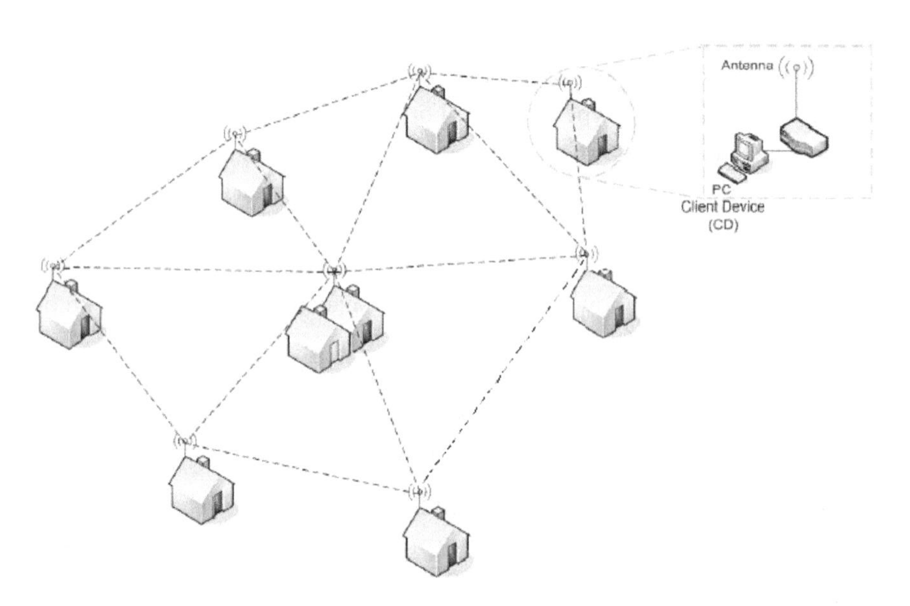

Fig. 7.7 Client WMNs architecture.

Nodes in this architecture use one radio because there is no other type of nodes in the networks (i.e., the mesh routers are the clients). Figure 7.7 illustrates a client WMN network.

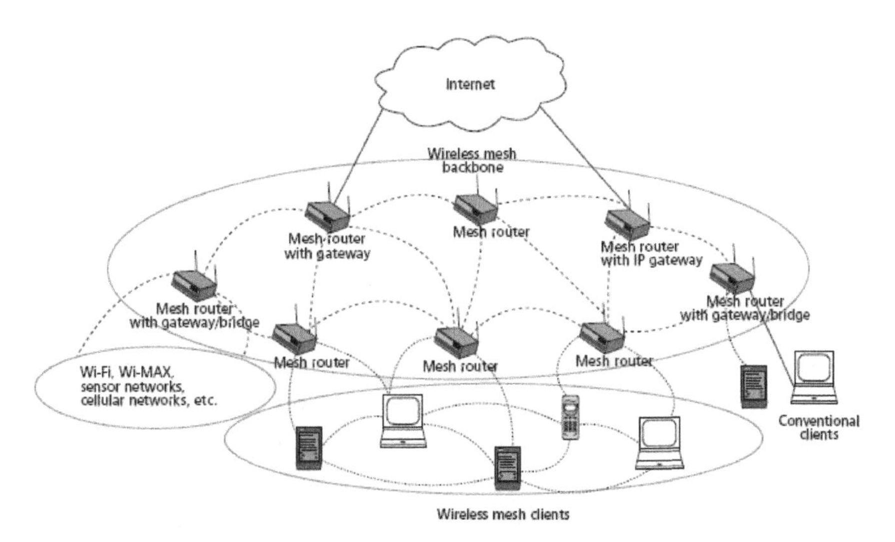

Fig. 7.8 Hybrid WMNs architecture.

Hybrid WMNs: This architecture is a combination of both infrastructure/backbone WMNs and client WMNs architecture. Client nodes can communicate directly through a peer-to-peer connection and through mesh routers too. This hybrid architecture requires different radios/channels to be used — at least one in mesh routers network and one in the client network in order to provide connectivity to different types of networks. Figure 7.8 shows a hybrid WMN network architecture formed by infrastructure/backbone WMNs and client WMNs. It also shows different radios used in each level of the network.

7.8 Routing in WMN and Cross Layers

Wireless mesh networks and ad hoc networks are very similar communication structures. So, most routing protocols used in ad hoc networks could be used in wireless mesh networks, to find the best path between the source and the destination and to disseminate routes information into the network. Mobile ad hoc routing protocols can be classified in two types: reactive and proactive protocols.

7.8.1 Reactive Protocol

This type of protocol can be known as on demand protocols, because information about route can be obtained when needed. This type of protocol suite wireless networks because the nature of wireless networks is dynamic, where information in routing tables could be changed before it has been used. Reactive protocols save networks bandwidth by reducing the number of unnecessarily control messages and routing updates. The most popular reactive protocols are ad hoc on demand distance vector protocol AODV and Dynamic source routing protocol (DSR).

7.8.2 Proactive Protocols

This type of protocols keep the routing table for each node in the network updated, by exchanging routes data periodically. This exchange results in the availability of routing information at all times. A proactive routing protocol represents a mechanism which reduces network latency inasmuch as there is no need to determine a route when data needs to be transmitted [5]. Sending route updates on periodic basis reduces the delay in packet delivery, because all routes to destination will be known. Performance in term of data delivery ratio and end-to-end delay is best in proactive protocols [6].

Exchanging periodic routes update, however, will produce a high overhead which will reduce the bandwidth utilisations, especially when the updates are not used. Examples of proactive protocols are Optimised Link State Routing Protocol for Ad Hoc Networks OLSR, Topology Dissemination based on Reverse-Path Forward (TPRPF), and Source-Tree Routing in Wireless Networks STAR.

7.8.3 Ad-Hoc on Demand Distance Vector Protocol (AODV)

Ad hoc on demand distance vector protocol (AODV) is a reactive protocol that supports both unicasting and multicasting traffic [7]. It does not update routing table entries on periodic basis but discovers routes and uses them when they needed. AODV uses a destination sequence number which is used to compare the freshness of route during route discovery. AODV uses route discovery/route reply. When a source node has no information about the route to the destination and it wants to send a message to the destination node, it sends a route request to neighbour nodes, if any of the intermediate nodes has a sufficiently fresh route, it sends it back to the sender. Otherwise, the intermediate node will pass the route request packet to other neighbours and then update its routing table. Freshness of the route is measured by comparing the

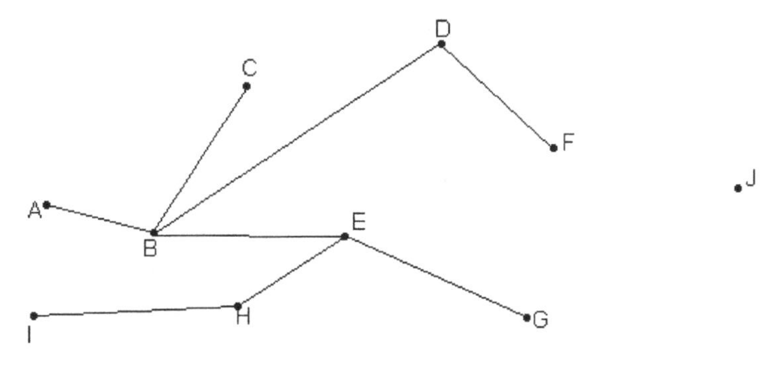

Fig. 7.9 AODV routing.

destination sequence number in the packet and the intermediate node routing table's sequence number. During communication, if a fresher route has been sent to the sender, the sender will use that route and discard the current one. AODV uses a timer for each routing table entry to maintain routes. If the timer reaches time out, the route is discarded from the routing table.

Figure 7.9 shows that node A wants to send a packet to node G. It first sends a route request packet to its neighbours (i.e. node B), Node B will send a route reply packet to node A if it has a fresh enough route to the destination node. If node B has not got a fresh enough route or has not got any route, it passes the request to neighbour nodes which are nodes C, D and E. Nodes C and D will either pass the packet to their neighbours or discard it if time out is reached. The packet will reach node G, so node G will reply with route reply. During this process, if any of the intermediate nodes knows the route, it will reply with a route reply packet.

7.8.4 Dynamic Source Routing Protocol (DSR)

Dynamic source routing protocol (DSR) is a reactive protocol. It contains two-phase route discovery and route maintenance. Each packet in DSR contains the full path from the source to the destination [8].

In route discovery phase, when the source node wants to send a packet and has not got any route available in the cache, it sends a route request to neighbour nodes only, and starts a timer, If the timer reaches time out before getting the route reply, the source sends a route request to the entire network [8]. The route request packet contains the following information: some identifiers, destination identity, source identity, and a unique ID which is used to check whether a node has seen this packet or not because, when the source node broadcasts the request, it is more likely that a node could receive the packet more than once.

When a node receives the request, it first checks if it has received the packet before, as a result the node either discards the packet if it has received it before, or adds its identity to the packet and retransmits it again, unless the node itself is the destination. In this case the node will reply by a route reply packet.

Nodes always check their cache for any available route to the destination and send a reply packet to the initial node before broadcasting the route to the network. This phase is continued until the route request packet reaches the destination node.

Figure 7.10 shows that node A wants to send a packet to node G. Node A sends a request to its neighbour nodes first, which is node B in this example. If B has a route, it will reply with route reply, node A will wait for node B's reply and start a timer. If the timer reaches time out, node A will broadcast the request to the network. When node E receives the request, it replies by route reply if it knows the route or adds its identity (node E identity) to the route request packet and broadcasts it again.

Nodes always verify the reception of the packet by acknowledgment. If any intermediate node can not verify the reception of the packet from next node, it will send back an error message to the previous node that received the packet. From then the intermediate node removes the next node that could not verify the transmission from the routing table.

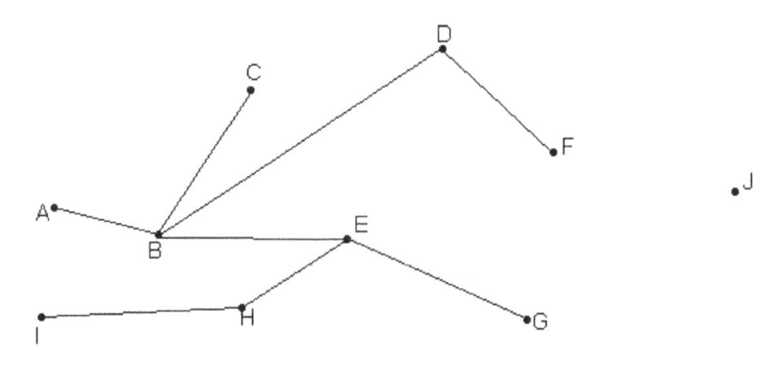

Fig. 7.10 DSR routing example.

In the above example, if the packet reaches node D and D broadcasts the packet but it did not receive acknowledgment from node F, Node D will send an error message to the node that sent the packet to node D which will be node B and delete node F from the table.

7.8.5 Topology Dissemination Based on Reverse-Path Forwarding (TBRPF)

Topology Dissemination Based on Reverse-Path Forwarding (TBRPF) is a proactive, link-state routing protocol designed for mobile ad hoc networks, which provides hop-by-hop routing along shortest paths to each destination [9]. Each node running TBRPF computes a source tree (providing paths to all reachable nodes) based on partial topology information stored in its topology table, using a modification of Dijkstra's algorithm. To minimise overheads, each node reports only part of its source tree to neighbours. TBRPF uses a combination of periodic and differential updates to keep all neighbours informed of the reported part of its source tree. Each node also has the option to report additional topology information (up to the full topology), to provide improved robustness in highly mobile networks. TBRPF performs neighbours discovery using 'differential' HELLO messages which report only the changes in the status of neighbours [9]. TRRPF consists of two modules: the first module is a neighbour discovery module, which is responsible for the discovery of neighbours, the second is a routing module, which is responsible for the discovery of the topology and route's computation.

Neighbour discovery model

The neighbour discovery model enables a node to detect neighbour nodes such that a direct bi-directional link between them exists. TBRPF sends hello messages which report the changes in neighbour nodes only; this small message makes it possible to transmit a hello message in more frequently. If a node has more than one interface, then a separate neighbour discovery messages run for each interface with a separate neighbour's table for each interface.

The status of the link can take one of the three possibilities 1-Way, 2-Way or lost, the hello message then can update these tables.

Routing module

Each node maintains a source tree that indicates the shortest paths to all reachable nodes in the network [9] using the shortest path algorithm. Each node periodically sends a portion of the source tree to its neighbours. These messages are transmitted less frequently than the hello messages; nodes combine these periodic topology updates and the hello messages information to compute the topology and the path to each node in the network.

7.8.6 Performance Metric Used in WMN Routing

Wireless mesh networks use radio links, so, the decision of the best path should not depend only on minimum hop count but it should also depend on the quality of the links. Different metrics should be used to determine the best available routes for packets from the source to the destination. This section briefly describes the most important metric which used to determine the routing path in wireless mesh networks. These metrics are used to find how good the path is and then determine the path between the source and destination.

7.8.6.1 Hop Count

Hop count is a very simple routing metric. It uses the shortest path to find the route regardless of the quality of the link. It does not require any calculation other than finding the shortest path given that the topology is known. This metric does not give a good performance when the nodes are static because it tends to include a wireless link between distance nodes [10]. This metric will consider a one-hop path over a slow link as a better option over a two-hop with a reliable fast link [11]. Hop count can be improved if the nodes in the network can estimate the topology of the network. This estimation may reduce the number of route discovery packets and reduce the network latency [12, 13].

7.8.6.2 Per-hop Round Trip Time

This metric measures the average delay by calculating the round trip time (RTT) seen by unicast probes between neighbour nodes [14]. A node sends a periodic probe packet which has a time stamps to its neighbours. Each neighbour receiving the packet responds with an acknowledgment, so the sending node will calculate the RTT and the average value of a moving window containing the recent received values of RTT.

The value of RTT determines which path the node will choose and it will also tell the node about the status of the link — whether it is good if the value of RTT is small or busy by having a large value, or unstable if no acknowledgment is received.

7.8.6.3 Pre-hop Packet Pair Delay

In a RTT metric measurement, part of the value of the RTT is the delay in the sending node. This means the metric is not giving an accurate status of the links. To overcome this shortage of RTT metric a packet pair metric is used.

In a packet pair delay a node sends two probe packets then the receiving neighbour calculates the delay between receiving the first and second packets and reports it back to the sending node.

7.8.6.4 Expected Transmission Count

Expected transmission count (EXT) estimates the number of transmissions to transfer a packet successfully over a wireless link [10], EXT is computed by sending a probe each second. This probe contains the number of probes received by the sending node from its neighbours in the last ten seconds. The probe will be used to calculate the expected transmission number according to the following equation:

$$ETX = \frac{1}{(1 - p_f)(1 - p_r)} \tag{7.8}$$

Where p_f is the frame loss ratio between the sending node to the receiving node, p_r is the frame loss ration between the receiving node to the sending one. By finding this value, the sending node will determine the number of retransmissions to transmit the packet successfully. As this number diminishes, then the quality of the link is better and the path for send data will be determined.

References

[1] D. Johnson, K. Matthee, D. Sokoya, L. Mboweni, A. Makan, and H. Kotze, "Building a Rural Wireless Mesh Network," Wireless Africa Team of the Meraka Institute, 2007.

[2] T. S. Rappaport, *Wireless Communications* vol. 1, second ed: Prentice-Hall, 2002.

[3] I. F. Akyildiz and W. Xudong, "A survey on wireless mesh networks," *Communications Magazine, IEEE*, vol. 43, pp. S23–S30, 2005.
[4] A. Raniwala and C. Tzi-cker, "Architecture and algorithms for an IEEE 802.11-based multi-channel wireless mesh network," presented at INFOCOM 2005. 24th Annual Joint Conference of the IEEE Computer and Communications Societies. Proceedings IEEE, 2005.
[5] G. Held, *Wireless Mesh Networks*. New York: Auerbach Publication Taylor & Francis Group, 2005.
[6] R. S. Wolff and G. Dawra, "A Terrain Based Routing Protocol for Sparse Ad-Hoc Intermittent Network (TRAIN)," presented at Wireless Networks and Emerging Technologies (WNET 2006) 2006.
[7] C. Perkins and E. Belding-Royer, "Ad hoc On-Demand Distance Vector (AODV) Routing," Network Working Group.ietf, 2003.
[8] D. B. Johnson and D. A. Maltz, "Dynamic Source Routing in Ad Hoc Wireless Networks," in *Mobile Computing*, T. Imielinski and H. F. Korth, Eds.: Springer US, 1996, pp. 153–181.
[9] R. Ogier, F. Templin, and M. Lewis, "Topology Dissemination Based on Reverse-Path Forwarding (TBRPF)," Network Working Group.ietf, 2004.
[10] D. S. J. De Couto, D. Aguayo, J. Bicket, and R. Morris, "A high-throughput path metric for multi-hop wireless routing," in *Proceedings of the 9th annual international conference on Mobile computing and networking*. San Diego, CA, USA: ACM, 2003.
[11] D. Richard, P. Jitendra, and Z. Brian, "Comparison of routing metrics for static multi-hop wireless networks," in *Proceedings of the 2004 conference on Applications, technologies, architectures, and protocols for computer communications*. Portland, Oregon, USA: ACM, 2004.
[12] M. Al-hattab and J. I. Agbinya, "Use of Street Maps to Aid Node Localization in Mobile Wireless Networks," presented at Parallel and Distributed Processing with Applications, 2008. ISPA '08. International Symposium on, 2008.
[13] M. Al-Hattab and J. I. Agbinnya, "Topology prediction and convergence for networks on mobile vehicles," presented at Computer and Communication Engineering, 2008. ICCCE 2008. International Conference on, 2008.
[14] A. Adya, P. Bahl, J. Padhye, A. Wolman, and L. Zhou, "A Multi-Radio Unification Protocol for IEEE 802.11Wireless Networks," Microsoft Research 2004.

8

Spectrum Sensing and Sharing for Cognitive Radio

Rania A. Mokhtar* and Rashid A. Saeed†

*Universiti Putra Malaysia (UPM), Serdang, Malaysia
†Telekom Malaysia (TM) Research and Development Innovation Centre, Malaysia

In recent years, cognitive radios have been introduced as a new paradigm for enabling much higher spectrum utilization, providing more personal and reliable radio services, reducing harmful interference, and facilitating the interworking or convergence of different wireless networks. In this chapter we study the intelligent open spectrum sharing using cognitive radios. This includes the discussion of the existing spectrum regulations and its challenges for spectrum scarcity. The advanced spectrum sharing and sensing is also discussed. In this context the cooperative spectrum sensing has been proposed to overcome the problem associated with the local sensing e.g. the hidden node problem due to noise uncertainty, fading, and shadowing. In this chapter we also present a hard decision auto-correction reporting scheme that directly correct the error in the reported bit, and further minimizes the average number of the reporting bits by allowing only the user with a detection information -binary decision 1- to report its result. The sensing performance is investigated and the numerical result shows great decrease in reporting bit without affecting the sensing performance.

8.1 Introduction

Cognitive Radio (CR) Definition

The definitions of CR are still being developed by industry and academia. At one extreme, a Full CR is assumed to be a fully re-configurable radio device that can "cognitively" adapt itself to both users' needs and its local environment. For example, a mobile handset may use cognitive reasoning to automatically reconfigure itself from a cellular radio to a PMR radio, or it may automatically power down when in a sensitive environment (such as a hospital, cinema or airport). This full CR is often referred to as a Mitola radio (named after the MITRE scientist Joseph Mitola) [2]. It is unlikely to be achieved in the next 20 years because it implies the availability of full software defined radio technologies coupled with cognitive capabilities. If flexibility of hardware and intelligence to control or configure the hardware, are two axes of a matrix (see Figure 6.1, then a full cognitive radio (Mitola radio) would be at the top right.

As we move back from this full Mitola radio implementation, we expect to see achievable forms of intelligent reconfigurable CRs within the next 5 years. These will be radios that can intelligently adapt at the physical layer, using software defined radio techniques coupled with basic intelligence. These intelligent radios may deliver significant benefits without the need to achieve a complete Mitola implementation. Many of today's radio systems already exhibit some characteristics of a cognitive radio (e.g. WLANs, military follower jammers); such as interference avoidance or adaptive modulation scheme selection to facilitate co-existence. Although Cognitive Radio technology, in its full form (i.e. a Mitola radio), holds much promise in maximising spectral efficiency,

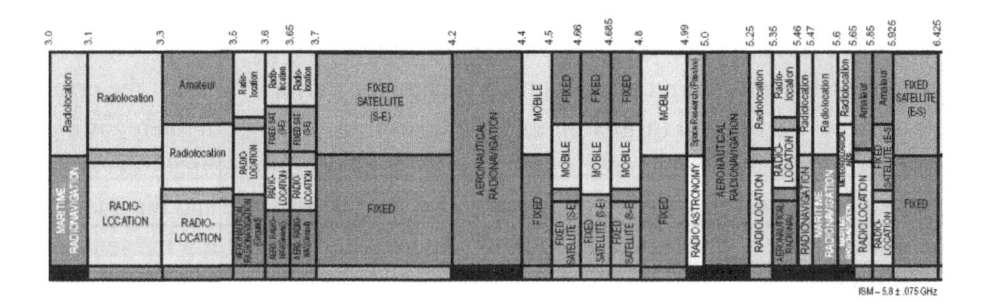

Fig. 8.1 The FCC frequency allocation from 3–6 GHz.

the technology implies a radical departure from existing methods of spectrum regulation. Many complications and challenges exist, however, from both a technical and a regulatory standpoint and this need to be understood before regulatory actions are considered for CR.

The Federal Communications Commission (FCC) has identified in (FCC, 2005a) the following (less revolutionary) features that cognitive radios can incorporate to enable a more efficient and flexible usage of spectrum:

- Frequency Agility — The radio is able to change its operating frequency to optimize its use in adapting to the environment.
- Dynamic Frequency Selection (DFS) — The radio senses signals from nearby transmitters to choose an optimal operation environment.
- Adaptive Modulation — The transmission characteristics and waveforms can be reconfigured to exploit all opportunities for the usage of spectrum
- Transmit Power Control (TPC) — The transmission power is adapted to full power limits when necessary on the one hand and to lower level on the other hand to allow greater sharing of spectrum.
- Location Awareness — The radio is able to determine its location and the location of other devices operating in the same spectrum to optimize transmission parameters for increasing spectrum re-use.
- Negotiated Use — The cognitive radio may have algorithms enabling the sharing of spectrum in terms of prearranged agreements between a licensee and a third party or on an ad-hoc/real-time basis.

One other comment is in order. A broadly defined cognitive radio technology accommodates a scale of differing degrees of cognition. At one end of the scale, the user may simply pick a spectrum hole and build its cognitive cycle around that hole. At the other end of the scale, the user may employ multiple implementation technologies to build its cognitive cycle around a wideband spectrum hole or set of narrowband spectrum holes to provide the best expected performance in terms of spectrum management and transmit-power control, and do so in the most highly secure manner possible.

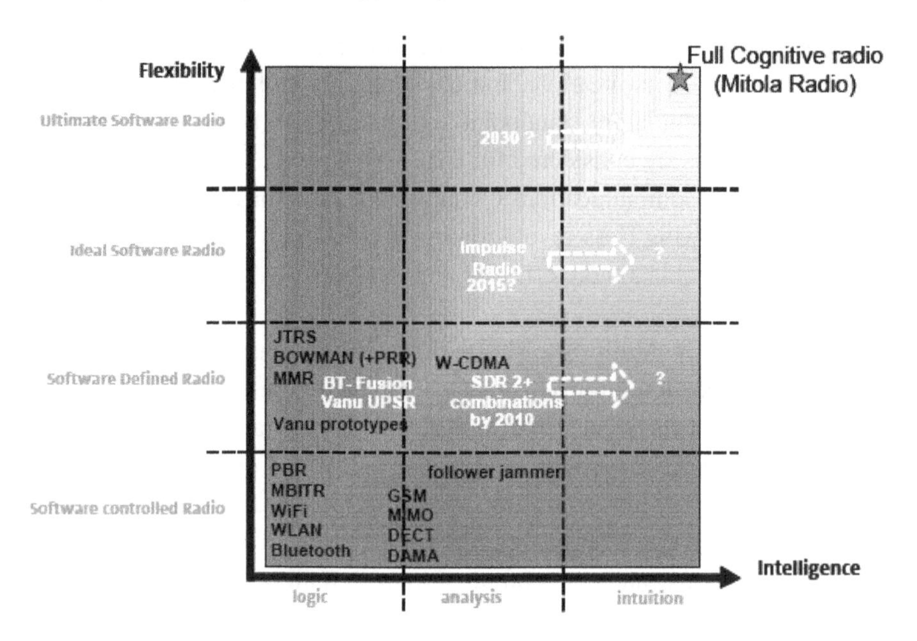

Fig. 8.2 Y-axis of RF flexibility and x-axis of Intelligent Signal Processing (ISP). A Mitola radio would be placed at the top right section. Current and planned radio systems would sit towards the left and below.

Scarcity and Waste of Spectrum

It is commonly believed that there is a crisis of spectrum availability at frequencies that can be economically used for wireless communications. This misconception has arisen from the intense competition for use of spectra at frequencies below 3 GHz. At higher frequencies, as seen by the snapshot of spectrum usage in an urban area shown in Figure 6.3, there is actually very little usage at the time, place and direction that this measurement was taken. Analysis of the snapshot in Figure 6.3 reveals that the actual utilization in the 3–4 GHz frequency band is 0.5% and drops to 0.3% in the 4–5 GHz band. This seems totally in contradiction to the concern of spectrum shortage, since in fact we have spectrum abundance, and the spectrum shortage is in part an artificial result of the regulatory and licensing process.

What is remarkable is that this low level of usage seems inconsistent with the FCC frequency chart from 3–6 GHz shown in Figure 6.1, which indicates that there are multiple allocations over all of the frequency bands. It is this discrepancy between FCC allocations and actual usage, which indicates that

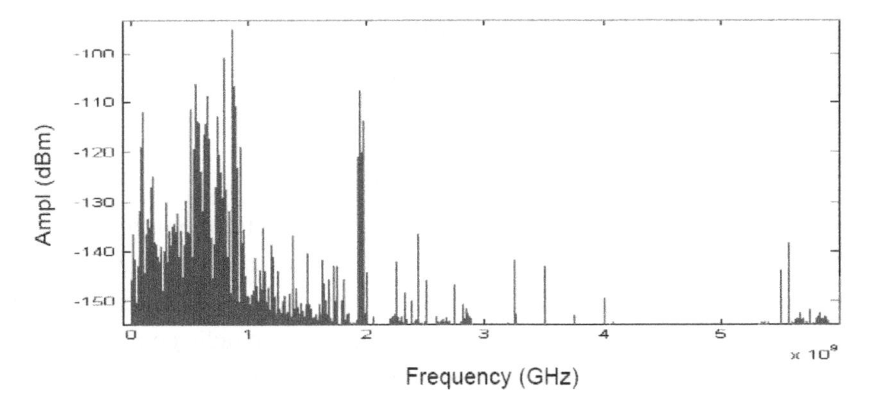

Fig. 8.3 A snapshot of the spectrum utilization up to 6 GHz in an urban area: taken at mid-day with 20 kHz resolution taken over a time span of 50 microseconds with a 30 degree directional antenna.

a new approach to spectrum licensing is needed. Part of the solution can be found by observing in Figure 6.3, that there is considerable usage in the upper 5 GHz band in this location. This corresponds to the unlicensed UNII spectra, which has only minimal constraints from the regulatory standpoint. What is clearly needed is an extension of the unlicensed usage to other spectral bands, while accommodating the present users who have legacy rights and also to insure that future requirements can be met.

An approach, which can meet these goals, is to develop a radio that is able to sense the spectral environment over a wide available band and use the spectrum only if communication does not interfere with licensed user. These un-licenses low priority Secondary Users (SU) would thus be using Cognitive Radio (CR) techniques, to ensure non-interfering coexistence with higher priority users and thus reduce concerns of a general allocation to unlicensed use [1].

The sensing should involve more than just determining the power in a frequency band as presented in Figure 6.3, since a wireless channel actually is built on multiple signal dimensions that include time, frequency, physical space, and user networks. The optimal CR operation will allow sensing of the environment and transmission optimized across all of the dimensions and thus allows a truly revolutionary increase in the ability to support new wireless applications. In a sense our cognitive radio discovers unused capacity and creates out of this unused capacity a "virtual unlicensed spectrum" to be used in a way not constraining the licensed owners.

8.2 Radio Spectrum Regulation

A brief overview about regulation bodies and global institutions related to the regulation of radio spectrum is given in this section.

Origin and Scope of Radio Spectrum Regulation

The loss of the Titanic in April 1912 was a milestone for the introduction of radio spectrum regulation: Many people were rescued due to the reception of the radioed SOS but many lives were lost due the absence of the radio operator of the nearest ship. This event indicated the relevance of radio communication for public safety and the dangers from unreliable communication. Regulation of radio spectrum has its origin in the economic regulation of railroads: In the US, the Communications Act from 1934 and its predecessors where principally concerned about control of monopoly power when the market is served by a single provider. Since the beginning of the 19th century, communication services of telephony, radio and television have been regulated according to this model to provide service to the public on a nondiscriminatory basis to fair and reasonable prices and conditions.

From the technical point of view, radio spectrum is a public resource that can be used without many limitations. The usage of spectrum implies always interference to neighboring radios sharing the same spectrum. Therefore, radio spectrum regulation is required to allow a reliable and efficient spectrum usage. Regulators approach spectrum regulation in determining how particular bands of spectrum can be used, make rights available to licensees or unlicensed users and define rules constraining the accesses to this spectrum. Ideally, the regulators' decision making targets at the increase of public welfare and it reflects the public interest.

The report of the FCC's Spectrum Policy Task Force defines spectrum regulatory mechanisms in a similar way. There, the assignment of spectrum rights is differentiated into an "exclusive use" model, a "command-and-control" model and a "commons" or "open access" model. The "command-and-control" is currently the preferred regulation model and refers to the "licensed spectrum for shared usage" and "unlicensed spectrum" models. Radio spectrum regulation has to take influence on the development of access protocols and standards to balance the following goals [6]:

- An adequate QoS should be possible to all radios depending on the supported applications

- No radio should be blocked from spectrum access and transmission for extended durations
- Spectrum management policies and standards should not slow down innovations in the economically significant, but rapidly changing, communication sector
- The limitedly available spectrum should be used efficiently, including special re-use of spectrum and solving the "tragedy of commons".
- Spectrum can be used in a dynamically adaptive way, taking the local communication environment like spectrum usage policies into account The costs of devices should be not increased significantly through techniques prescribed by regulation

Licensed and Unlicensed Spectrum

Licensed Spectrum

Large parts of the radio spectrum are allocated to licensed radio services in a way that is often referred to as "command-and-control". Licensing spectrum covers the exclusive access to spectrum and the spectrum sharing of the licensed spectrum through strictly regulated devices. In case of exclusive spectrum usage, a license holder pays a fee to have this privilege. Exclusive access rights have the advantage of preventing potential interference which implies dangers to a reliable and thus chargeable communication. In case of spectrum scarcity, licensed spectrum is highly valuable leading to economic profits, as consumers need to pay for using it. Having an immense commercial impact, spectrum licenses can be bounded to requirements which are to be fulfilled like a concrete transmission technology allowed in this spectrum or a certain percentage of population to be reached by the network when licensing the spectrum. The UMTS auctions in Europe are an example for this. Licenses in recent times are time bound. Today's most often used licensing model in Europe is to license spectrum for shared usage restricted to a specific technology. Emission parameters like the transmission power and interference to neighboring frequencies like out of band emissions are restricted. Regulation takes care for protection against interference and for a limited support of coexistence capabilities like Dynamic Channel Selection (DCS) in DECT that are mandatory and part of the standard. Licensing spectrum takes a lot of time and is very difficult. Therefore, licensing is also very expensive. The licensing process constrains innovation as it forms a barrier which is difficult to overcome when introducing new technologies. The inflexibility of exclusive usage rights from licensing spectrum leads to inefficient spectrum utilization, as the license prohibits the usage of the spectrum if it is under-utilized or even unused by the license holder.

Unlicensed Spectrum

The access to unlicensed spectrum is open but its utilization is strictly regulated. An unlimited number of users are sharing the same unlicensed spectrum. Spectrum usage is allowed to all devices that satisfy certain technical rules or standards in order to mitigate potential interference. Examples for these technical rules are the limitation of transmission power or advanced coexistence capabilities. The usage rights at unlicensed spectrum are flexible and no concrete methods to access spectrum are specified.

Part 15 Regulation in the United States

The Part 15 rules of the FCC (2005b) in the "Code of Federal Regulations, Title 47 Telecommunication" describe the regulations under which a transmitter may be operated without requiring an individual license. It also contains the technical specifications and administrative requirements of Part 15 devices. ISM low-power devices like garage openers for instance are allowed to transmit at 1 Watt when using spread spectrum technologies. Three basic principles describe in general the rules of Part 15 regulation: (i) "Listen before talk", (ii) "when talking, make frequent pauses and listen again" and (iii) "don't talk too loud". When detecting a busy channel, either another unused channel is chosen or the radio waits until the channel is idle again. These simple etiquettes do neither require interoperability nor information exchange between spectrum sharing devices. IEEE 802.11 is designed for operation in frequency bands subject to Part 15 regulation and realizes spectrum access according to these three basic principles. An analysis of the QoS capabilities of 802.11 indicates that with the current Part 15 regulation a QoS support is impossible in the case of coexistence and additional coordination is required. The accidental background noise emitted from consumer electronics, like personal computers operating with an internal clock at 2–3 GHz, are also restricted in the Part 15 regulation.

Tragedy of Spectrum Regulation

The success of unlicensed spectrum draws to a close, as the severe QoS constrains to spectrum access imposed by the upcoming multimedia applications cannot be fulfilled with today's means for coexistence. In case of short-distance wireless communication, spectrum demand is extremely localized and often sporadic. In such a scenario, the competition for shared spectrum is limited. Therefore, the regulatory instrument of restricting transmission, e.g., limiting the maximum emission power, is successful. In all other deployment scenarios, as for instance WLANs, unlicensed spectrum usage is a victim of its own success: Too many parties and different technologies are using the same unlicensed spectrum so that it is getting overused and thus less usable for all. In economics the phenomenon is referred to as the "tragedy of commons". Hazlett [11] additionally introduces the "tragedy of the anticommons": Contrary to the over-use of spectrum due to missing regulation of spectrum access, the "tragedy of the anticommons" refers to inefficient spectrum utilization because of too restrictive regulation. The "tragedy of commons" and the associated inefficient over-use of spectrum results to an under-investment into technology and questions thus the "open access" licensing. Therefore, to anticipate the "tragedy of commons", regulators impose restrictions like transmission power. As consequence, many alternative systems are not allowed to operate in such a spectrum which leads again to inefficient under-utilization of spectrum. In [12] it is concluded, that limiting spectrum sharing through spectrum regulation is the only way out of this tragedy.

Open Spectrum

Open spectrum allows anyone to access any range of spectrum without any permission under consideration of a minimum set of rules from technical standards or etiquettes that are required for sharing spectrum [13]. Open spectrum targets at the complete liberalization of radio communication in overcoming regulatory roadblocks. The core concept of open spectrum is that technologies and standards are able to dynamically manage the spectrum access and sharing, replacing the static spectrum assignments resulting from bureaucratic licensing [14]. New, upcoming radio transmission technologies suggest that radio spectrum can be treated as open accessible common property rather than a collection of stringed together access rights [15]. Nevertheless, squeezing much more wireless communication out of a given bandwidth solves not the tragedy of commons [16]. But rules and etiquettes on the one hand and technologies allowing more than one user simultaneously using the same frequency on the other hand enable open spectrum. As a consequence, spectrum is similar to a public highway: Traffic rules facilitate a continuous traffic flow and help to avoid collisions. Further on may the overall throughput be increased through changing the means of transportation from single used cars to car-pooling or public transportation and congestion can be mitigated. Three primary technologies which have seen great development progress in the recent years can be identified to realize open spectrum:

- Low-power, ultra-wide band underlay spectrum usage.
- Cognitive, frequency agile radios based on SDR
- Cooperation-based, self-configuring meshed networks

As open spectrum considerably impacts the way spectrum is regulated, regulation authorities have to face the new challenges. The promising advantages of open spectrum impose a fundamental rethinking of regulating spectrum [17]:

- Overcoming the scarcity of spectrum: A dynamic spectrum sharing through cooperative techniques improving efficiency makes regulatory limitation of spectrum access obsolete.
- The competition through innovation while sharing spectrum as a common resource is in many cases superior compared to auctioning licensed spectrum at the market.
- Investment costs for exploiting spectrum are essentially reduced as no tremendous fees for temporal licenses are required resulting into essentially lower service costs.
- Wireless broadband technologies are a solution for the last-mile bottleneck. The special challenges of the last-mile access can be solved in using long-range communications, wideband underlay or meshed architectures.
- Wide-area 3rd Generation (3G) systems can be complemented by more efficient short-range and meshed communication technologies of unlicensed operation.

8.3 Spectrum Sharing and Flexible Spectrum Access

For the rest of this tutorial it is differed between primary (incumbent) and secondary users of spectrum, where as secondary users defer to primary users in utilizing spectrum. Regardless of the regulatory model, flexibility and efficiency need to be reflected in spectrum access. Spectrum sharing plays thereby an important role to increase spectrum utilization, especially in the context of open spectrum. Techniques that sense and adjust to the radio environment are essentially required as for instance in unlicensed bands and to enable secondary access to spectrum.

Underlay and Overlay Spectrum Sharing

The open access to most of the radio spectrum, even if the spectrum is licensed for a dedicated technology, is permitted by radio regulation authorities only for radio systems with minimal transmission powers in a so-called underlay sharing approach. Thereby techniques to spread the emitted signal over a large band of spectrum are used so that the undesired signal power seen by the incumbent licensed radio devices is below a designated threshold. Spread spectrum, Multi-Band Orthogonal Frequency Division Multiplex (OFDM) or Ultra-Wide Band (UWB) as introduced below are examples for such techniques. The transmission power is strictly limited in underlay spectrum sharing to reduce the possibility for a potential interference. The Spectrum Policy Task Group of the FCC suggested in [27] the Interference Temperature Concept for underlay spectrum sharing to allow low power transmissions in licensed (used) bands. The FCC proposes there to allow secondary usage of shared spectrum if the interference caused by a device is below a sufficient threshold. The FCC identifies for this a well defined space between the original noise floor and the licensed signal of the incumbent radios. This space is branded as "new opportunities for spectrum use" [28] and it is illustrated in Figure 6.4. The space refers to the power level of the signals at the receiver in a specific band at a certain geographical location.

Only a small fraction of the radio spectrum is available as open frequency band for unlicensed operation. Nevertheless, these bands have stimulated an immense economic success of wireless technologies like the popular WLAN IEEE 802.11. On the other hand, the actual availability of new spectrum is a seemingly intractable problem. Cognitive radios use flexible spectrum access techniques for identifying under-utilized spectrum and to avoid harmful

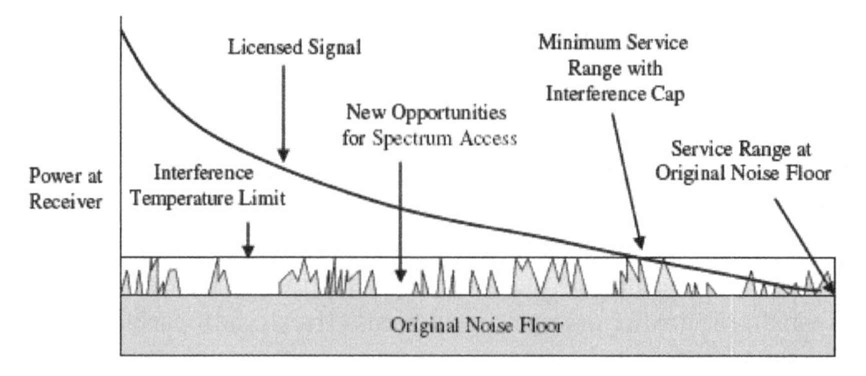

Fig. 8.4 Underlay spectrum sharing corresponding to the Interference Temperature Concept of the FCC [26].

interference to other radios using the same spectrum. Such an opportunistic spectrum access to under-utilized spectrum, whether or not the frequency is assigned to licensed primary services, is referred as overlay spectrum sharing. Overlay sharing requires new protocols and algorithms for spectrum sharing. Additionally, spectrum regulation is impacted, especially in case of vertical spectrum sharing as introduced below: The operation of licensed radios systems may not interfere when identifying spectrum opportunities and during secondary operation in licensed spectrum. DFS is a simple example for how unlicensed spectrum users (IEEE 802.11a) share spectrum with incumbent licensed users (radar stations) using overlay sharing.

Opportunistic Spectrum Usage

Under-utilized spectrum is in the following referred to as spectrum opportunity. The terms “white spectrum” and “spectrum hole” can be used equivalently. To use spectrum opportunities with overlay sharing, cognitive radios adapt their transmission schemes such that they fit into the identified spectrum usage patterns, as illustrated in Figure 6.5. Thus spectrum opportunities have to be identified in a reliable way. Additionally, their usage requires coordination especially in distributed environments. A spectrum opportunity is defined by location, time, and frequency and transmission power. It is a radio resource that is either not used by licensed radio devices, and/or it is used with predictable patterns such that idle intervals can be detected and reliably predicted. The accurate identification of spectrum opportunities is a challenge, as it depends on the predictability and the dynamic nature of spectrum usage.

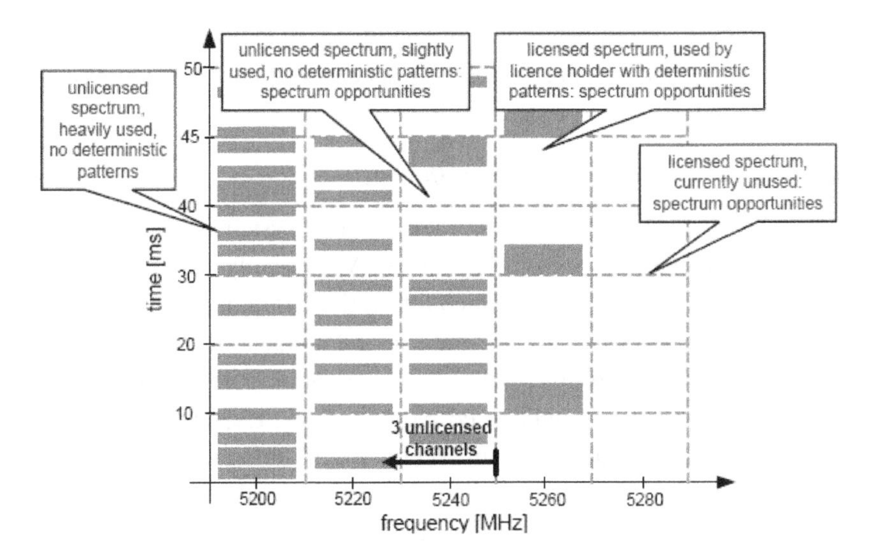

Fig. 8.5 Spectrum usage example at 5 GHz. Three 802.11a/e channels and the frequencies above are depicted. A cognitive radio identifies spectrum opportunities.

The frequency and predictability of spectrum usage by primary radio devices is decisive for the success of opportunity identification and the efficiency of its usage by cognitive radios. Therefore, a less frequent and predictable spectrum usage can be regarded as contribution to cooperation. In summary, deterministic patterns in spectrum usage and characteristic signal features of the radio signals transmitted by primary radio systems facilitate an improvement of spectrum opportunity identification. Different spectrum usage patterns and their classification as spectrum opportunity are shown in Figure 6.8. In this figure, time is progressing from bottom to top, frequency increases from right to left. Patterns of random spectrum usage of IEEE 802.11a/e are depicted (in three channels of the unlicensed 5 GHz band and the frequencies above), in parallel to a predictable, deterministic spectrum usage of licensed spectrum.

IEEE 802.11k

A new type of measurements improving spectrum opportunity identification is being developed in the standardization group of IEEE 802.11k [29] which provides means for measurement, reporting, estimation and identification of characteristics of spectrum usage. Spectrum awareness for distributed resource sharing in IEEE 802.11e/k is described in [30] while radio resource measurements for opportunistic spectrum usage on the basis of 802.11k are analyzed in [31]. The improvement of confidence in radio resource measurements as approach to judging reliability in spectrum opportunity identification is discussed in [32].

Vertical and Horizontal Spectrum Sharing

The overlay spectrum sharing with licensed radio systems requires not only fundamental changes in spectrum regulation. Additionally, new algorithms for sharing spectrum are necessary, which reflect the different priorities for spectrum usage of the licensed, i.e., incumbent, and unlicensed radio systems. To reflect this priority, the terms primary and secondary radio systems are often used for the licensed and unlicensed radio systems, respectively. Cognitive Radios will have to share spectrum (i) either with unlicensed radio systems with limited coexistence capabilities enabling them to operate in spite of some interference from dissimilar radio systems or (ii) with licensed radio systems designed for exclusively using spectrum. The sharing of licensed spectrum with primary radio systems is referred to as vertical sharing and the sharing between equals as for instance in unlicensed bands is referred to as horizontal sharing. These terms of horizontal and vertical spectrum sharing are first mentioned in [33]. Another example for horizontal spectrum sharing is the usage of the same spectrum by dissimilar cognitive radios that are not designed to communicate with each other directly. These dissimilar cognitive radio systems have the same regulatory status, i.e., similar rights to access the spectrum, comparable to the coexistence of devices operating in unlicensed spectrum. Vertical spectrum sharing promises to have the advantage that neither a lengthy and expensive licensing process nor a reallocation of spectrum is required. Vertical and horizontal sharing requires the capability to identify spectrum opportunities as introduced above. Cognitive radios are able to operate without harmful interference in sporadically used licensed spectrum requiring no modifications in the primary radio system. Nevertheless, in order to protect their transmissions, licensed radio systems may assist cognitive radios to identify spectrum opportunities in vertical sharing scenarios. This help is referred to as "operator assistance" in the following. In horizontal sharing, the cognitive radios autonomously identify opportunities and coordinate their usage with other cognitive radios in a distributed way. To avoid chaotic and unpredictable spectrum usage as in today's unlicensed bands, advanced approaches such as "spectrum etiquette" are helpful.

Coexistence, Coordination and Cooperation

In literature about spectrum sharing, the terms coexistence, coordination and cooperation are often used in different ways. A definition of these terms is given therefore in the following, as these terms are especially important in the context of QoS support. Means for coexistence target at interference avoidance in a distributed communication environment. Consequently, no communication among coexisting devices is required and possible. In case of a less utilized shared spectrum, coexistence capabilities suffice to enable a reliable communication. In the recent years, coexistence has been very successful but is now victim of its own success: Two many often dissimilar radio systems are coexisting in the unlicensed frequency bands like Wi-Fi and Bluetooth. Under such a severe competition for accessing the shared spectrum, no QoS support is possible due to missing coordination. Today's implemented approaches to coexistence have limited interference prevention and their spectrum utilization is very ineffective as coexistence implies only little incentive to conserve spectrum. Mutual coordination, either centralized or decentralized, is required in spectrum sharing to enable the support of QoS. QoS support refers in this context to the exclusive usage of spectrum at a predictable point of time for certain duration. Under cooperation, altruistic devices delimitate their spectrum usage and carry each other's traffic in the hope for benefiting from a potential cooperation when all radios participate. Cooperation comes along with the danger of being exploited by selfish, myopic radios resulting into a disadvantage for the cooperating radios. Cooperation is required in building up one self-configuring network of mutually coordinated radios in a distributed communication environment. The usage of deterministic patterns when allocating spectrum can also be regarded as cooperation. These deterministic patterns help to increase accuracy for other radios for identification of spectrum opportunities and enable a distributed coordination on the basis of observation. In distributed environments, cooperation can be created and enforced through protocols, either as part of a standard or realized as spectrum sharing etiquette. The enforcement of cooperation is difficult for regulation authorities but may be easier for a license holder.

Coexistence-based Spectrum Sharing

The amount of flexibility in the design of access protocols which is imposed by regulating authorities is decisive for the ability of radios to coexist, share spectrum or even interoperate for mutual coordination. [38] defines therefore "Rules of Coexistence" considering the grade of flexibility from the least to the most restrictive one:

- Spectrum usage is not constrained. This approach has the advantage of stimulating innovations but the big disadvantage of being only applicable in scenarios where less spectrum utilization can be expected. No QoS guarantee can be given and efficient spectrum usage is not encouraged.
- Spectrum usage is constrained by the transmission power. A maximum power level for radio emissions is defined. Finding the adequate level may be difficult and depends on the application scenarios of the envisaged radios operating in the considered frequency band. Limiting transmission power increases the potential re-use of spectrum but more radios are required for covering a certain area.
- Constrains on access protocols are imposed, that require no communication among coexisting devices. DFS in the U-NII band, as introduced below, is an example for this. Suitably defined etiquettes for spectrum usage can facilitate the support of QoS, increase spectrum efficiency and add fairness to spectrum sharing. These etiquettes enable distributed coordination and require cooperation as introduced above. The design of such etiquettes is a complex challenge especially in the context of spectrum sharing.
- A minimal standard is required for operation. A common signaling channel used by all radios operating in a shared frequency band is an example for this.
- The common signaling channel enables mutual coordination and thus increases the level of supportable QoS.
- The interoperation of dissimilar radio networks is part of their standard. The Base Station Hybrid Coordinator (BSHC) concept for integrating IEEE 802.11(e) in the frame structure of IEEE 802.16 (WiMAX).
- In the following section several approaches to coexistence are introduced that do neither require nor support communication among spectrum sharing devices.

Dynamic Frequency Selection

The FCC demands in [39] that devices operating in the U-NII band use Dynamic Frequency Selection (DFS) to protect radar systems from interference. DFS lets the transmitter dynamically switch to another channel whenever a certain condition is met. This condition is for instance a threshold level like -62 dBm for devices with a maximum EIRP less than 200 mW. When a radio signal is detected, the channel must not be utilized for a certain period of time. Further, a DFS using device senses the available spectrum for unused channels before initiating transmission and accesses only these channels. The FCC prescribes that the DFS aims at a uniform spreading of load over the available channels. Additionally, a continuous monitoring of spectrum during operation is demanded. The possibilities and limitations of DFS for cognitive radios and policies are discussed in [40]. For fundamental details on DFS and radio networks using DFS see [41].

Transmit Power Control

Transmit Power Control (TPC) is a mechanism that adapts the transmission power of a radio to certain conditions, e.g., a command signal from a communication target when the received signal strength falls below a predefined threshold. In the U-NII band for instance a radio reduces it's transmission power by 6 dB when TPC is triggered according to [42]. Only low-power radios operating at power levels higher than 500 mW require TPC in the U-NII band.

Ultra-Wide Band

Ultra-Wide Band (UWB) enables underlay spectrum sharing. UWB is a pulse positioning transmission technique with a very short pulse time duration using a very large frequency band in spectrum. Contrary to many other Radio Frequency (RF) communication techniques, UWB does not use RF carriers. Instead UWB uses modulated high frequency pulses of low power with a duration less than one nanosecond. From the perspective of other communication systems the UWB transmissions are part of the low power background noise. Therefore, UWB promises to enable the usage of licensed spectrum without harmful interference to primary communication systems. Figure 6.6 shows the spectrum crossover of the narrowband interferers in UWB systems.

IEEE 802.16.2

Coexistence in Fixed Broadband Wireless Access (FBWA) Systems of IEEE 802.16 operating in licensed bands is standardized in the Working Group

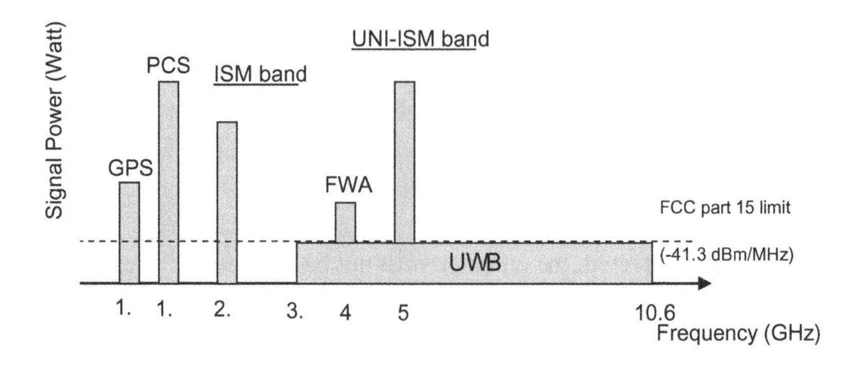

Fig. 8.6 Spectrum crossover of the narrowband interferers in UWB systems.

IEEE 802.16.2. It (IEEE, 2004c) provides recommendations for the design and coordination of FBWA systems in order to avoid interference and to facilitate their coexistence. 802.16.2 Suggests threshold parameters, like the distance between two interfering base stations, to assess the necessity for inter-operator coordination. These parameters are used to define guidelines for geographical spacing and frequency re-use. A concept of using Power Spectral Flux Density (PSFD) values is introduced in order to trigger different levels of initiatives taken by an operator to give notifications to other operators. Maximum values are defined for the PSFD that can be tolerated as a result from co-channel interference originating from adjacent operators. Frequency guard bands, recognition of cross-polarization differences, antenna angular discrimination, spatial location differences, the use of adaptive antennas and frequency assignment substitution are suggested to reduce the probability of interference.

IEEE 802.16h

The IEEE 802.16h License-Exempt Task Group is developing improved mechanisms for enabling coexistence among license-exempt systems based on IEEE 802.16. The standardization efforts target for instance at medium access control enhancements and policies. Additionally, 802.16h also focuses on the coexistence of such systems with primary radio systems when hierarchically sharing spectrum. A distributed architecture for radio resource management is suggested [49] that enables communication and exchange of parameters between different networks formed by one 802.16 Base Station (BS) and its associated Subscriber Stations (SSs). Each BS has a Distributed Radio

Resource Management (DRRM) entity to execute the spectrum sharing policies of 802.16h and to build up a data base for sharing information related to actual and intended future usage of radio spectrum. 802.16h proposes a coexistence protocol to realize all functions required for coexistence as for example detecting the neighborhood topology, to register to the data base or the negotiation for sharing radio spectrum. The DRRM uses the coexistence protocol to communicate with other BSs or license-exempt data bases when interacting with MAC or PHY.

IEEE 802.19

The IEEE 802.19 working group calls itself Coexistence Technical Advisory Group. 802.19 is aiming at the development and maintenance of policies defining the responsibilities of IEEE standardization efforts to consider coexistence with existing standards and other standards under development. If demanded, 802.19 evaluates the conformance of the developed standard to the coexistence policies and offers a documentation of the coexistence capabilities to the public.

Coordination-based Horizontal Spectrum Sharing

Common Spectrum Coordination Channel

A widely accepted approach to spectrum sharing is the usage of a Common Spectrum Coordination Channel (CSCC). The basic idea of CSCC is to standardize a simple common protocol for periodically signaling radio and service parameters [50] regarded as spectrum etiquette mechanism. The CSCC enables coordination through mutual observability between different neighboring radio devices via a simple common protocol. As shown in [51] at the example of contending 802.11b and Bluetooth devices, the CSCC approach impacts the complete protocol stack (e.g., physical layer, MAC layer, packet formats). In general, a common control channel as part of the shared spectrum is highly vulnerable to interference so that the complete network might be disrupted though deliberated jamming or through coexisting radios operating in the same spectrum. A permanently available signaling channel shared by several networks is also suggested in [52]. A Network Access and Connectivity Channel (NACCH) is introduced to enable communication between different networks. In this way they form a larger network where users have universal access and roaming support. This approach extends the exchange of control information between networks for coordinating spectrum usage with the aspect of transferring user data among networks over a common radio channel.

Dynamic Spectrum Allocation

In Dynamic Spectrum Allocation (DSA), spectrum allocations are changed over time depending on network loads in assigning continuous spectrum quantities to different Radio Access Networks (RANs) [53]. DSA aims at the exploitation of spatial and temporal variations of traffic loads in RANs. In the frequency domain for instance, a flexible guard band that separates two adjacent frequency bands used by different RANs can be adapted to shift bandwidth between these RANs. In DSA, an exclusive usage of spectrum by one operator using one RAN is assumed.

a. Brokerage-based Spectrum Sharing

Spectrum can be regarded as economic good which is traded by a broker as introduced above. Auctions are an efficient way to determine the value of spectrum among many parties. An auction can be regarded as a partial information game in which the real valuation that a bidder gives to spectrum is hidden to the broker and the other bidders. Auction theory is a very well developed field of research in economics as well as in wireless communication. A technology independent introduction and theoretical results on spectrum auctioning are given for example in [54]. The different treatment of spectrum in the context of a brokerage-based spectrum sharing leads to dissimilar approaches. The auctioning mechanisms introduced in [55] consider spectrum as public resource. Price and demand functions characterize there the optimal response functions of the bidders leading to a unique Nash equilibrium for an arbitrary number of agents with heterogeneous quasi-linear utilities. There, spectrum sharing between different operators providing their spectrum to a common spectrum pool is discussed. An operator temporarily leases spectrum from a broker who controls a pool of spectrum and performs the auctions. An analytical investigation of the problem and an approach to solve it is described in [56]: An operator of a CDMA cell populated by delay-tolerant terminals operating at various data rates tries to optimize its revenue depending on its own pricing policies. Brokerage-based spectrum sharing implies a substantial periodical signaling that scales with the number of participating parties. The automatically bidding through agents located in the MAC layer is also discussed in [57] for an Orthogonal Frequency Division Multiple Access (OFDMA)/Time Division Duplex

(TDD) system like IEEE 802.16. Piggybacked multi-unit, sealed-bid auctions are suggested to take heavy time and signaling constraints into account. It can be questioned in general, if a QoS guarantee can be given when using spectrum based on auctions. Therefore, a customer-interested operator might not be willing to provide its spectrum to the pool even if in a short term the selling of its capacity leads to higher revenues.

b. Inter Operator Spectrum Sharing

Another approach to DSA is introduced in [58] as Dynamic Inter Operator Spectrum Sharing. There, based on spectrum shared in UTRA Frequency Division Duplex (FDD), each operator deploys its own independent radio access network. This work is initiated in [59] where a shared UMTS FDD RAN is assumed. From the regulatory perspective, Inter Operator Spectrum Sharing is limited to one Radio Access Technology (RAT) and a frequency band licensed for being used by this single RAT is dynamically shared between several operators. The consideration of a synchronous inter-operator system requires heavy cooperation which is very unlikely for competing operators. Further it has been shown in [60] that only partial spectrum sharing leads to favorable capacity gains. These gains depend heavily on network parameters like cell radii or transmission powers.

IEEE 802.11y

IEEE 802.11y is the youngest Study Group of 802.11. It is working towards a standard for contention-based protocols operating in a 50 MHz frequency band at 3.650-3.700 GHz. The frequency band was opened to unlicensed services in [61]. A contention- based listen before talk protocol is there demanded to operate in this frequency band. Therefore, the FCC defines in [62] a contention-based protocol as: "A protocol that allows multiple users to share the same spectrum by defining the events that must occur when two or more transmitters attempt to simultaneously access the same channel and establishing rules by which a transmitter provides reasonable opportunities for other transmitters to operate. Such a protocol may consist of procedures for initiating new transmissions, procedures for determining the state of the channel (available or unavailable), and procedures for managing retransmissions in the event of a busy channel." Some of these characteristics are satisfied by currently available IEEE 802.11a systems operating in the U-NII band at 5 GHz: They are frequency agile, have the ability to sense signals from neighboring transmitters, offer adaptive modulation and use TPC. 802.11y aims at the definition of an extension to 802.11 OFDM, such that fixed and wireless devices can operate in the 3650–3700 MHz band.

Coordination-based Vertical Spectrum Sharing

All approaches to horizontal spectrum sharing can be used for vertical spectrum sharing when being combined with an accurate identification of spectrum opportunities. When multiple radios regard simultaneously the same spectrum as unused by the incumbent radio system, the access of the secondary radios needs to be coordinated (centralized or distributed) to enable a support of QoS on the one hand and to increase efficiency of spectrum usage on the other hand.

Common Control Channel

In the context of vertical spectrum sharing, [61] introduce opportunistic spectrum usage in creating a "virtual unlicensed band", there also referred to spectrum pool, with the help of a common control channel. A hierarchical control channel structuring is suggested in differing between a universal control channel, used by all groups for coordination and separated group control channels are used by members of a group. Preceding, the need for sophisticated techniques for spectrum sensing and the establishment of spectrum sensing as cross-layer functionality is motivated in [50]. For shadowing and multipath environments, cooperative decision making is suggested in order to reduce the probability of interference to primary users. A dedicated control channel located in licensed spectrum is suggested by the DARPA XG Program [11] to enable coordination in shared spectrum. This simple approach has several disadvantages, as a fixed licensed channel is required. The licensing of such a channel is a long-winded and expensive process. The licensed channel has a fixed bandwidth and is therefore limited in scalability. Nevertheless, this approach promises a high visibility and reliability.

IEEE 802.22

The IEEE 802.22 working group is targeting the standardization of a cognitive radio interface for fixed, point-to-multipoint, Wireless Regional Area Networks (WRANs) of 40 km or more operating on unused channels in the VHF/UHF TV bands between 54 and 862 MHz [12]. A wireless broadband replacement of DSL and cable modem services in less populated areas with unused VHF/UHF TV bands is envisaged. In not causing harmful interference to the incumbent TV broadcasting system, 802.22 realizes vertical spectrum sharing as introduced above and illustrated in Figure 6.7. Especially areas where wired networks are too expensive to deploy due to sparse population are in the focus of IEEE 802.22. Recent publications indicate that the 802.22 working group is concentrating on the slight modification of existing wireless communication protocols without taking up the fundamental ideas of cognitive radios. The 802.22 architecture, its requirements and applications are for instance discussed in [31], where IEEE 802.16 is assumed as basis for an 802.22 MAC. The performance of ultra sensitive feature TV detectors for measurements of analog and digital TV signals is investigated in [34]. The ability to detect TV signals below the thermal noise level, as one critical requirement for secondary usage of the television band, is evaluated. The collaborative spectrum sensing for detecting TV transmissions by a group of unlicensed devices is considered in [15]. It is shown that collaboration among spectrum sharing devices leads to efficient spectrum utilization and the individual computational complexity of the detection algorithms of each node can be decreased.

Spectrum Pooling

In [26] an OFDM-based approach to secondary usage in overlay spectrum sharing is developed referred to as Spectrum Pooling. In an 802.11 like scenario the spectrum measurements of mobile terminals are gathered centrally by an access point. Unused spectrum of different owners is merged into a common pool optimized for a given application. A so-called "licensed system public rental" hosts this common spectrum pool and users can temporarily rent spectrum during idle periods of the licensed users. Although [14] introduce a speed-up protocol to bypass the MAC layer and just use the physical layer for signaling, an essential weakness of spectrum pooling is only mitigated: The central gathering of measurement information takes considerable time and management effort.

Value-Orientation

Spectrum sharing among different radio systems can be understood as a scenario in which a society of independent decision-makers is formed [33]. Therefore, basic concepts to classify social action that are taken from social science can be applied to define system strategy rules. The rules represent algorithms

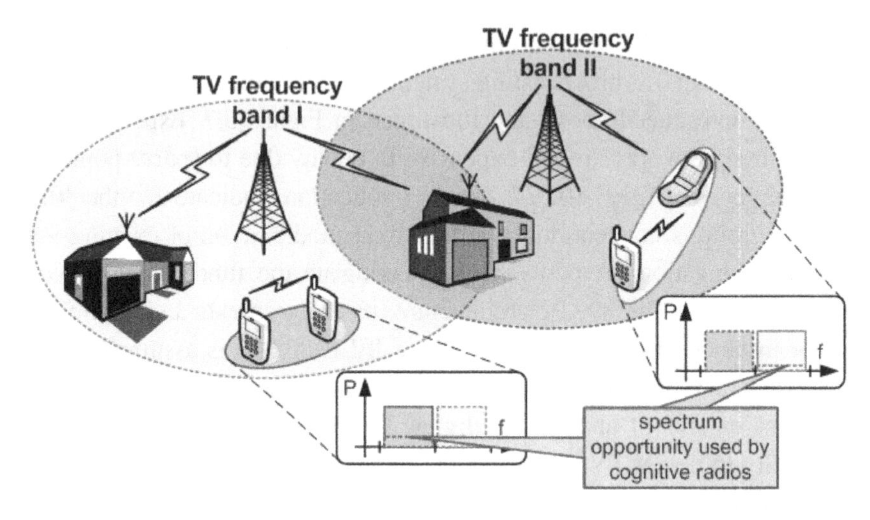

Fig. 8.7 Cognitive radios operating in frequency bands of TV and radio broadcasts. At different locations the cognitive radios identify different frequencies as unused and regard them as spectrum opportunities. This example of vertical spectrum sharing is the basis for IEEE 802.22.

for decision-making entities (referred to as actors) that reside in the radio systems. For a simple scenario of spectrum sharing, the need for regulation as opposed to voluntary rules is investigated. A spectrum sharing scenario of a contention-based medium access is analyzed under the assumption that the radio systems do not communicate with each other, but operate using the same radio resources. Voluntary standards and social concepts to mitigate the two main problems of open spectrum access are addressed: Incumbent protection and fair coexistence.

Spectrum Load Smoothing

The application of water-filling in the time domain enables a decentralized and coordinated opportunistic usage of spectrum. This is referred to as Spectrum Load Smoothing (SLS). In applying SLS, competing radio systems simultaneously aim at an equal overall utilization of the spectrum. In observing the past usage of the radio resource, the radio systems interact and redistribute their allocations of the spectrum under consideration of their individual QoS requirements. Due to the principle of SLS, these allocations are redistributed to less utilized or unallocated spectrum taking the QoS requirements of the coexisting networks into account. Further, SLS allows an optimized usage of the available spectrum: An operation in radio spectrum originally licensed to other communication systems is facilitated: The SLS implicitly achieves usage of unused spectrum by secondary radios and releases it again when needed by primary radios.

Spectrum sharing challenges

In the previous sections, the theoretical findings and solutions for spectrum sharing in CR networks are investigated. Although there already exists a vast amount of research in spectrum sharing, there are still many open research issues for the realization of efficient and seamless open spectrum operation. In the following, we detail the challenges for spectrum sharing in CR networks along with some possible solutions.

Common Control Channel (CCC)

Many spectrum sharing solutions, either centralized or distributed, assume a CCC for spectrum sharing [10]. It is clear that a CCC facilitates many spectrum sharing functionalities such as transmitter receiver handshake [12], communication with a central entity [19], or sensing information exchange. However, due to the fact that CR network users are regarded as visitors to the spectrum they allocate, when a primary user chooses a channel, this channel has to be vacated without interfering. This is also true for the CCC. As a result, implementation of a fixed CCC is infeasible in CR networks. Moreover, in a network with primary users, a channel common for all users is shown to be highly dependent on the topology, hence, varies over time [10]. Consequently, for protocols requiring a CCC, either a CCC mitigation technique needs to be devised or local CCCs need to be exploited for clusters of nodes [11]. On the other hand when CCC is not used, the transmitter receiver handshake becomes a challenge. For this challenge, receiver driven techniques proposed in [12] may be exploited.

Dynamic Radio Range

Radio range changes with operating frequency due to attenuation variation. In many solutions, a fixed range is assumed to be independent of the operating spectrum [63]. However, in CR networks, where a large portion of the wireless spectrum is considered, the neighbors of a node may change as the operating frequency changes. This affects the interference profile as well as routing decisions. Moreover, due to this property, the choice of a control channel needs to be carefully decided. It would be much efficient to select control channels in the lower portions of the spectrum where the transmission range will be higher and to select data channels in the higher portions of the spectrum where a localized operation can be utilized with minimized interference. So far, there exists no work addressing this important challenge in CR networks and we advocate operation frequency aware spectrum sharing techniques due the direct interdependency between interference and radio range.

Spectrum Unit

Almost all spectrum sharing techniques discussed in the previous sections consider a channel as the basic spectrum unit for operation. Many algorithms and methods have been proposed to select the suitable channel for efficient operation in CR networks. However, in some work, the channel is vaguely defined as "orthogonal non-interfering" [14], "TDMA, FDMA, CDMA, or a combination of them" [15], or "a physical channel as in IEEE 802.11, or a logical channel associated with a spectrum region or a radio technology" [61]. In other work, the channel is simply defined in the frequency dimension as frequency bands [11]. It is clear that the definition of a channel as a spectrum unit for spectrum sharing is crucial in further developing algorithms. Since a huge portion of the spectrum is of interest, it is clear that properties of a channel may not be constant due to the effects of operating frequency.

On the contrary, a channel is usually assumed to provide the same bandwidth as other channels [18]. Furthermore, the existence of primary users and the heterogeneity of the networks that are available introduce additional challenges to the choice of a spectrum unit/channel. Hence, different resource allocation units such as CSMA random access, TDMA time slots, CDMA codes, as well as hybrid types can be allocated to the primary users. In order to provide seamless operation, these properties need to be considered in the choice of a spectrum unit. In [11], the necessity of a spectrum space for a spectrum unit is also advocated. The possible dimensions of the spectrum space are classified as power, frequency, time, space, and signal. Although not orthogonal, these dimensions can be used to distinguish signals [20]. For this purpose, we describe a three dimensional space model for modeling network resources that has been proposed in [13]. Although this work focuses on heterogeneity

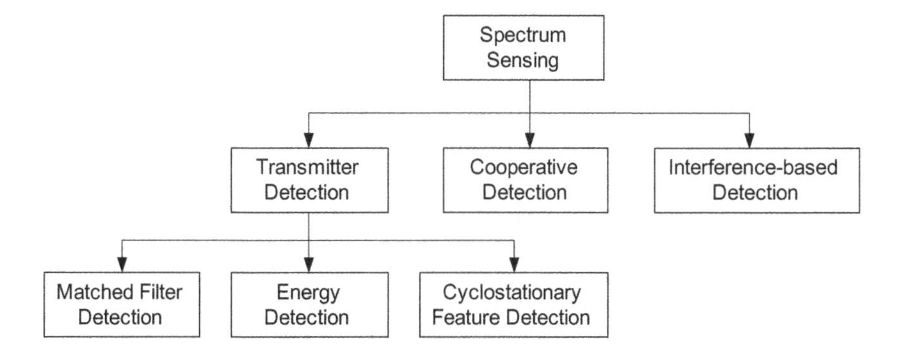

Fig. 8.8 Classification of spectrum sensing techniques.

in next generation/fourth generation (NG/4G) networks, as discussed in [22], it can be easily incorporated into CR networks.

Spectrum Sharing Life Cycle

Spectrum Sensing

An important requirement of the CR network is to sense the spectrum holes. As explained earlier, a cognitive radio is designed to be aware of and sensitive to the changes in its surrounding. The spectrum sensing function enables the cognitive radio to adapt to its environment by detecting spectrum holes. The most efficient way to detect spectrum holes is to detect the primary users that are receiving data within the communication range of a CR user. In reality, however, it is difficult for a cognitive radio to have a direct measurement of a channel between a primary receiver and a transmitter. Thus, the most recent work focuses on primary transmitter detection based on local observations of CR users. Generally, the spectrum sensing techniques can be classified as transmitter detection, cooperative detection, and interference-based detection, as shown in Figure 6.9. In the following sections, the spectrum sensing methods for CR networks and the open research topics in this area are discussed.

Transmitter Detection (non-cooperative detection)

The cognitive radio should distinguish between used and unused spectrum bands. Thus, the cognitive radio should have capability to determine if a signal from primary transmitter is locally present in a certain spectrum. Transmitter detection approach is based on the detection of the weak signal from a primary transmitter through the local observations of CR users. Basic hypothesis model for transmitter detection can be defined as follows [19]:

$$x(t) = \begin{cases} n(t) & H_0 \\ hs(t) + n(t) & H_1 \end{cases} \tag{8.1}$$

where $x(t)$ is the signal received by the CR user, $s(t)$ is the transmitted signal of the primary user, $n(t)$ is the AWGN and h is the amplitude gain of the channel. $H0$ is a null hypothesis, which states that there is no licensed user signal in a certain spectrum band. On the other hand, $H1$ is an alternative hypothesis, which indicates that there exist some licensed user signals. Three schemes are generally used for the transmitter detection according to the hypothesis model [10]. In the following subsections, we investigate matched filter detection, energy detection and cyclostationary feature detection techniques proposed for transmitter detection in CR networks.

Matched filter detection: When the information of the primary user signal is known to the CR user, the optimal detector in stationary Gaussian noise is the matched filter since it maximizes the received signal-to-noise ratio (SNR) [61]. While the main advantage of the matched filter is that it requires less time to achieve high processing gain due to coherency, it requires a priori knowledge of the primary user signal such as the modulation type and order, the pulse shape, and the packet format. Hence, if this information is not accurate, then the matched filter performs poorly. However, since most wireless network systems have pilot, preambles, synchronization word or spreading codes, these can be used for the coherent detection.

Energy detection: If the receiver cannot gather sufficient information about the primary user signal, for example, if the power of the random Gaussian noise is only known to the receiver, the optimal detector is an energy detector [22]. In order to measure the energy of the received signal, the output signal of band-pass filter with bandwidth W is squared and integrated over the observation interval T. Finally, the output of the integrator, Y, is compared with a threshold, k, to decide whether a licensed user is present or not [33]. If the energy detection can be applied in a non-fading environment where h is the amplitude gain of the channel as shown in (1), the probability of detection P_d

and false alarm P_f are given as follows [4]:

$$P_d = P\left\{Y > \frac{\lambda}{H_1}\right\} = Q_m(\sqrt{2\gamma}, \sqrt{\gamma}) \tag{8.2}$$

$$P_f = P\left\{Y > \frac{\lambda}{H_0}\right\} = \frac{\Gamma(m, \lambda/2)}{\Gamma(m)}, \tag{8.3}$$

where γ is the SNR, $\lambda = \text{TW}$ is the time bandwidth product, $\Gamma(m)$ and $\Gamma(n, m)$ are complete and incomplete gamma functions and $Q_m()$ is the generalized Marcum Q-function. From the above functions, while a low P_d would result in missing the presence of the primary user with high probability which in turn increases the interference to the primary user, a high P_f would result in low spectrum utilization since false alarms increase the number of missed opportunities. Since it is easy to implement, the recent work on detection of the primary user has generally adopted the energy detector [5]. In [66], the shadowing and the multi-path fading factors are considered for the energy detector. In this case, while P_f is independent of λ, when the amplitude gain of the channel, h, varies due to the shadowing/fading, P_d gives the probability of the detection conditioned on instantaneous SNR as follows:

$$P_d = \int_x Q_m(\sqrt{2\gamma}, \sqrt{\gamma}) f_\lambda(x) dx \tag{8.4}$$

where $f_\lambda(x)$ is the probability distribution function of SNR under fading. However, the performance of energy detector is susceptible to uncertainty in noise power. In order to solve this problem, a pilot tone from the primary transmitter is used to help improve the accuracy of the energy detector in [67]. Another shortcoming is that the energy detector cannot differentiate signal types but can only determine the presence of the signal. Thus, the energy detector is prone to the false detection triggered by the unintended signals.

Cyclostationary feature detection: An alternative detection method is the cyclostationary feature detection [8]. Modulated signals are in general coupled with sine wave carriers, pulse trains, repeating spreading, hopping sequences, or cyclic prefixes, which result in built-in periodicity. These modulated signals are characterized as cyclostationarity since their mean and autocorrelation exhibit periodicity. These features are detected by analyzing a spectral correlation function. The main advantage of the spectral correlation function is that it differentiates the noise energy from modulated signal energy, which is a result of the fact that the noise is a wide-sense stationary signal with no correlation, while modulated signals are cyclostationary with spectral correlation due to the embedded redundancy of signal periodicity. Therefore, a cyclostationary feature detector can perform better than the energy detector in discriminating against noise due to its robustness to the uncertainty in noise power [9]. However, it is computationally complex and requires significantly long observation time. For more efficient and reliable performance, the enhanced feature detection scheme combining cyclic spectral analysis with pattern recognition based on neural networks is proposed in [6]. Distinct features of the received signal are extracted using cyclic spectral analysis and represented by both spectral coherent function and spectral correlation density function. The neural network, then, classifies signals into different modulation types.

Cooperative Detection

The assumption of the primary transmitter detection is that the locations of the primary receivers are unknown due to the absence of signaling between primary users and the CR users. Therefore, the cognitive radio should rely on only weak primary transmitter signals based on the local observation of the CR user [7]. However, in most cases, a CR network is physically separated from the primary network so there is no interaction between them. Thus, with the transmitter detection, the CR user cannot avoid the interference due to the lack of the primary receiver's information as depicted in Figure 10(a). Moreover, the transmitter detection model cannot prevent the hidden terminal problem. A CR transmitter can have a good line-of-sight to a receiver, but may not be able to detect the transmitter due to the shadowing as shown in Figure 10(b). Consequently, the sensing information from other users is required for more accurate detection.

Spectrum sensing is a base aspect of cognitive radio to insure no interference for the primary (licensed user), however the spectrum sensing can be done locally in the node, where it will be susceptible to shadowing and fading that could cause a hidden node problem and will degrade the sensing performance. The cooperative sensing is proposed [2] to overcome the problems associated with the local sensing, different cooperative method is discussed that exploit the multiuser diversity in sensing process [20]. Different protocols can be employed in order to report the local sensing result to other secondary user or a common server in centralized or decentralized architecture respectively, the

Amplify and forward (AF) in which the relay transmits the signal obtained from the transmitter without any processing achieved full diversity [2]. In [25] the AF is been used in two cognitive user scenario, where each user considered as a relay for other user signal in the next time slot, the scheme show a reduction in detection time and increase agility.

In cooperative sensing architectures, the control channel can be implemented using different methodologies. These include a dedicated band, unlicensed band such as ISM, and underlay ultra wide band (UWB) system [20]. In order to minimize communication overhead, different quantization of the local obtained signal is introduced. It was shown that two or three bits quantization was most appropriate without noticeable loss in the performance [2]. In [8] a hard decision (binary quantization) is proposed for arbitrary large node population. However, the total number of sensing bits transmitted to the central is still very huge. Further to minimize reporting bandwidth a two level quantization method was recently proposed in [9], the method identify the users with a reliable information only to report a binary decision (0,1) to the common server as shown in Figure 6.12, however the method reduce the number of reporting bits but with a degradation in sensing performance Their result show that misdetection probability P_m is degraded by the imperfect channel and the false alarm probability P_f is bounded by the reporting error probability. This means that spectrum sensing cannot be successfully conducted when the desired P_f smaller than the bound $\bar{P}_f$. If the channels between cognitive users and the central server are perfect the local decision will send to central server without error, in practice, the reporting channels may experience fading which will deteriorate the performance of the cooperative spectrum sensing. A cluster based method was proposed in [7] where the most favorable user in each cluster is selected to report to central server, the method improved the sensing performance comparing to conventional sensing.

Based on Figure 8.10, every cognitive user conducts a local sensing and if a primary user detected, a hard decision '1' is sent to central server, otherwise no action is taken, If the server receives a local decision '0' due to imperfect reporting channel, according to a pre-knowledge, it is able to auto correct the reported error, to make it '1'. For simplicity energy detection based sensing is assumed for the local sensing method where the output of the integrator, Q, is compared with a threshold, λ, to decide the presence of a primary user. If Q exceed the threshold, a reporting decision, R, is taken and binary decision

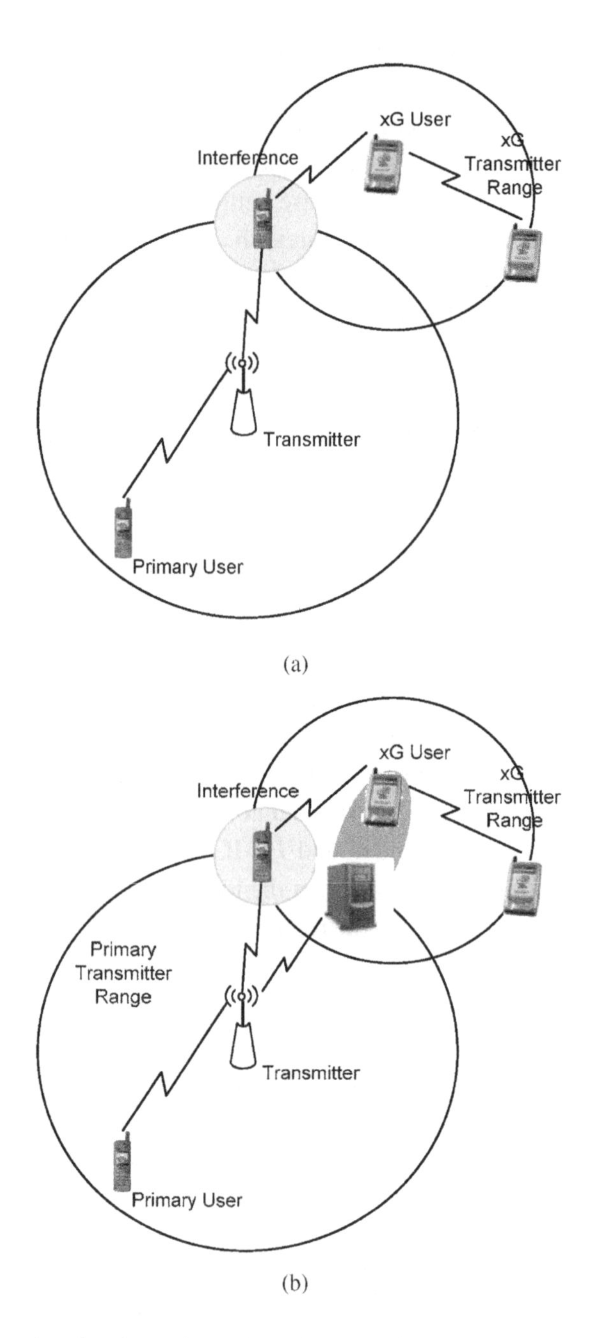

Fig. 8.9 Transmitter detection problem: (a) Receiver uncertainty and (b) shadowing uncertainty.

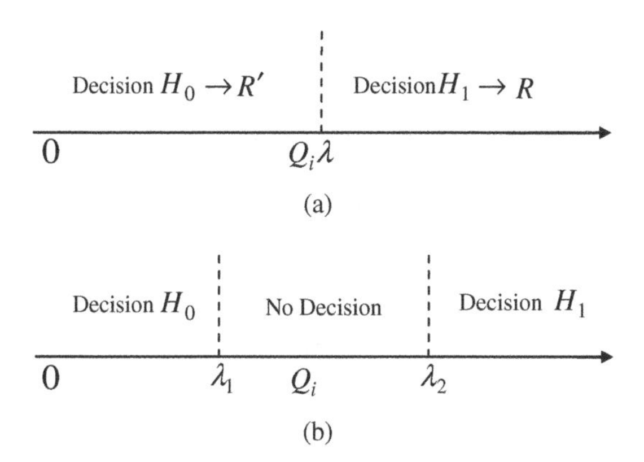

Fig. 8.10 (a) Auto_correction reporting method with one threshold for the ith user, (b) Censoring detection method with bi-thresholds for the ith user.

'1' is sent to central server otherwise "no report" decision, R', is taken. This is given by:

$$R = \begin{cases} 0 & Q < \lambda \quad R' \\ 1 & Q > \lambda \quad R \end{cases} \tag{8.5}$$

The considered system model of our interest is shown in Figure 6.12(a)

Following the work of [28], where the white noise and the signal term are modeled as zero mean Gaussian random variable, the decision metric R of the energy detector follows the following distribution:

$$R \sim \begin{cases} X_{2m}^2 & H_0, R' \\ X_{2m}^2(2\gamma) & H_1, R \end{cases} \tag{8.6}$$

where m is the time bandwidth product. X_{2m}^2 and $X_{2m}^2(2\gamma)$ represents the central and non-central chi-square distribution with $2m$ degree of freedom respectively and 2γ non centrality parameter for the later. The SNR γ is exponentially distributed with mean value $\bar{\gamma}$ for Rayleigh fading channel. Assume that the receiver receives K (where $K = 0, 1, \ldots, N$) out of N local decision 1 reported for the cognitive users. The final decision H at the server is done based on K. if the server receives any local decision 1 or 0 a final decision $H = 1$ is taken. If no local decision is reported to the server, then a final decision $H = 0$ is taken. H is given by

$$H = \begin{cases} 1 & K \geq 1 \\ 0 & K = 0 \end{cases} \tag{8.7}$$

Let K' denote the normalize average number of reporting bits

$$K' = \frac{K_{\text{avg}}}{N} \tag{8.8}$$

where N is the maximum number of reporting bits (for maximum N users in the system).

If R_K represents that Kcognitive users has reported, then the number of users without report is $N - K$ can be represented as $\bar{R}_{N-K}$. This can be modeled as:

$$P\{R_K\} = 1 - P\{Q < \lambda\} \tag{8.9}$$

$$P\{R_{N-K}\} = P\{Q < \lambda\} \tag{8.10}$$

Further, let $P_0 = P\{H_0\}$, $P_1 = P\{H_1\}$. Then the average number of reporting bits is:

$$K_{\text{avg}} = P_0 \sum_{K=0}^{N} K \binom{N}{K} P\{\bar{R}_{N-K}|H_0\} + P_1 \sum_{K=0}^{N} K \binom{N}{K} P\{R_K|H_1\} \tag{8.11}$$

$$K' = 1 - P_0 R_0' - P_1 R_1' \tag{8.12}$$

Where R_0', R_1' represent the probability of 'no report' under hypothesis H_0, H_1 respectively.

$$R_0' = P\{Q < \lambda|H_0\}, R_1' = P\{Q < \lambda|H_1\} \tag{8.13}$$

From equation (12) it can be shown that, the normalized average number of reporting bits k is always smaller than 1. If the channel between the cognitive users and the central server are perfect, and a full reporting is employed, the detection probability, P_d, the false alarm probability, P_f, and the misdetection probability, P_m, are given by [10] as:

$$P_d = 1 - \prod_{k=1}^{N} (1 - P_{d,k}), \tag{8.14}$$

$$P_f = 1 - \prod_{k=1}^{N} (1 - P_{f,k}), \tag{8.15}$$

and

$$P_m = \prod_{k=1}^{N} P_{m,k} \tag{8.16}$$

where $P_{d,k}$, $P_{f,k}$, $P_{m,k}$ are the detection probability, the false alarm probability, and the misdetection probability for the kth cognitive user, respectively. Under Rayleigh fading, γ, would have an exponential distribution. In this case, the Cumulative Distribution Function (CDF) of collected energy, Q, under hypothesis H_0, H_1 is

$$F(\lambda) = \int_0^{\lambda} f(Q|H_0)dQ = 1 - \frac{\Gamma\left(m, \frac{\lambda}{2}\right)}{\Gamma(m)}, \tag{8.17}$$

and

$$G(\lambda) = \int_0^{\lambda} f(Q|H_1)dQ = 1 - e^{-\frac{\lambda}{2}} \sum_{k=0}^{m} \frac{1}{k!}\left(\frac{\lambda}{2}\right)^k + \left(\frac{1+\bar{\gamma}}{\bar{\gamma}}\right)^{m-1}$$
$$\times \left(e^{-\frac{\lambda}{2(1+\bar{\gamma})}} - e^{-\frac{\lambda}{2}} \sum_{k=0}^{m-2} \frac{1}{k!}\left(\frac{\lambda\bar{\gamma}}{2(1+\bar{\gamma})}\right)^k\right) \tag{8.18}$$

Where $\Gamma(.,.)$, $\Gamma(.)$ are incomplete and complete gamma functions respectively [21]. In case $K = 0$, no report is sent to the server, that is, no primary user is active in the frequency band. Let β_0, β_1 denote the probability of 'no report' under hypothesis H_0, H_1, respectively.

$$\beta_0 = P\{K = 0|H_0\} = (R_0')^N \tag{8.19}$$
$$\beta_1 = P\{K = 0|H_1\} = (R_1')^N \tag{8.20}$$

Here the detection probability P_d, the false alarm probability P_f and the misdetection probability P_m are given as follows:

$$P_d = P\{H = 1, K \geq 1|H_1\} = 1 - \beta_1 \tag{8.21}$$
$$P_f = P\{H = 1, K \geq 1|H_0\} = 1 - \beta_0 \tag{8.22}$$

and

$$P_m = 1 - P_d \tag{8.23}$$

Figure 6.13 shows the tradeoff between the spectrum sensing performance and the average number of reporting bits, i.e., P_M vs. $\bar{K}$, for given false alarm

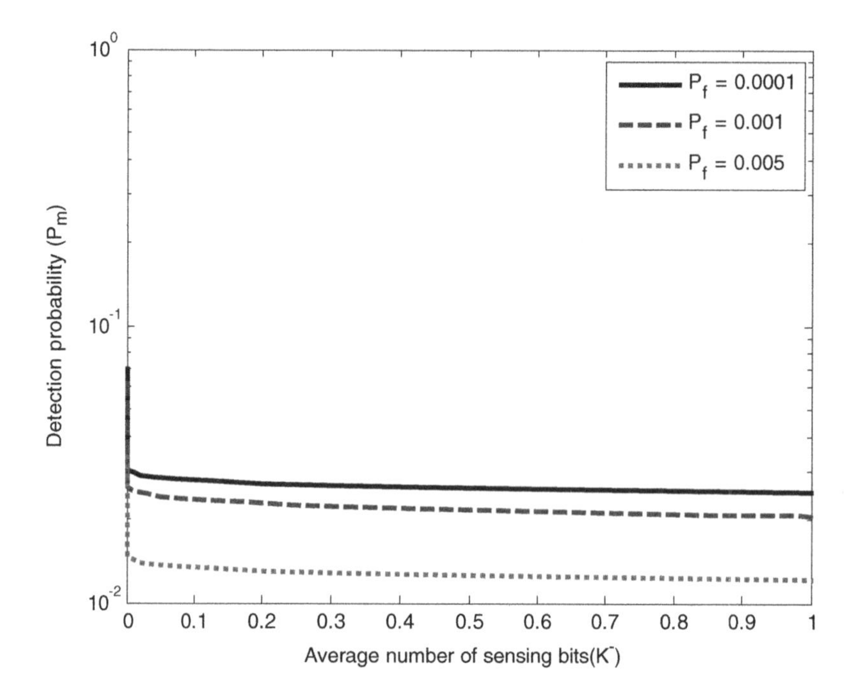

Fig. 8.11 P_m versus K' for $P_f = 0.0001, 0.001, 0.005$, $N = 10$ SNR $= 10$ dB.

probability, $P_F = 0.0001, 0.001, 0.005$, respectively. It can be observed that, for a fixed false alarm probability, the missing probability P_m changes a little when K' varies from 0.5 to 1, which means that we can achieve a large reduction of number of sensing bits at a very little expense of performance loss. In the case of non-cooperative detection the CR users detect the primary transmitter signal independently through their local observations. Cooperative detection refers to spectrum sensing methods where information from multiple CR users is incorporated for primary user detection. Cooperative detection can be implemented either in a centralized or in a distributed manner [51]. In the centralized method, the CR base-station plays a role to gather all sensing information from the CR users and detect the spectrum holes.

On the other hand, distributed solutions require exchange of observations among CR users. Cooperative detection among unlicensed users is theoretically more accurate since the uncertainty in a single user's detection can be minimized [42]. Moreover, the multi-path fading and shadowing effect are the main factors that degrade the performance of primary user detection methods [33]. However, cooperative detection schemes allow mitigating the multi-path fading and shadowing effects, which improves the detection probability in a heavily shadowed environment [24]. In [25], the limitation of non-cooperative spectrum sensing approaches is investigated. Generally, the data transmission and sensing function are collocated in a single CR user device. However, this architecture can result in suboptimal spectrum decision due to possible conflicts between data transmission and sensing. In order to solve this problem, in [6], two distinct networks are deployed separately, i.e., the sensor network for cooperative spectrum sensing and the operational network for data transmission. The sensor network is deployed in the desired target area and senses the spectrum. A central controller processes the spectrum information collected from sensors and makes the spectrum occupancy map for the operational network. The operational network uses this information to determine the available spectrum.

While cooperative approaches provide more accurate sensing performance, they cause adverse effects on resource-constrained networks due to the additional operations and overhead traffic. Furthermore, the primary receiver uncertainty problem caused by the lack of the primary receiver location knowledge is still unsolved in the cooperative sensing. In the following section, we explain interference-based detection methods, which aim to address these problems.

Interference-based Detection

Interference is typically regulated in a transmitter-centric way, which means interference can be controlled at the transmitter through the radiated power, the out-of-band emissions and location of individual transmitters. However, interference actually takes place at the receivers. Therefore recently, a new model for measuring interference, referred to as interference temperature shown in Figure 6.3 has been introduced by the FCC [17]. The model shows the signal of a radio station designed to operate in a range at which the received power approaches the level of the noise floor. As additional interfering signals appear, the noise floor increases at various points within the service area, as indicated by the peaks above the original noise floor. Unlike the traditional transmitter-centric approach, the interference temperature model manages interference at the receiver through the interference temperature limit, which is represented by the amount of new interference that the receiver could tolerate. In other words, the interference temperature model accounts for the cumulative RF energy from multiple transmissions and sets a maximum cap on their aggregate level. As long as CR users do not exceed this limit by their transmissions, they can use this spectrum band. However, there exist some limitations in measuring the interference temperature. In [67], the interference is defined as the expected fraction of primary users with service disrupted by the CR operations. This method considers factors such as the type of unlicensed signal modulation, antennas, ability to detect active licensed channels, power control, and activity levels of the licensed and unlicensed users. However, this model describes the interference disrupted by a single CR user and does not consider the effect of multiple CR users. In addition, if CR users are unaware of the location of the nearby primary users, the actual interference cannot be measured using this method.

8.4 Conclusions

The unsatisfied demand for freely available radio spectrum indicates that the necessary radio spectrum will not be available in the near future. The present regulation framework constrains the fast changing of the spectrum's status. The limited nature of radio resources in the current unlicensed frequency bands is the main reason for this scarcity. Many radio systems that support consumer electronics and personal high data-rate networks operate in these unlicensed frequency bands. Even worse, the demand for additional spectrum is growing faster than the technology is able to increase spectral efficiency, although latest research has had tremendous success to increase spectral efficiency and capacity in radio communication. Multiple Input Multiple Output (MIMO) and Space Time Division Multiple Access (SDMA) are just two examples for the recent advance in communication technology. In comparison to other licensed radio systems, today's widely deployed radio systems operating in the unlicensed frequency bands are already efficiently using these bands due to local re-use. Only a small fraction of radio spectrum is open for free, unlicensed operation. Many radio systems require rigorous protection against interference from other radio devices. Nowadays, such protection against interference is guaranteed in licensing radio spectrum for exclusive usage. Most of the radio spectrum is therefore licensed to traditional communication systems resulting into inefficient spectrum usage.

References

[1] J. Mitola III and G. Q. Maguire Jr., "Cognitive Radio: Making Software Radios More Personal," IEEE Pers. Commun., 6(4), August 1999, pp. 13–18.
[2] M. Vilimpoc and M. McHenry, 2006, Dupont Circle Spectrum Utilization During Peak Hours; http://www. newamerica.net/files/archive/Doc_File_183_1.pdf
[3] C. Ting, S. S. Wildman, and J. M. Bauer, "Government Policy and the Comparative Merits of Alternative Governance Regimes for Wireless Services," Proc. IEEE DySPAN '05, Baltimore, MD, November 8–11, 2005, pp. 401–19.
[4] C. Jackson, "Dynamic Sharing of Radio Spectrum: A Brief History," Proc. IEEE DySPAN '05, Baltimore, MD, November 8–11, 2005, pp. 445–66.
[5] B. A. Fette, Ed., Cognitive Radio Technology, Elsevier, 2006.
[6] IEEE 1900 Std. Committee,IEEE SCC41 home page: http://www.scc41.org.
[7] IEEE, "Standard Definitions and Concepts for Spectrum Management and Advanced Radio Technologies," P1900.1 draft std., V. 0.31, June 2007.
[8] P. Pawelczak, G. J. Janssen, and R. V. Prasad, "Performance measures of Dynamic Spectrum Access Networks," Proc. IEEE GLOBECOM '06 San Francisco, CA, Nov. 27–Dec. 1, 2006; http://dutetvg.et.tudelft.nl/~przemyslawp/files/ prfrmnc.pdf
[9] M. Bronzel, D. Hunold, G. Fettweis, "IBMS featuring Wireless ATM", In Proc. of ACTS Mobile Communication Summit 97, pp. 641–646, Aalborg, Denmark, October 1997.
[10] Fatih Capar, Ihan Martoyo, Timo Weiss, and Friedrich Jondral. Comparison of bandwidth utilization for controlled and uncontrolled channel assignment in a spectrum pooling system. In Proceedings of the VTC Spring 2002, pp. 1069–1073, Birmingham (AL), 2002.
[11] URL http://www.fcc.gov/oet/info/database/spectrum/
[12] FCC. Et docket No. 02–155. Order, May 2002.
[13] FCC. Et docket No. 03–322. Notice of Proposed Rule Making and Order, December 2003.
[14] G.-P. Fettweis, K. Iversen, M. Bronzel, H. Schubert, V. Aue, D. Maempel, J. Voigt, A. Wolisz, G. Walf, and J.-P. Ebert, "A Closed Solution for an Integrated Broadband Mobile System (IBMS)", ICUPC'96, pp. 707–711, Cambridge, Massachusetts, USA, October 1996.
[15] Joseph Mitola III. Cognitive radio for flexible mobile multimedia communications. In MoMuC'99, San Diego, CA, 1999.
[16] Joseph Mitola III. Cognitive Radio An Integrated Agent Architecture for Software Defined Radio. PhD thesis, KTH Royal Institute of Technology, Stockholm, Sweden, 2000.
[17] Jospeh Mitola III. Software radios-survey, critical evaluation and future directions. In IEEE National Telesystems Conference, pages 13/15–13/23, New York, 1992.
[18] Timo Weiss and Friedrich Jondral. Spectrum pooling: An innovative strategy for the enhancement of spectrum efficiency. IEEE Communications Magazine, 42:S8–S14, March 2004.
[19] Darpa XG working group. The xg architectural framework rfc v1.0, 2003. URL http://www.darpa.mil/ato/programs/XG/rfc_af.pdf.
[20] Darpa XG working group. The xg vision rfc v1.0, 2003. URL
[21] http://www.darpa.mil/ato/programs/XG/rfc_vision.pdf.

[22] C. Doan, S. Emami, A. Niknejad and R. Brodersen. Design of CMOS for 60 GHz Applications. IEEE International Solid-State Circuits Conference Digest of Technical Papers, 2004.
[23] S. Y. Poon, D. N. C. Tse, and R. W. Brodersen. An adaptive multiple-antenna transceiver for slowly flat-fading channels. IEEE Trans. on Communications, Vol. 51, pp. 1820–1827, November 2003.
[24] Ada S. Y. Poon, Robert W. Brodersen and David N. C. Tse. Degrees of Freedom in Spatial Channels: A Signal Space Approach. submitted to IEEE Trans. On Information Theory, August 2003.
[25] Chunyi Peng, Haitao Zheng "Utilization and fairness in spectrum assignment for opportunistic spectrum access" Mobile Networks and Applications 11(4), pp. 555–576, August 2006.
[26] Haitao Zheng, Chunyi Peng, "Collaboration and Fairness in Opportunistic Spectrum Access" 2005 IEEE International Conference on Communications (ICC 2005), Volume 5, pp. 3132–3136 Vol. 5, May 2005.
[27] Economides, A.A.; Silvester, J.A.; "A game theory approach to cooperative and non-cooperative routing problems" International Telecommunications Symposium, (ITS 1990) SBT/IEEE, Page(s): 597–601, September 1990.
[28] Chakravarthy, V.D. Wu, Z. Shaw, Temple, M.A. Kannan, R. Garber, F., "A general overlay/underlay analytic expression representing cognitive radio waveform" International Waveform Diversity and Design Conference, 2007, page(s): 69–73, June 2007.
[29] Anon., "Gartner Dataquest says PC market experienced slight upturn in 2002, but industry still shows no strong rebound," Gartner press release, January 16, 2003.
[30] Gartner Inc., as cited in Bray, Hiawatha, "Intel and rivals put their chips on table," Boston Globe, March 13, 2003.
[31] Anon., "IDC sees solid PC market in second half of 2002 and projects growth of more than 8% in 2003," IDC press release, December 6, 2002.
[32] M. Burrows, M. Abadi, and R. Needham, "A Logic of Authentication", ACM Transaction on Computer Systems, Vol. 8(1), pp. 18–36, USA, February 1990.
[33] Fenton, Andrew. "Security at Your Fingertips – New Notebooks Offer Biometric Protection." PC World Magazine, October 2003.
[34] "MPC: Presentation. Bio-Metric Fingerprint Scanner." MPC corporate web site. URL: http://www.buympc.com/presentations/biometrics.html, June 2003.
[35] "Learn about IBM Embedded Security Subsystem." IBM corporate web site. URL: http://www.pc.ibm.com/us/think/thinkvantagetech/security.html June 2003.
[36] "CryptCard product information". URL: http://www.gtgi.com/products_cryptcard.php, June 2003.
[37] R. I. Sharp, "User authentication". Technical university of Denmark, Spring 2004.
[38] A. Kahate, Cryptography and Network Security, 1st Edition, Tata McGraw-Hill Company, India, 2003.
[39] Security and Palm OS www.palmsource.com.
[40] A. clark. "Do You Really Know who is using Your System". Technology of Software Protection Specialist Croup, February 1990.
[41] B. D. Noble, M. D. Corner, "The Case for Transient Authentication", In Proceeding of 10th ACM SIGOPS European Workshop, Saint-Emillion, France, September 2002.

[42] Microsoft. "Encrypting File System for Windows 2000". http://www.microsoft.com/windows2000/techinfo/howitworks/security/encrypt.asp.

[43] Intel. "Biometric User Authentication Fingerprint Sensor Product Evaluation Summary", 2002.

[44] "Targus Defcon Fingerprint Security Authenticator." Electronic Gadget Depot corporate web site. URL: http://www.electronicgadgetdepot.com/t/ Targus/index4.htm, October 2003.

[45] M. D. Corner, B. D. Noble. "Protecting Applications with Transient Authentication". MOBICOM'02, Atlanta, Georgia, USA, September 2002.

[46] P. J. Phillips, A. Martin, C. L. Wilson, and M. Przybocki. "An introduction to evaluating biometric systems". IEEE Computer, February 2000.

[47] M. Blaze. "Key management in an encrypting file system". In Proceedings of the summer 1994 USENIX Conference, pp. 27–35, Boston, MA, June 1994.

[48] M. Blaze. "A cryptographic file system for UNIX". In Proceedings of the 1st ACM Conf. on Computer and Communications Security, pp. 9–16, Fairfax, VA, November 1993.

[49] "Most reported laptop thefts occur inside the office", news release by Kensington, January 26, 1999.

[50] Anon., "Warning: value of lost computers underestimated", Security, February 2002.

[51] Anon., "Security software uses Internet, e-mail to track down a vanished laptop", Washington Post, June 7, 2002.

[52] Junnarkar, Sandeep, "Setting a trap for laptop thieves", August 23, 2002.

[53] Targus Defcon MDP — Motion Data Protection. Targus corporate web site. October 2003 URL: http://www.targus.com/us/product_details.asp?Sku =PA480U

[54] Specification of the Bluetooth System, Core v1.1, Bluetooth SIG, February 2001 URL: http://www.bluetooth.com/ pdf/Bluetooth_11_ Specifications _Book.pdf

[55] Specification of the Bluetooth System, Profiles v1.1, Bluetooth SIG, February 2001.

[56] J. K. W. Chang, "An Interaction of Bluetooth Technology for Zero Interaction Authentication" Honours Project, School of Computer Science, Carleton University, April 2003.

[57] R. Mettälä, "Bluetooth Protocol Architecture v1.0", Bluetooth SIG, August 1999.

[58] Y. Hu, A. Perrig, and D. B. Johnson, "Wormhole detection in wireless ad hoc networks", Technical report, Department of Computer Science, Rice University, June 2002.

[59] J. Daemen and V. Rijmen, AES proposal: Rijndael. Advanced Encryption Standard Submission, 2nd version, March 1999.

9

Cooperative Networks: Optimization Through Game Theory and Mobility Prediction

Dimitris E. Charilas*, Theodore B. Anagnostopoulos†,
Athanasios D. Panagopoulos* and Christos B. Anagnostopoulos†

National Technical University of Athens, Greece
†*National and Kapodistrian University of Athens, Greece*

Planning and optimization of modern wireless networks can contribute drastically to the enhancement of Quality of Service (QoS) provided to end users. While the QoS may be enhanced through innovative protocols and new technologies, future trends should focus, also, on the efficiency of resource reservation and allocation, and network/terminal cooperation. First steps are currently being undertaken in this direction with the emergence of cognitive radios and cooperative relaying techniques and technologies. Cooperative communication networks, in which wireless nodes cooperate with each other in transmitting information, promise significant gains in overall throughput and energy efficiency and should therefore be considered during the planning of wireless networks. Certain cooperation schemes involve the dynamic formation of user groups so that resources can be exchanged and common interests can be served. The performance of such schemes requires efficient mobility prediction algorithms so that decisions based on users' mobility behavior and path prediction are taken instantly and correctly. Path prediction allows the network and services to further enhance the QoS levels the user enjoys. Such mechanisms are mostly meaningful in infrastructures like wireless LANs. In this way, mobile terminals should be able to choose the best cooperation tactics in an environment populated by a grand number of mobile users and heterogeneous networks. The goal of this chapter therefore is to summarize (i) cooperation trends, mainly based on the principles of game theory and (ii) refer to the adoption of mobility prediction mechanisms that enables network management.

9.1 Planning Cooperative Networks

Introduction

Generally speaking, the wide access wireless networks have larger coverage and support better mobility but have lower data rates and require higher power consumption on mobile terminals, for instance cellular networks GPRS and UMTS; the local access wireless networks have higher data rates and consume much less power on the terminals but have smaller coverage with limited mobility, for instance nomadic wireless networks WLAN and Bluetooth. The tradeoff between coverage and data rate is due to the relation between radio signal attenuation and distance.

With increasingly congested frequency bands and an ever-growing demand for higher data-rates, innovative approaches are needed to increase the spectral efficiency and hence cost per bit/s/Hz over the wireless link. Cooperative systems and radios, which are capable of intelligently forming mutually cooperative entities, are a promising way to achieve this increase in capacity. A comprehensive survey on cooperation in communication systems is given in [1] and in [2], where cooperation issues at the Physical, Medium Access Control, and Network layers are addressed.

A feasible solution to implement the envisioned wireless networks and sophisticated terminals is twofold. On the one side cellular and nomadic (short-range) heterogeneous networks should cooperate with each other and on the other side the terminals forming a cluster should also collaborate. The envisioned cooperative networks architecture is based on cellular reception of data which is then forwarded or shared among mobile devices within each others' proximity over the short-range link. To implement such cooperative networks, it is required that all the coexisted heterogeneous (wide access/local access) wireless networks can be designed to cooperate efficiently. And the multi-modality terminals can exploit the highly cooperative heterogeneous networks to cooperate with peers to realize the envisioned network.

As Frank Fitzek et al. state in [3], cooperation can be divided into two categories: **Macro cooperation** and **Micro cooperation**. In the macro cooperation the cooperating entities are wireless terminals, virtual access points, wireless routers and other macroscopic wireless system parts. The potential of macro cooperation is exploiting shorter radio propagation distance to reduce interference and to increase data rate. As a result, lower transmission and reception power is required. Additionally, cooperative diversity (or cooperation gain) can be exploited to form virtual antenna array (or virtual MIMO) to achieve network performance gain. In the micro cooperation the cooperating entities are microscopic components including processing units, functional parts and algorithms. The basic idea of micro cooperation is virtually sharing these microscopic components to obtain gain by designed cooperation mechanism.

Cooperative Services in 4G

Instead of the traditional peer to peer communication between base station and each terminal using a cellular link, where each terminal is operating autonomously, cooperation among terminals is introduced. In such a scenario the terminals are communicating over a short range communication in parallel to the cellular communication. Such architecture offers virtual high data rate, lower energy consumption and new business models. An essential precondition for the functionality of cooperative access is a highly flexible capacity distribution between the cellular and the short range communication. The capacity needed is a function of the number of cooperative users.

The success of 4G will consist in the combination of network and terminal heterogeneity, as stated in [4–6]. Network heterogeneity guarantees ubiquitous connection and provision of common services to the user, ensuring at least the same level of QoS when passing from one network's support to another one. Moreover, due to the simultaneous availability of different networks, heterogeneous services are also provided to the user [3].

The concept of node cooperation introduces a new form of diversity that results in an increased reliability of the communication, leading both to the extension of the coverage and the minimization of the power consumption [7, 8]. In fact, mobile terminals are less susceptible to the channel variations and shadowing effects and can transmit at lower power levels in order to achieve a certain throughput, thus increasing their battery life. Furthermore, cooperative transmission strategies may increase the end-to-end capacity and hence the spectral efficiency of the system.

By sharing energy resources, the cooperating terminals inherently share processing as well as spectrum resources. This makes cooperation a paradigm of gaining energy efficiency by sharing processing and spectrum resources among cooperating terminals. It is apparent that proper incentives, e.g., in terms of energy efficiency has to be in place before each terminal agrees to participate in a cooperating network and to spend energy resources to serve the needs from other cooperating terminals.

The authors of [7] and [8] demonstrate such an example, where terminals in the same multicast group, thus in physical proximity, can communicate with each other using high-speed wireless links. The basic idea behind this mechanism is illustrated in Figure 9.1. Users belonging to the same multicast group may download data and at the same time establish Bluetooth connections to share downloaded data with less fortunate users. The decision about sharing resources may involve a reputation matrix, as it will be explained in

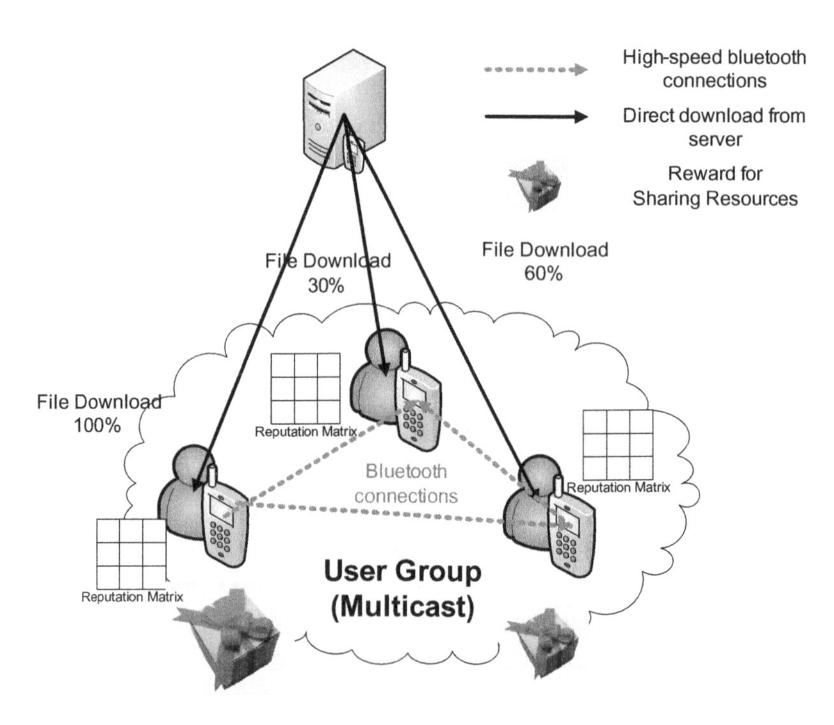

Fig. 9.1 User cooperation through data sharing.

the following paragraph. Finally, resource sharing users are rewarded according to their contribution. In [9] the authors describe a hybrid (e.g., between ad-hoc and infrastructure based) architecture called MobiShare that enables mobile devices to share their data encapsulated in services.

9.2 Incentive Mechanisms

In several cases, optimization issues require collaboration among nodes. However, some nodes may refuse to cooperate in order to conserve their limited resources (for example, energy), resulting in traffic disruption or overall QoS degradation. Nodes exhibiting such behaviour are termed selfish. Incentive mechanisms intend to provide a framework that forces players to cooperate for the best interest of all participants. In other words, they provide a **motive** so that each individual prefers to work along with others, sometimes sacrificing their own resources and sometimes benefiting from the resources of others. For example, the authors of [10] propose an incentive mechanism called bandwidth exchange where a node can delegate a portion of its bandwidth to another node in exchange for relay cooperation. Two types of incentive mechanisms are distinguished:

- Credit exchange systems
- Reputation based systems

A significant problem arising when dealing with node behavior issues is that sometimes, due to packet collisions and interference, cooperative nodes will be perceived as being selfish, which will trigger a retaliation situation. Such poor judgment may lead nodes to stop cooperating and thus degrade the overall network performance. In other words, disadvantaged nodes that are inherently selfish due to their precarious energy conditions shouldn't be excluded from the network using the same basis as for malicious nodes.

Pricing-Based Systems

In pricing-based systems, a node receives payment (reimbursements) for forwarding packets for others which can then be used for transmitting one's own data in the future. Buttayan and Hubaux [11] and Zhong et al. [12] develop pricing-based protocols where the amount charged per packet is determined exogenously and is the same for each node in the network. The drawback with the above approaches is the assumption of a simplified channel model — the energy required to forward a packet is assumed to be constant regardless of transmission distance. Stimulation mechanisms that take into account the fading channel have been developed during the last years. Results show that in this case cooperation is highly dependent on network geometry. The authors of [13] focus their work on stimulating nodes to participate in cooperative diversity transmissions in wireless ad hoc networks. More specifically, they consider that the source and relay may interact through reimbursement prices, transmitter power control and forwarding/protocol preferences such that their utilities are maximized.

Reputation-Based Systems

In wireless networks nodes can be thought of as members of a community that share a common resource. The key to solve problems related to node misbehavior derives from the strong binding between the utilization of a common resource and the cooperative behavior of the members of the community. Thus, all members of a community that share resources have to contribute to the community life in order to be entitled to use those resources. However, the members of a community are often unrelated to each other and have no information on one another's behavior. Generally, *reputation* is defined as the amount of trust inspired by a particular member of a community in a specific setting or domain of interest. Members that have a good reputation, because they helpfully contribute to the community life, can use the resources while members with a bad reputation, because they refused to cooperate, are gradually excluded from the community. Reputation is formed and updated along time through direct observations and through information provided by other members of the community. Pietro Michiardi and Refik Molva in [13] classify reputation in three types:

- *Subjective reputation* is calculated directly from a subject's observation. A subjective reputation at time t from subject s_i point of view is calculated using a weighted mean of the observations' rating factors, giving more relevance to the past observations. The reason why more relevance is given to past observations is that a sporadic misbehavior in recent observations should have a minimal influence on the evaluation of the final reputation value: as a result, it is possible to avoid false detections due to link breaks and to take into account the possibility of a localized misbehavior caused by disadvantaged nodes.

- *Indirect reputation* adds the information provided by other members of the community to the final value given to the reputation of a subject.
- *Functional reputation* allows for the possibility to calculate a global value of a subject's reputation that takes into account different observation/evaluation criteria.

Each type of reputation is obtained as a combination of different observations made by a subject over another subject with respect to a defined function. Furthermore, the above types of reputation information can be combined. A common approach is followed in [14], where the authors propose a reputation-based mechanism as a means of building trust among nodes. A node may autonomously evaluate its neighbours based on the completion of the requested service(s). Each node can maintain a reputation table, where a reputation index is stored for each of the node's immediate neighbours. For each successfully delivered packet, each node along the route increases the reputation index of its next-hop neighbour that forwarded the packet. Conversely, packet delivery failures result in a penalty applied to such neighbours by decreasing their reputation index. Once a node's reputation, as perceived by its neighbours, falls below a pre-determined threshold all packets forwarded through or originating at that node are discarded by those neighbours and the node is isolated.

9.3 Cross-Layer Optimization

Cross-layer optimization determines a general concept of useful interactions among different layers of the protocol stack with a view to improving wireless network performance. There are many approaches for cross-layer design such as network routing based on physical layer information, TCP performance based on channel modelling prediction etc [15]. For example, the performance of a wireless network can be greatly improved by taking into account the properties of the dynamic properties of physical channel and adapt efficiently the modulation scheme, the coding rate and the power control loop.

The parameters that should be exchanged among the layers are specific figure of merits at each layer, or particular information that can be used to estimate other quantities and determine accordingly the cross-layer optimization scenario. There are two approaches for the evaluation of the cross-layer design [16]: the evolutionary approach also referred in the literature as layer-related approach that focuses mainly on the improvement of the interoperability among the layers with a view to optimizing the overall network performance. The second approach is the revolutionary approach that tries to optimize the network performance without binding to the general concept layering. It is a philosophical extension of the network functionalities thus the revolutionary approach may lead in problems with compatibility among the layers. There are other two design cross-layer approaches taking into account the exchange of signalling during the operation. These are [17]: the implicit cross-layer design, where there is no exchange of information among different layers during operation, but the design takes into consideration all the layer interactions, and the explicit cross-layer design, where signalling interactions among adjacent or not protocol layers are employed during the phase of operation. Dynamic adaptations can be simultaneously performed at different layers [18].

Here we are going to present some physical and MAC layer-related interactions that are very important for the optimization of resource management schemes of wireless networks. It is substantial for the resource allocation schemes to be aware of the physical layer behavior. Some relevant metrics that describe the physical layer features are the Bit-Error-Ratio (BER), the Signal-to-Noise Ratio (SNR), the Signal-to-Noise-plus-Interference Ratio (SNIR). The knowledge of the Channel State Information (CSI) enables the appropriate choice of modulation and coding schemes in order to compensate the fading. Many authors have investigated the impact of the knowledge of SNR and of the required transmit power in order to achieve the minimum consumed energy. Generally, the information from the physical layer is very important for the upper-layers design [18–20]. As far as Data Link/MAC layer's relevant information that can be offered to higher are: current FEC scheme, number of retransmitted frames and packet length. The effect of an optimal packet length has been investigated in [21].

In the rest of the subsection, we will concentrate on cross-layer optimization for OFDM wireless networks [22, 23]. OFDM (Orthogonal Frequency Division Multiplexing) is one of the techniques that have been chosen to become the basic scheme for broadband wireless communication networks. OFDM is a combination of modulation and multiplexing. The general concept is to divide the available spectrum into several subcarriers. It has been shown

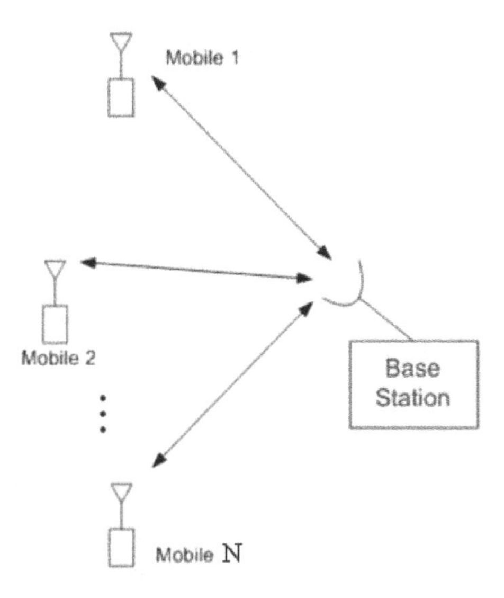

Fig. 9.2 Downlink/Uplink Multiuser system.

that OFDM is an effective technique to combat frequency selective multipath fading. Furthermore, in an OFDM wireless network, different subcarriers can be allocated to provide a flexible multiuser access scheme and to take advantage multiuser diversity. Here we will present briefly a utility based cross-layer optimization of OFDM systems, considering a network with a Base Station and N users as shown in Figure 9.2. The quality of each user can be quantified by the SNR which is defined as

$$SNR_i(f) = \frac{|H_i(f)|^2}{N_i(f)} \tag{9.1}$$

when the transmission power density is unity, and $H_i(f)$ is the instantaneous frequency response of channel impulse response, and $N_i(f)$ is the noise power density function. The Base Station is aware of the channel state information of each user and this can be exploited for utility-based cross-layer optimization. The achievable throughput of user i for a given BER and transmission power $p(f)$ is expressed as [24]:

$$T_i(f) = \log_2\left(1 - \frac{1.5}{\ln(5 \cdot BER)} \cdot p(f) \cdot SNR_i(f)\right) \tag{9.2}$$

In order to obtain the performance bounds of the cross-polarization, the fundamental assumption is that the bandwidth of each orthogonal subcarrier is infinitesimal. Let us consider a single cell of N user and that the frequency bandwidth B is divided in N non-overlapping frequency sets. If we define BW_i the frequency set assigned to user i, the transmission throughput of each user is expressed:

$$Tr\_throughput_i = \int_{BW_i} T_i(f)df \tag{9.3}$$

The network performance can also be improved through the adjustment of the transmission power with a total transmission power constraint by:

$$\frac{1}{B}\int_0^B p(f)\cdot df \leq 1 \tag{9.4}$$

The utility functions map the network resources that a user utilizes into a real number. The utility function should be a non-decreasing function of the data throughput. There are two choices regarding the utility function:

$$U(Tr\_throughput) = \begin{cases} Tr\_throughput \\ 0.16 + 0.8\ln(Tr\_throughput - 0.3) \end{cases} \tag{9.5}$$

The latter is used to capture the user's feeling such as the level of satisfaction for assigned certain resources [25]. The utility-based cross-layer optimization is to assign network resources (including subcarrier frequency and power density) and to maximize the average utility of the network:

$$\frac{1}{N}\sum_{i=1}^{N} U_i(Tr\_throughput_i) \tag{9.6}$$

where $U_i(Tr\_throughput_i)$ is the i user utility if it has transmission throughput $Tr\_throughput_i$.

On the other hand, cooperative diversity schemes consisting of multiple nodes that share resources create multiple diversity channels and thereby improve system performance, typically in terms of availability, range and throughput increase. A fundamental building block for cooperative diversity systems is the relaying channel [26], which has been studied in the context of fading channels in recent years [27].

9.4 Cooperative Relaying

Relaying is a significant means to meet the demanding requirements of next generation wireless networks in terms of system coverage and capacity. The

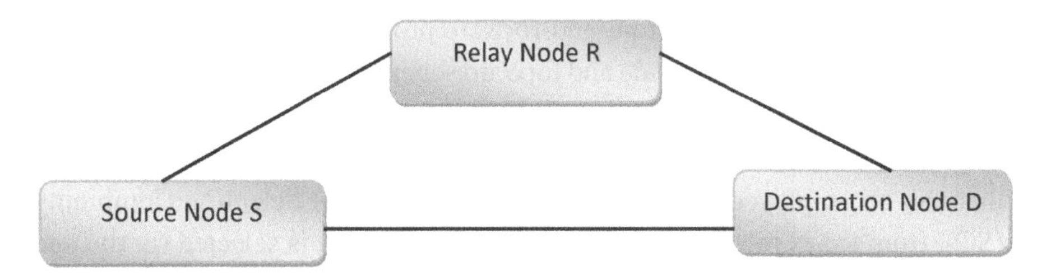

Fig. 9.3 A simple cooperative relaying scenario.

cooperative relay technique can further enhance the performance of a relay-based network by employing the antennas of different relay stations as a part of an antenna array, and applying the multi-antenna techniques among several relays. Coordinating the transmitted signals properly, the link quality, data rate and coverage of the relay-based system can be greatly improved. Here we are going to present the basics of cooperative relays.

We consider a cooperative scenario as depicted in Figure 9.3, which includes a single relay and where all nodes feature a single antenna. According to their forwarding strategy, the cooperative protocols can be categorized as [28]: (a) Amplify and Forward, where the relay acts as an analogue repeater (transparent repeater, bent-pipe system), (b) Decode and Forward, where the relay fully decodes, encodes and retransmits the received signal (regenerative), (c) Decode and Re-encode, where the relay fully decodes the received message, but constructs a codeword that is different from the source codeword. The transmission schemes based on protocol nature are further classified [28], (a) Fixed protocols, where the relay always forwards a processed version of the received message, (b) Adaptive protocols, where the relay uses a threshold rule to make a decision autonomously, whether to forward or not and (c) Feedback protocols, where the relay has an assistant role on the transmission only when the destination requires extra information. As a further classification, relays in non-regenerative systems can be categorized into [29]: (a) channel state information assisted relays, and (b) blind relays. Non-regenerative systems with CSI assisted relays use instantaneous CSI of the first hop to control the gain introduced by the relay.

We are now going to present some cooperative relay protocols [29]: (a) the well-known Alamouti scheme [30], (b) the transmission takes place in two time slots, the first time slot is dedicated for the information transmission from source to relay and the second slot is employed from the relay to retransmit the data to the destination (store-and-forward protocol). It is used for coverage extension. (c) Adaptive decode and forward schemes, (d) Adaptive decode and re-encode schemes and (e) Distributed hybrid ARQ.

In the destination node, there are various diversity combining techniques that can be employed and these are distinguished: (a) Selection Combining (SC), from the N received signals, the strongest signal is selected (or the one with the minimum attenuation), (b) Switched combining, where the receiver switches to another signal when the current SNR falls below a predefined threshold. This is a less efficient technique than selection combining (c) Equal Gain Combining (EGC), where all the received signal are added coherently, (d) Maximal Ratio Combing (MRC), where the received signals are weighted with respect to their SNR and then summed. There are many papers in the literature that are dealing with the performance of the cooperative relays under various fading channels.

The idea of cooperation has also been utilized to the design of higher layer protocols e.g., MAC [31, 32]. In [33], the general concept of cooperation has been employed to the routing problem of ad hoc networks. Multiple nodes are cooperating in sending the information to a single receiver node, and they can precisely delay their transmitted signal to achieve perfect phase synchronization. The problem of finding the optimal cooperative route from a source node to a destination node can be formulated as a Dynamic Programming problem.

Very recently, a new approach to cross layer design of multihop wireless networks with cooperative diversity has been proposed and succeeds in maximizing the gain in the physical layer and hitting off the interactions with higher layers [34]. In this paper the authors have applied the nonlinear optimization techniques to develop the optimization frameworks. The first framework has to do with joint routing and cooperative resource allocation which minimizes the total power consumption, while the second framework uses congestion control through a utility function to break down a balance between maximizing the sum rate utility and minimizing total cost assumption.

9.5 Introduction to Game Theory

Game theory is a mathematical tool developed to understand competitive situations in which rational decision makers interact to achieve their objectives, aimed at modeling situations in which decision makers have to make specific actions that have mutual, possibly conflicting, consequences [35].

It has been used primarily in economics, in order to model competition between companies. Game theory techniques have recently been applied to various engineering design problems in which the action of one component has impact on (and perhaps conflicts with) that of any other component. As a tool, it may be used for forming cooperation schemes among entities such as nodes, terminals or network providers. During the last years, game theory has widely been applied to networking, in most cases to solve routing and resource allocation problems in a competitive environment. A great number of references are included in [36]. The most fundamental concepts of non-cooperative game theory, as well as the modeling of several examples is included in [37] by Mark Felegyhazi and Jean-Pierre Hubaux.

Basic Principles of Game Theory

Game theory is related to the actions of decision makers who are conscious that their actions affect each other. A game consists of a principal and a finite set of players $N = \{1, 2, \ldots, N\}$, each of which selects a strategy $s_i \in S_i$ with the objective of maximizing his utility u_i. The utility function $u_i(s) : S \rightarrow R$ characterizes each player's sensitivity to everyone's actions. In **non-cooperative** games, each player selects strategies without coordination with others. The strategy profile **s** is the vector containing the strategies of all players: $s = (s_i)$, $i \in \mathbf{N} = (s_1, s_2, \ldots, s_N)$. On the other hand, in a **cooperative game**, the players cooperatively try to come to an agreement, and the players have a choice to bargain with each other so that they can gain maximum benefit, which is higher than what they could have obtained by playing the game without cooperation.

The *equilibrium strategies* are those that the players pick while trying to maximize their individual payoffs. In game theory, the Nash equilibrium is a solution concept of a game involving two or more players, in which no player has anything to gain by changing only his own strategy unilaterally. If each player has chosen a strategy and no player can benefit by changing his strategy while the other players keep theirs unchanged, then the current set of strategy choices and the corresponding payoffs constitute a *Nash equilibrium*. A pure strategy s_i is strictly dominated for player i if there exists $s_i' \in S_i$ such that $u_i(s_{i'}, s_{-i}) > u_i(s_i, s_{-i}) \forall s_{-i} \in S_{-i}$. It is customary to denote by s_{-i} the collective strategies of all players except player i.

When a player makes a decision, he can use either a pure or a mixed strategy. If the actions of the player are deterministic, he is said to use a pure strategy. If probability distributions are defined to describe the actions of the player, a mixed strategy is used.

In strategic or *static games*, the players make their decisions simultaneously at the beginning of the game. On the other hand, the model of an *extensive game* defines the possible orders of the events. The players can make decisions during the game and they can react to other players' decisions. Extensive games can be finite or infinite. A class of extensive games is repeated games, in which a game is played numerous times and the players can observe the outcome of the previous game before attending the next repetition.

9.6 Optimization of Wireless Networks Through Game Theory

Applications of non-cooperative games in 4G

As a first example of a situation for which game theory is an appropriate analysis tool, we consider **random access** to a communications channel, in this case slotted Aloha. A more analytical explanation of this mechanism may be found in [38, 39]. Users who wish to transmit typically wish to do so as soon as possible. If multiple users try to transmit simultaneously, though, all accesses fail; in addition, unsuccessful attempts to transmit may be costly. The users trying to transmit have conflicting objectives. In slotted Aloha, time is divided into slots and via some method of synchronization, all users are presumed to know where the slot boundaries are located. When a user wishes to access the shared channel the user waits until the next slot boundary and then begins attempting to transmit. If two or more users try to transmit in the same slot, the users become "backlogged" and must attempt to transmit again in a future slot.

A key field where non-cooperative games can be successfully applies is **admission control**. The basic goal of an admission control algorithm in cellular networks is to control the admission of new sessions within the network with the goal of maintaining the load of the network within some boundaries. Admission control takes place each time a new session request is received and decides whether it should be allocated resources or be rejected due to lack of resources. In this kind of games the providers may compete with each other over handling certain requests, or compete with the customers so that both parties increase their utility and thus their level of satisfaction.

In the first type of such games the networks constitute the players. As individual players in the game, the access networks will therefore try to maximize their own payoff by choosing the best available strategy in a rational manner, meaning that they will try to choose the request that best fits their characteristics. Such a game may be played in rounds. In each round of the game the networks must decide which request will maximize their payoff and then select it. Once a request is selected it is removed from the set of service requests and the game is repeated, until all requests have been selected. A typical example can be found in [40]. The proposed game is non-zero-sum and non-cooperative, since a player is unable to bind and enforce agreements with other players.

In the customer vs. provider schemes, the main goal is to maximize not only the QoS offered to customers, but also the provider's gain, therefore balancing the interests of both parties. Such an attempt has been modeled in [41] and [42]. The authors there consider that each customer has a contract with a specific service provider, thus him being the default network choice ("home" provider); nevertheless, if case of insufficient resources, the customer is free to pursue higher QoS at another provider, given that there is some kind of federation agreement between the visited and the home provider as in roaming (possibly under a small monetary penalty). Each user-provider combination in these schemes is considered as a two-player game $G_{j,}$, as illustrated in Figure 9.4 The decision of each player may be based on measurements extracted from on-going sessions of the all service types [42]. We assume that the service provider has two choices: either admit or reject the request. The customer also possesses two strategies: either leave or stay with the service provider, leaving us with four possible strategy combinations.

As far as the solution of the game in this scenario is concerned, two cases are distinguished. Assuming the case where the system is not full, the user request will be accepted and the probability that a customer leaves is near to 0. In this case, there is a Nash equilibrium when the service provider accepts the request while the user remains with the provider. Assuming now the case where the system is loaded to a certain extent or even overloaded, the user request may be not accepted and the probability that a customer leaves is non zero. Even in this case, there is also a pure strategy Nash equilibrium at the

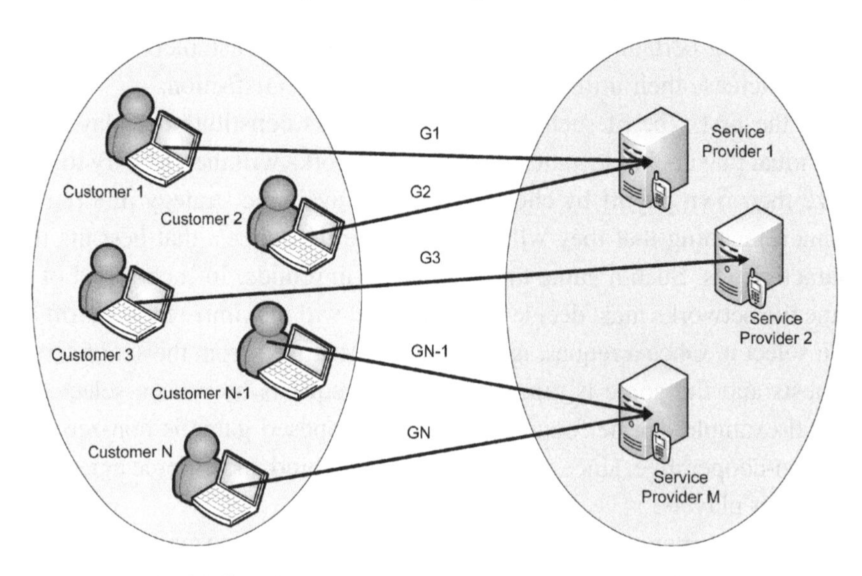

Fig. 9.4 Game formulation between customers and service providers.

pair, which depends on the relation between some terms in the payoffs. The new request is accepted if the revenue generated from admitting the request is greater than the possible revenue loss is the user leaves. Otherwise, the provider is better to reject the request.

Using game theory to model cooperation in wireless networks

Since in 4G the most efficient allocation of resources is required, user cooperation may be modeled according to the principles of game theory. We envision a novel future architecture where users may form freely and dynamically resource sharing groups, where users are expected to share as many resources they see best for their own interests. Game-theoretic backgrounds can easily fabricate mechanisms for rewarding generous users or punishing selfish ones. Another possible application of game theory in 4G involves the resource or even client exchange among different networks or even providers.

A cooperative game is a game in which the players have the option of planning as a group in advance of choosing their actions. Let $N = \{1, 2, \ldots, N\}$ be a set of n players. Non-empty subsets of N, $S, T \subseteq N$ are called a coalition. The coalition form of an n-player game is given by the pair (N, u), where u is the characteristic function. A coalition that includes all of the players is called a grand coalition. The characteristic function assigns each coalition S

its maximum gain, the expected total income of the coalition denoted $u(S)$. The core is the set of all feasible outcomes that no player or coalition can improve upon by acting for themselves. The objective is to allocate the resources so that the total utility of the coalition is maximized. In wireless networks the formation of coalitions involves the sharing of certain resources; however, as the costs of such resource sharing outweigh the benefits perceived by the nodes, users are less likely to participate, compromising overall network goals.

As a first example, we consider the **power control problem**. Each user's utility is increasing in his signal-to-interference-and-noise ratio (SINR) and decreasing in his power level. If all other users' power levels were fixed, then increasing one's power would increase one's SINR. However, when a user raises her transmission power, this action increases the interference seen by other users, driving their SINRs down, inducing them to increase their own power levels. If the user's transmit power is too high, then he is squandering precious battery power while having little impact on his bit error rate. The users will therefore attempt to make the best possible choices, taking into account that the other users are doing the same thing. By assumption, the users have complete information about each other. Then, according to game theory, the users will choose an operating point which is a Nash Equilibrium. MacKenzie and Wicker in [43] formulate a power control game for a CDMA system.

The **spectrum sharing problem** addresses the issue of how to allocate the limited available spectrum among multiple wireless devices. The problem has two important goals: efficiency and fairness. The allocation of spectrum should utilize as much of the resource as possible. However, when utilization is maximized, fairness can be compromised. The available bandwidth can be divided equally into multiple channels. Each node can transmit in any combination of channels at any time and can set its transmit power on each channel.

The most common application of cooperative game theory in wireless networks is the **routing problem**. In the routing problem, the source nodes can be viewed as the players in the game. The action set available to each player is the set of all possible paths from the source to the destination. In wireless ad hoc networks for example, nodes communicate with far off destinations using intermediate nodes as relays. Since wireless nodes are energy constrained, it may not be in the best interest of a node to always accept relay requests. On the other hand, if all nodes decide not to expend energy in relaying, then network throughput will drop dramatically. For this reason, ad hoc and peer-to-peer networks sometimes operate as voluntary resource sharing networks, relying on users' willingness to spend their own resources for the common good [44–46]. In [45] the utility function of such a game is modelled as $U_j(s) = a_j(s) + \beta_j(s)$ where $a_j(s) = a_j(\sum_{i \in N, i \neq j} s_i)$ is the benefit accrued by a user from others' sharing of their resources and $\beta_j(s) = \beta_j(s_j)$ is the benefit (or cost) accrued by sharing one's own resources with others, s being the joint action taken by all players ($s = 0$ stands for sharing and $s = 1$ for not sharing). The latter may be negative, since there may be a cost to participating in the network (such as faster depletion of a node's energy resources) or positive, if there exist financial incentives for participation or if the user derives satisfaction in doing so.

9.7 Optimization Through Mobility Prediction

Introduction to Mobility Prediction

In order to render mobile context-aware applications intelligent enough to support users everywhere/anytime and materialize the so-called *ambient* intelligence, information on the present context of the user has to be captured and processed accordingly. A well-known definition of *context* or *contextual information* is the following: *context is any information that can be used to characterize the situation of an entity. An entity is a person, place or object that is considered relevant to the integration between a user and an application, including the user and the application themselves* [47]. Context refers to the current values of specific ingredients that represent the activity of an entity and environmental state (e.g., time, location, a user is attending meeting, neighbouring mobile terminals, network access points). One of the more intuitive capabilities of mobile context-aware applications is their *pro-activity*. Predicting contextual ingredients and thus user actions enables a new class of applications to be developed. Nowadays, the most important contextual ingredient is *location*. Estimation of the current location and prediction of the future location of a user enables the introduction of advanced Location-based Services (LBS) [48]. In our case, location prediction can be used to improve resource reservation in wireless networks.

Mobile context-aware applications gather user location information, process and learn user mobility behaviour in order to predict the mobile user's new movement and context. Hence, such applications operate proactively, deliver advanced service in advance, and manage network resources (e.g., packets, proxy-cache content retrieval, and network resources reservation). In this section we report context models and Machine Learning (ML) methods for representing and predicting the location of mobile users.

The representation of context in each mobile context-aware application poses a significant challenge. Modeling context is complicated due to the fact that different mobile applications value differently their pieces of context. Especially, in a mobile computing environment different types of context might appear, for instance, the context of each node in a group (individual context), the context of the group itself (collaborative context) [49, 50]. In our case, we are interested in representing context with which we can infer/predict the future user movement. Context representation in mobility prediction includes a model of representing current and historical spatio-temporal information of a unique entity (user) — *spatial context*, e.g., user identifier, position coordinates, observed time, network cell identifier, user preferred areas, routes and itineraries. Based on the corresponding positioning technology and the provided facility, we study two basic models for representing spatial context: the *Cellular* and *Coordinate* model. The reported prediction algorithms represent location information through such models.

Location (or context) prediction is quite similar to pattern *classification* and *prediction*, in the sense that, the values of certain contextual ingredients (e.g., direction, location, speed) are determined and predicted. Moreover, Machine Learning (ML) is primarily concerned with the design and development of algorithms for pattern classification and prediction. Specifically, ML refers to *the study of algorithms that improve automatically through experience* [51]. ML provides algorithms for learning a system to classify observations and predict unknown situations based on a history of patterns. Hence, if we are able to trace and learn the mobility patterns of a user, i.e., the observed movement trajectories, then we can classify and, consequently, predict the future mobility behaviour through appropriate ML methods.

ML methods can be roughly categorized as follows: *supervised learning* (classification/regression) and *unsupervised learning* (clustering/knowledge induction). We report classification methods in order to predict the future location of a mobile user, i.e., *spatial context*. Each training trajectory $\mathbf{h}_i$ is assigned to a specific class g_i from a fixed set of classes (e.g., symbolic locations like home, university campus, train station). Hence, the goal of spatial classification is to predict the class g_0 of an unseen trajectory $\mathbf{h}_0$ based on a history of N trajectories $[\mathbf{h}_i]_{i=1,\ldots,N}$. We study certain classification models, *predictors*, to deal with location prediction. The contextual information for location prediction can be the history of the trajectories of a mobile user $[\mathbf{h}_i]_{i=1,\ldots,N}$, along with the current location and direction. Prediction results, i.e., the future user location, are provided through efficient, low complexity algorithms that require low classification time. Context prediction based on ML is quite promising and efficient as it achieves increased accuracy at reasonable computing/storage cost. We perform a comparative analysis of several proposed schemes found in the literature, e.g., [52, 53], and report prior work.

Figure 9.5 depicts the main idea of a predictor $\mathcal{P}$. Actually the user starts his/her movement from the *start* point. After certain time he/she walked a *trajectory* in the movement space. Consider that, at the time being, the user is at the *now* point. The predictor is used for predicting a point (the *prediction* point) as close as to the future point (*after* point) having certain *accuracy* of that prediction, e.g., the *prediction* point and the *after* point are within a certain area. The size of the area is determined by the corresponding mobile context-aware application that exploits the predictor $\mathcal{P}$.

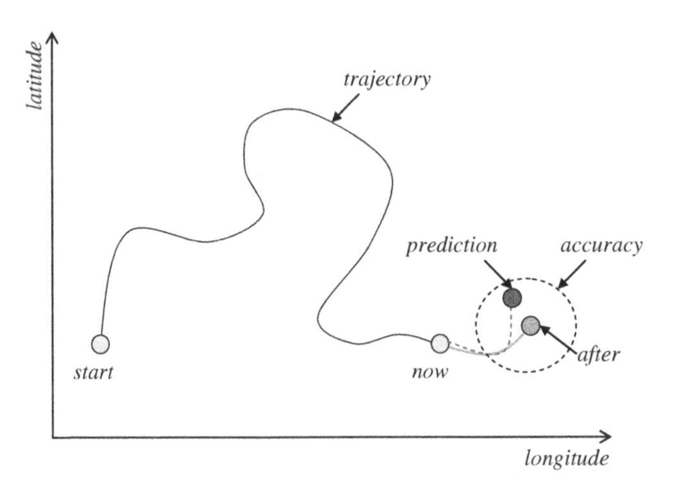

Fig. 9.5 The main idea of the location prediction problem.

9.8 Context Representation

Ultimately, the contextual information that can represent the user mobility behaviour is *location* $\mathbf{u}$ and *time* t. However, location information can be differently captured by different positioning systems. For instance, by adopting a GPS receiver, location $\mathbf{u}$ are the longitude x and latitude y coordinates, thus, at time t, $\mathbf{u} = [x, y]$. In addition, a user movement in a cellular system can be represented through the transition between neighbouring network cells, thus, at time t, $\mathbf{u} = [x]$ denoting that user locates in cell x. We deal with both representations of location and refer to prediction algorithms adopting such representations.

Moreover, the contextual information considered for prediction refers to the *history* of user movements $\mathbf{h}$. Specifically, the history of movements reflects the mobility behaviour for a user, which is not necessary the same for all time and periods. For instance, a mobile user might follow different routes at weekends or on holidays than his/her daily program. Such history refers to the user movement trajectory and is represented by a m-dimensional vector $\mathbf{h}$ of m time-ordered visited locations $\mathbf{u}_i$, that is, $\mathbf{u}_j < \mathbf{u}_i$ if the user visited location $\mathbf{u}_j$ before $\mathbf{u}_i$, $i, j = 1, \ldots, m$. Such locations are used for predicting future location notated as $\mathbf{u}_0$.

Cellular Model

By adopting cellular topology, the user roams through a number of cells in a cellular mobile network. Each *cell* has a unique identifier (Cell ID, CID) x_i. A cellular network consists of a set $X = \{x_i : i = 1, \ldots, |X|\}$ of cells. We assume that, at any given time, the user is within a cell $x_i \in X$. We formulate the learning problem for location prediction $\mathbf{u}_0$ as follows: *given a cellular representation model derived from the movement history of a user, construct a predictor* $\mathcal{P}$ *that predicts the next cell* $\mathbf{u}_0 = [x_0]$ *to which that user is going to move in the immediate future.* We want to predict which cell the user is going to move to in the next transition. Let us assume that, $\mathcal{P}$ deals only with the mobility behaviour of a single user and does not cover groups of users. Specifically, each training example — pair $(\mathbf{h}_i, g_i)$ consists of the trajectory $\mathbf{h}_i = [\mathbf{u}_{ij}] = [x_{i1}, \ldots, x_{im}]$ of dimension m, where x_{ij} is a CID and g_i is the CID of the predicted cell, which is called the *class label* for that training pair. We adopt the notation $t(x_i)$ to denote the time of user entrance to cell x_i.

For each cell in trajectory x_i it holds strictly that $t(x_{i1}) < \cdots < t(x_{im})$ and the cell g_i is the antecedent of the cell x_{im}, i.e., $t(x_{im}) < t(g_i)$ — the next user movement. Then $T = \{(\mathbf{h}_1, g_1), \ldots, (\mathbf{h}_N, g_N)\}$ is the training set with N training pairs.

Sliding Window Cellular Model

The *sliding window cellular* model is a specification of the cellular model where the training set T of assigned user trajectories contains tuples of $(\mathbf{h}_i, g_i)$ such that the g_i class label of the ith tuple is the mth element of the $\mathbf{h}_{i+1}$ vector for $i = 1, \ldots, m$, $m > 1$. This means that, $\mathbf{h}_{i+1}$ is an expansion of $\mathbf{h}_i$ with the class label g_i for m overlapped trajectories. The parameter $w = m + 1$ is called *window* of m trajectories. We explain the considered model in mode details below:

The $\mathbf{h}_i = [x_{i1}, \ldots, x_{im}]$ vector in the cellular model does not model time; instead, it models cell transitions. Then, from the complete sequence of $m - 1$ transitions of a user, x_{im} is the cell in which the user is currently positioned (i.e., $t(x_{im}) = now$) and x_{im+1} is the cell to which he/she is going to move in the next transition. The $x_{im+1} = x_0$ location constitutes the prediction target and $\mathbf{h}_i$ is the sequence of cells from which the user passed before reaching x_{im+1}. Hence, the set $\{\mathbf{h}_i\}$, $i = 1, \ldots, m$, stores the last $m$movements and the training set T is *expanded* with mN pairs including for each location the previous and the next location of the user. Consider a set of cells $X = \{x_1, x_2, x_3, x_4, x_5\}$ and assume a user that had the following sequence of transitions $\langle x_1, x_2\rangle$, $\langle x_2, x_3\rangle$, $\langle x_3, x_4\rangle$, $\langle x_4, x_5\rangle$ ($\langle x_k, x_l\rangle$ denotes the transition from cell x_k to x_l). Applying a sliding window $w = 3$, we derive the following three trajectories $\mathbf{h}_1 = [x_1, x_2]$, $\mathbf{h}_2 = [x_2, x_3]$, and $\mathbf{h}_3 = [x_3, x_4]$. Obviously the corresponding class labels are $g_1 = x_3$, $g_2 = x_4$, and $g_3 = x_5$. With this model we can exact that the user transited from x_2 to x_4 passing through x_3. Figure 9.6 depicts an area divided into cells in which several sliding windows of a user movement are shown.

Coordinate Model

In the coordinate model, we make use of time-stamped longitude and latitude, which represents more fine-grained location information than a single cell identifier (see *Cellular Model*). It is worth noting that such type of representation model requires exact position information regardless the network morphology. Specifically, in the coordinate model the movement history is

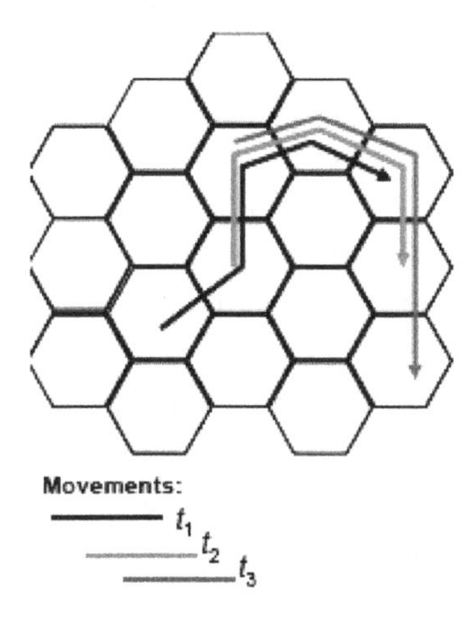

Fig. 9.6 The three trajectories h_1, h_2, and h_3 of a sliding window length $m = 4$. Note that the sliding windows are overlapping in a cellular network.

represented through a sequence of 3-D points (3DPs) visited by the moving user, i.e., time-stamped trajectory points in a 2D surface. The spatial attributes in that model denote longitude x and latitude y. Hence, $\mathbf{u} = (x, y, t)$ is a 3DP. We formulate the learning problem for location prediction $\mathbf{u}_0 = (x_0, y_0, t_0)$ as follows: *given a coordinate representation model derived from the movement history of a user, construct a predictor $\mathcal{P}$ that predicts the next 3DP $\mathbf{u}_0$ to which that user is going to move in at time t_0.*

The user trajectory $\mathbf{h}$ consists of m time-ordered 3DPs, $\mathbf{h} = [\mathbf{u}_i] = [\mathbf{u}_1, \ldots, \mathbf{u}_m]$, $i = 1, \ldots, m$. It holds true that $t(\mathbf{u}_1) < t(\mathbf{u}_2) < \cdots < t(\mathbf{u}_m)$, i.e., time-stamped coordinates; $t(\cdot)$ returns the time coordinate of $\mathbf{u}$. Time t assumes values between 00:00 and 23:59. To avoid state information explosion, trajectories may contain time-stamped points sampled at specific time instances. For instance, the movement of each user is sampled at $1.66 \cdot 10^{-3}$ Hertz, i.e., once every 10 minutes. Sampling at very high rates (e.g., in the order of a Hertz) is meaningless as the derived points will be highly correlated. If we assume that $\mathbf{h}$ is a finite sequence of m 3DPs, i.e., $\mathbf{h}$ is a $3 \cdot m$ dimension vector, and adopt a value of $m = 6$ then, we estimate the future position of a mobile terminal from a movement history of 50 minutes (i.e., 5 samples). The aim is to query the system with a $m - 1$ 3DPs sequence so that the predictor $\mathcal{P}$ returns a 3DP, which is the predicted location of the mobile terminal.

Query Trajectory

In order to conclude from a prediction algorithm the future user movement we have to construct a query. Such query refers to the historical spatial context of the mobile user for certain period of time. Specifically, a prediction algorithm refers to $n + l$ predictor $\mathcal{P}$ in the sense that, we monitor the spatial context of a given mobile user for n time units and require to predict the l future movements. In this chapter we deal with $n + 1$ predictors, that is, we query about the next user movement in time once the user has been monitors for the last n time units. We define *query trajectory* $\mathbf{q}$ as the $m - 1$ dimensional vector of $[\mathbf{u}_i]$, $i = 1, \ldots, m - 1$. The $\mathbf{u}_i$ locations are the *actual* locations of the user for the last ith movement. Given a $\mathbf{q}$ with a $m - 1$ history we predict the mth user movement thus obtaining a $(m - 1) + 1$ predictor $\mathcal{P}$.

Mobility Profile Model

We introduce the parameter *degree of movement randomness*, $\delta \in [0, 1]$, in order to express the randomness of the mobility behaviour of a user, i.e., the way a user transits between cells/coordinates and changes directions. Such degree is used for assessing the performance of the predictors under various levels of uncertainty and unpredictability w.r.t. mobility behaviour. The δ degree denotes the possible transition patterns of a user trajectory between locations.

A certain trajectory can derive either from a *deterministic* movement (assuming a low value of δ) or a *random* movement (assuming a high value of δ), thus, the adoption of δ provides an objective criterion for assessing movement prediction algorithms. It allows a correct interpretation of the performance evaluation results (e.g., a high accuracy may not necessarily indicate an efficient algorithm if the testing patterns were quite deterministic). The deterministic trajectories represent regular movements (e.g., the route from home to work). On the other hand, random trajectories represent purely random movements between predefined locations (e.g., a quick detour for a coffee after leaving home and before getting to work). Therefore, a value of $\delta \sim 1.0$ does not mean explicitly non-deterministic mobility behaviour. Instead, such movement is constrained by obstacles in the examined space. In this chapter, we adopted the mobility pattern generator discussed in [54]. Through this generator we obtained trajectories with specific δ values in the set $\{0.0, 0.25, 0.5, 0.75, 1.0\}$. The five discrete values of δ range from the regular pattern ($\delta = 0.0$) to completely random pattern ($\delta = 1.0$). It should be noted that, the value of δ influences the size of the training patterns (movement history) since the more random the movement is the more transitions are generally required for a certain itinerary (i.e., moving from a given origin to a given destination).

9.9 Location Prediction Algorithms

A prediction algorithm is capable of learning the user mobility behaviour through specific training (i.e., exploiting the training set T) and inferring the future user location $\mathbf{u}_0$. During the learning phase, the predictor $\mathcal{P}$ builds a Knowledge Base (KB) using the T set, KB(T), through which one can conclude, retrieve or revise the depiction of the mobility behaviour of a mobile user. The predictor based on KB(T) can act as follows:

- estimates $\mathbf{u}_0$ with certain probability $p(\mathbf{u}_0)$, and,
- updates and/or expands the KB(T) based on a reinforcement mechanism [55].

In this chapter, we study predictors that are trained once and are able to estimate $\mathbf{u}_0$ without revising the built KB(T).

Learning a Predictor

A number of different learning paradigms have been developed over the years for classification and prediction. Since there is no single classification algorithm that is better than all the others irrespective of the application domain, each time we face a new classification problem we have to assess the suitability of the algorithms. We experiment with several classification algorithms trying to cover a range, as broad as possible, of different learning paradigms for location prediction. In the next paragraphs, we give a description of the different types of learning paradigms that we use for location prediction.

- *Bayesian learning*. Bayesian classification algorithms are statistical learning algorithms based on the Bayes theorem. The Naïve Bayes algorithm (the simplest Bayesian classifier) [56] assumes that the effect of the value of an attribute on the class attribute is independent of the values of the other attributes given the value of the class attribute (*conditional independence*). That is, each value of $\mathbf{u}_i$ in $\mathbf{h}$ is independent of the value of g in the training tuple $(\mathbf{h}, g)$.

- *Decision Tree learning*. A decision tree consists of decision nodes (i.e., the value of $\mathbf{u}_i$) and leaves (i.e., the value of g). A leave is usually associated with a single class (e.g., g) that is the majority class of the training examples that arrive to that leave (for the training tuple $(\mathbf{h}, g)$). Splits are introduced in the building of the tree according to the outcome of a function f (information gain ratio). When examples are classified, a function is used in each split to determine the downstream path to be followed [51]. An indicative decision tree-based algorithm is the C4.5 classifier [57].
- *Rule-Induction learning*. Rule-induction performs a depth-first search in a graph $G(V, E)$ generating one path, that is, it constructs the appropriate pair $(\mathbf{h}, g)$, $\forall \mathbf{h} \in T$. V is a set of attributes and E is the set of edges denoting dependencies among attributes. Such path represented as a *classification rule* [58] is a conjunction of conditions with discrete or numeric attributes. In our case, the tuple $(\mathbf{h}, g)$ is represented as the rule: $\mathbf{u}_1 \wedge \mathbf{u}_2 \wedge \cdots \wedge \mathbf{u}_m \rightarrow g$ or as the rule IF $\mathbf{u}_1$ AND $\mathbf{u}_2$ AND ... AND $\mathbf{u}_m$ THEN $\mathbf{u}_0 = g$. A rule is said to *cover* an example (i.e., a query trajectory $\mathbf{q}$) if the later fulfils all the conditions of that rule. A representative rule-induction algorithm is the RIPPER (Repeated Incremental Pruning to Produce Error Reduction) [59].
- *Instance-based learning*. An instance-based algorithm uses a distance function $\|\mathbf{h}_i - \mathbf{q}\|$ in order to determine which example vector $\mathbf{h}_i$ of the training set T is kth closest to an un-classified example $\mathbf{q}$ (i.e., the query trajectory), $k > 0$. Once the nearest example vector $\mathbf{h}_i$ has been determined, its class label g_i is selected as the class label for $\mathbf{q}$. A representative instance-based algorithm is the k-Nearest Neighbour (kNN) classifier [51].
- *Ensemble-learning*. The ensemble learning algorithms combine a number of base classifiers/predictors, produced by different learning algorithms (as described above) or by different training sets, in order to achieve better classification performance than their constituents. Base models can be combined in different ways in

order to generate ensemble-learning algorithms. Popular ensemble-learning algorithms are the following:

- *Learning by Voting*: Each base classifier k predicts, *votes*, a class $g(k)$. The final class g_0 is that, which assumes the greatest number of votes, that is, $g_0 = \arg\max\{votes(g(k)), \forall k\}$.
- *Learning by Bagging*: Several training sub-sets T_i are formed from the initial training set T by random re-sampling with replacement. A base classifier is trained with each training sub-set T_i, i.e., it builds a KB(T_i). The final class is determined by voting of the base classifiers.
- *Learning by Boosting*: In boosting, the diversity of the base classifiers is a result of different training sets T_i. The method works in an iterative manner by re-sampling the current training dataset and by giving higher resample weights to tuples that are hard to classify. The final class is determined by a weighted voting of the base classifiers where the weights are determined on the basis predictive performance of the base classifiers. A typical example of a boosting algorithm is AdaBoost M1 boosting [51].

Analyzing a Predictor

We discuss in details the prediction algorithms based on their capacity of dealing with the cellular or coordinate representation models. For that, we distinguish the predictors into non-parametric and non-metric.

Non-parametric Predictor

The k-Nearest Neighbour (kNN)-predictor [60] is *memory-based* and requires no model to be fit. The characterization of memory-based indicates that *ts* stores all tuples from T in KB(T). Given a query trajectory $\mathbf{q}$, we find the k trajectories $\mathbf{h}_{(r)}, r = 1, \ldots, k$ from the tuples $(\mathbf{h}_{(r)}, g_{(r)})$ that are closest in distance to $\mathbf{q}$. This means that $\mathbf{u}_0 = g_{(r)}$. Then, the kNN predictor classifies $\mathbf{q}$ using majority voting among the k neighbours that is, $g_0 = \arg\max_{(r)}\{P(g_{(r)}|\mathbf{q})\}$. The value $P(g_{(r)}|\mathbf{q})$ is the probability for classifying $\mathbf{q}$ with class $g_{(r)}$. Thus, kNN classifies $\mathbf{q}$ by assigning to it the label most frequently appearing among the k nearest trajectories, i.e., the $\mathbf{u}_0 = g_0$.

When adopting the cellular model, a trajectory $\mathbf{h}$ assumes nominal values, i.e., CIDs. If $\gamma(\cdot, \cdot)$ denotes a generalization of the Kronecker delta function then $\gamma(x_i, x_q) = 0$ if the two arguments (CIDs) match and 1 otherwise, with $[x_i] = \mathbf{u}_i \in \mathbf{h}$ and $[x_q] = \mathbf{u}_q \in \mathbf{q}$. The distance in the feature space is then $\|\mathbf{h}_{(r)} - \mathbf{q}\| = \Sigma_j\{\gamma_j(x_{(r)j}, x_{qj})\}$, $j = 1, \ldots, m$.

For the kNN, the error on the training data is an increasing function of k and is always zero for $k = 1$. In fact, the 1-nearest-neighbor (1NN) classifier is typically adopted for low-dimensional problems [60] (in our case we could use $m = 3$ or $m = 4$ last positions). The 1NN partitions the movement space into convex hulls consisting of all trajectories closer to a $\mathbf{h}_i$ than to any other training trajectory. All trajectories of a convex hull are labelled by the corresponding g_i. Hence, the movement space is abstracted as a Voronoi tessellation. In addition, the bias estimate for the 1NN is low as long as it uses only the training trajectory closest to the query trajectory. The error rate of the 1NN, asymptotically, is never worse than twice the Bayes rate [61]. That is, when the minimum misclassification error assumed by the Bayes selection is small then the corresponding misclassification error for the 1NN is also small.

Non-metric Predictor

Decision trees are a common approach for classifying nominal data. Classification proceeds from top to bottom; each node denotes a particular cell x of the trajectory $\mathbf{q}$. A path from the root to a leaf corresponds to a specific trajectory that leads to a possible g class label, which appears in the tree as a leaf node. Each decision outcome at a node $r$is called a *split* and corresponds to splitting a subset $T(r)$ of T. Such decision is affected by specific entropy measures of the subtrees of the current node. However, rather than splitting each node into just two subset of T at each stage, we might consider multi-way splits into more than two subsets. The problem is that multi-way splits fragment the T set quickly, leaving insufficient data at the next level down, i.e., we defy the time sequence of cells of the user mobility pattern.

As long as, the multi-way split can be achieved by a series of (recursive) binary splits, the latter way is preferred. Hence, the predictor classifies a movement pattern in node r to the majority class $g_{(r)}$ that is, $g_{(r)} = \arg\max_g\{P_{(r)g}\}$. The class $g_{(r)}$ maximizes the quantity $P_{(r)g} = N_{(r)}^{-1} \cdot \Sigma_i\{I(g_i = g)\}$, where, $N_{(r)}$ is the number of observations in $T(r)$ with $\mathbf{h}_i \in T(r)$, $(\mathbf{h}_i, g_i) \in T$; I is unit if it holds true that $g_i = g$; otherwise zero. A particular case of such method is the C4.5 algorithm [57], which is quite popular. In C4.5 each leaf node has an associated *rule* — the conjunction of the decisions leading from root to that leaf, i.e., the timed sequence of a recorded trajectory. C4.5 also classifies movement patterns with missing values, i.e., unrecorded cells and even trajectories with gaps. Finally, the used C4.5 with binary splits has the capability of pruning/deleting redundant antecedents (e.g., cell transitions) in the formed tree.

9.10 Assessment of Mobility Prediction Algorithms

Once we train a $(m-1)+1$ predictor $\mathcal{P}$ with a KB(T), we are able to query $\mathcal{P}$ in order to obtain future user movements. This means that a $\mathcal{P}$ predictor has to:

- decide on the future location $\mathbf{u}_0$ with certain accuracy,
- make a decision on the future location $\mathbf{u}_0$ within an acceptable time frame, and,
- demand space resources as a little as possible.

Evidently, a $\mathcal{P}$ predictor decides on $\mathbf{u}_0$ with probability $p(\mathbf{u}_0) \in [0, 1]$ that is, we obtain that $\mathbf{u}$ from the movement space, which maximizes $p(\mathbf{u})$.

Probability of Correctness

We assess a $\mathcal{P}$ predictor with a series of test. Specifically, during the testing phase of the predictor, we employ a test set Y which was not used in the training phase, i.e., $Y \cap T = \emptyset$. The produced predictor with KB(T) is then applied on Y and the prediction accuracy is assessed. The prediction accuracy for the $\mathcal{P}$ predictor is quantified through the *probability of correctness*, $a(\mathcal{P})$. The $a(\mathcal{P})$ indicator is calculated by the correct predictions achieved by KB(T) over the test set Y. The value of a refers to the ratio of the correctly predicted trajectories in $Y' \subseteq Y$:

$$a(\mathcal{P}) = |Y'|/|Y|.$$

A popular method to estimate the a value for a $\mathcal{P}$ is the re-sampling method called *cross-validation* [62, 63]. In n fold cross-validation, the training set T is divided into n subsets T_i, $i = 1, \ldots, n$, of equal size. A predictor with KB(T) is trained n times, each time setting aside one $Y = T_i$, which will be used as the test set for the determination of the a value. The reported value is the average value for the n repetitions of the learning and testing phases. Usually, $n = 10$.

Statistical Significance

A predictor $\mathcal{P}$ depends heavily on the training set T thus there is a need for *statistical testing* in order to:

- assess the expected error rate of the predictor, and,
- compare the expected error rates of two predictors, thus, reasoning about their relative efficiency.

It is important to determine whether the difference $|a(\mathcal{P}_1) - a(\mathcal{P}_2)|$ for $\mathcal{P}_1$ and $\mathcal{P}_2$ predictors is *statistically significant*, i.e., whether the difference in the a values for $\mathcal{P}_1$ and $\mathcal{P}_2$ is not attributed to statistical errors. The statistical significance $ss(\mathcal{P}_1, \mathcal{P}_2) \in \{0, 1\}$ is determined through the *McNemar* test [64]. By adopting such test, we are interested either in the proportion of the correctly classified trajectories by $\mathcal{P}_1$ and in the proportion of the incorrectly classified trajectories by $\mathcal{P}_2$ for the same training set or the opposite. Both $\mathcal{P}_1$ and $\mathcal{P}_2$ are trained and tested with the same T and Y sets. Hence, the McNemar test indicates whether a correct prediction by $\mathcal{P}_1$ and a false prediction by $\mathcal{P}_2$ is *more* or *less* likely than the reverse. In the case that, $\mathcal{P}_1$ and $\mathcal{P}_2$ are statistically insignificant, i.e., $ss(\mathcal{P}_1, \mathcal{P}_2) = 0$, the difference of their probabilities of correctness is meaningless and both predictors can be safely considered as equivalent; otherwise $ss(\mathcal{P}_1, \mathcal{P}_2) = 1$. The threshold level for the statistical significance is typically set to 0.05 [51]. This value derives from the chi-square distribution with one degree of freedom.

Experimentation

We make use of the ML classifiers implemented in Java in the Weka[1] workbench [65]. Specifically, we experiment with the abovementioned learning types of location predictors adopting the cellular and the sliding window cellular model. We use five discrete categories of randomness from the regular pattern ($\delta = 0.0$ with 500 training tuples) to completely disordered trajectories ($\delta = 1.0$ with 1000 training tuples). We experiment with the following classifiers from the Weka workbench:

- the Naïve Bayes classifier,
- the J48 Decision Tree-based classifier (an implementation of the C4.5 algorithm),

[1]Download Weka from http://www.cs.waikato.ac.nz/ml/weka/

- the Jrip Classification Rule classifier (an implementation of the RIPPER algorithm), and,
- the Ibk incremental algorithm (an implementation of kNN algorithm).

From the category of the Ensemble learning algorithms we experimented with:

- Voting with the base classifiers constructed by J48 and Ibk,
- Bagging were the base classifiers were learned by IBk, and, finally,
- AdaBoost M1, where the base classifiers were also learned by IBk.

The prediction accuracy a for each predictor algorithm is estimated by 10-fold cross-validation.

At first, we compare the predictive performance of each prediction algorithm with the default accuracy, i.e., the prediction accuracy obtained from a naïve classifier that always predicts the majority class. A predictor is considered appropriate for a certain prediction problem if its probability of correctness is significantly better than the default accuracy. Table 9.1 depicts the values of default accuracy w.r.t., the different levels of randomness δ. Figure 9.5 depicts the probability of correctness a of the IBk, J48, Naïve Bayes, JRip, Vote, Bagging and AdaBoost M1 predictors for the cellular representation. The probability of correctness of all algorithms is significantly better than the default accuracy for any given δ (see Table 9.1). The algorithm that achieves the top performance (for all levels of randomness) is the Vote algorithm constructed by the J48 and Ibk algorithms.

An important factor is the selection of the sliding window size m. The appropriate value of m is estimated by experimentation. In Figure 9.7, we can observe the probability of correctness a (Vote) of the Vote predictor adopting the sliding window cellular model having $2 \leq m \leq 20$ and a training set of

Table 9.1. The value of the majority class w.r.t degree of randomness γ

Degree of Randomness (δ)	Majority Class ratio
0%	9.06%
25%	6.53%
50%	5.48%
75%	3.81%
100%	3.69%

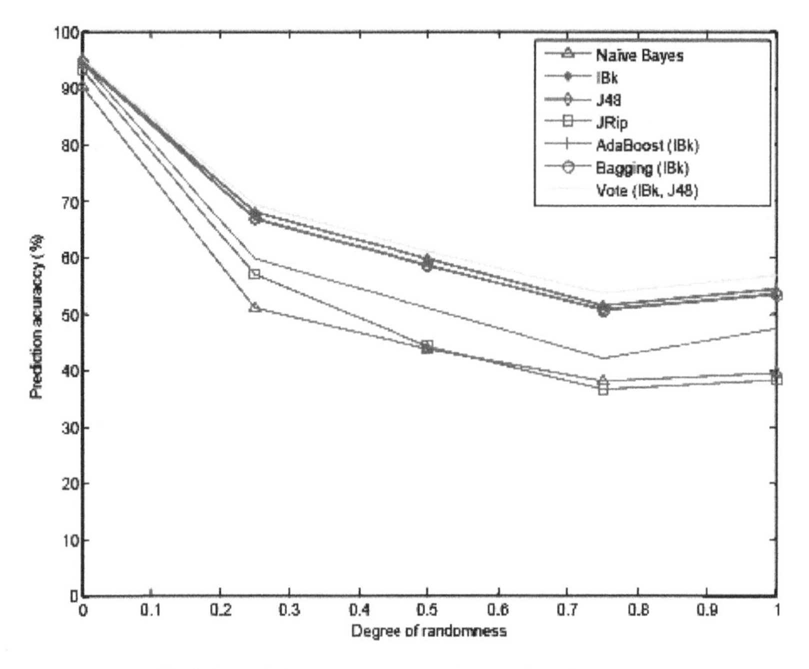

Fig. 9.7 Prediction accuracy (a) vs. degree of randomness δ.

tuples T for 40 days. The best results are obtained for $m = 4$. The case $m = 4$ assumes better prediction results even when the user exhibits a high degree of randomness. When the size m is small the considered vectors point at many different class attributes (almost equiprobable) hence prediction accuracy falls. When the size m raises, the considered vectors of cell identifiers is quite large to achieve a close match (between sample and test data) when randomness increases. The size $m = 4$ reflects the regularity of a common user movement.

9.11 Summary of Chapter

The goal of this chapter was, on the one hand, to present new trends on cooperation among entities in wireless networks, and on the other hand to provide an insight on tools that enable a more efficient implementation of cooperation schemes. These tools are game theory, which models networks entities as rational players competing with each other for a better distribution of resources, and mobility prediction, which estimates nodes' future location so that resource exchange and group formation can be more accurate. More specifically, we have attempted to demonstrate how game theory can

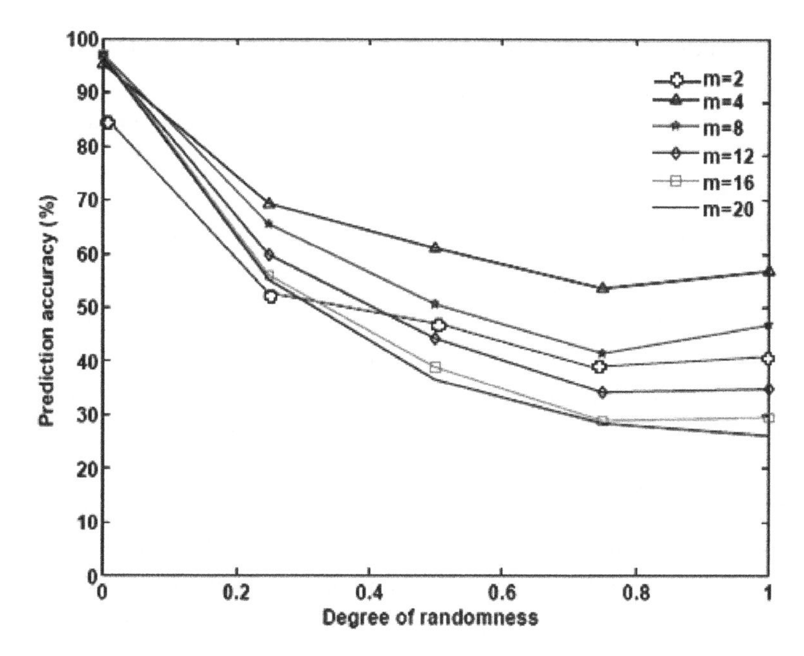

Fig. 9.8 Prediction accuracy a (Vote) vs. window length m.

be applied to wireless networking in order to optimize resource allocation. Throughout the modeled games highlighted, is should be clear how game-theoretic solutions may effectively predict/simulate realistic user behavior in competitive or cooperative scenarios. In addition, we study issues related to mobility prediction based on Machine Learning algorithms. We report certain representation models adopted by classifiers and examine their applicability in mobility prediction. Furthermore, we refer methods that assess the prediction algorithms like probability of correctness and certain statistical measures.

Problems on Game Theory

Problem 9.1

Which elements does one have to specify in order to model a problem as a game?

Solution

A game can be modeled as $G = (P, A, S_i, \pi_{ij})$ where:

- $P = \{1, \ldots, n\}$ denotes the set of players
- $A = \{1, \ldots, n\}$ denotes the available resources in the game (action set)

- S_i denotes the set of strategies for player i, i.e., all possible choices of a specific request from set A.
- π_{ij} denotes the payoff assigned to player i after choosing resource j.

Problem 9.2

Consider the following network selection problem. Two WLAN networks are available, each one offering an overall data rate of 100 Mbps. Furthermore, four different types of services are considered, all of the same priority. The goal is to distribute six requests optimally between the two networks. The utility function for each network is calculated by $U = \mathrm{NP} \times \frac{B_{\mathrm{av}}}{B_{\mathrm{req}}}$ where NP is a constant with values between 0 and 1, B_{av} is the available bandwidth and B_{req} is the bandwidth requested by the service request. The objective is to model the addressed problem as a non-cooperative game between the available networks played in rounds. In each round the networks will select the most appropriate request, as indicated by the utility function. For simplicity reasons, assume that in such cases, where two equivalent options are available, the networks will choose the first one, since it possesses chronological priority. The following table sums up the information needed for the calculation of utilities. Do not forget that the available bandwidth for each network changes after a new request is accepted! Which requests are handled by WLAN1 and which ones by WLAN2?

	Web browsing	Access to corporate database	Video download	Peer-to-peer file sharing
Required Data Rate	512 kbps	512 kbps	5 Mbps	5 Mbps
NP for WLAN 1	0.578	0.615	0.76	0.673
NP for WLAN 2	0.789	0.875	0.584	0.545
Service Requests	R1, R4	R3, R6	R2	R5

Solution

WLAN 1 \ WLAN 2												
Services	R1		R2		R3		R4		R5		R6	
R1			113	12	113	106	113	154	113	18	113	106
R2	15	154			15	106	15	154	15	18	15	106
R3	131	154	131	12			131	154	131	18	131	106
R4	113	154	113	12	113	106			113	18	113	106
R5	12	154	12	12	12	106	12	154			12	106
R6	131	154	131	12	131	106	131	154	131	18		

WLAN 1 \ WLAN 2								
Services	R2		R4		R5		R6	
R2			15	153	15	17	15	106
R4	112	12			112	17	112	106
R5	12	12	12	153			12	106
R6	131	12	131	153	131	17		

WLAN 1 \ WLAN 2				
Services	R2		R5	
R2			15	17
R5	12	12		

WLAN 1: R3, R6, R2

WLAN 2: R1, R4, R5

Problems on Mobility Prediction

Problem 9.3

Consider the cellular model and a network of X cells. Assume that $m = 5$, that is, the predictor monitors five movements. Build an extended representation cellular model, say *directional model*, which in each trajectory **h** stores the eight possible directions of the mobile user, w.r.t. hexagonal cell shape, having also the last known position. Specifically, the **h** trajectory consists of m attributes that is $\mathbf{u}_i = [d_i], i = 1, \ldots, m$ and the corresponding future direction is g_i. In this model, we can predict the future x_0 once we have predicted the next direction g_i and we have stored the last known visited cell x_m. Hence, $\mathbf{h} = [d_1, d_2, \ldots, d_m]$. The possible directions d_i can be the North (N), North-West (NW), West (W), South-West (SW), South (S), South-East (SE), East (E), and North-East (NE). A possible trajectory for $m = 4$ is $\mathbf{h}_1 =$ [N, N, NW, W] and the user is located in x_1 directed from NW to W (the last direction transition in $\mathbf{h}_1$). If that trajectory is assigned the class label $g_1 = W$, then the user will be located to the western cell of x_1.

- Create a directional predictor $\mathcal{P}_d$ adopting the Naïve Bayes classifier from Weka[2], which is trained with data represented through the considered directional model. Then, test the predictor and assess its probability of correctness.
- Would the directional model be appropriate for highly disordered paths (the user changes directions with a high frequency)?
- Consider those users that are driving to high ways. Would this model be appropriate for prediction?
- A multiple appearance of the trajectory [NW, NW, NW, NW] what could mean for the Naïve Bayes classifier?

Problem 9.4

Consider the sliding window cellular model and a network of X cells. Assume that a predictor monitors the mobile user for the m lastly visited cells and records the corresponding time stamps. Hence, a possible representation model, say *temporal cellular* model could be: $\mathbf{h} = [x_i, t_i, x_{i+1}, t_{i+1}, \ldots, x_m, t_m]$, where t_i is the time-stamp that the user locates

[2] The Weka Naïve Bayes classifier is implemented in the Java class: weka.classifiers.bayes.NaiveBayes

in x_i cell. The class label for such model is the future cell x_0. The time-stamp feature assumes value in two different sets: The Nominal set $N = \{morning, noon, afternoon, evening, night\}$ and the Concrete set $C = [00:00, 23:59]$, i.e., the time in seconds.

- Create a temporal predictor $\mathcal{P}_t(N)$ adopting the J48 (an implementation of the C4.5 classifier) classifier from Weka[3], which is trained with data represented through the considered model with the Nominal set. Then, test the predictor and assess its probability of correctness.
- Repeat the first problem with a temporal predictor $\mathcal{P}_t(C)$ built on the Concrete set. Then, compare both predictors $\mathcal{P}_t(N)$ and $\mathcal{P}_t(C)$ w.r.t. the probability of correctness.
- Examine the behaviour of the probability of correctness of the $\mathcal{P}_t(N)$ predictor for diverse values of the m parameter. For instance, $m = 4$, $m = 8$, $m = 16$, and $m = 32$.
- Which is the behaviour of the $\mathrm{P}_t(\mathrm{N})$ predictor for large values of m?

Problem 9.5

Consider the coordinate model. Implement the kNN classifier $\mathcal{P}_{k,m}$ using the Euclidean distance φ between locations $\mathbf{u}_i$ and $\mathbf{u}_j$, i.e., $\varphi^2(\mathbf{u}_i, \mathbf{u}_j) = (x_i - x_j)^2 + (y_i - y_j)^2 + (t_i - t_j)^2$. The entire distance Φ between trajectories $\mathbf{h}_p$ and $\mathbf{h}_q$ is: $\Phi^2(\mathbf{h}_p, \mathbf{h}_q) = m^{-1/2}(\Sigma_{l=1,\ldots,m}\varphi^2(\mathbf{u}_{il}, \mathbf{u}_{jl}))$. Assume $m = 4$ and $m = 8$ and experiment with $k = 1$, $k = 4$ and $k = 10$.

- Examine the behaviour of the probability of correctness of the $\mathcal{P}_{k,m}$ predictor for diverse value of k and m. Which is the behaviour of the $\mathcal{P}_{k,m}$ predictors for large value of k and low value of m and vice versa? Use time-stamped training data sets with GPS coordinates from the site: http://www.openstreetmap.org/traces.
- Repeat the same problem adopting the distance: $\varphi^2(\mathbf{u}_i, \mathbf{u}_j) = (x_i - x_j)^2 + (y_i - y_j)^2$ if $|t_i - t_j| > \varepsilon$; 0, otherwise. The value of ε is $30'$ that is two trajectories that have occurred within a half hour are considered the same. The value of ε denotes the time-granularity for considering two trajectories equivalent in time.

---

[3]The Weka Id3 tree-based classifier is implemented in the Java class: weka.classifiers.trees.j48

References

[1] F. H. P. Fitzek and M. Katz, "Cooperation in Wireless Networks: Principles and Applications — Real Egoistic Behavior is to Cooperate!", Springer, April 2006.

[2] L. Militano, F. H. P. Fitzek, A. Iera, and A. Molinaro, "On the beneficial effects of cooperative wireless peer to peer networking. In Tyrrhenian International Workshop on Digital Communications (TIWDC'07), Ischia, Italy, September 2007.

[3] Q. Zhang, F. H. P. Fitzek, and M. Katz, "Evolution of Heterogeneous Wireless Networks: Cooperative Networks". 2006. In 3nd International Conference of the Center for Information and Communication Technologies (CICT) — Mobile and wireless content, services and networks — Short-term and long-term development trends, Copenhagen, Denmark, November 2006.

[4] S. Frattasi, H. Fathi, and F. H. P. Fitzek, "4G: A User-Centric System," 2006. Kluwer — Wireless Personal Communications Journal (WPC) — Special Issue on Advances in Wireless Communications: Enabling Technologies for 4G.

[5] S. Frattasi, H. Fathi, F. H. P. Fitzek, M. Katz, and R. Prasad, "Defining 4G Technology from the User Perspective," *IEEE Network Magazine* 20(1): 35–41, 2006.

[6] M. Katz and F. H. P. Fitzek, "Cooperative Techniques and Principles Enabling Future 4G Wireless Networks," 2005. In The International Conference on EUROCON 2005, pp. 21–24.

[7] S Frattasi, B Can, F Fitzek, and R Prasad, "Cooperative Services for 4G," *Proceedings of the 14th IST Mobile & Wireless Communications*, 2005.

[8] G. P. Perrucci, F. H. P. Fitzek, A. Boudali, M. Canovas Mateos, P. Nejsum, and S. Studstrup, "Cooperative Web Browsing for Mobile Phones," In International Symposium on Wireless Personal Multimedia Communications (WPMC'07), India, 2007.

[9] E. Valavanis, C. Ververidis, M. Vazirgianis, G. C. Polyzos, and K. Nørvåg, "MobiShare: Sharing Context-Dependent Data and Services from Mobile Sources," *Proc. IEEE/WIC Int'l Conference on Web Intelligence (WI 2003)*, Halifax, Canada, October 2003.

[10] Dan Zhang Ileri, O. Mandayam, N., "Bandwidth exchange as an incentive for relaying," 42nd Annual Conference on Information Sciences and Systems, 2008 (CISS 2008).

[11] N. Shastry and R. S. Adve, "Stimulating Cooperative Diversity in Wireless Ad Hoc Networks through Pricing," *IEEE International Conference on Communications*, 2006 (ICC '06).

[12] L. Buttyan and J. P. Hubaux, "Stimulating Cooperation in self-organizing mobile *ad hoc* networks," ACM/Kluwer MONET, vol. 8, pp. 579–592, October 2003.

[13] P. Michiardi and R. Molva, "Core: A cooperative reputation mechanism to enforce node cooperation in mobile ad hoc networks," In Communications and Multimedia Security Conference (CMS), 2002.

[14] M. Tamer Refaei, Vivek Srivastava, Luiz DaSilva, Mohamed Eltoweissy, "A Reputation-based Mechanism for Isolating Selfish Nodes in Ad Hoc Networks," Second Annual International Conference on Mobile and Ubiquitous Systems: Networking and Services (MobiQuitous'05).

[15] X. Lin, N. B. Shroff, and R. Srikant, "A Tutorial on Cross-Layer Optimization in Wireless Networsk," IEEE Journal on Selected Areas in Communications, 24(8), August 2006.

[16] F. Aune, "Cross-Layer Design Tutorial, Norwegian University of Science and Technology, Department of Electronics and Telecommunications, Trondheim, Norway, published under Creative Common License, November 2004.

[17] G. Giambenne and S. Kota, "Cross-layer protocol optimization for satellite communication networks: A survey," International Journal of Satellite Communications and Networking, 24, pp. 323–341, 2006.

[18] J. P. Ebert and A. Wolisz, "Combined tuning of RF power and medium access control for WLANs," Mobile Networks and Applications, 6(5), Special Issue on Mobile Multimedia Communications, 2001.

[19] J. Zhao, Z. Guo and W. Zhu, "Power efficiency in IEEE 802.11a WLAN with cross-layer adaptation," *IEEE ICC2003*, 2003.

[20] R. Ferrus, L. Alonso, A. Umbert et al., "Cross layer scheduling strategy for UMTS downlink enhancement," *IEEE Communications Magazine*, pp. S.24–S.28, 2005.

[21] V. Atanasovski and L. Gavrilovska, "Throughput performance of contenting IEEE 802.11a Systems in Rayleigh Fading Channels," WPMC 2004, pp. 2925–1929.

[22] G.-C. Song and Y. (G.) Li, "Cross-layer optimization for OFDM wireless networks — Part I: theoretical framework," *IEEE Transactions on Wireless Communications*, 4(2), pp. 614–624, March 2005.

[23] G.-C. Song and Y. (G.) Li, "Cross-layer optimization for OFDM wireless networks — Part II: algorithm development," *IEEE Transactions on Wireless Communications*, 4(2), pp. 625–634, March 2005.

[24] X. Qiu and K. Chawla, "On the performance of adaptive modulation in cellular systems," *IEEE Transactions on Communications*, 47(6), pp. 884–895, June 1999.

[25] Z. Jiang, Y. Ge, and Y. (G.) Li, "Max-utility wireless resource management for best effort traffic," *IEEE Transactions on Wireless Communications*, 4(1), pp. 100–111, January 2005.

[26] A. Sendonaris, E. Erkip, and B. Aazhang, "User cooperation diversity. Part I. System description," *Communications, IEEE Transactions on*, 51(11), pp. 1927–1938, November 2003.

[27] J. N. Laneman, D. N. C. Tse, and G. W. Wornell, "Cooperative diversity in wireless networks: Efficient protocols and outage behavior," *Information Theory, IEEE Transactions on*, 50(12), pp. 3062–3080, December 2004.

[28] E. Zimmermann et al., "On the performance of cooperative relaying protocols in wireless networks," *European Transactions on Telecommunications*, 16, pp. 5–16, 2005.

[29] M. Hasna, and M. Alouini, "A performance study of dual-hop transmissions with fixed gain relays," *IEEE Transactions on Wireless Communications*, 3(6), 1963—1968, 2004.

[30] S. M. Alamouti "A simple transmit diversity technique for wireless communications" *IEEE Journal on Selected Areas in Communications*, 16(8): 1451–1458, October 1998.

[31] H. Zhu and G. Cao, "rDCF: a relay enabled medium access control protocol for wireless ad hoc networks," in Proceedings of IEEE INFOCOM 2005.

[32] P. Liu, et al., "Cooperative wireless communications: A cross-layer approach," *IEEE Wireless Communications Magazine*, 13(4), pp. 84–92, August 2006.

[33] Amir Khandani, Jinane Abounadi, Eytan Modiano, Lizhong Zhang, "Cooperative Routing in Wireless Networks," Allerton Conference on Communications, Control and Computing, October, 2003.

[34] Le Long, Ekram Hossain, "Cross-Layer Optimization Frameworks for Multihop Wireless Networks Using Cooperative Diversity," *IEEE Transactions on Wireless Communications*, 7(7), July 2008.

[35] D. Fudenberg and J. Tirole, *Game Theory*. MIT Press, 1991.

[36] E. Altman, T. Boulogne, R. El-Azouzi, T. Jimenez, and L. Wynter, "A survey on networking games in telecommunications," *Computers and Operations Research*, 33(2), pp. 286–311, February 2006.

[37] Mark Felegyhazi and Jean-Pierre Hubaux, "Game Theory in Wireless Networks: A Tutorial," in EPFL technical report, LCA-REPORT-2006-002, February, 2006.

[38] Allen B. MacKenzie and Luiz A. DaSilva, "Game Theory for Wireless Communications," Morgan and Claypool Publishers, 2006.

[39] Allen B. MacKenzie and Stephen B. Wicker, "Game Theory and the Design of Self-Configuring, Adaptive Wireless Networks," *IEEE Communications Magazine*, pp. 126–131, November 2001.

[40] D. Charilas, O. Markaki, E. Tragos, "A Theoretical Scheme for applying game theory and network selection mechanisms in access admission control," International Symposium on Wireless Pervasive Computing (ISWPC) 2008, May 2008.

[41] H. Lin et al., "ARC: An Integrated Admission and Rate Control Framework for Competitive Wireless CDMA Data Networks Using Noncooperative Games," *IEEE Transactions on Mobile Comp.*, 4(3), pp. 243–258, May–June 2005.

[42] P. Vlacheas, D. Charilas, E. Tragos, and O. Markaki, "Maximizing Quality of Service for Customers and Revenue for Service Providers through a Noncooperative Admission Control Game," ICT Mobile Summit 2008, June 2008, Stockholm

[43] Allen B. MacKenzie, Stephen B. Wicker, "Game Theory in Communications: Motivation, Explanation and Application to Power Control," IEEE Global Telecommunications Conference 2001, vol. 2, pp. 821–826.

[44] Luiz A. DaSilva and Vivek Srivastava, "Node Participation in Ad Hoc and Peer-to-Peer Networks : A Game-Theoretic Formulation" in Workshop on Games and Emergent Behavior in Distributed Computing Environments, Birmingham, U K, September 2004.

[45] A. Economides and J. Silvester, "Multi-objective routing in integrated services networks: A game theory approach," in IEEE INFOCOM Networking in the 90s/Proceedings of the 10th Annual Joint Conference of the IEEE and Communications Societies, vol. 3, pp. 1220–1227, 1991.

[46] V. Srivastava, J. A. Neel, A. B. MacKenzie, J. E. Hicks, L. A. DaSilva, J. H. Reed, and R. P. Gilles, "Using Game Theory to Analyze Wireless Ad Hoc Networks," *IEEE Communications Surveys and Tutorials*, 7(5), Fourth Quarter 2005.

[47] A. Dey, "Understanding and using context," *Personal and Ubiquitous Computing*, 5(1), pp. 4–7, 2001.

[48] J. Hightower, G. Borriello, "Location Systems for Ubiquitous Computing," *IEEE Computer*, 34(8), August 2001.

[49] A. Salkham, R. Cunningham, and A. Cahill, "A taxonomy of collaborative context-aware systems," Ubiquitous Mobile Information and Collaboration Systems, 2006, pp. 899–911.

[50] P. Brezillon, M. Borges, J. Pino, and J. Pomerol, "Context-awareness in group work: Three case studies," *Decision Support Systems*, 2004, pp. 115–124.

[51] E. Alpaydin, *Introduction to Machine Learning*. The MIT Press, 2004.

[52] G. Liu and G. Maguire Jr. "A Class of Mobile Motion Prediction Algorithms for Wireless Mobile Computing and Communications," MONET, 1, pp. 113–121, 1996.

[53] A. Bhattacharya, S. Das, LeZi "Update: An Information Theoretic Approach to Track Mobile Users In PCS Networks," Proceedings of ACM/IEEE Mobicom, 1999.

[54] M. Kyriakakos, N. Frangiadakis, S. Hadjiefthymiades, and L. Merakos, RMPG: "A Realistic Mobility Pattern Generator for the Performance Assessment of Mobility Functions," Simulation Modeling Practice and Theory, 12(1), Elsevier, 2004.

[55] K. Narendra and M. A. L. Thathachar, "Learning Automata — An Introduction," Prentice Hall, 1989.

[56] J. Han and M. Kamber, "Data Mining: Concepts and Techniques," Morgan Kaufmann Series in Data Management Systems, 2001.

[57] J. R. Quinlan, "C4.5: Programs for Machine Learning," Morgan Kaufmann Series in Machine Learning, 1993.

[58] M. Mahmood, M. Zonoozi, and P. Dassanayake, "User Mobility Modeling and Characterization of Mobility Patterns," IEEE Communications, 15(7), 1997.

[59] W. Cohen, A. Prieditis, and S. J. Russel, "Fast Effective Rule Induction," Proc. Int. Conf. on Machine Learning, 1995.

[60] T. Hastie, R. Tibshirani, and J. Friedman, "The Elements of Statistical Learning; Data mining," Inference and Prediction, Springer, NY, 2001.

[61] T. Cover, and P. Hart, "Nearest neighbor pattern classification," *IEEE Trans. On Information Theory*, 13(1), 1967.

[62] M. Weiss and C. Kulikowski, "Computer Systems That Learn: Classification and Prediction" Methods from Statistics, Neural Networks, Machine Learning and Expert Systems, MK in Machine Learning, 1991.

[63] M. Plutowski, S. Sakata, and H. White, "Cross-validation estimates integrated mean squared error," Advances in Neural Information Processing Systems, 6, 1994.

[64] T. Dietterich, "Approximate Statistical Tests for Comparing Supervised Classification Learning Algorithms," Neural Computation, 10: 1895–1923, 1998.

[65] I. Witten and E.Frank, "Data Mining: Practical Machine Learning Tools and Techniques," Morgan Kaufmann Series in Data Management Systems, 2005.

10

On the Sensor Placement Problem in Directional Wireless Sensor Networks

Yahya Osais, Marc St-Hilaire and F. Richard Yu

Carleton University, Canada

The Sensor Placement (SP) problem is a fundamental planning problem in directional Wireless Sensor Networks (WSNs). It manifests itself in the desire of network designers to minimize the network cost while meeting the requirements of coverage, lifetime and connectivity. If the cost, sensing region and sensing range of each sensor are predetermined, the SP problem can be formulated as either a minimum SP problem or a minimum cost SP problem. On the other hand, if the optimal configuration of each sensor is to be determined, the SP problem can be posed as an optimal sensor configuration problem. The goal of this chapter is to introduce the reader to the SP problem in directional WSNs and its various mathematical formulations. The necessary motivation and background will be provided to enable the reader to acquire the essence of the SP problem. Also, examples will be given to illustrate the viability and effectiveness of the mathematical models. Finally, directions for further research will be suggested.

10.1 Introduction

Wireless Sensor Networks (WSNs) are built from sensor nodes which are tiny devices equipped with multiple on-board sensors, a microcontroller and a wireless interface. Sensors collect information about different phenomena (e.g., temperature, concentration of a substance and light intensity) and send it to one or more base stations. The collected information provides the necessary situational awareness required by decision makers.

The limited energy supply and communication capacity of individual sensors create many technical challenges for designers of WSNs. One such challenge is sensor placement in which the types, numbers, locations and configurations of sensors are determined so that the total network cost is minimized while the requirements of coverage, connectivity and lifetime are satisfied. Sensor placement is an important task in planning WSNs because it can have a significant influence on the overall network performance. Performance of WSNs is based on the levels of coverage and connectivity they provide for the longest possible duration of time.

Planning of traditional WSNs has received excellent attention from the WSNs community. However, when a traditional WSN is equipped with directional sensors, network planning problems cannot be treated in the same way as in traditional WSNs. Also, the algorithms designed for traditional WSNs behaves differently when used in directional WSNs (for example, see [1, 2]). In addition to that, the mathematical models developed for traditional WSNs cannot directly be used for optimizing directional WSNs due to the different parameters involved.

Directional WSNs inherit all the technical challenges introduced by traditional WSNs. In addition, they introduce new ones that are unique to them. For example, they are expected to be capable of carrying video data which typically requires much higher bandwidth than that required by traditional WSNs whose main purpose is to observe and sample the surrounding environment. Another example is the directionality of the sensors which has significant impact on the overall network coverage.

There is a pressing need for efficient and effective directional WSN planning techniques because of the following reasons:

1. The current cost of an average wireless sensor node is around several hundred US dollars. For example, an ordinary Crossbow sensor node consisting of the MICAz mote and MTS400 multi-sensor board costs over US$300 [3]. In addition, if we take into consideration the cost of batteries and networking devices, the

cost of deploying a large-scale (> 3000 nodes) WSN would easily exceed a million US dollars.

2. There might be a limit on the maximum number of sensor nodes to be used for performing a mission. This could be due to a limited budget or payload.
3. Deterministic placement may be favored over random placement. This is because random placement can introduce non-efficient spending of energy. Also, it may result in limited coverage and connectivity, e.g., some sensors may not survive the air drop.
4. The wireless sensor environment is often hostile and contains complex terrains. It is also an interference-rich environment. Therefore, sensing and communication capabilities of sensor nodes are greatly affected. Also, network communication increases, which results in more interference and contention for medium access.
5. A large part of the job of a WSN is to collect and convey data. Redundancy is naturally present in data due to the high density of sensor nodes. Therefore, it is desirable to minimize redundancy but ensures that the ensemble of the sensed data contains sufficient information for the sink to subsequently query a small number of sensors for detailed information.

This chapter presents three network planning problems that naturally arise during the planning phase of directional WSNs. They are called the Minimum Sensor Placement (MSP), Minimum Cost Sensor Placement (MCSP) and Optimal Sensor Configuration (OSC) problems. These problems have received very little attention in the literature. Up to now, most of the published works concentrate on optimizing directional WSNs after they are randomly deployed in a sensor field. The proposed techniques involve organizing the directions of sensors to maximize the coverage and lifetime of the deployed WSNs.

The chapter is organized as follows. In Section 2, the necessary background and notation are explained. In Section 3, the relevant literature is reviewed and critiqued. In Section 4, the problem statement is presented. In Section 5, the mathematical formulations of the MSP, MCSP and OSC problems are described. In Section 6, numerical results are discussed. Finally, in Section 7, some conclusions and possible future directions are given.

10.2 Background

We envision the following problem. A directional sensor network is to be deployed in a field. An engineer is assigned the task of studying the field and coming up with a floorplan for it. The floorplan consists of a set of placement sites for directional sensors, set of control points to be monitored by the sensors and location of the base station. Control points may represent regions of interest in the sensor field or points at which a target occurs (or is anticipated to occur). Figure 10.1 shows a typical floorplan for a 2D sensor field.

The behavior of a sensor is dictated by many parameters. However, we consider only parameters that play a role in the issues relevant to this work. A sensor can completely be characterized by the following parameters (see Figure 10.2):

1. (x_i, y_i): the Cartesian coordinates that denote the location of the sensor in a two-dimensional plane,
2. v_{id}: a unit vector that cuts the sensing sector into half. This parameter defines the direction (or orientation) of the sensor,
3. φ_i: the maximum angle of sensing that can be achieved by the sensor. It is also called the Field of View (FOV), and

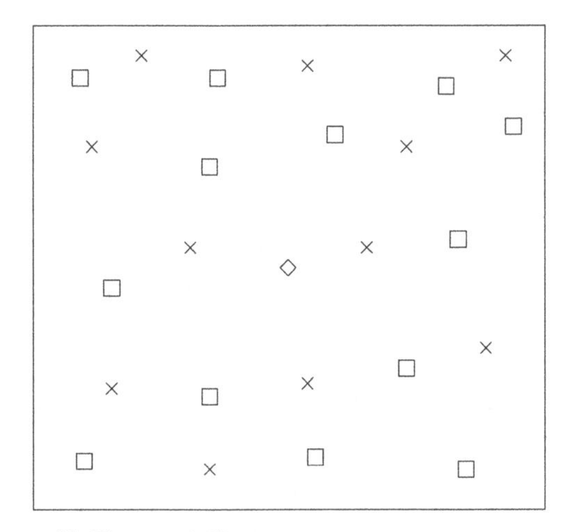

Fig. 10.1 A typical floorplan for a 2D sensor field.

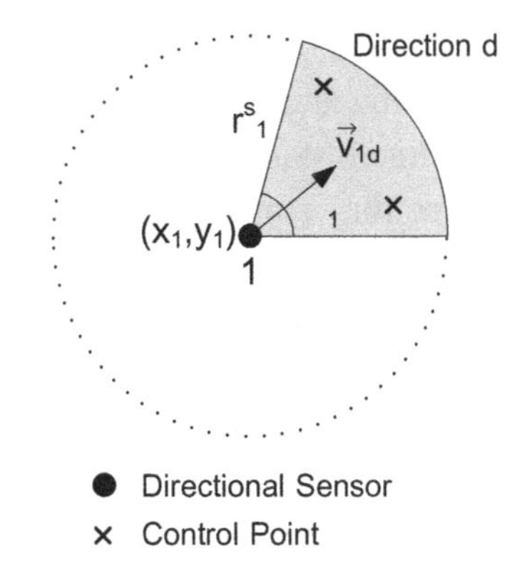

Fig. 10.2 A directional sensor monitoring two control points.

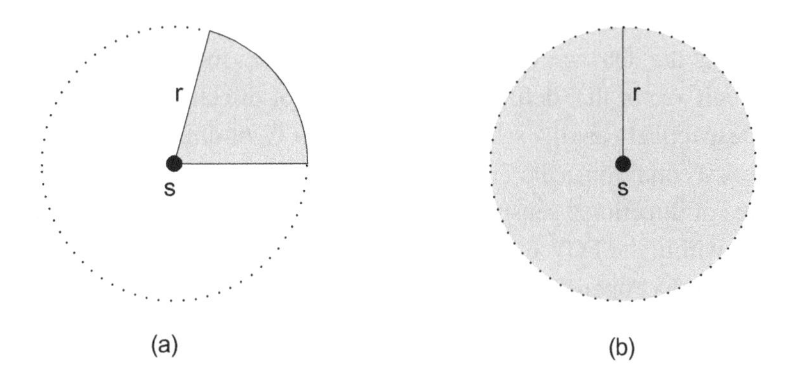

Fig. 10.3 Sensing region of (a) a directional sensor and (b) an isotropic sensor.

4. r_i^s: the maximum sensing range of the sensor beyond which a control point cannot be monitored.

Sensors can be distinguished by their FOV. Those whose FOV is 360° are called *isotropic*. On the other hand, those whose FOV is less than 360° are called *directional*. Figure 10.3(a) and 10.3(b) respectively illustrate the sensing region of a directional sensor and isotropic sensor. The shadowed area is the current sensing region of the sensor. As can be seen from the figure, the sensing region of an isotropic sensor is represented as a disk around the sensor node and that of a directional sensor is represented as a sector.

The FOV of a directional sensor is defined as the angle whose sides define the range of vision of the sensor. The set of possible FOVs include the angles from 0° to 360°. On the other hand, the direction of a directional sensor is defined by the angle between the positive x-axis and its unit vector. The set of possible directions include the angles from 1° to 360°. A sensor can only choose one direction at any time instant.

With each choice of direction, sensing range and FOV, a certain subset of control points will be covered by the directional sensor. The relationship between a directional sensor and a control point is determined by the Target In Sector (TIS) test [4]. This test states that a control point j is covered by a directional sensor i if the following two conditions are true:

$$\|\vec{d}_{ij}\|_2 \leq r_i^s \tag{10.1}$$

$$\vec{v}_{id} \cdot \vec{d}_{ij} \geq \|\vec{d}_{ij}\|_2 * \cos\left(\frac{\phi_1}{2}\right) \tag{10.2}$$

where $\vec{d}_{ij}$ is the distance vector from directional sensor i to control point j, $\vec{v}_{id}$ is a unit vector that defines the direction d of directional sensor i and r_i^s and φ_i respectively are the sensing range and FOV of directional sensor i.

The first condition in the TIS test checks if control point j is within the sensing range of directional sensor i. The second condition checks if the distance vector is within the FOV of directional sensor i. This is done by performing the inner product operation with equality when control point j is along one of the two edges of the sensing sector of directional sensor i.

It should be pointed out that in a real network design task, the sets of possible sensing ranges, FOVs and directions will be finite and contain only discrete values. That is because the first two sets are dictated by the available sensors. The last set, however, is dictated by the network designer who typically requires a small number of directions.

In the mathematical formulations of the MSP and MCSP problems, the number of possible directions a sensor can take is assumed to be equal to $\frac{360^\circ}{\phi_i}$. The first direction is always at an angle of $\frac{\phi_i}{2}$ degrees measured from the positive x-axis. The next direction is at an angle of ϕ_i degrees from the previous one. In this way, the whole circular area around the sensor is accounted for and there is no overlapping between sensing regions.

10.3 Related Work

The directional sensor placement problem is similar to the art gallery problem which is the theoretical study on how to place guards (or devices with vision capabilities) in an arbitrarily shaped polygon so as to cover the entire area. Chvatal [5] proved that for a polygon P with n sides, at most $\lfloor \frac{n}{3} \rfloor$ guards with a 360° range of vision are needed such that every point of P is visible from at least one of them. Since this fundamental result, many variations of the art gallery problem with alternative definitions for visibility have been studied. For example, Toth [6] considered guards whose range of vision is 180°. He showed that a polygon with n sides can also be monitored by at most $\lfloor \frac{n}{3} \rfloor$ guards whose range of vision is 180°. In a subsequent work, he showed that $\left(\frac{2n}{3} - 1\right)$ guards with a $(180 - \epsilon)$ where $\epsilon > 0$ are possibly necessary to monitor a polygon with n sides [7]. Ntafos and Tsoukalas [8] considered the k-guard placement problem which seeks a placement of k guards in a polygon such that the total area or the portion of the perimeter of the polygon visible to the guards is maximized. They showed that this placement problem is NP-hard and presented polynomial time algorithms for optimally placing a single guard. Czyzowicz et al. [9] considered rectangular polygons which are further divided into smaller rectangles such that any two rectangles sharing a side can be monitored by a single guard. They showed that $\lceil \frac{n}{3} \rceil$ guards are always sufficient to protect all rectangles in a rectangular polygon.

Clearly, from the above discussion, the directional sensor placement problem might not be reduced to an art gallery problem. This is because of the following reasons:

1. The sensing region (or visibility) of a directional sensor is characterized by two parameters which are the range of vision and sensing range,
2. The polygon formed by connecting the points corresponding to the placement sites may not enclose all the control points in the sensor field,
3. Sensors can be of different types (e.g., different ranges). In the art gallery problem, however, guards are assumed to be of the same type (i.e., similar capabilities), and
4. The location of sensors is restricted to a finite number of sites. Also, the number of points to be covered is finite.

The directional sensor placement problem is also similar to the camera placement problem in computer vision where the goal is to determine the optimal number and location of cameras for a region to be monitored. Horster and Lienhart [10, 11] developed a binary integer programming model by discretizing the region into a grid. Each grid point represents a potential placement site for a camera. They used a simple visibility model where they described the range of vision of a camera by a triangle. The proposed model is focused on coverage with respect to a predefined sampling rate. It guarantees that an object in the monitored region will be imaged at a minimum resolution.

Inspired by the above work, Zhao and Cheung [12] developed an iterative grid based binary integer programming model for the visual tagging problem where the goal is to identify distinctive visual features of objects in two or more camera views. They also presented a comprehensive visibility model for computing the visibility of a single tag. The proposed model finds the minimum number of cameras, their poses and their locations in the monitored region in order to achieve a desired level of visibility.

Clearly, the models developed for the camera placement problem might not be used to solve the directional sensor placement problem for the following reasons:

1. Cameras are mounted on walls and ceilings. Sensors, however, are placed on the ground. This requires new visibility models,
2. Sensors have a limited energy supply. Cameras do not have this problem since they are typically powered from wall outlets. Thus, energy must be considered in optimization,
3. Sensors must be connected among themselves. Thus, connectivity is another element that must be considered in optimization, and
4. Sensors can process data before it is delivered to the base station. This gives rise to issues such as routing and redundancy reduction.

Several models exist in the literature for the placement problem of isotropic sensors. Chakrabarty et al. [13] were the first to study this problem on sensor fields comprising discrete points that are grid points. A 2D or 3D grid of points is to be covered. Sensors can be placed only at grid points. Each grid point is to be covered by at least α sensors, where $\alpha \geq 1$. There are $|\Upsilon|$ sensor types. A sensor of type k costs c_k dollars and has a sensing range of r_k^s meters. Only one sensor can be placed at any grid point. They made an assumption such that all sensors can directly communicate with the base station. Therefore, the connectivity constraints were not considered. The objective is to find the least cost sensor placement that provides the required α-coverage. The problem is formulated as an Integer Linear Programming (ILP) problem with $\mathcal{O}(|\Upsilon|(|\Omega| + |\Pi|)^2)$ variables and $\mathcal{O}(|\Upsilon|(|\Omega| + |\Pi|)^2)$ constraints, where $|\Omega|$ and $|\Pi|$ are the number of control points and placement sites, respectively. Control points and placement sites are combined into one set corresponding to the grid points.

Sahni and Xu [14] proposed another model that reduces the number of variables to $\mathcal{O}(|\Upsilon|(|\Omega| + |\Pi|))$ and the number of constraints to $\mathcal{O}(|\Omega| + |\Pi|)$. That is possible because the set of neighbors for each grid point is computed a priori and given to the model as an input. The proposed model can be applied to any set of discrete points.

The above two models are practically solvable only for a small number of points. Wang and Zhong [15] compared the runtime of the two models. The models are implemented using *lpsolve 5.5* [16]. It was observed that when the sensor field is bigger than 9×9 (i.e., 81 grid points), execution of the Chakrabarty et al.'s model is suspended with an out-of-memory error message. On the other hand, for a 20×20 sensor field, the runtime of Sahni and Xu's model for $\alpha = 1$ surges to 14 hours.

Sahni and Xu [17] proposed another ILP formulation for finding the minimum cost placement of sensors. The model handles the problem of placing sensors at a subset of preselected sites so as to minimize the sensor cost while providing a specified degree of coverage of the control points. The sets of placement sites and control points are merged together to form a single set. Each sensor is capable of directly communicating with the base station which is situated within the communication range of all sensors. The total number of variables is $\mathcal{O}(|\Upsilon|(|\Omega| + |\Pi|))$ and constraints is $\mathcal{O}((|\Upsilon| + m)(|\Omega| + |\Pi|))$, where m is the number of modalities. A modality is the quantity to be monitored, such as temperature, humidity and sound.

The above models might not be used to solve the directional sensor placement problem because they assume that sensors have a circular sensing range. Also, they only consider coverage. In our case, we consider coverage and connectivity. In addition, our model captures the limited visibility of sensors and their ability to change their orientation.

There are some works that especially deal with directional sensor networks. They all assume that directional sensors are randomly deployed in the sensor field and the goal is to organize their directions to achieve some objective. For example, the maximization of target coverage while minimizing the number of active sensors was studied in [4]. Area coverage was considered in [18]. The goal was to maximize the covered area by scheduling the directions of sensors. Prolonging the network lifetime was discussed in [19]. The proposed solution is to organize the directions of sensors into a group of non-disjoint cover sets. Each cover set covers all the targets in the sensor field. One cover set is activated at a time.

10.4 Problem Statement

Dropping sensors from an aircraft or randomly throwing them into a sensor field may not be acceptable because of the post-deployment work required in order to reach a desired level of performance. Besides, human instinct and expertise are insufficient to determine the best trade-off between the conflicting objectives of a mission. For example, reducing the number of sensors will minimize the deployment cost. However, it may yield poor coverage and shorten the lifetime of the WSN. Therefore, WSN designers need effective and efficient planning techniques to help them make sure that their designed networks will perform as best as possible on the ground.

A sensor field can be approximated by a set of points where sensors can be installed and a set of critical points that must be covered. These two sets of points can be regular or random. They can also be merged into one set. For example, in a fire detection application in which the monitored area is represented as a grid, each grid point represents a control point as well as a placement site.

Given the above representation of a sensor field, we wish to address the following fundamental questions:

1. What is the minimum number of directional sensors necessary to cover all control points in the sensor field?
2. What is the minimum possible cost if more than one directional sensor type is available?
3. What is the optimal configuration (i.e., sensing range, field of view and direction) of each directional sensor to be installed in the sensor field?

Engineers face these questions during the design and planning stages of directional WSNs. They arise because of the need to reduce the deployment and installation costs by minimizing the number of sensors and optimally configuring them.

In the next section, we present optimization models to answer the above questions. Such an approach is necessary to understand the fundamental performance bounds and how they are affected by different design parameters. It also enables the comparison of heuristics with respect to the optimal solutions.

10.5 Problem Formulations

Notation

The notation is composed of sets, decision variables, parameters and constants.

Sets

- Ω Control (or critical) points
- Π Placement sites to install sensors
- Υ Sensor types
- $\mathcal{S}$ Sensing ranges
- $\mathcal{F}$ Field of views
- $\mathcal{D}$ Directions
- $\mathcal{D}_k$ Directions for a type-k sensor

Decision Variables

- x_{id} A 0-1 variable such that $x_{id} = 1$ if and only if the sensor installed at a site $i \in \Pi$ is oriented toward direction $d \in \mathcal{D}$
- x_{id}^{k} A 0-1 variables such that x_{id}^{k}=1 if and only if a type-k sensor ($k \in \Upsilon$) is installed at a site $i \in \Pi$ and oriented toward direction $d \in \mathcal{D}_k$
- x_i^{sfd} A 0-1 variable such that $x_i^{sfd} = 1$ if and only if the sensor installed at $i \in \Pi$ has a sensing range $s \in \mathcal{S}$, a FOV $f \in \mathcal{F}$ and is oriented toward direction $d \in \mathcal{D}$
- σ_{ij}^{k} A 0-1 variable such that $\sigma_{ij}^{k} = 1$ if and only if a control point $j \in \Omega$ is covered by a type-k sensor ($k \in \Upsilon$) installed at a site $i \in \Pi$
- σ_{ij}^{d} A 0-1 variable such that $\sigma_{ij}^{d} = 1$ if and only if a control point $j \in \Omega$ is covered by a sensor installed at a site $i \in \Pi$ and oriented toward direction $d \in \mathcal{D}$

- y_1^{il} A 0-1 variable such that $y_1^{il} = 1$ if and only if a sensor at a site $i \in \Pi$ can communicate with a sensor at site $l \in \Pi/\{i\}$
- y_2^{ib} A 0-1 variable such that $y_2^{ib} = 1$ if and only if a sensor at a site $i \in \Pi$ can communicate with the base station at location b
- f_1^{il} The flow from a sensor node $i \in \Pi$ to a sensor node $l \in \Pi/\{i\}$
- f_2^{ib} The flow from a sensor node $i \in \Pi$ to the base station b

Parameters

- c Base cost of a sensor (not including the cost of sensing range and FOV)
- c_b Cost of the base station
- c_k Cost of a type-k sensor
- r_k^s Sensing range of a type-k sensor
- ϕ_k FOV of a type-k sensor
- c_s Cost of one meter of sensing range
- c_f Cost of one degree of FOV

Constants

- Δ Adjacency matrix between placement sites where $\delta_{il} = 1$ if the Euclidean distance between two placement sites $i, l \in \Pi$; $i \neq l$ is less than or equal to the uniform transmission range of sensor nodes
- Θ Adjacency matrix between control points and a sensor node where $\Theta_{ij}^{d} = 1$ if a sensor node installed at site $i \in \Pi$ and oriented toward direction $d \in \mathcal{D}$ covers a control point $j \in \Omega$
- Θ_k Adjacency matrix between control points and a type-k sensor where $\Theta_{ij}^{kd} = 1$ if a type-k sensor installed at site $i \in \Pi$ and oriented toward direction $d \in \mathcal{D}_k$ covers a control point $j \in \Omega$
- Γ Adjacency matrix between placement sites and the base station location where $\gamma_{ib} = 1$ if the Euclidean distance between the base station location b and a placement site $i \in \Pi$ is less than or equal to the reception range of the base station
- Λ Distance matrix $[d_{ij}]$ where d_{ij} is the Euclidean distance between a placement site $i \in \Pi$ and a control point $j \in \Omega$
- C_{tl} Capacity of the wireless link between a sensor node at site $i \in \Pi$ and a sensor node at site $l \in \Pi/\{i\}$

- C_{ib} Capacity of the wireless link between a sensor node at site $i \in \Pi$ and the base station at location b
- $C_{\max}$ Maximum amount of data that a sensor node can handle (transmit and/or receive) per time unit (i.e., node capacity)
- R_i Rate at which data is generated by a sensor node at site $i \in \Pi$
- r_e Uniform transmission range for all sensors
- r_r^b Reception range of the base station

Minimum Sensor Placement

In the MSP problem, the goal is to find the minimum number of directional sensors such that every control point is covered by at least one sensor and the resulting network is connected. All sensors have the same FOV and set of directions.

Formulation

The objective function is given by the following equation:

$$F = \min \sum_{\{i \in \Pi\}} \sum_{\{d \in \mathcal{D}\}} x_{id} \tag{10.3}$$

which is the possible total number of sensors that can be installed in the sensor field. The goal is to minimize this function subject to the following set of requirements and constraints.

The coverage requirement is represented by the following inequality:

$$\min \sum_{\{i \in \Pi\}} \sum_{\{d \in \mathcal{D}\}} x_{id} \cdot \Theta_{ij}^{d} \geq 1 \quad \forall j \in \Omega \tag{10.4}$$

which states that a control point must be covered by at least one sensor. A control point is assigned to a sensor only if it can be covered by one of its possible directions.

The following constraint captures the fact that a sensor can be oriented toward one direction only.

$$\sum_{\{d \in \mathcal{D}\}} x_{id} \leq 1 \quad \forall i \in \Pi \tag{10.5}$$

The connectivity requirement is captured by the following flow conservation constraint which is generated for every placement site.

$$R_i \sum_{\{d \in \mathcal{D}\}} x_{id} + \sum_{\{l \in \Pi; l \neq i\}} f_1^{li} = \sum_{\{l \in \Pi; l \neq i\}} f_1^{li} + f_2^{li} \quad \forall\, i \in \Pi \tag{10.6}$$

Since the base station does not generate any traffic, only flows between placement sites and between placement sites and the base station are considered.

A flow between two placement sites is possible only if two sensors are installed at those sites and they can communicate with each other (i.e., a wireless link can be established between them). This is ensured by the following constraint.

$$y_1^{il} \leq \frac{\delta_{il}}{2}\left(\sum_{\{d\in\mathcal{D}\}} x_{id} + \sum_{\{d\in\mathcal{D}\}} x_{ld}\right) \quad \forall\, i,l \in \Pi;\ i \neq l \tag{10.7}$$

Similarly, a wireless link can be established between the base station and a placement site only if there is a sensor installed at that site and can communicate with the base station. This is ensured by the following constraint.

$$y_2^{ib} \leq \gamma_{ib} \sum_{\{d\in\mathcal{D}\}} x_{id} \quad \forall\, i \in \Pi \tag{10.8}$$

The following constraints are the capacity constraints. They ensure that the capacity of the wireless links and nodes is not exceeded. Also, they ensure that a flow can be realized only if there is a wireless link to carry it.

$$f_1^{il} \leq C_{il} \cdot y_1^{il} \quad \forall\, i,l \in \Pi;\ i \neq l \tag{10.9}$$

$$f_2^{ib} \leq C_{ib} \cdot y_2^{il} \quad \forall\, i \in \Pi \tag{10.10}$$

$$\sum_{\{l\in\Pi; l\neq i\}} f_1^{il} + f_2^{ib} \leq C_{\max} \cdot \sum_{\{d\in\mathcal{D}\}} x_{id} \quad \forall\, i \in \Pi \tag{10.11}$$

Finally, the following constraints indicate that decision variables are binary and flow variables are positive integer.

$$x_{id}, y_1^{il}, y_2^{ib} \in \mathbb{B} \quad \forall\, i,l \in \Pi;\ i \neq l;\ d \in \mathcal{D} \tag{10.12}$$

$$f_1^{il} f_2^{ib} \in \mathbb{N}^+ \quad \forall\, i,l \in \Pi;\ i \neq l; \tag{10.13}$$

The number of variables in the above ILP formulation is $\mathcal{O}(|\Pi|(|\mathcal{D}| + |\Pi|))$ and the number of constraints is $\mathcal{O}(|\Pi|^2 + |\Omega|)$

Example

Figure 10.4 shows an example about the MSP problem. There are five placement sites, five control points and one base station. Since the FOV of each sensor is 45°, a sensor has eight directions. The set of possible directions has

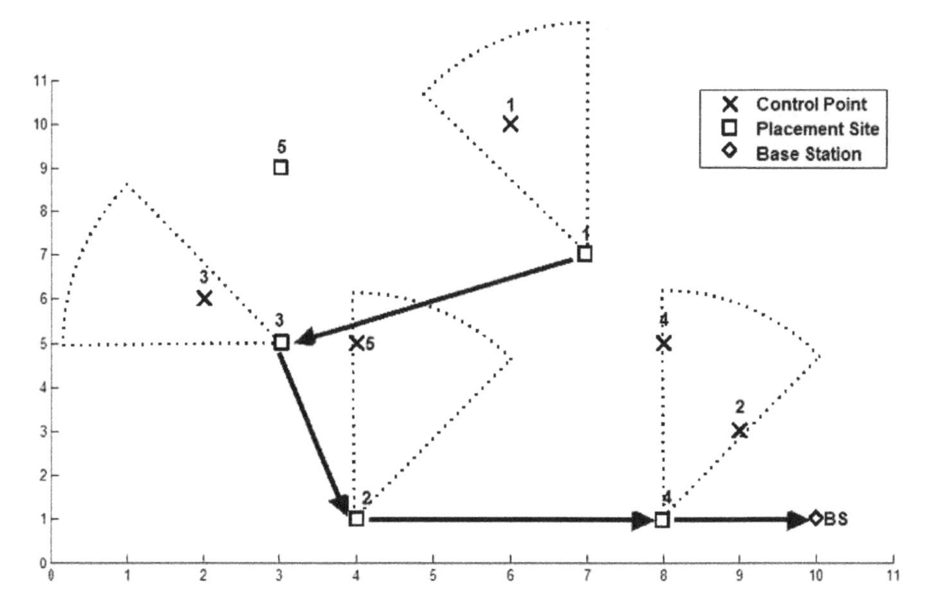

Fig. 10.4 Network layout generated by the MSP model for a five placement sites and five control points instance.

eight elements which are 22.5°, 67.5°, 112.5°, 157.5°, 202.5°, 247.5°, 292.5° and 337.5°. After building and solving the model, four sensors are required to cover the five control points. They are sensors 1, 2, 3 and 4. Sensor 1 is oriented toward direction 112.5°. Sensor 3 is oriented toward direction 157.5°. Sensors 2 and 4 are both oriented toward direction 67.5°. Four wireless links are established to carry traffic from the sensor nodes to the base station. They are shown by the bold arrows.

Minimum Cost Sensor Placement

In the MCSP problem, the goal is to minimize the cost of a directional WSN by appropriately selecting sensors from a set different sensor types. Each sensor type is characterized by its cost, FOV and set of directions. The requirements are that each control point must be covered by at least one sensor and the resulting network must be connected.

Formulation

The objective function is given by the following equation:

$$F = \min \sum_{\{k \in \Upsilon\}} C_k \sum_{\{i \in \Pi\}} \sum_{\{d \in \mathcal{D}_k\}} x_{id}^{k} \tag{10.14}$$

which represents the total cost of sensors. The goal is to minimize this function subject to the following set of requirements and constraints.

A control point can be assigned to a type-k sensor only if it can be covered by one of its directions. This is represented by the following inequality.

$$\sigma_{ij}^{k} \leq \sum_{\{d \in \mathcal{D}_k\}} x_{id}^{k} \cdot \Theta_{ij}^{kd} \quad \forall\, i \in \Pi;\; j \in \Omega; k \in \Upsilon \tag{10.15}$$

The coverage requirement is that a control point must be covered by at least one sensor. This is captured by the following inequality.

$$\sum_{\{i \in \Pi\}} \sum_{\{k \in \Upsilon\}} \sigma_{ij}^{k} \geq 1 \quad \forall\, j \in \Omega \tag{10.16}$$

A type-k sensor can be oriented toward one direction only. The following constraint captures this requirement.

$$\sum_{\{k \in \Upsilon\}} \sum_{\{d \in \mathcal{D}_k\}} x_{id}^{k} \leq 1 \quad \forall\, i \in \Pi \tag{10.17}$$

The following constraints are similar to constraints (10.6)–(10.13) in the formulation of the MSP problem.

$$R_i \sum_{\{k \in \Upsilon\}} \sum_{\{d \in \mathcal{D}_k\}} x_{id}^{k} + \sum_{\{l \in \Pi; l \neq i\}} f_1^{il} = \sum_{\{l \in \Pi; l \neq i\}} f_1^{il} + f_2^{ib} \quad \forall\, i \in \Pi \tag{10.18}$$

$$y_1^{il} \leq \frac{\delta_{il}}{2} \left(\sum_{\{k \in \Upsilon\}} \sum_{\{d \in \mathcal{D}_k\}} x_{id}^{k} + \sum_{\{k \in \Upsilon\}} \sum_{\{d \in \mathcal{D}_k\}} x_{ld}^{k} \right) \quad \forall\, i, l \in \Pi; i \neq l \tag{10.19}$$

$$y_2^{ib} \leq \gamma_{ib} \sum_{\{k \in \Upsilon\}} \sum_{\{d \in \mathcal{D}_k\}} x_{id}^{k} \quad \forall\, i \in \Pi \tag{10.20}$$

$$f_1^{il} \leq C_{il} \cdot y_1^{il} \quad \forall\, i, l \in \Pi; i \neq l \tag{10.21}$$

$$f_2^{ib} \leq C_{ib} \cdot y_2^{ib} \quad \forall\, i \in \Pi \tag{10.22}$$

$$\sum_{\{l \in \Pi; l \neq i\}} f_1^{il} + f_2^{ib} \leq C_{\max} \cdot \sum_{\{k \in \Upsilon\}} \sum_{\{d \in \mathcal{D}_k\}} x_{ld}^{k} \quad \forall i \in \Pi \tag{10.23}$$

$$x_{id}^{k}, \sigma_{ij}^{k}, y_1^{il}, y_2^{ib} \in \mathbb{B} \quad \forall\, i, l \in \Pi; i \neq l; j \in \Omega; k \in \Upsilon; d \in \mathcal{D}_k \tag{10.24}$$

$$f_1^{il} f_2^{ib} \in \mathbb{N}^{+} \quad \forall\, i, l \in \Pi; i \neq l \tag{10.25}$$

The number of variables in the above ILP formulation is $\mathcal{O}(|\Pi||\Upsilon||\Omega|)$ and the number of constraints is $\mathcal{O}(|\Pi||\Upsilon||\Omega|)$.

Table 10.1. Types and directions of sensors for a 40 placement sites and 30 control points instance of the MCSP problem.

Sensor No.	Type	Direction
1	1	337.5°
2	2	330°
3	2	210°
4	2	30°
5	2	90°
6	2	90°

Example

Consider a sensor field with 40 placement sites and 30 control points. There are two sensor types. A type-1 sensor has a 45° FOV and eight directions. A type-2 sensor has a 60° FOV and six directions. Six sensors are required to cover all the control points in the sensor field. Table 10.1 gives the placement sites where these sensors are to be installed. It also gives the type and direction of each sensor.

Optimal Sensor Configuration

Unlike MSP and MCSP, in the OSC problem, the adjacency matrix between control points and sensors (i.e., Θ) is not available as an input for the optimization model. However, the sets of possible sensing ranges, FOVs and directions a sensor can take are given. Then, the goal is to minimize the network cost by appropriately configuring each sensor to be installed in the sensor field. This is accomplished by choosing the optimal sensing range, FOV and direction for each sensor.

Formulation

The objective function is given by the following equation:

$$F = \min \sum_{\{s \in \mathcal{S}\}} \sum_{\{f \in \mathcal{F}\}} (c + c_s r^s + c_f \phi^f) \sum_{\{d \in \mathcal{D}\}} \sum_{\{i \in \Pi\}} x_i^{sfd} \tag{10.26}$$

which represents the cost of the WSN. The cost of a directional sensor is determined by its sensing range and FOV and the sensor base cost. The goal is to minimize this function subject to a set of requirements and constraints.

The following two inequalities are the conditions for coverage of a control point by a sensor. They are derived from the TIS test. If the solver decides to cover a control point j by installing a sensor at placement site i, it assigns a sensing range, FOV and direction to sensor i such that the following two conditions are true.

$$\sum_{\{s\in\mathcal{S}\}} r^s \sum_{\{f\in\mathcal{F}\}} \sum_{\{d\in\mathcal{D}\}} x_i^{sfd} \geq d_{ij} \sum_{\{d\in\mathcal{D}\}} \sigma_{ij}^d \quad \forall\, i \in \Pi;\, j \in \Omega \tag{10.27}$$

$$\sum_{\{f\in\mathcal{F}\}} \phi^f \sum_{\{s\in\mathcal{S}\}} x_i^{sfd} \geq 2\cos^{-1}\left(\frac{\vec{v}_{id}\cdot\vec{d}_{ij}}{\|\vec{d}_{ij}\|_2}\right)\sigma_{ij}^d \quad \forall\, i \in \Pi;\, j \in \Omega;\, d \in \mathcal{D} \tag{10.28}$$

The following constraints ensure that each control point is covered by at least one sensor and only one sensor can be installed at each placement site.

$$\sum_{\{i\in\Pi\}} \sum_{\{d\in\mathcal{D}\}} \sigma_{ij}^d \geq 1 \quad \forall\, j \in \Omega \tag{10.29}$$

$$\sigma_{ij}^d \leq \sum_{\{s\in\mathcal{S}\}} \sum_{\{f\in\mathcal{F}\}} x_i^{sfd} \quad \forall\, i \in \Pi;\, j \in \Omega;\, d \in \mathcal{D} \tag{10.30}$$

$$\sum_{\{s\in\mathcal{S}\}} \sum_{\{f\in\mathcal{F}\}} \sum_{\{d\in\mathcal{D}\}} x_i^{sfd} \leq 1 \quad \forall\, i \in \Pi \tag{10.31}$$

The following constraints are similar to constraints (10.6)–(10.13) in the formulation of the MSP problem.

$$R_i \sum_{\{s\in\mathcal{S}\}} \sum_{\{f\in\mathcal{F}\}} \sum_{\{d\in\mathcal{D}\}} x_i^{sfd} + \sum_{\{l\in\Pi;l\neq i\}} f_1^{li} = \sum_{\{l\in\Pi;l\neq i\}} f_1^{il} + f_2^{ib} \quad \forall\, i \in \Pi \tag{10.32}$$

$$y_1^{il} \leq \frac{\delta_i l}{2}\left(\sum_{\{s\in\mathcal{S}\}} \sum_{\{f\in\mathcal{F}\}} \sum_{\{d\in\mathcal{D}\}} x_i^{sfd} + \sum_{\{s\in\mathcal{S}\}} \sum_{\{f\in\mathcal{F}\}} \sum_{\{d\in\mathcal{D}\}} x_l^{sfd}\right) \quad \forall\, i,l \in \Pi;\, i \neq l \tag{10.33}$$

$$y_2^{ib} \leq \gamma_{ib} \sum_{\{s\in\mathcal{S}\}} \sum_{\{f\in\mathcal{F}\}} \sum_{\{d\in\mathcal{D}\}} x_i^{sfd} \quad \forall\, i \in \Pi \tag{10.34}$$

$$f_1^{il} \leq C_{il} \cdot y_1^{il} \quad \forall\, i,l \in \Pi;\, i \neq l \tag{10.35}$$

$$f_2^{ib} \leq C_{ib} \cdot y_2^{ib} \quad \forall\, i \in \Pi \tag{10.36}$$

$$\sum_{\{l\in\Pi;l\neq i\}} f_1^{il} + f_2^{ib} \leq C_{\max} \cdot \sum_{\{s\in\mathcal{S}\}} \sum_{\{f\in\mathcal{F}\}} \sum_{\{d\in\mathcal{D}\}} x_i^{sfd} \quad \forall\, i \in \Pi \tag{10.37}$$

$$x_i^{sfd}, \sigma_{ij}^d, y_1^{il}, y_2^{ib} \in \mathbb{B} \quad \forall\, i, l \in \Pi; i \neq l; j \in \Omega; d \in \mathcal{D}; s \in \mathcal{S}; f \in \mathcal{F} \tag{10.38}$$

$$f_1^{il} f_2^{ib} \in \mathbb{N}^+ \quad \forall\, i, l \in \Pi; i \neq l \tag{10.39}$$

The number of variables in the above ILP formulation is $\mathcal{O}(|\Pi||\mathcal{S}||\mathcal{F}||\mathcal{D}|)$ and the number of constraints is $\mathcal{O}(|\Pi||\Omega||\mathcal{D}|)$.

Example

Figure 10.5 shows the network layout generated by the above ILP model for a 40 placement sites and 30 control points instance of the OSC problem. The arrows indicate the directions of the sensors. Eleven sensors are required to cover the all the control points in the sensor field.

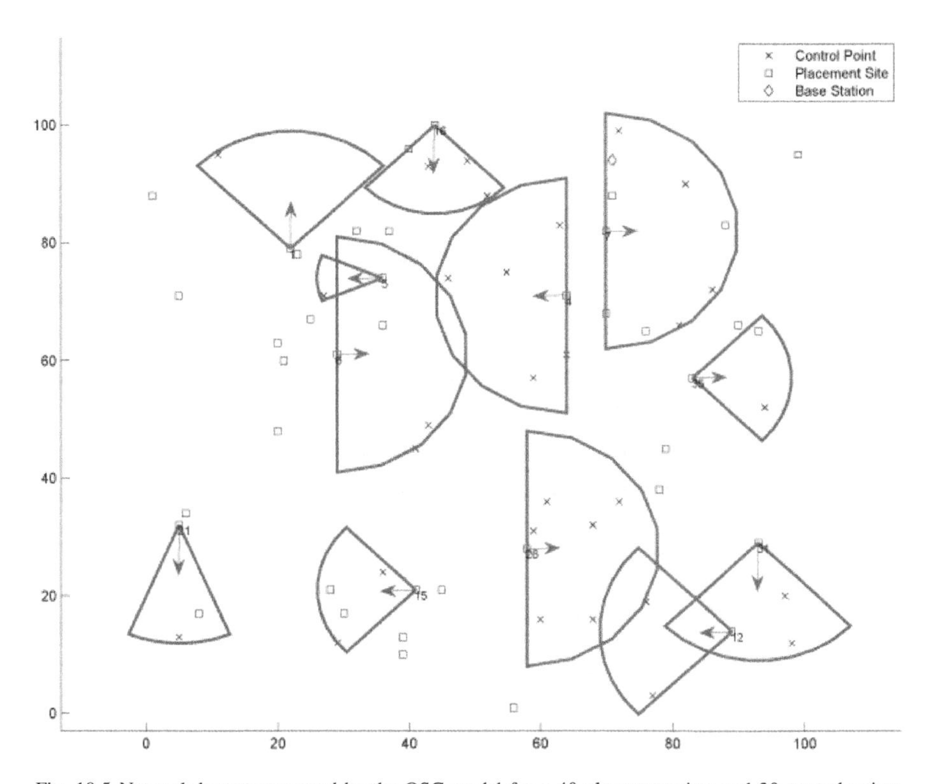

Fig. 10.5 Network layout generated by the OSC model for a 40 placement sites and 30 control points instance.

10.6 Numerical Results

In this section, numerical results are presented to assess the performance of the MSP, MCSP and OSC models. The models are implemented using *CPLEX 10.1.1* [20]. For the computing platform, a Unix workstation equipped with a 3 GHz CPU and 1 GB memory is used.

First, the MSP model is used to solve the grid coverage problem presented in [13–15]. This is possible because it is a general model that can handle both isotropic and directional sensors. Therefore, the FOV is set to 360°. Since only the coverage is considered, the resulting ILP model is composed of an objective function (Eq. 10.3) and two constraints (Ineqs. 10.4 and 10.5). The model has $\mathcal{O}(|\Pi|)$ variables and $\mathcal{O}(|\Omega|)$ constraints.

Table 10.2 shows the results. The first column is the grid size. The second column is the number of grid points which is also the number of placement sites and control points in the sensor field. The next three columns respectively give the number of sensors needed to cover the control points, the percentage of reduction with respect to the initial number of placement sites and the runtime of the model. The table shows a small improvement in the runtime over that reported in [15]. Since the two models have the same complexity, the improvement in the runtime may be explained by the fact that different computers and different solvers (CPLEX vs. lpsolve) were used.

Secondly, we study the effect of the number of placement sites on the runtime of the MSP, MCSP and OSC models and the percentage of reduction in the number of sensors required to cover all the control points. The experiments are setup as follows. Locations of placement sites, control points and the base station are uniformly generated. There are 50 control points that need to be covered and one base station where all the traffic must be sent. The node capacity of each sensor is 40 Kbps and the link capacity of each wireless link is 10 Kbps. The rate of data generation is 512 bps for each sensor. For each

Table 10.2. Percentage of reduction and runtime for the grid coverage problem.

Grid Size	#Grid Points	#Sensors	%Reduction	Runtime (sec.)
9×9	81	12	85.12	0.55
10×10	100	14	86	1.57
11×11	121	16	86.78	1.83
12×12	144	18	87.5	6.13
13×13	169	20	88.16	26.5

problem size (i.e., number of placement sites), five instances of the problem were randomly generated and the average was calculated.

For the MSP problem, the size of the sensor field is 20 m × 20 m. Each sensor has four directions (i.e., FOV = 90°) and a transmission range of 10 m. Table 10.3 shows the results. The first column indicates the number of possible placement sites in the sensor field. The next three columns respectively represent the minimum number of sensors required in order to cover the 50 control points, the percentage of reduction with respect to the initial number of placement sites and the runtime of the MSP model.

As can be seen from Table 10.3, the execution time is increasing with respect to the problem size. Even with 200 potential placement sites, CPLEX only took, on average, 504 seconds. We also noticed that from one instance to another, the execution time is variable. This can be explained by the fact that CPLEX is using the branch and bound algorithm. As far as the reduction in the number of sensors is concerned, we can see that as the number of potential placement sites increases, the reduction percentage also increases. Since more placement sites are available, better locations can be selected. As a result, one sensor node will be able to cover more control points, thus reducing the total number of sensors.

For the MCSP problem, the size of the sensor field is 100 m × 100 m and has 30 control points. There are two sensor types which both have a transmission range of 25 m. A type-1 sensor costs $20 and has a 20 m sensing range and 45° FOV. A type-2 sensor costs $40 and has a 50 m sensing range and 60° FOV.

Table 10.3. Percentage of reduction achieved by the MSP model along with the runtime.

#PSs	#Sensors	%Reduction	Runtime (sec.)
30	14	53.33	0.31
40	14	65	1.36
50	13	74	3.77
60	13	78.33	3.97
70	12	82.86	8.58
80	12	85	17.13
90	12	86.67	23
100	12	88	50.18
150	11	92.67	99.12
200	11	94.5	503.61

Table 10.4. Percentage of reduction achieved by the MCSP model along with the runtime.

	#Sensors					
# PSs	Type-1	Type-2	Total	% Reduction	Cost ($)	Runtime (sec.)
40	1	5	6	85	220	4.26
50	3	5	8	84	260	4.93
60	3	5	8	86.67	260	11.71
70	2	4	6	91.43	200	20.43
80	1	6	7	91.25	260	33.12
90	1	5	6	93.33	220	49.31
100	1	5	6	94	220	71.72
150	6	6	12	92	240	209.61
200	3	5	8	96	260	372.17

Table 10.4 shows the results. The first column indicates the number of possible placement sites in the sensor field. The fourth column gives the minimum total number of sensors required to cover the control points. The number of sensors from each type is shown in the second column for type-1 and third column for type-2. The next three columns respectively represent the percentage of reduction with respect to the initial number of placement sites, total cost of sensors and runtime of the MCSP model.

As can be seen from Table 10.4, the execution time is increasing with respect to the problem size. Even with 200 potential placement sites, CPLEX only took, on average, 372 seconds. Similar to MSP, we can see that as the number of potential placement sites increases, the reduction percentage generally increases.

For the OSC problem, the size of the sensor field is $50\,\text{m} \times 50\,\text{m}$ and has 50 control points. Each sensor has a transmission range of $10\,\text{m}$. The costs of $1\,\text{m}$ of sensing range and 1° of FOV are set to \$0.1 and \$0.2, respectively. The sets of possible FOVs, directions and sensing ranges are as follows:

- $\mathcal{F} = \{45^\circ, 90^\circ, 180^\circ\}$,
- $\mathcal{D} = \{90^\circ, 180^\circ, 270^\circ, 360^\circ\}$, and
- $\mathcal{S} = \{5\,\text{m}, 10\,\text{m}, 15\,\text{m}, 20\,\text{m}\}$.

Table 10.5 shows the results. The first column indicates the number of possible placement sites in the sensor field. The next four columns respectively represent the minimum number of sensors required in order to cover the 50 control points, the percentage of reduction with respect to the initial number of placement sites, the total cost of the WSN and the runtime of the OSC model.

Table 10.5. Percentage of reduction achieved by the OSC model along with the runtime.

#PSs	#Sensors	%Reduction	Cost ($)	Runtime (sec.)
30	9	70	150.5	20.98
40	10	75	138.2	78.21
50	9	82	128.62	115.46
60	8	86.7	130	157.75
70	9	87.1	134.1	407.32
80	10	87.5	130.4	415
90	9	90	125.5	757.3
100	9	91	135	962.12
150	10	93.3	115.25	2739.32
200	9	95.5	113	4326.65

As can be seen from Table 10.5, the execution time is increasing with respect to the problem size. This is as expected since the complexity of the problem grows with the number of potential placement sites. Another interesting aspect is that the number of placement sites, the cost of the network and the reduction percentage are closely related. In fact, if the number of potential placement sites increases, the total cost of the WSN will generally decrease (up to a certain limit) and the reduction percentage will increase. This is because better locations can be selected for sensors when more placement sites are available.

Finally, we study the effect of varying the number of control points on the number of sensors obtained by the MCSP model. The number of placement sites is fixed at 50 and the solution is the average of five randomly generated problem instances. Our main observations are the following. First, as the number of control points increases, the sensor cost increases. This behavior is as expected since the directions of the existing sensors may not cover the newly added control points. Therefore, more sensors are added to the sensor field. Secondly, when the number of control points reaches a certain threshold (500 in our experiments), the number of placement sites becomes insufficient. That is, even if a sensor is installed at every placement site, the resulting set of directions would not be enough to cover all the control points. Therefore, more placement sites must be introduced into the floorplan of the sensor field.

10.7 Summary and Further Research

The planning of directional WSNs aims to determine the optimal placement and configuration of directional sensors that meet the requirements of coverage and connectivity while reducing the installation and deployment costs. To that end, this chapter has presented the sensor placement problem in directional WSNs. Also, it has critiqued the current works in some relevant areas. After that, it has described three planning problems that naturally arise when designing directional WSNs. The viability and effectiveness of the mathematical models have been illustrated through numerical results.

More work is still needed in this area. The following directions are suggested:

(1) Coverage is not a single notion that fits all applications. Different definitions of coverage exist in the literature (e.g., see [21] and [22]). Planning models incorporating these notions of coverage need to be developed and evaluated against their proposed applications.
(2) Network lifetime is another important metric for developing more efficient directional WSNs. The lifetime of a directional WSN can be estimated in the planning phase and thus the necessary resources can be determined. The MSP, MCSP and OSC models can be extended to include network lifetime. Besides, new models might be needed to account for the new emerging technologies and applications.
(3) Some applications require the deployment of a large number of sensor nodes, making the MSP, MCSP and OSC impractical. This necessitates the use of heuristic techniques like genetic algorithms and tabu search. This avenue should be investigated, especially when performance and cost become an issue.

10.8 Problems

Problem 14.1

Describe and give examples of directional sensors.

Problem 14.2

Discuss three possible applications of directional WSNs.

Problem 14.3

Distinguish between the MSP, MCSP and OSC problems.

Problem 14.4

Explain how the directional MSP problem can be transformed into an isotropic MSP problem.

Problem 14.5

Extend the MSP model to account for the situation in which there is more than one base station location.

Problem 14.6

Consider a 10×10 grid in which each grid point represents a potential placement site and a control point. Let the coordinates of the point at the lower left corner of the grid be (0, 0). Also, assume that the grid has a resolution (i.e., distance between adjacent points) of 1 meter. Answer the following questions:

(a) Give the sets of possible sensing ranges, FOVs and directions a sensor can take.
(b) Using a sensor type with a 2 m sensing range and a 180° FOV, how many sensors of this type are needed to cover all the 100 control points?
(c) Assume that you are allowed to use only two sensors. What would be their optimal location and configuration?

References

[1] S. Soro and W. B. Heinzelman, "On the coverage problem in video-based wireless sensor networks," In Proc. *International Conference on Broadband Networks*, IEEE, 2005, pp. 932–939.

[2] J. Adriaens, S. Megerian, and M. Potkonjak, "Optimal worst-case coverage of directional field-of-view sensor networks," In Proc. *International Conference on Sensor and Ad Hoc Communications and Networks*, IEEE, 2006, pp. 336–345.

[3] The WSN Starter Kit, Crossbow Technology, Inc. [Online]. http://www.xbow.com

[4] J. Ai and A. A. Abouzeid, "Coverage by directional sensors in randomly deployed wireless sensor networks," *Journal of Combinatorial Optimization*, 11(1), pp. 21–41, February 2006.

[5] V. Chvatal, "A combinatorial theorem in plane geometry," *Journal of Combinatorial Theory Series B*, 18, pp. 39–41, 1975.

[6] C. D. Toth, "Art Gallery Problem with Guards whose Range of Vision is 180," *Computational Geometry — Theory and Applications*, 17, pp. 121–134, 2000.

[7] C. D. Toth, "Art Galleries with guards of uniform range of vision," *Computational Geometry — Theory and Applications*, 21, pp. 185–192, 2002.

[8] S. Ntafos and M. Tsoukalas, "Optimum Placement of Gaurds," *Information Sciences*, 76, pp. 141–150, 1994.

[9] J. Czyzowicz, E. Rivera-Campo, N. Santoro, J. Urrutia, and J. Zaks, "Guarding rectangular art galleries," *Discrete Applied Mathematics*, 50, pp. 149–157, 1994.

[10] E. Horster and R. Lienhart, "Approximating optimal visual sensor placement," In Proc. *International Conference on Multimedia and Expo*, IEEE, 2006, pp. 1257–1260.

[11] E. Horster and R. Lienhart, "On the optimal placement of multiple visual sensors," In Proc. *International Workshop on Video Surveillance and Sensor Networks*, ACM, 2006, pp. 111–120.

[12] J. Zhao and S.-C. S. Cheung, "Multicamera surveillance with visual tagging and generic camera placement," In Proc. *International Conference on Distributed Smart Cameras*, ACM/IEEE, 2007, pp. 259–266.

[13] K. Chakrabarty, S. Sitharama, H. Qi, and E. Cho, "Grid coverage for surveillance and target location in distributed sensor networks," *IEEE Transactions on Computers*, 51(12), pp. 1448–1453, 2002.

[14] S. Sahni and X. Xu, "Algorithms For Wireless Sensor Networks," *International Journal of Distributed Sensor Networks*, 1(1), pp. 35–56, January 2005.

[15] J. Wang and N. Zhong, "Efficient point coverage in wireless sensor networks," *Journal of Combinatorial Optimization*, 11(3), pp. 291–304, May 2006.

[16] M. Berkelaar, P. Notebaert, and K. Eikland, "LP Solve: Linear Programming Code, version 5.5," Eindhoven University of Technology, 2005.

[17] X. Xu and S. Sahni, "Approximation algorithms for sensor deployment," *IEEE Transactions on Computers*, 56(12), pp. 1681–1695, December 2007.

[18] W. Cheng, S. Li, X. Liao, S. Changxiang, and H. Chen, "Maximal coverage scheduling in randomly deployed directional sensor networks," In Proc. *International Conference on Parallel Processing Workshops*, IEEE, 2007, pp. 68–73.

[19] Y. Cai, W. Lou, M. Li, and X.-Y. Li, "Target-oriented scheduling in directional sensor networks," In Proc. *26th International Conference on Computer Communications*, IEEE, 2007, pp. 1550–1558.

[20] ILOG Inc., Incline Village, NV. [Online]. http://www.ilog.com/products/cplex

[21] S. Kumar, T. H. Lai, and A. Arora, "Barrier coverage with wireless sensors," *Wireless Networks*, 13(6), pp. 817–834, December 2007.

[22] S. Meguerdichian, F. Koushanfar, G. Qu, and M. Potkonjak, "Exposure in wireless Ad-Hoc sensor networks," In Proc. *International Conference on Mobile Computing and Networking*, IEEE, 2001, pp. 139–150.

11

Layered Architecture for Mobility Models — Lemma

Alexander Pelov and Thomas Noel

Laboratoire des Sciences de l'Image, de l'Informatique et de la T´el´ed´etection & University Louis Pasteur, Strasbourg, France

This chapter presents the generic layered architecture for mobility models (*LEMMA*), which can be used to construct a wide variety of mobility models, including the majority of models used in wireless network simulations. The fundamental components of the architecture are described and analyzed, in addition to its benefits. One of the core principles stipulates that each mobility model is divided in five distinct layers that communicate via interfaces. This allows their easy replacement and recombination, which we support by reviewing 19 layers that can form 480 different mobility models. Some of the advanced features provided by the architecture are also discussed, such as layer aggregation, and creation of hybrid and group mobility models. Finally, some of the numerous existing studies of the different layers are presented.

11.1 Mobility Models

Motivation

Today, due to the exceptional success of wireless networks worldwide, it is difficult to imagine our lives without them — be it satellite communications or mobile phones. Yet, only a couple of decades ago they were far from being indispensable for the majority of us. During that time, a tremendous amount of efforts has been put in the improvement of these networks, which resulted in the creation of a wide variety of protocols and technologies fulfilling many different requirements, ranging from Bluetooth and ZigBee, to UWB, Wi-Fi, WiMAX, GSM, and UMTS.

The major differences between wired and wireless networks are the shared communication medium and terminals' ability to change their positions.[1] When addressing the latter issue, the majority of wireless technologies aim at maintaining any ongoing communications as long as possible by defining mechanisms to manage node mobility. Multiple such solutions have been proposed in regard to the various environments and particularities — e.g., routing packets in a centralized network such as UMTS is very different from the same task in a MANET. All these propositions were first tested in simulations in order to compare their performances and select the best solution, thus avoiding the development of full-featured implementations of all candidates. Because of this, it is very important to be able to draw valid conclusions from the simulations, by insuring that each subsystem and the simulation as a whole are as realistic as possible.

About This Chapter

In this chapter we are going to be addressing the part of the simulation governing nodes' movements — the *mobility model.* We are going to study microscopic mobility models (Bettstetter, 2001) used in wireless network simulations. Many models have been proposed generating various types of movement patterns (e.g., vehicle in a city, pedestrian in a shop). Some of them are purely synthetic, while others are based on diverse collections of data or have well-established theoretical grounds. With that multitude of possibilities, it is still astounding to discover that the most popular mobility model by far is the Random Waypoint (Kurkowski et al., 2005). One of the major reasons for this is that the synthetic models are much simpler and fit equally well or equally bad all scenarios. Complex models, on the other hand, are difficult to compare to each other and are generally specialized in a limited set of use cases. This does not incite the non-specialist to get acquainted with all their peculiarities in order to decide to what extent they may be useful and what possible adjustments could be made. Furthermore, following the classical model presentations, it is impossible to easily recombine different parts of several

[1]Throughout this work we are going to use the terms mobile *terminals*, *nodes*, *mobile nodes* (*MN*) and *users* interchangeably.

mobility models, which means that the general users are limited to the set of predefined scenarios, which may or may not correspond to their needs.

Examples

As an illustration, we are going to present two mobility models used in multiple simulation scenarios. Even though some of the details have been omitted for the sake of clarity, it is difficult to see right away how can one combine the different traits in order to obtain new models, or even — to what extent do these models differ. Later, we are going to give the LEMMA representation of these two models.

First, we are going to introduce the Random Waypoint Mobility Model (*RWP*) (Johnson & Maltz, 1996). All nodes are confined in a 2D rectangle area. Node movement is generated by first selecting a destination point, and then moving at a constant speed until reaching it. The speed is uniformly distributed in a predefined range $[V_{min}; V_{max}]$. Once the destination point is reached, the node *pauses* for a time period selected from the range $[P_{min}; P_{max}]$. Afterwards, the procedure is repeated, until the end of the simulation. The execution of this model is illustrated in Figure 11.1.

The Manhattan Mobility Model (Bai et al., 2003) is also defined in a 2D rectangle area. Node movement is restricted to an idealized street map containing only vertical and horizontal streets, each street having two lanes

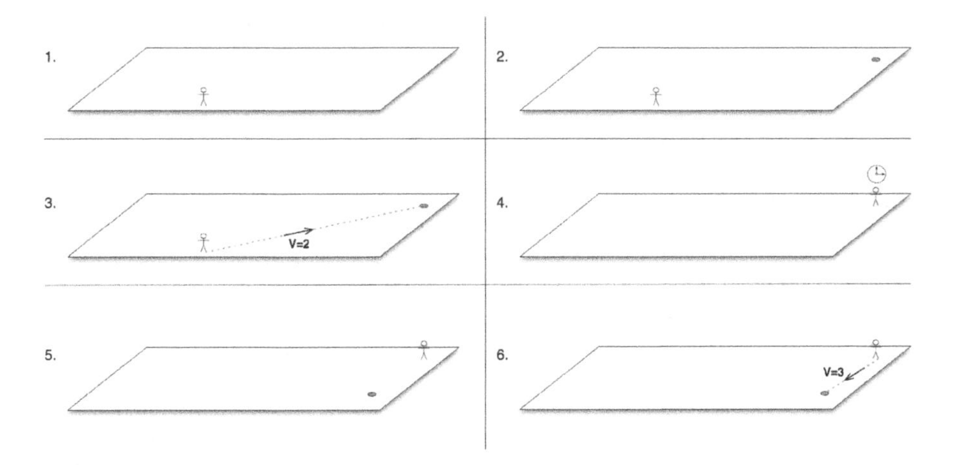

Fig. 11.1 Execution of the Random Waypoint Mobility Model. First, the node selects the destination point, then moves constantly with a selected speed, and finally pauses. The process is repeated until the end of the simulation.

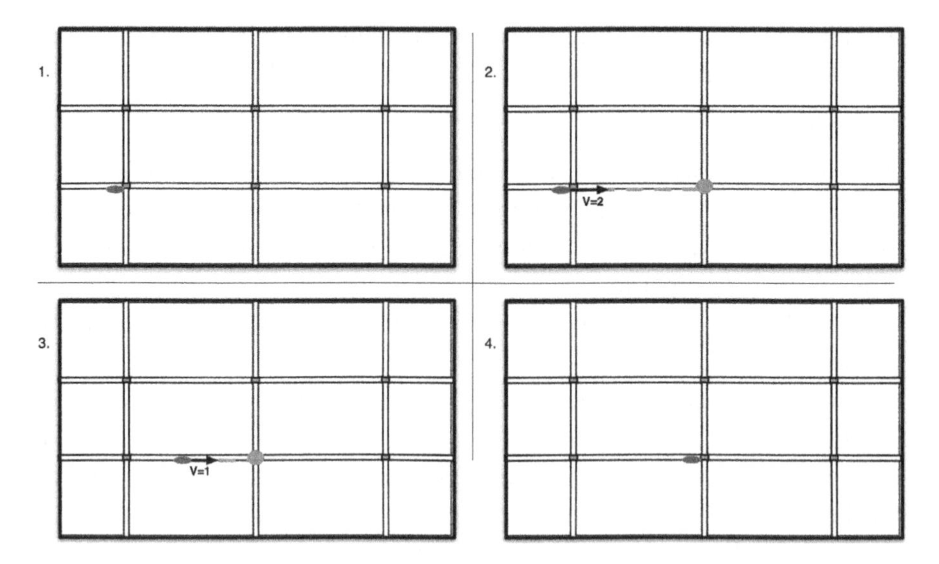

Fig. 11.2 Execution of the Manhattan Mobility Model. First the node (given as a filled ellipse) probabilistically decides which direction it should take. Then, it moves by randomly changing its speed during its movement, until reaching the next crossroad, where the process repeats.

(one for each of the directions). Upon reaching an intersection, a node decides probabilistically whether to continue moving on the same street, or to turn left or right. The speed changes randomly on each time slot and depends on the maximal node acceleration, and the speed of the node preceding it on the same lane of the street (Figure 11.2).

11.2 Architecture Description

In this section we are going to describe in details the generic layered architecture for mobility models. The architecture stipulates that a model is divided into several *layers* (Figure 11.4), each layer having distinct, well-defined functions. Each layer exposes an interface, which can be used only by the layer directly above it, and its output is fed to the layer directly below it.

The usefulness and the viability of this approach has been proven in many existing systems, such as the TCP/IP family of protocols. Individual layers are less complex than the whole model itself, which helps simplify the development, the validation and the usage of new mobility models. The abstract description of the different layers and their interactions allows new propositions to be made and studied independently of the rest. Afterwards, they can be used in conjunction with any combination of existing layers. Several layers can be aggregated in order to fine-tune the behavior of the nodes, while sharing layer implementations across a set of nodes may simulate group behavior. Furthermore, this separation allows gaining a better understanding of the influence of the different layers on the final result. Additionally, presenting an elaborate model in this form helps increase its readability, underline the major contributions, and insure that all necessary details are given, which is not always the case. Most importantly, because layers are functionally and semantically distinct, one can define specialized validation routines on a per-layer basis, e.g., by using study results and analyses performed in other research areas given later in this chapter.

The movement of a node in a given environment may be regarded as a result of the interaction of a set of spatiotemporal processes. The environment is common for all nodes and contains all objects, properties and constraints, which may affect the movements of the nodes, such as points of interest and obstacles. The movement processes specify node's movements as a function of their environment and simulation parameters (Figure 11.3). In the vast majority of cases, there is only one movement process, which governs the movement of all nodes in a simulation. However, one may find scenarios where each node has its own movement process (e.g., as in multi-agent simulations), or processes that govern the movement of all nodes within a given subset of the simulation area, etc.

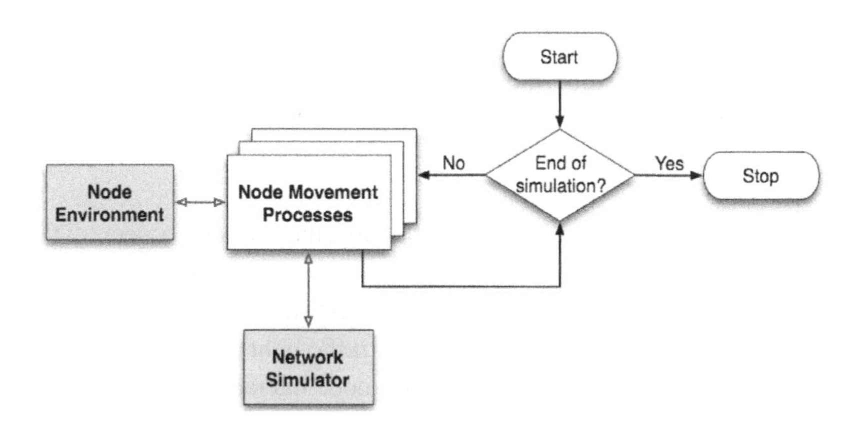

Fig. 11.3 Movement process — node environment interaction.

Environment

The node environment is constituted of four types of entities, namely: *simulation area*, *zones*, *constraints* and *movement influencing factors*.

- *Simulation Area*: The *simulation area* is the universe where the nodes "live" and move. It may be a one-, two- or three- dimensional space, or some other (user-defined) space. Each point P in this space is characterized by a set of coordinates $(p_0, \ldots, p_N)$. The action to be taken if a node tries to move outside the simulation area depends on its *border behavior*. In (Bettstetter, 2001) Bettstetter summarized the bounce-back, wrap-around and delete-and-replace behaviors.
- *Zones*: The simulation area provides a sort of "low-level" coordinate-based positioning. However, people rarely think in terms of latitude and longitude — one would rather use names of places, cities, streets, buildings, etc. A zone is a connected set of points with an optional set of attributes. The zones form a high-level addressing space on top of the simulation area. They are a generic construct that can be used to represent a wide range of scenario entities, such as buildings, lakes, districts, desks, walls, etc. This allows the movement process layers to be defined in a scenario independent way. Moreover, the definition of zone operations is straightforward (union, intersection, ...), which enhances their expressiveness (e.g., picking all buildings from a given district).
- *Constraints*: Node movement is directly affected by the movement *constraints*, which restrict the possible movement trajectories. They depend on the movement process itself, e.g., a constraint may be in the form of a graph, a set of rectangles.
- *Movement Influencing Factors*: Finally, the low-level aspects of the movement depend on the *movement influencing factors*. They are also dependent on the movement process and may include, amongst others, various traffic regulations (traffic lights, minimal/maximal speed, ...), rules that specify inter-node interactions (e.g., no collisions, speed matching), etc.

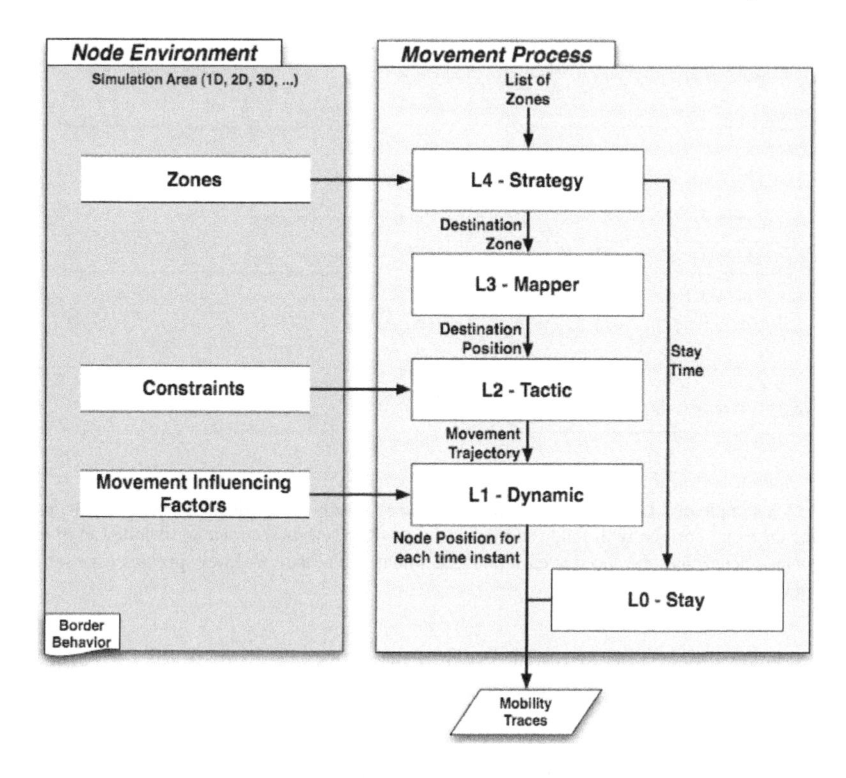

Fig. 11.4 Node environment entities and movement process layers.

Movement Subprocesses

According to the layered architecture, a general node movement process can be divided into five layers called *strategy*, *mapper*, *tactic*, *dynamics*, and *stay* (as shown in Figure 11.4).

A typical process execution has five stages (Figure 11.5). First, the *strategy* selects the next zone to be visited. Then the *mapper* chooses a specific point from that zone, which allows the *tactic* to generate the trajectory. Finally, the *stay* layer determines the node movement during the stay period. The whole process is repeated until the end of the simulation, or until it is replaced (i.e., the node is assigned another mobility model). The purpose of the individual layers is detailed hereafter. There exist an abundance of studies on various mobility-related phenomena, which map directly to the different layers of our architecture. Some of them are also outlined in the following paragraphs.

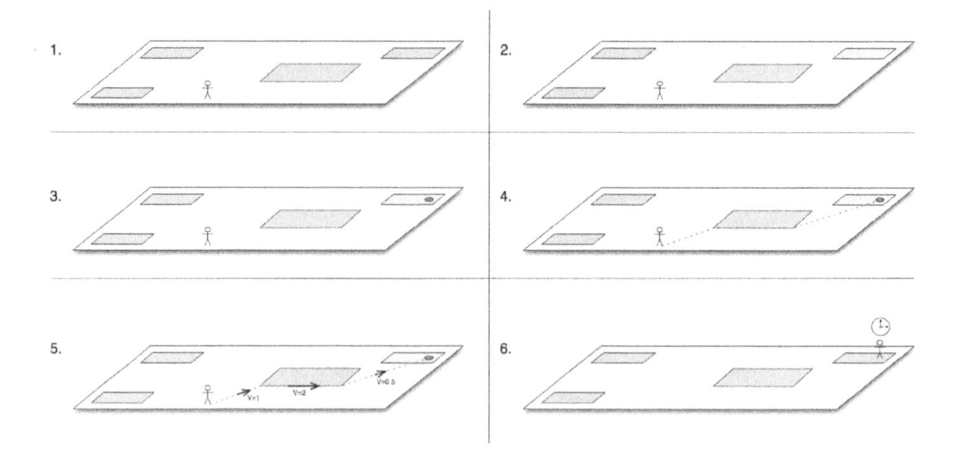

Fig. 11.5 Example of a LEMMA execution. The *strategy* selects the destination zone in 2. The *mapper* chooses the exact position in 3, followed by the *tactic*, which selects the route to be followed in 4. In 5, the *dynamic* determines the speed at each point, and finally in 6, the *stay* layer specifies what should be done in this zone (e.g., pause).

- *Strategy*: The *strategy* layer represents the high-level movement decision-making process. It determines node's destination zones, but does not specify the movement trajectory itself. A strategy takes as input a list of zones it should choose from. The result of the execution of a strategy is the pair {*next_zone*, *stay_time*}, where *next_zone* is the next zone to be visited by the given node, and *stay_time* is the time it should stay in the selected zone after reaching it. Most often the nodes simply pause during the stay time, but one can specify a different movement process to govern their movement during the zone stay time.

  Location prediction studies are strongly related to the strategy layer. Indeed, correctly modeling the high-level movement patterns (determined by the strategy) is the goal of many of them. Extensive analyses have been performed in the area of cellular networks, and because of the direct correspondence *cell* ⇔ *zone*, the majority of them can be considered as direct implementations of the strategy layer. For example, Bhattacharya and Das (Bhattacharya & Das, 1999) have proposed to create movement profile for each user, which is implemented as a Lempel-Ziv tree, whose alphabet holds one letter per zone. This idea is further developed and studied by

Song et al. (Song, 2008) and compared to Markov-based predictors.

Another class of relevant work includes the Origin-Destination studies performed by the traffic engineering researchers, e.g., (Abrahamsson, 1998). Other methods used to create and/or validate strategy layer implementations include user activity modeling, and the related activity-based approaches, such as the ones discussed in (Axhausen & Gärling, 1992).

- *Mapper*: Having the next zone to be visited, the zone-to-coordinates *mapper* translates the high-level zone addresses generated by the strategy to "low-level" coordinates, which are then passed on to the tactic layer.
- *Tactic*: The *tactic* layer is the trajectory-generating process. It generates a route from point *A* (the current node position) to point *B* (the position supplied by the *mapper*), which satisfies the set of constraints set by the user.

  Selecting routes between two given locations is also a subject of extensive studies, which can be regarded as studies of the tactic layer. The shortest path is not always the one being selected, as there may be other factors that affect people's decisions, e.g., traffic conditions, route attractiveness. It has even been shown that people may take different routes when doing a round trip (Golledge, 1995).
- *Dynamic*: Finally, the movement *dynamics* layer specifies the speed and the acceleration, and possibly some small deviations from the trajectory that has been defined by the *tactic*.

  Dynamics have also been thoroughly studied from many different points of view. If we concentrate on the case of vehicle movement, we will find multiple models of driver behavior, such as the velocity-difference model (Jiang et al., 2001) or the Intelligent Driver Model (Treiber et al., 2000). Pedestrian movement requires other types of models, e.g., emergency evacuations (Kuligowski, 2005).
- *Stay*: Upon reaching the end of its trajectory, and hence its destination zone, the node may be required to stay in it for a certain amount of time. During this time, the movement process defined

in the stay layer governs its movement. In most of the models, the node simply pauses (i.e., it does not move), but there also exist models that define no zone stay time, the Random Waypoint (e.g., as in (Blazevic et al., 2001)), etc. The process set in this layer can be described with the same layered architecture, thus making any model described here an implementation of this layer.

11.3 Existing Layers

Here are presented several examples of the environment components and mobility layers defined in the previous section. Most of them have been taken from existing mobility models, which helps illustrate the flexibility and the general character of the architecture.

Environment Components

- *Simulation Area*: In the huge majority of mobility scenarios, the *simulation area* is a 2D rectangle with a wrap-around or bounce-back border behavior (all studies using the Random Waypoint or the Random Walk mobility models fall in this category). Amongst the rarely used alternatives we can note the 3D parallelepiped area (Tuduce & Gross, 2005), and the fish bowl and Swiss flag (Le Boudec & Vojnovic, 2006).
- *Zones*: Zones with various shapes have been used in simulation scenarios — mainly rectangles, but also cubes and circles. These zones contain the points enclosed by the geometric figures, or their borders. We can also distinguish the singleton zone, which contains a single point from the simulation area.
- *Constraints*: The set of constraints used for mobility scenarios is limited. Almost all constraints can be expressed in terms of graphs embedded in the simulation area, zone avoidance or zone confinement. Indeed, most of the papers only describe the ways these constraints are configured, e.g., building graphs by using Voronoi paths (Jardosh et al., 2003), Delaunay triangulation (Huang, 2005), synthetic maps (Bai et al., 2003), real-world maps (Nousiainen et al., 2002; Saha & Johnson, 2004; Naumov et al., 2006; Yoon et al., 2006; Bratanov, 1999).

Almost all constraints can be expressed in terms of graphs embedded in the simulation area, zone avoidance or zone confinement. The embedded graph is a graph whose vertices are assigned a location (a point or a zone). A node obeying this constraint is restrained to move only on the edges and in the vertices of the graph. Zone avoidance restrains the conforming nodes from entering into zones marked to be avoided. This may be used to simulate walls, impenetrable buildings, etc. On the contrary, zone confinement keeps the nodes from leaving the zones marked to be their habitat. An example scenario may include rescue operation where different teams work in distinct areas of the simulation (as in (Aschenbruck et al., 2004)).

- *Movement influencing factors*: The factors influencing the movement dynamics vary, with the majority of cases using a simple constant speed movement. However many elaborated propositions exists, which change the speed of the nodes as a function of the speed or the number of the surrounding ones (Tan et al., 2002; Saha & Johnson, 2004; Stepanov, 2002), or add traffic signalization compliance (Markoulidakis, 1997; Kim & Bohacek, 2005; Ilyas & Radha, 2005; Potnis & Mahajan, 2006).

Mobility Model Layers

A small sample of the layer implementations found in the literature is reviewed in the following paragraphs. The parameters for the *strategies* and the *dynamics* are given, while for the *tactics* this is the case for the constraints they enforce. The presented *mappers* are parameterless. All these layers are given in a succinct, yet unabridged manner. In the typical use case, a scenario creator will pick one member from each layer and use it directly in his/her scenario.

- *Strategies*
    1. ***Manually Specified***
       Zones — List of zones to visit, in the order to be visited.
       Stay time — An ordered list of stay times.
       The entire sequence of zones to be visited with the corresponding stay times is fixed in the scenario (e.g., as the movement of groups' logical centers in (Hong et al., 1999)).

2. ***Uniform Random Zone***

   Zones — List of zones to choose from.

   $P_{\min}$, $P_{\max}$ — Minimum and maximum stay times.

   The next zone to be visited is chosen from the list at random (with uniform distribution), and the stay in the zone is for a randomly selected amount of time (uniformly distributed in $[P_{\min}; P_{\max}]$) (Johnson & Maltz, 1996).

3. ***User-distributed Random Zone***

   Zones — List of zones to choose from.

   $P_D(z)$ — The distribution of the stay times, which depends on the zone.

   $Z_D(z, t)$ — The distribution from which the next zone is selected, as a function of time and zone.

   The next zone to be visited is randomly chosen from the list, with a user-specified distribution $Z_D(z, t)$, which depends on the current simulation time and zone. The stay time at the zone is drawn from the user-specified distribution $P_D(z)$, which depends on the selected destination zone. This strategy may be seen in the WWP (Hsu et al., 2005).

4. ***Mathematical Function***

   Zones — List of zones to choose from.

   $F(t)$ — Vector function of the time t defining the motion.

   For each moment of the simulation t, the smallest zone containing the point $F(t)$ is chosen and the stay time is fixed to 0, e.g., (Tolety, 1999; Bergamo & others, 1996).

- *Mappers*

  1. ***Fixed***

     The point to be selected is drawn from a list, which is specified by the creator of the scenario (e.g., (Minder et al., 2005)).

  2. ***Random***

     Randomly select a point in the given zone, as in (Blazevic et al., 2001).

3. ***Random Border***

   Randomly select a point belonging to the border of the given zone, as in (Hong & Rappaport, 1986).

4. ***Gravity Center***

   Select the point of gravity of the given zone. It can be defined for an arbitrary zone as the sum of the radius vectors of all points in this zone, divided by the number of summands. Used in (Scourias & Kunz, 1999).

- *Tactics*

  1. ***Linear***

     Does not enforce any constraints.

     The node should move in a straight line from its current position to the destination point. Used in a wide variety of models, such as (Hong & Rappaport, 1986; Basagni et al., 1998; Johansson et al., 1999; Blazevic et al., 2001; Tuduce & Gross, 2005).

  2. ***Zone Avoiding — Shortest Route***

     Enforces zone avoidance constraints.

     The movement is linear, with nodes never crossing zones marked as inaccessible. Instead, the shortest possible route is taken. If more than one shortest path exist, the one to be taken is randomly chosen.

  3. ***Zone Avoiding — Border Route***

     Enforces zone avoidance constraints.

     The movement is linear, with nodes never crossing zones marked as inaccessible. Instead, the node moves in straight line directly towards its endpoint until it reaches an inaccessible zone, which it surrounds tightly following the borders in a predefined direction (e.g., counterclockwise in 2D) until it is able to move on the straight line to its endpoint.

  4. ***Zone Confining — Shortest and Border Route***

     Enforces zone confinement constraints.

Analogical to the *Zone Avoiding* cases, with the difference that nodes are allowed to move only on the zones marked as accessible.

5. ***Graph-constrained — Shortest Route***

   Enforces embedded graph constraints.

   Nodes obeying the constraints of an embedded graph can move only on its edges and in its vertices. That is, the movement is linear, following the edges of the graph. If a vertex is represented by a zone, the node takes the shortest path to the next edge. If the graph is oriented, then the movement follows the orientation of the edges. Finding a route in the graph is done by ignoring the size of the vertices and then selecting the shortest route (if the edges have associated weights, they are taken into account). This tactic is met in many of the map-based models (both synthetic and real-world based).

- *Movement Dynamics*

  1. ***Constant Movement***

     V_{min}, V_{max} — Minimal and maximal velocity.

     The node moves at a constant velocity, drawn uniformly from the interval $[V_{min}, V_{max}]$ for each trajectory. Very frequently used, as in (Hong & Rappaport, 1986; Basagni et al., 1998; Johansson et al., 1999; Tuduce & Gross, 2005; Hsu et al., 2005). Setting $V_{min} = V_{max}$ will result in a frequently used scenario, where the nodes move at a constant speed throughout the simulation.

  2. ***Smooth Random (Bettstetter, 2001)***

     V — Set of velocities.

     The speed selection is characterized by the use of a set of target speeds V (the speed a node intends to achieve) and a linear acceleration. The node moves with constant speed v until a random process draws a new target speed from the set of possible target velocities V. The node then accelerates (or decelerates) linearly until this desired speed is achieved (or a new target speed is chosen in the mean time).

3. ***Edge-limited (Breyer et al., 2004)***

   The speed of a node depends on the number of nodes on the same edge of the graph. If there are many nodes on a given edge, the speed of all of them is reduced, which may be used to simulate overcrowding/congestion with little complexity.

4. ***Acceleration-constant-deceleration***

   $V_{\min}, V_{\max}$ — Minimal and maximal velocity.

   The node accelerates for a given time until reaching a randomly drawn speed in $[V_{\min}, V_{\max}]$, continues moves at a constant speed, and finally decelerates before reaching its destination.

5. ***Random (Bai et al., 2003)***

   $V_{\min}, V_{\max}$ — Minimal and maximal velocity.

   $a_{\min}, a_{\max}$ — Minimal and maximal acceleration.

   Node velocity changes on each time slot, with uniformly selected acceleration in $[a_{\min}, a_{\max}]$. The speed is always kept in the range $[V_{\min}, V_{\max}]$.

6. ***Preceding-node-limited (Bai et al., 2003)***

   D — Base dynamic.

   SD — Safety distance.

   The node is governed by the based dynamic **D**, given as parameter. If the node is moving on a graph, and there is another node on the same edge of the graph, and the distance between the two nodes is less than **SD**, the speed of the base dynamic is limited to be at most the same as the preceding node.

Examples

If we return to the examples given beforehand — the Random Waypoint Mobility Model (*RWP*) and the Manhattan Mobility Model (*MMM*) — we can demonstrate the way they can be formed with the help of *LEMMA*. For both scenarios we choose a 2D simulation area with wrap-around border behavior. For the RWP we choose a single zone with the size of the whole simulation area. For the MMM we create one zone per intersection, all zones being vertices in a planar graph, one edge per street lane. At this point the environment for the two scenarios is completely described.

To represent the RWP movement process we may use the *Uniform Random Zone* strategy, the *Random* mapper, the *Linear* tactic and the *Constant Movement* dynamic, with a simple pause as a stay layer. For the MMM, we need the *User-distributed Random Zone* strategy with *Random Border* mapper, *Graph-constrained Shortest Route* tactic and *Preceding-node-limited* dynamic based on the *Random* dynamic.

This representation not only allows us to see the essence of each model clearly, but also provides the possibility to create new models based on the existing ones by simply changing a given layer, layer parameter or environment setup. For example, by replacing the strategy of the *MMM* to *Uniform Random Zone* strategy we obtain the City Section Mobility Model (Camp et al., 2002).

11.4 Strategy Aggregation, Hybrid and Group Mobility Models

With the given architecture and the specified layers it is possible to create a wide variety of mobility models. However, there is more potential lying in the chosen approach. Indeed, it is possible to create advanced mobility scenarios, which use aggregated layers, or produce group or hybrid motion. In this section, we are going to describe the ways to achieve these powerful features.

Combining Multiple Strategies

Defining relations between zones is straightforward, as they are sets of points. As a consequence, since strategies take a set of zones as input, and output a single zone and stay time, it is possible to combine two or more strategies into a single one with the help of some inter-layer "glue" modules, which we are going to call *strategy adaptors*. Aggregating strategies may be done with the help of two simple types — zone and time adaptors (as shown in Figure 11.6).

Constructing Heterogeneous Models

A node may be required to change its movement process during the course of the simulation, as for example pedestrians getting on and off busses, taxies,

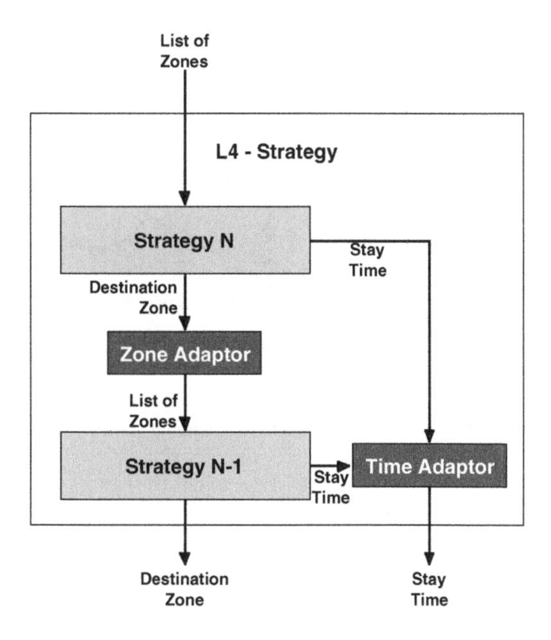

Fig. 11.6 Using adaptors to combine strategies.

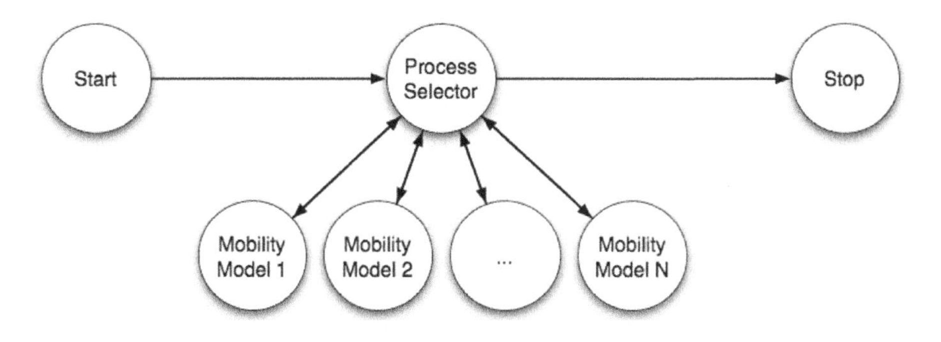

Fig. 11.7 Constructing a new heterogeneous model by combining several mobility models.

etc. These types of scenarios may be created by combining several mobility models, each representing a single kind of movement pattern. The movement of a node is then governed by a single "simple" model at a time, having a *process selector* to chose the active model according to some conditions, like current time, zone, surrounding nodes, etc. (see Figure 11.7). The same principle is illustrated in Figure 11.8 from the perspective of an individual node.

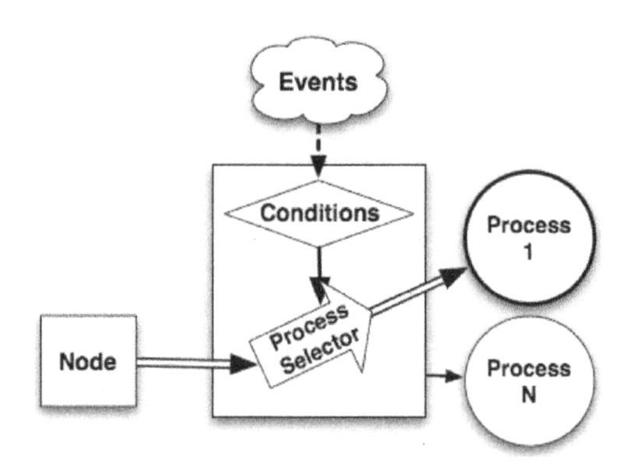

Fig. 11.8 Process Selector schematic functionning — node perspective.

The active mobility process for the node may be changed upon the occurrence of an event, and depending on the specified conditions. For example, one may use two movement processes — one to simulate daytime activities, and the other for the night movements, changing them every 12 hours.

Furthermore, it is possible to change a single layer during the model execution, which once again shows the flexibility of the selected approach. One can imagine a case, where the tactic is changed as a function of the destination point chosen by the *mapper* layer, as shown in Figure 11.9. Using this technique, it is easy to specify scenarios where nodes travel the long distances at high speeds, while handling the short distances at lower speeds (thus simulating getting on/off vehicular transport), and all this — without changing the individual layer definitions and implementations.

Creating Group Mobility Models

Having defined the interfaces between all layers it is easy to share the instances of some of the high-level layers across several nodes, as shown in Figure 11.10. This facilitates the creation of a group-like behavior, e.g., we may define the Reference Point Group Mobility Model (*RPGM*) (Hong et al., 1999) as having a shared strategy, mapper and tactic layers and using different parameters for the dynamic layer of each node (i.e., the specific reference point). Because, the RPGM is the most used group mobility model (most of the other models producing group movement patterns are its subset (Camp et al., 2002)), by following the example given in this paragraph, it is possible to obtain the majority of group-based movement traces used for wireless network research.

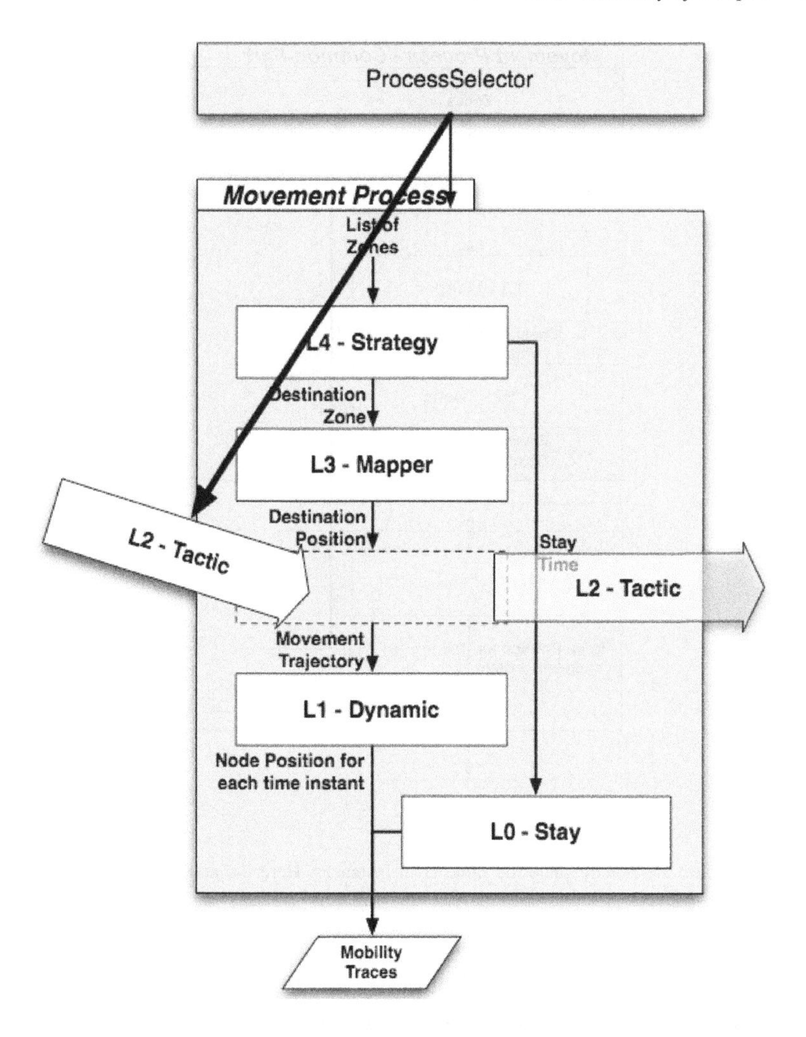

Fig. 11.9 Fine-grained hybrid node behaviour may be achieved by changing a single layer of its process.

11.5 Summary of Chapter

The layered mobility model architecture was presented in this chapter. It can be used to construct a wide variety of models, including the majority of microscopic models used in wireless network simulations. It divides the movement process in five layers, each having distinct, strictly defined functions, as well as the ways these layers interact. Additionally, the major entities of the simulation environment have been described, along with the way the different layers

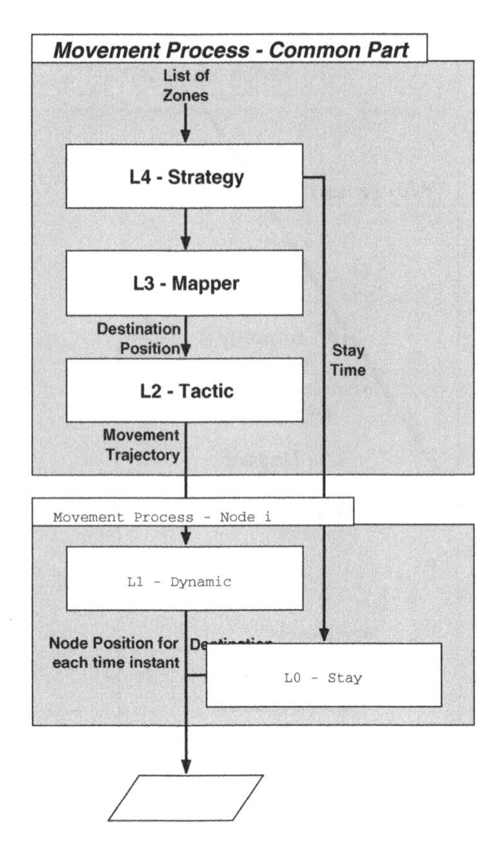

Fig. 11.10 Several nodes may share the same layer instances. Here the nodes have common *strategy*, *mapper* and *tactic* layers.

interact with them. The means of building heterogeneous and group models, and aggregating multiple layers have also been described.

References

[1] Bettstetter, C. (2001) Smooth is better than sharp: A random mobility model for simulation of wireless networks, 19–27.

[2] Kurkowski, S., Camp, T., & Colagrosso, M. (2005) MANET simulation studies: The incredibles. *SIGMOBILE Mob. Comput. Commun. Rev.* 9, 50–61.

[3] Johnson, D. B. & Maltz, D. A. (1996) Dynamic Source Routing in Ad Hoc Wireless Networks, 353.

[4] Bai, F., Sadagopan, N., & Helmy, A. (2003) IMPORTANT: A framework to systematically analyze the impact of mobility on performance of routing protocols for adhoc networks, 2, 825–835.

[5] Bettstetter, C. (2001) Mobility modeling in wireless networks: Categorization, smooth movement, and border effects. *SIGMOBILE Mob. Comput. Commun. Rev.* 5, 55–66.

[6] Bhattacharya, A. & Das, S. K. (1999) LeZi-update: An information-theoretic approach to track mobile users in PCS networks. 1–12.

[7] Song, L. (2008) Evaluating Mobility Predictors in Wireless Networks for Improving Handoff and Opportunistic Routing.

[8] Abrahamsson, T. (1998) Estimation of origin — destination matrices using traffic count–a literature survey.

[9] Axhausen, K. W. & Gärling, T. (1992) Activity-based approaches to travel analysis: Conceptual frameworks, models, and research problems. *Transport Reviews* 12, 323–341.

[10] Golledge, R. (1995) Path selection and route preference in human navigation: A progress report. 988/1995, 207–222.

[11] Jiang, R., Wu, Q., & Zhu, Z. (2001) Full velocity difference model for a car-following theory. *Physical Review E* 64.

[12] Treiber, M., Hennecke, A., & Helbing, D. (2000) Congested Traffic States in Empirical Observations and Microscopic Simulations. *Physical Review E* 62, 1805.

[13] Kuligowski, E. D. (2005) Review of 28 Egress Models, 68–90.

[14] Blazevic, L., Giordano, S., & Boudec, J. L. (2001) Self organized terminode routing simulation, 81–88.

[15] Tuduce, C. & Gross, T. (2005) A mobility model based on WLAN traces and its validation. 1, 664–674.

[16] Le Boudec, J. Y. & Vojnovic, M. (2006) The random trip model: Stability, stationary regime, and perfect simulation. *Networking, IEEE/ACM Transactions on* 14, 1153–1166.

[17] Jardosh, A., Belding-Royer, E. M., Almeroth, K. C., & Suri, S. (2003) Towards realistic mobility models for mobile ad hoc networks, 217–229.

[18] Huang, D. (2005) Using Delaunay triangulation to construct obstacle detour mobility model. 3, 1644–1649.

[19] Nousiainen, S., Kordybach, K., & Kemppi, P. (2002) User distribution and mobility model framework for cellular network simulations, 518–522.

[20] Saha, A. K. & Johnson, D. B. (2004) Modeling mobility for vehicular ad-hoc networks, 91–92.

[21] Naumov, V., Baumann, R., & Gross, T. (2006) An evaluation of inter-vehicle ad hoc networks based on realistic vehicular traces, 108–119.

[22] Yoon, J., Noble, B. D., Liu, M., & Kim, M. (2006) Building realistic mobility models from coarse-grained traces, 177–190.

[23] Bratanov, P. (1999) User Mobility Modeling in Cellular Communications Networks.

[24] Aschenbruck, N., Frank, M., Martini, P., & Tolle, J. (2004) Human mobility in MANET disaster area simulation — a realistic approach, 668–675.

[25] Tan, D. S., Zhou, S., Ho, J., Mehta, J. S., & Tanabe, H. (2002) Design and Evaluation of an Individually Simulated Mobility Model in Wireless Ad Hoc Networks.

[26] Stepanov, I. (2002) Integrating Realistic Mobility Models in Mobile Ad Hoc Network Simulation.

[27] Markoulidakis, J. G. (1997) Mobility modeling in third generation mobile telecommunication systems. *IEEE PCS.*

[28] Kim, J. & Bohacek, S. (2005) A survey based mobility model of people for urban mesh networks. *MeshNets'05.*

[29] Ilyas, M. U. & Radha, H. (2005) The influence mobility model: A novel hierarchical mobility modeling framework. 3, 1638–1643.

[30] Potnis, N. & Mahajan, A. (2006) Mobility models for vehicular ad hoc network simulations, 746–747.

[31] Hong, X., Gerla, M., Pei, G., & Chiang, C. (1999) A group mobility model for ad hoc wireless networks, 53–60.

[32] Hsu, W., Merchant, K., Shu, H., Hsu, C., & Helmy, A. (2005) Weighted waypoint mobility model and its impact on ad hoc networks. *SIGMOBILE Mob. Comput. Commun. Rev.* 9, 59–63.

[33] Tolety, V. (1999) Load reduction in ad hoc networks using mobile servers.

[34] Bergamo, M. et al. (1996) System Design Specification for Mobile Multimedia Wireless Network MMWN.

[35] Minder, D., Marrón, P. J., Lachenmann, A., & Rothermel, K. (2005) Experimental construction of a meeting model for smart office environments.

[36] Hong, D. & Rappaport, S. S. (1986) Traffic model and performance analysis for cellular mobile radio telephone systems with prioritized and nonprioritized handoff procedures. *Vehicular Technology, IEEE Transactions on* 35, 77–92.

[37] Scourias, J. & Kunz, T. (1999) Activity-based mobility modeling: realistic evaluation of location management schemes for cellular networks. *Wireless Communications and Networking Conference, 1999. WCNC. 1999 IEEE* 1.

[38] Basagni, S., Chlamtac, I., Syrotiuk, V. R., & Woodward, B. A. (1998) A distance routing effect algorithm for mobility (DREAM), 76–84.

[39] Johansson, P., Larsson, T., Hedman, N., Mielczarek, B., & Degermark, M. (1999) Scenario-based performance analysis of routing protocols for mobile ad-hoc networks, 195–206.

[40] Breyer, T., Klein, M., Obreiter, P., & König-Ries, B. (2004). Activity-based user modeling in service-oriented ad-hoc-networks.

[41] Camp, T., Boleng, J., & Davies, V. (2002) A survey of mobility models for ad hoc network research. *Wireless Communications & Mobile Computing (WCMC): Special Issue on Mobile Ad Hoc Networking: Research, Trends and Applications* 2, 483–502.

12

Optimisation of CDMA-Based Radio Networks

Moses Ekpenyong and Joseph Isabona

University of Uyo, Nigeria

In this chapter we study the optimisation methods of 3G (CDMA-based) networks. We provide background issues on optimisation, discuss various existing methods, process and type of problems they typically address and appraise optimization techniques that can be utilized to solving these problems.

12.1 3G Network Optimisation: Background of Issues

The high price of third generation (3G) licenses and growing competition in the telecommunication market are putting enormous pressure on manufacturers and operators to configure the next-generation mobile network in the most cost-effective way possible and provide high quality 3G services. Because of recent demands to improve overall network performance and ensure efficient utilisation of network resources in 3G networks, radio network optimisation has become even more important than in previous generation networks.

Network optimisation is concerned with providing a network configuration process that improves the performance of a network, allowing the translation from desired quality of service (QoS) targets to measured network performance and finding the best possible network design. The techniques of optimisation borders on best design practices and amounts to finding different kinds of planning-related actions that provides reasonable trade-off between the network model complexity and performance reality.

3G radio networks are based on the Code Division Multiple Access Technology (CDMA) that uses spread spectrum techniques. 3G may also be referred to as Universal Mobile Telecommunication System (UMTS). There are three standards accepted by the International Telecommunications Union (ITU). They are Wide-band CDMA (WCDMA), CDMA2000 and Time Division Synchronous CDMA (TD-SCDMA). Although these three standards exist in 3G networks (each used in different parts of the world), the general planning and optimisation process and overall objectives are the same.

Due to the complexity of the 3G network, its size and the necessity of having to deal with many factors and control parameters, we will in this chapter treat optimization under three broad categories namely topology optimization, automated optimization and network optimization.

12.2 Topology Optimisation

The goal of radio network topology planning is to provide a configuration that offers the required coverage for different services, and simultaneously maximises the system capacity. This section appraises the impact of:

(i) coverage overlap
(ii) selection of site location
(iii) sectoring and antenna beamwidth
(iv) antenna downtilt

on the Wideband Code Division Multiple Access (WCDMA) network coverage and system capacity. Moreover, the impact of the aforementioned elements is appraised on pilot pollution that reflects partly on the expected functionality of the network, and on site evolution from a 3-sectored site to a 6-sectored site [1].

Coverage Overlap

In any cellular network, coverage overlap is required in order to combat the harmful effect of slow fading of signals, and furthermore, to enable the provision of indoor coverage with an outdoor network (building penetration loss). Consequently, in cellular networks, most of the other-cell interference is produced by the coverage overlap requirements.

Generally, coverage overlap is affected by the link budget, antenna configuration, average site spacing, and propagation environment. The first three elements borders strictly on topology related factors, while the last element is defined by the planning environment, which also defines the propagation slope [2]. The impact of the maximum allowable path loss on the coverage overlap is obvious; a higher allowable path loss provides better coverage and can thus increase the coverage overlap. The maximum allowable path loss is naturally affected by the base station antenna gain (connection to antenna horizontal and vertical beamwidth). Secondly, a higher antenna position decreases the propagation slope, and therefore increases the cell coverage and resulting coverage overlap. Moreover, a higher antenna position increases the probability of line-of-sight (LOS) connections. Thirdly, the closer the base station sites are to each other, the larger the resulting coverage overlap. Lastly, the propagation environment impacts on the propagation slope, thereby affecting the amount of coverage overlap.

In summary, a small coverage overlap might reduce the network performance through too low network coverage, whereas too high coverage overlap reduces the network performance, increases other-cell-interference level, and reduces system capacity. This is essentially the foundation of radio network topology planning, which requires optimised coverage overlap. Hence, its impact on system capacity must be understood when selecting sites during the topology planning phase. The aim of this section is to achieve optimum coverage overlap that maximises the system capacity. A similar approach from roll-out optimised network configuration viewpoint is treated in [3].

Site Spacing

Site spacing (i.e., average distance between sites) is defined either by the coverage or capacity requirements for a planning area. Coverage requirements define the site spacings typically in rural areas, where the capacity does not constrain the system performance and observable QoS. Capacity requirements (expected customer density) define site spacings in capacity-limited environment; though, the coverage requirements for indoor users also affect the site density of an urban planning area. If high indoor coverage probabilities (80–90%) are required, the average site density grows, which automatically results in large coverage overlap areas. This easily increases the risk of observing higher other-cell-interference levels as well. Therefore, antenna height optimisation and antenna downtilt are strongly required.

Antenna Height

The selection of antenna height is typically performed according to the planning environment. In a microcellular planning environment, antennas are systematically deployed under the average roof top level for capacity purposes. For an urban macrocellular network layer, antenna heights follow the average roof top levels.

Correspondingly in suburban areas, the propagation occurs most of the time clearly above roof-top level due to relatively higher antenna position with respect to the average roof top levels. On the other hand, the propagation loss in rural areas is dominated by the undulation of the terrain. This implies that the propagation slope varies between 20 dB/dec (free space) and 45 dB/dec (dense urban) depending on the propagation environment [2, 4, 5].

From a radio network topology optimisation viewpoint, the selection of the antenna height also depends on the site location. A choice of a low antenna installation height increases the number of required sites in a planning area. Besides, a lower antenna position in an urban area reduces service coverage probabilities in the network, and might decrease the QoS. Conversely, sectors become more isolated from each other, which results in lower other-cell-interference levels. If a higher antenna position is selected, coverage probabilities can be enhanced. However, signals are propagated for longer distances (referred to as over shooting), which exposes the network to higher other-cell-interference levels. Furthermore, higher antenna position may increase Soft Handover (SHO) areas at the cell edges and result in higher overhead for SHO connections.

Coverage Overlap Index

Here, we present the effect of coverage overlap on the system capacity with extensive set of system level simulations. Moreover, an optimum empirical value for coverage overlap is evaluated with coverage overlap index (COI). All relevant simulation parameters and the simulation environment descriptions can be found in [6].

The coverage overlap index (COI) is defined here as

$$COI = \frac{1 - \text{length of dominance area}}{\text{length of actual coverage area}} \qquad (12.1)$$

where length of dominance area is the length of the geographical area, where the cell is intended to be the most probable server. The length of actual coverage area is the cell range defined by the maximum allowable path loss towards the horizontal plane of an antenna and can be calculated with an adequate propagation model. In the context of multi-service WCDMA network, the maximum allowable path loss is defined by the service with the highest path loss (typically, speech/voice). If COI $\rightarrow 0$, the cells in a network would not have sufficient overlap, and the network would most probably be unable to provide continuous network coverage (without planning margins such as slow fading margin). However, the other-cell interference level would definitely be low as well. Hence, in practice, COI must be higher than zero in order to tolerate slow fading and to achieve indoor coverage.

In order to provide an idea of the range of COI, let us consider an example with link budget values presented in [7]. The isotropic path loss (i.e., without any margins) equals 157.1 dB in the downlink. This can be mapped into a cell range of 3.16 km by using the Okumura-Hata propagation model with 35 dB/dec propagation slope, 25 m antenna height, and 2100 MHz frequency. However, by taking into account standard deviation of slow fading ($\sigma_{SF} = 7$ dB), SHO gain (3 dB), and outdoor location probability requirement (95%), the resulting maximum allowable path loss would be 149.8 dB, and the corresponding cell range 1.95 km. If the network were deployed based on outdoor coverage, the resulting COI would be, by the definition, COI $= 1 - (1.95/3.16) = 0.383$. On the contrary, if 90% indoor location probability were required ($\sigma SF = 9$ dB and building penetration loss (BPL) $= 15$ dB), the indoor path loss would be 135.6 dB, and the cell range, 0.77 km. Finally, the COI would equal to: $1 - (0.77/3.16) = 0.756$. Note that this example does not include the impact of antenna downtilt that effectively decreases the path loss towards antenna boresight, and hence reduces the actual coverage area. Nevertheless, due to planning margins, coverage overlap always exists, which on the contrary, increases other-to-own-cell interference in WCDMA. Therefore, a certain level of other-to-own-cell interference has to be accepted (e.g., $i = 0.5$ [8]).

Technically, the utilisation of COI has two different approaches. Taking an academic approach, a possible network together with site and antenna configuration could be selected freely, and hence optimisation of COI should also be based on all these parameters. This kind of scenario could arise in a case where planned site density was smaller than the density candidate site locations, which is a rather hypothetical assumption. In a more practical approach, site locations (and correspondingly site spacings) would be fixed (or pre-defined), and, moreover, antenna heights could not be significantly changed. In this kind of scenario, the optimisation of COI would be based almost solely on optimising the antenna configuration (mostly downtilt).

Empirical Optimum COI

In the following, we present an empirically evaluated optimum COI based on the simulations presented in [6]. Parameters in the evaluation include:

(i) maximum allowable isotropic path loss in DL (157.55 dB, 160.55 dB, 163.55 dB)
(ii) simulated site spacings (1.5 km, 2.0 km, 2.5 km)
(iii) simulated antenna heights (25 m, 35 m, 45 m)
(iv) different antenna types (65°/6°, 65°/12°, 33°/6°)
(v) optimum downtilt angles for each scenario

Three different antenna radiation patterns are considered. They include 3-sectored sites with 65°/6°, 3-sectored sites with 65°/12°, and 6-sectored sites with 33°/6°, where xy°/z° denotes the half-power beamwidth in the horizontal (xy)/vertical plane (z). The maximum allowable isotropic path loss differs only due to different antenna gain. For the evaluation of COI, antenna electrical downtilt is taken into consideration by decreasing the maximum path loss in the direction of horizontal plane of the antenna using the Okumura-Hata propagation model at a frequency of 2100 MHz. The propagation slope is derived as a function of antenna height. The average area correction factor (−14 dB) was evaluated based on the exact land use in the simulation area. Furthermore, as the network was based on the hexagonal grid with nominal antenna directions, the length of the dominance areas for 3-sectored sites were 2/3 of the site spacing and for 6-sectored sites 1/2 of the site spacing. Figures 12.1 and 12.2 gather the average sector throughput as a function of COI for different network, site, and antenna configurations. The average sector throughput is derived based on the average BS TX power of 39 dBm of all cells.

Figure 12.1(a) shows the sector throughput as a function of COI for 3-sectored network with 65°/6°. The squared dots are for optimum downtilt scenarios and the circular dots are for non-tilted scenarios. Most of the scattering

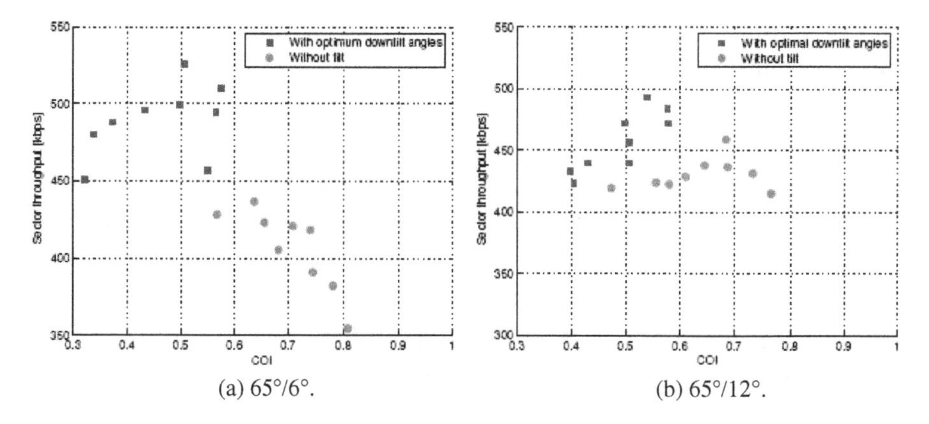

(a) 65°/6°. (b) 65°/12°.

Fig. 12.1 Average sector throughput as a function of COI for 3-sectored sites.

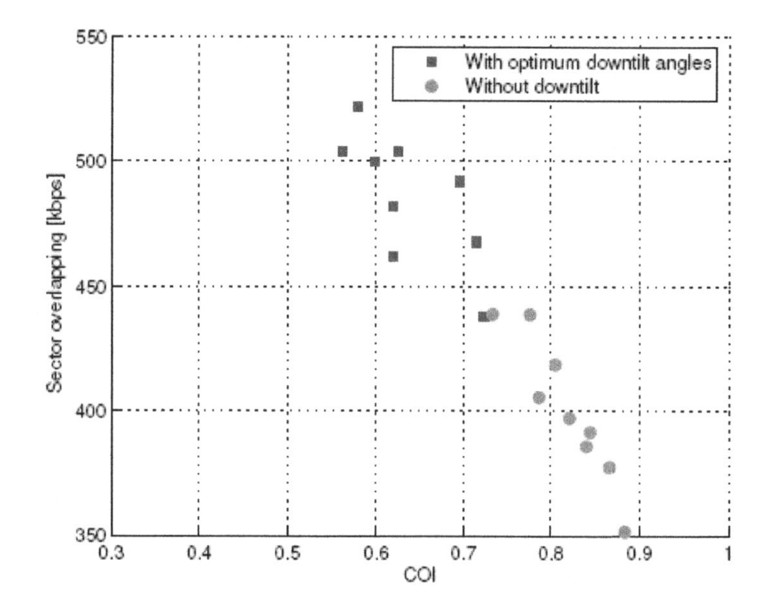

Fig. 12.2 Average sector throughput as a function of COI for 6-sectored network; 33°/6°.

between samples are caused by the utilisation of a digital map (the propagation environment was different from those with all site spacing) and practical antenna radiation patterns.

The results indicate that maximum sector throughput can be achieved with COI $\approx$ 0.5 (Figure 12.1(a)), and this can only be achieved by downtilting. The configuration that provides the highest throughput is 1.5 km site spacing with 45 m antenna height.

Nevertheless, the importance of the results is that the system capacity can be maximised with higher antenna position and shorter site spacing. Without any downtilt, the resulting COI is naturally higher with a correspondingly lower capacity. The results with and without downtilt cannot be directly compared because of the fact that the definition of COI does not address the enhanced coverage near the base station due to antenna downtilt. With different antenna type (larger vertical beamwidth and smaller antenna gain), the changes in the throughput as a function of COI are not dramatic (Figure 12.1(b)). Moreover, optimum COI is obviously close to 0.5. The optimisation process of COI should be based in practice on automatic optimisation due to different dominance areas and antenna heights.

Figure 12.2 shows the corresponding relation of COI and sector throughput for 6-sectored sites. However, the selected site spacings and antenna heights are not able to achieve COI < 0.5. Nevertheless, the trend of the results indicates that optimum COI would be achieved around 0.5.

Study of Site Locations and Sector Directions

The site location selection is performed during the topology planning phase. However, the requirements for site density are defined either by coverage or capacity requirements, and are provided from configuration planning (or from initial topology planning). In practice, the amount of candidate site locations is rather limited, and hence an operator is not always able to utilise wanted site locations.

A set of candidate site locations is reduced, e.g., due to topographical irregularities of the terrain, or authority constraints or government regulations that could prevent an operator for deploying a base station to an optimal place from a network performance point of view. Also, in urban environments, where base stations are often located on the top of buildings, the physical space required for hardware of the planned site solution may not be enough. A rather extensive and practical view of possible problems during site location selection and site acquisition is provided in [9].

Another approach for WCDMA site locations is the opportunity to reuse the site locations of an existing cellular network (e.g., GSM). This is economically a beneficial approach due to co-locating and co-siting opportunity, if only the costs are considered. However, from an optimum performance viewpoint, site locations of the existing network could be derived from the ones planned for a new network.

To allow for network coverage, existing GSM locations could be selected also for WCDMA with certain coverage limitations. In summary, the selection process of WCDMA site location has several aspects and also several constraints, and hence, the solutions might be somewhat non-optimal from a technical standpoint.

Another limitation that might occur during site acquisition is the selection of sector direction (or antenna direction). Due to external problems, the antenna directions might not have been directed as planned. Moreover, because of obstacles close to the base station site location, or due to errors, e.g., in the base station antenna implementation, the antenna directions may change, and thus affect the network performance. Hence, the effect of antenna direction deviation on the network performance should also be considered.

The following section provides simulation results and analysis of two different irregular configurations: namely, irregular site location[s] and irregular sector directions. First, a comparison between hexagonal grid planning and irregular network configuration is presented in order to observe the impact of non-hexagonal site locations on the network performance. Secondly, the impact of irregular antenna directions on the WCDMA network performance is addressed.

Irregular Site Locations

In a homogenous and totally flat environment, a hexagonal grid planning with equal site spacings would be the most efficient strategy to deploy a cellular network, and assuming further that the traffic distribution were homogeneous. Even in a generally flat environment, local terrain and clutter cause dramatic spatial variations in the received power. This phenomenon is observed as slow fading, where the mean level of the received signal tends to have a log-normal distribution [10, 11]. In addition, the regularity of traffic distribution is broken by different building distributions. Hence, in practice, a hexagonal network is not an optimum network layout for a cellular network. The displacement of base station locations from the ideal hexagonal grid has been found to have a negligible impact on carrier-to-interference (C/I) values in cellular systems through simulations in a homogenous environment and with uniform traffic distribution. Moreover, in [12] it is shown that especially a CDMA network is rather robust for moderate base station location changes with respect to an ideal hexagonal grid. The results are of great importance owing to the fact that small deviations do not affect cellular system performance. Hence, during the site selection process, most importance could be put on economical constraints. In contrast, [13] concentrates especially on UMTS system, the base station location is concluded to have a notable impact on downlink and uplink performance with a relatively large cell range.

Another set of simulation results regarding irregular base station site locations is considered in [14]. However, the aim was to study the impact of a small deviation in the base station site location on the top of a digital map on WCDMA system performance. Irregular network configurations (also called non-hexagonal) were formed by introducing a random deviation (i.e., an error vector) for each hexagonal site location. The maximum allowed deviation of the site location was 1/4 of the site spacing, and hence with 1.5 km site spacing for instance, the maximum deviation was 375 m. Figure 12.3 illustrates part of

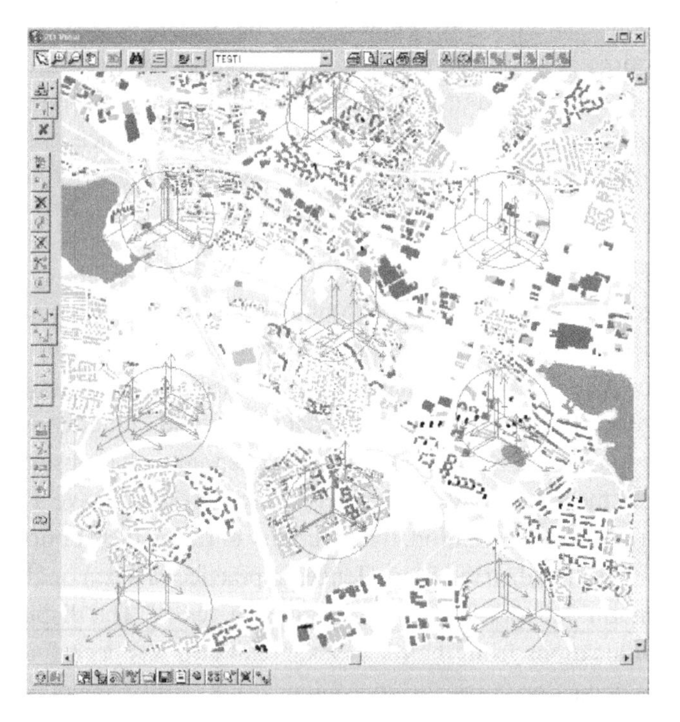

Fig. 12.3 Non-hexagonal site locations together with the hexagonal reference locations with 1.5 km site spacing. The ring around each hexagonal site location illustrates the maximum allowable deviation. [14]

Table 12.1. Example simulation results with 1.5 km and 3.0 km site spacings. The performance indicators for the non-hexagonal grids are averaged over all five non-hexagonal network layouts, and the values in parenthesis are their standard deviations. DL capacity value is estimated based on average TX power of 39 dBm of all base station sectors. [14]

		Reference grid		Non-hexagonal grids	
Parameter		1.5 km	3.0 km	1.5 km	3.0 km
Service probability	[%]	98.1	96.2	98.3(0.68)	94.0(1.08)
SHO probability	[%]	24.9	17.2	25.0(0.56)	18.0(0.5)
SfHO probability	[%]	3.9	4.6	3.9(0.19)	4.9(0.1)
UL_i		0.79	0.58	0.80(0.02)	0.60(0.02)
DL capacity	[kbps/sector]	360	360	359(2)	359(3)

a network with hexagonal and non-hexagonal site locations. Altogether, five different non-hexagonal grids were formed and simulated.

The simulation results in Table 12.1 show selected performance indicators for hexagonal and non-hexagonal network layouts of 1.5 km and 3.0 km site spacings. The results are rather pessimistic as no optimisation of antenna configuration was performed after randomisation of the network layout. With 1.5 km site spacing, the non-hexagonality is hardly seen in any of the performance indicators, and actually the average service probability is slightly larger for non-hexagonal grids. This minute increase comes from an improved network coverage. However, the network of 3.0 km site spacing is more coverage-limited, and hence the irregularity begins to affect the network performance. As a result, the average service probability is 2% smaller. However, randomisation of site locations does not change the network capacity as indicated by the results. In this scenario also, the increase of SHO probability is slightly larger. As is shown in [14], a higher indoor location probability does not affect the inter-related simulation results of a non-hexagonal grid. Moreover, the deviation does not have an impact on the system capacity with selected maximum location deviation and site spacings.

Irregular Sector Directions

The target of this subsection is to study the impact of errors in the nominal antenna direction on the network performance. Information on the exact sector orientation is crucial, e.g., in planning tools. Therefore, the simulation results with irregular sector directions were compared with the nominal sector

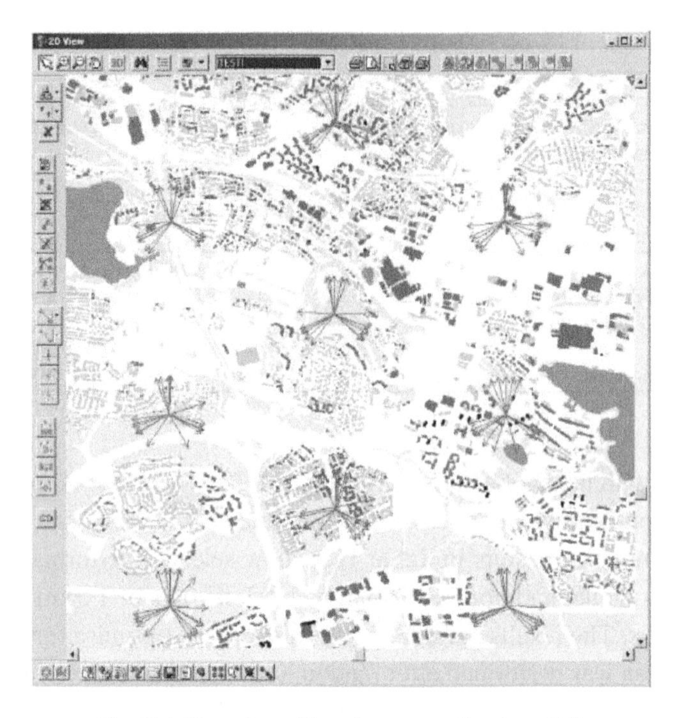

Fig. 12.4 Illustrations of irregular antenna directions. [14]

directions (i.e., 0°, 120°, and 240°) of the hexagonal grid in order to solve the impact on system performance [14].

To provide random antenna directions to each base station, an error is added using the normal distribution. Figure 12.4 shows an example of a part of the network with irregular antenna directions at the base stations. The resulting average error from the nominal antenna direction was 9.1° in the first scenario and 18.2° in the second one. Altogether, five different networks with irregular antenna directions were simulated.

Table 12.2 provides a set of simulation results with 1.5 km site spacing for nominal antenna directions. Due to increased irregularity of the antenna directions, the network performance with irregular sector directions is slightly reduced with respect to nominal sector directions. However, with an average deviation smaller than 20°, there is not much degradation in the system performance, if sector overlap is around 0.5 (3-sectored sites with 65° antennas). According to the simulation results in Table 12.2, the maximum system

Table 12.2. Example simulation results of 1.5 km site spacing. Random directions 1 is with 9.1° and random directions 2 is with 18.2° average error in the antenna direction. [14]

Parameter		Norminal directions	Random directions 1	Random directions 2
Service probability	[%]	98.1	98.0(0.3)	98.0(0.27)
SHO probability	[%]	24.9	25.1(0.18)	25.0(0.5)
SfHO probability	[%]	3.9	4.1(0.12)	4.9(0.37)
UL_i		0.79	0.80(0.01)	0.81(0.01)
DL capacity	[kbps/sector]	360	357(2)	355(3)

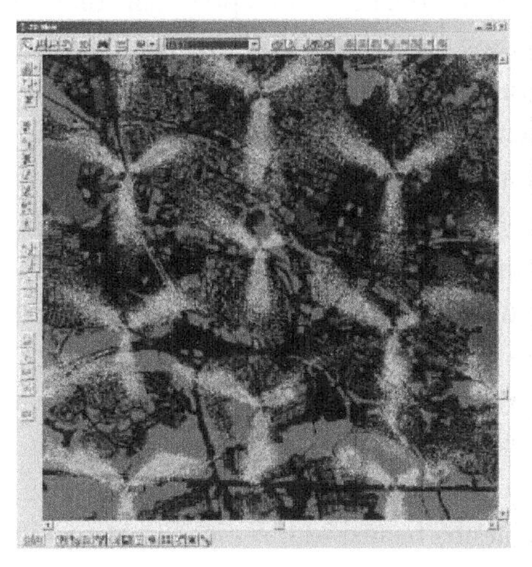

(a) Nominal antenna directions.

(b) Randomised antenna directions.

Fig. 12.5 Visualisation of Softer Handover (SfHO) probability from part of the network with (a) nominal antenna directions and (b) random antenna direction (with larger deviation). Black color indicates 0% SfHO probability and lighter color higher probability.

capacity degradation is less than 2%. As expected, the most visible impact of direction deviation is seen on the SfHO probability (see also Figure 12.5).

Sectoring and Antenna Beamwidth

Sectoring is generally identified as a convenient method to increase network coverage and system capacity in a cellular network. However, especially in WCDMA, the selection of the antenna beamwidth plays an important and crucial role in sectoring. The coverage enhancement with sectoring is based on the improvement of the power budget, and also on the fact that more antennas are implemented at the base station site. The capacity enhancement can be logically observed due to the increasing number of sectors. Ideally, doubling the number of sectors of a base station site would mean doubling the offered capacity of a particular site. However, an ideal sectoring efficiency can not be normally achieved due to nonoptimal antenna radiation patterns. The capacity gain of a 3-sectored site compared to a 1-sectored site is around 2.5–2.7, and 1.7–1.8 between a 6-sectored site and a 3-sectored site [15]. However, in [6] it was shown that a capacity gain of close to two could be achieved by using correct downtilt angles.

In WCDMA networks, 3-sectored sites probably offer a practical solution in the beginning of the network evolution. However, along with increasing capacity demands, higher order sectoring (such as 6-sectored sites) could be deployed in order to provide better network coverage and system capacity. However, there are several practical limitations in the deployment of 6-sectored sites such as the increasing need of cabling and RF amplifiers. Moreover, the need for capacity has to exceed a certain threshold before the deployment of a 6-sectored site or upgrading of an existing 3-sectored site to a 6-sectored site becomes economically viable.

The simulation results for WCDMA networks have indicated that with a so called cloverleaf 3-sectored configuration, the optimum antenna beamwidth varies between 35° and 65° [15]. However, with a so-called wide-beam 3-sectored configuration, the optimum antenna beamwidth is close to 80° [16]. On the other hand, for 6-sectored sites narrower antenna beamwidth is required (from 33° to 40°) [15]. Naturally, the importance of the selection of correct antenna beamwidth increases with higher order sectoring. In this chapter, the impact of antenna beamwidth (or more generally, sector overlap) on the WCDMA system capacity is addressed with different degree of coverage overlap. The presented results here are extended from the ones published in [15].

Fundamentally, they differ only in the sense that in the results presented here, 19 base station hexagonal grid configuration was used instead of 10 base station configuration.

Sector Overlap Index

The sector overlap index (SOI) is defined by the area that is covered by the halfpower beamwidth of all sector antennas belonging to a site:

$$SOI = \frac{angle\ covered\ by\ \theta_{-3dB}}{360^\circ} \tag{12.2}$$

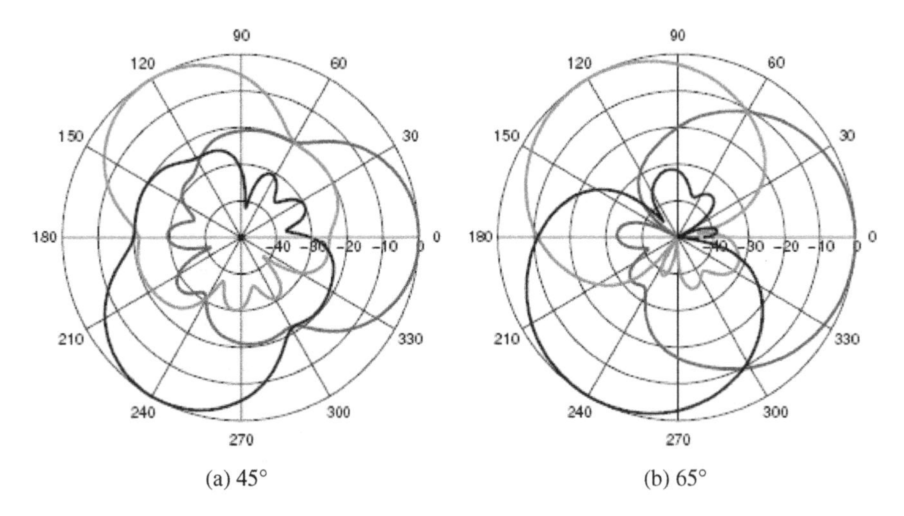

(a) 45°

(b) 65°

Fig. 12.6 Illustration of radiation patterns for a 3-sectored base station site with (a) 45° and (b) 65° antennas.

where angle covered by $\theta_{-3\mathrm{dB}}$ is the fractional angle of 360° covered by $\theta_{-3\mathrm{dB}}$ antenna horizontal beamwidths. Figure 12.6 provides an example of how different antenna horizontal beamwidths 'occupy' a whole circular area if the site is deployed in a 3-sectored manner with 45° antennas or 65° antennas. As seen from Figure 12.6, the 'switching point' between sectors is at the level of 20 dB and 10 dB with respect to antenna gain in the main beam direction, respectively. As an example, the corresponding SOI for these configurations is 0.375 and 0.54.

Optimum Antenna Beamwidths

Table 12.3 presents the achievable system capacities (served users per site) with a 95% service probability target for different configurations. For the 3-sectored configurations, the observed capacity values are moderately equal within the range of simulated antenna beamwidths. In most of the cases, the 65° antenna results in the highest capacity, but only with a marginal difference to the 45° beamwidth. Actually, for higher degree of coverage overlap, the capacity with 45° beamwidth is the same (or even higher) than the system capacity with 65° beamwidth. Hence, the results indicate that the importance of the selection of antenna beamwidth becomes more crucial if coverage overlap is increased. Moreover, the results indicate that coverage overlap and sector overlap have to be optimised simultaneously.

Table 12.3. The number of served users per site with 95% service probability target in different 3-sectored and 6-sectored configurations.

	Beamwidth					
	3-sectored			6-sectored		
Site spacing/antenna height (km/m)	45°	65°	90°	33°	45°	65°
1.5/25	161	164	155	312	290	225
1.5/45	125	115	107	225	210	165
2.0/25	160	165	156	320	305	240
2.0/45	155	155	140	300	245	218
2.5/25	155	160	150	310	280	225
2.5/45	155	155	145	305	260	223

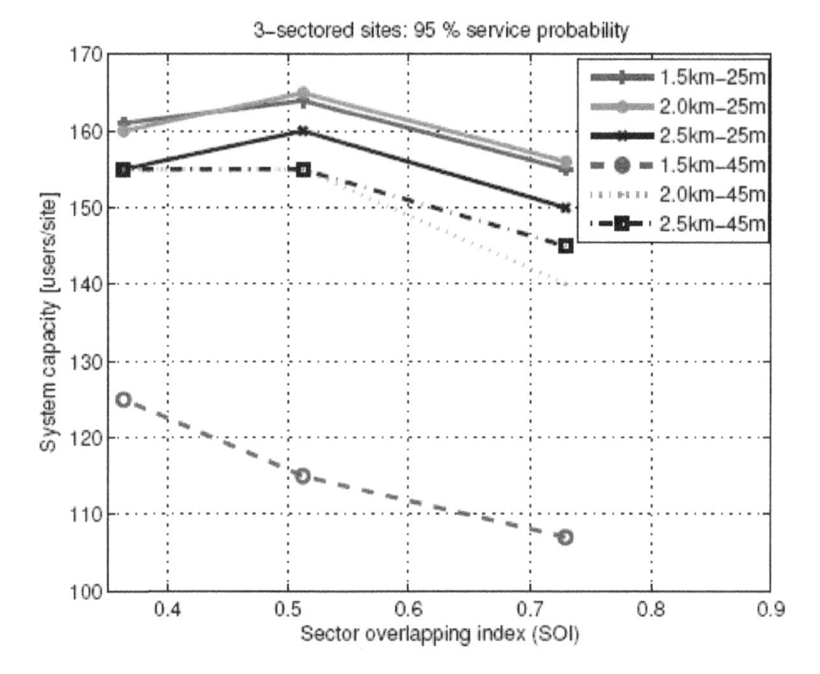

Fig. 12.7 Sector overlap for 3-sectored sites with 65°/6° antennas.

The system capacity values for different network configurations can be compared in terms of sector overlap. Figure 12.7 yields that with 25 m antenna height, SOI between 0.5 and 0.6 results in the optimum system capacity. However, with 45 m antenna heights (equal to higher COI), the optimum SOI < 0.5. Moreover, with 1.5 km and 45 m configuration, the optimum does not even exist within the range of simulated antenna beamwidths. Hence, it seems that if the COI is for some reason higher, the optimum sector overlap is smaller. Note, however, that these results do not yet include the impact of downtilting.

For the 6-sectored configurations, the selection of antenna beamwidth is more crucial in the whole range of selected antenna beamwidths (Table 12.3). The 33° antenna results in the highest capacity values irrespective of the network configuration. Whereas in the 3-sectored configurations, the achieved capacity enhancements vary from 6% to 17% between the best and the worst antenna beamwidth, the corresponding capacity enhancements of 35% or higher are observed with the 6-sectored configuration. Clearly, the selection of antenna beamwidth becomes more vital among higher order sectoring. In terms of SOI, it seems that the optimum beamwidth for 6-sectored sites would be even lower than 33° (Figure 12.8).

It can be seen from this study that sectoring obviously improves the system capacity. However, the evolution strategy for sectoring plays an important role for the final sectoring efficiency. For instance, maintaining 65° beamwidth also in the 6-sectored configuration results in a capacity gain close to 40% between the 6-sectored and 3-sectored configurations. In contrast, sectoring efficiency

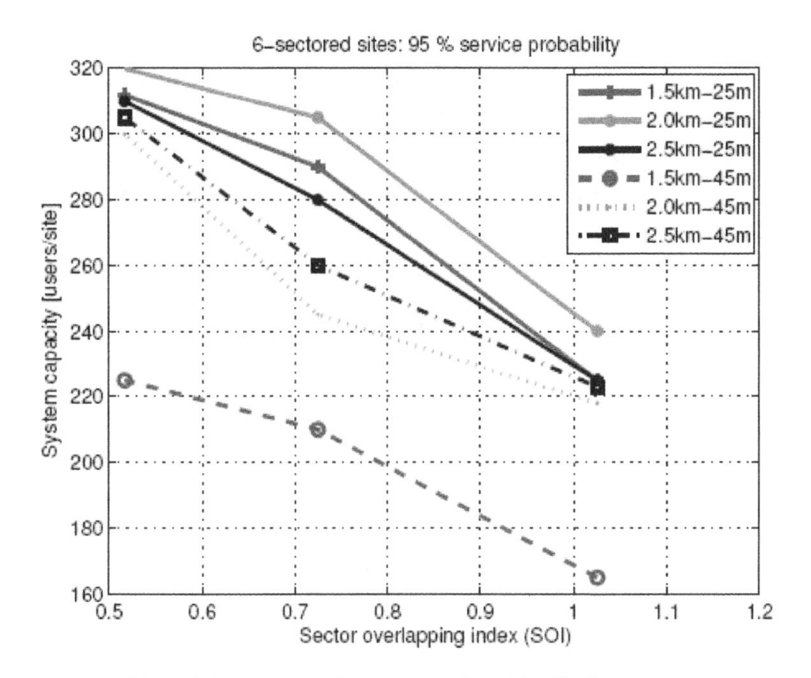

Fig. 12.8 Sector overlap for 6-sectored sites with 33°/6° antennas.

can be increased up to 95% if antennas are changed to 33° beamwidth (without downtilt). Thus, the achievable sectoring gain in capacity (between 6-sectored 33° and 3-sectored 65° configuration) can be around 1.8 and 1.95 — slightly depending on the network configuration.

Moreover, an optimum SOI seems to locate around 0.5, and seems to be connected with coverage overlap (index). The sensitivity of system capacity with respect to SOI is naturally higher (approximately twice) in a 6-sectored network due to double number of overlap sectors. Hence, further analysis should be performed in order to find an optimum SOI as a function of COI.

Antenna Downtilt

Antenna downtilt comprises the final optimisation method of radio network topology. With antenna downtilt, the vertical radiation pattern is directed towards the ground in order to control the radiation towards other cells. In general, the impact of antenna downtilt on system capacity is widely known for macrocellular [6, 17] as well as for microcellular environments [18–20].

With mechanical downtilt (MDT), the antenna element is physically directed towards the ground. Naturally, the areas near the base station experience better signal level due to the fact that the antenna main lobe is more precisely directed towards the intended dominance (serving) area. However, the effective downtilt angle corresponds to the physical one only exactly in the main lobe direction, and decreases as a function of horizontal direction in such a way that the antenna radiation pattern is not downtilted at all in the side lobe direction. Nevertheless, interference towards other cells is reduced mostly in the main lobe direction, which provides capacity enhancements in the downlink in GSM and in WCDMA. In the uplink, the capacity enhancement is typically smaller, as observed in [17]. Naturally, higher capacity gain in the downlink can be observed with larger coverage overlap [6].

Antenna electrical downtilt (EDT) is carried out by adjusting the relative phases of antenna elements of an antenna array in such a way that the radiation pattern can be downtilted uniformly in all horizontal directions. Changing the relative phases of different antenna elements slightly changes the vertical radiation pattern depending on the chosen EDT angle (i.e., depending on the relative phase differences between different antenna elements). By using EDT, the achievable downlink capacity gain is slightly higher than with MDT [6].

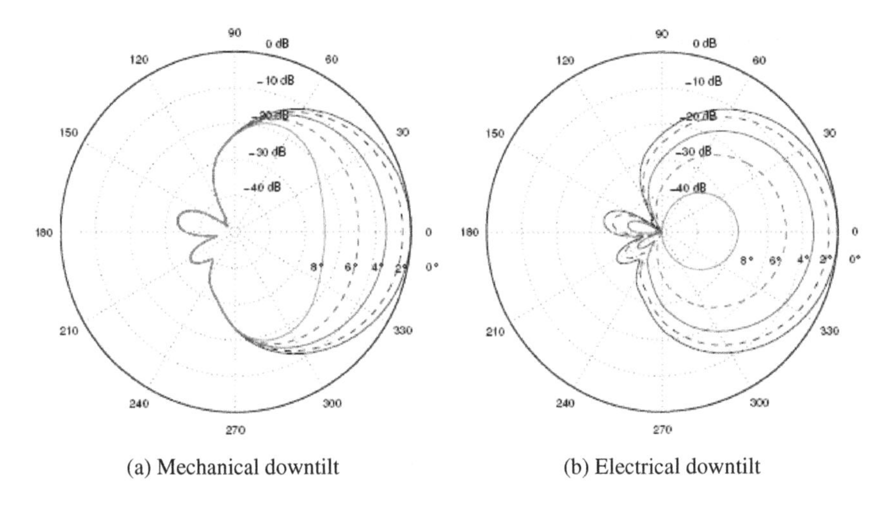

(a) Mechanical downtilt (b) Electrical downtilt

Fig. 12.9 The impact of antenna (a) mechanical and (b) electrical downtilt on the horizontal (azimuthal) radiation pattern in the horizontal plane. Antenna gain is normalised to zero and the scale is in decibels. The 'uptilt' of back lobe direction for mechanical downtilt is not illustrated. [6]

The fundamental differences between MDT and EDT are seen in the horizontal radiation patterns if antennas are downtilted (Figure 12.9). The relative widening of the horizontal radiation pattern is illustrated in Figure 12.9(a) for a horizontally 65° and vertically 6° wide antenna beam as a function of increasing downtilt angle. The reduction of the antenna gain towards the boresight, e.g., with an 8° downtilt angle, is as large as 25 dB, whereas towards a 60° angle the reduction is less than 10 dB. In contrast, Figure 12.9(b) illustrates the behavior of the horizontal radiation pattern for 65° and vertically 6° with EDT. It reduces radiation efficiently also towards the adjacent sectors, since all directions are downtilted uniformly. However, the coverage in the side lobe direction reduces rapidly as well, which lowers the network performance if antennas are downtilted excessively.

As the downtilt angle increases (with MDT and EDT), the soft handover (SHO) probability in the cell border areas decreases. However, in MDT, the relative widening of the horizontal radiation pattern increases the overlap between adjacent sectors, which makes softer handovers (SfHO) more attractive. This increase of softer handovers as a function of downtilt angle depends on sector overlap (i.e., sectoring and antenna horizontal beamwidth).

In summary, the utilisation of EDT antennas might become more attractive due to better ability to change the tilt angle remotely or automatically.

This is actually the idea in remote-controlled electrical tilt (RET) concept, or, continuously adjustable electrical downtilt (CAEDT). By using a sophisticated control mechanism (or system), tilt angles can be changed remotely even without any site visit [21]. Moreover, the usage of RET facilitates significantly the effort required for antenna downtilt optimisation.

A simple and automatic algorithm could be deployed in order to further automate the optimisation process as in the CAEDT concept. The need for a common interface for base station equipment to support RET has also been recognised within the 3GPP specification body, which is currently specifying RET concept for the Release 6 [22].

The latest research results have indicated the potential capacity gains that could be achieved by utilizing remote electrical downtilt. In [23], an optimum downtilt angle, which was evaluated based on the minimisation of uplink transmit power, is observed to change even with an homogenous traffic distribution according to traffic load. In [24], the idea is extended to include the impact of non-homogeneous traffic distribution. Moreover, with simultaneous usage of RET and pilot power adjustment, capacity gain is achieved: mainly through traffic load balancing. In [25], a method for load balancing with tilt angle control is presented. The tilt angle optimisation criterion was based on the minimisation of uplink load for a cluster of three cells. Achieved capacity gains are approximately 20–30% with respect to constant, network-wide tilt angle. However, as all these methods are based on the uplink, a downlink assessment would still be required.

We present in the following, the impact of downtilt in WCDMA with system level simulations in [6] and measurements in [26]. A feasibility study is also provided to assess the need for rapid change of the downtilt angle due to change in traffic conditions [17]. Table 12.4 gathers optimum downtilt angles (ODA) for simulated site and antenna configurations for a macrocellular suburban environment. The variation of optimum downtilt angles within the simulated configurations is from 3.4° up to 10.3°. The optimum downtilt angles for the 3-sectored configurations with 6° and 12° vertical antenna beamwidth varies between 4.3°–8.1° and 3.5°–10°, respectively. One reason for lower ODAs for 12° beamwidth is the interference conditions (i.e., lower coverage overlap) that differ due to lower antenna gain. Moreover, wider vertical spread of antenna pattern makes the use of downtilt not so beneficial. For 6-sectored configurations, the observed ODAs (4°–7°) are very close to the values of the corresponding 3-sectored configurations.

Table 12.4. Optimum downtilt angles for mechanically and electrically downtilted antennas for all simulated site and antenna configurations. Evaluation of an optimum downtilt angle is based on a simple algorithm that utilises the simulation results with two different traffic loads. [6]

Site Spacing (km)	Antenna height (m)	EDT 3-sec 6	EDT 3-sec 12	EDT 6-sec 6	MDT 3-sec 6	MDT 3-sec 12	MDT 6-sec 6
1.5	25	5.1°	7.3°	5.4°	5.7°	5.9°	4.9°
	35	6.1°	9.1°	6.3°	7.3°	8.1°	5.9°
	45	7.1°	10.3°	7.1°	8.1°	9.1°	7.0°
2.0	25	4.3°	5.6°	3.8°	5.1°	4.3°	3.8°
	35	5.8°	7.9°	5.1°	6.7°	7.5°	4.8°
	45	6.3°	9.3°	6.1°	6.9°	8.2°	5.9°
2.5	25	4.5°	5.2°	4.6°	5.1°	3.4°	3.7°
	35	5.4°	7.6°	5.3°	6.1°	4.4°	4.5°
	45	5.9°	8.3°	5.7°	6.9°	6.9°	5.8°

Empirical Equation for Selection of Downtilt Angle

Based on the simulated optimum downtilt angles, an empirical equation is derived [6]:

$$v_{opt} = 3[\ln(h_{bs}) - d_{dom}^{0.8}]\log_{10}(\theta_{-3dB}^{ver}) \tag{12.3}$$

Equation (12.3) relates the topological factors such as the base station antenna height (h_{bs} in meters), the intended length of the sector dominance area (d in kilometers), and also the half-power vertical beamwidth ($\theta - 3$ dB in degrees). The equation is derived with a simple curve fitting method. It provides a zero mean error with 0.5° standard deviation with respect to simulated optimum downtilt angles for all simulated scenarios. As the error of equation (12.3) is rather small, it could be embedded into a radio network planning tool. Thereafter, the tool would automatically provide a suggestion of downtilt angle for a planner based on the information of antenna vertical beamwidth, antenna height (also ground height level could be utilised), and expected dominance area of a particular sector. Hence, it could provide an initial downtilt angle setting for each antenna depending on the sector configuration.

Excessive Downtilt Angles Identification

The excessive downtilt angle using two different downtilt schemes have also been identified in [6]. With Electrical Downtilt (EDT) due to uniform reduction of the horizontal radiation pattern, a too large EDT angle could be identified by larger proportion of mobiles with higher uplink TX (transmit) power (or large uplink TX power before the cell edges). Naturally, depending on the actual power budget and selected services, the uplink or downlink may limit the coverage. In contrast, the increase of SfHO connections due to effective widening of horizontal beamwidth affects most on the downlink capacity degradation with too high Mechanical Downtilt (MDT) angles.

Expected Capacity Gains

Tables 12.5 and 12.6 provide the capacity gains and the corresponding maximum DL throughputs with selected network and antenna configurations for EDT and MDT. The downlink capacity gains vary from 0% up to 60%, depending heavily on the network configuration [6].

Generally, the capacity gain becomes larger if the coverage overlap increases, i.e., either the antenna height increases or the site spacing decreases. Considering an urban macrocellular environment, where the network is typically very dense due to requirements of higher coverage probabilities and capacity, the utilisation of antenna downtilt is mandatory. An interesting

Table 12.5. Capacity gains for electrical downtilt. Maximum sector throughput [kbps] in the downlink with optimum downtilt angles and corresponding capacity gains with respect to non-tilted scenario for all simulated network configurations. The maximum capacity values are based on 0.5 average DL load. [6]

Site Spacing (km)	Antenna height (m)	EDT 3-sec 6	EDT 3-sec 12	EDT 6-sec 6
1.5	25	494(18.1%)	472(2.8%)	492(27.5%)
	35	510(33.5%)	484(12.1%)	504(43.8%)
	45	526(48.4%)	493(18.7%)	522(58.1%)
2.0	25	457(8.0%)	440(1.4%)	438(9.4%)
	35	496(17.8%)	457(4.3%)	468(22.5%)
	45	499(27.7%)	472(8.0%)	500(37.5%)
2.5	25	451(5.3%)	423(0.8%)	462(6.3%)
	35	480(9.9%)	433(2.1%)	482(16.9%)
	45	488(20.4%)	440(2.6%)	504(31.3%)

Table 12.6. Capacity gains for mechanical downtilt. [6]

Site Spacing (km)	Antenna height (m)	MDT 3-sec 6	MDT 3-sec 12	MDT 6-sec 6
1.5	25	489(17.0%)	466(1.6%)	458(18.8%)
	35	500(30.8%)	475(9.9%)	474(35.3%)
	45	516(45.6%)	479(15.4%)	480(45.5%)
2.0	25	459(8.5%)	440(1.4%)	438(9.3%)
	35	494(17.4%)	453(3.3%)	458(20.0%)
	45	495(26.8%)	466(6.6%)	471(29.4%)
2.5	25	451(5.3%)	424(1.1%)	456(5.0%)
	35	487(11.5%)	433(2.0%)	464(12.5%)
	45	479(18.3%)	437(2.0%)	463(20.6%)

observation in [6] is the increase of absolute sector capacity as a function of higher antenna position. Geometrically thinking, it is obvious that the achievable capacity gain is higher with higher antenna position. This is due to better ability to aim the antenna beam towards the intended dominance area. However, a higher antenna position requires more precise adjustment of the antenna downtilt angle. Hence, from the radio network planning point of view, placing antennas higher and using larger downtilt angles provides better system capacity. This planning approach might not be applicable for an urban network if the antennas are placed on the top of buildings, since the probability of distant interferers easily increases with higher antenna positions. The achieved isolation between cells with rooftop antenna installation could be lost, and thus in practice the net effect could be close to zero.

SHO Probabilities

An example of reduction of soft handover (SHO) probability is shown in Figure 12.10 for 3-sectored 1.5 km configurations (25 m and 45 m antenna heights) as a function of EDT and MDT angle.

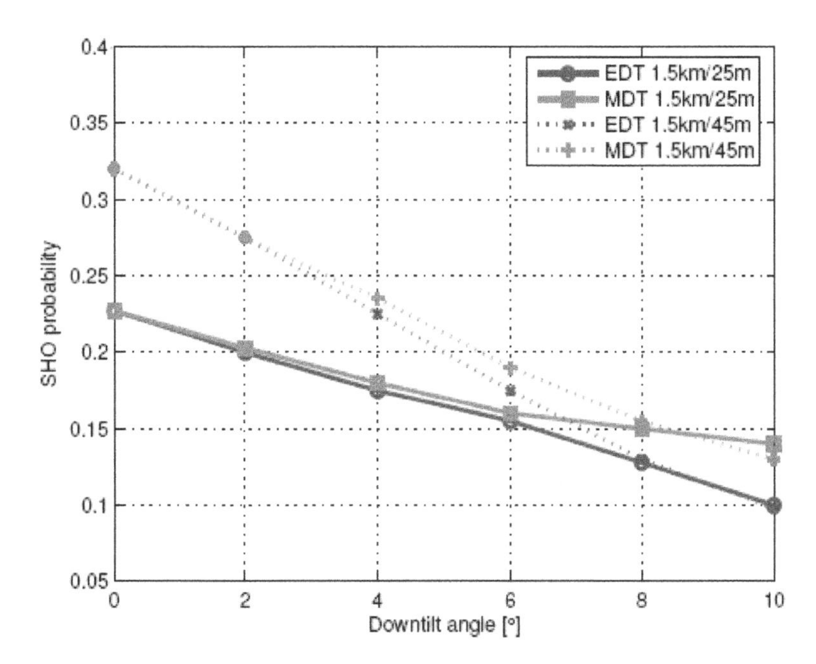

Fig. 12.10 An example of SHO reduction with EDT and MDT downtilt schemes with different antenna heights of 3-sectored sites and antenna vertical beamwidth of 6°.

Clearly, EDT is able to reduce the SHO probability more efficiently due to more efficient antenna pattern control. This is partly the reason for higher capacity gains for EDT. Moreover, it is observed that with optimum downtilt angles, SHO probabilities are systematically around 17%, which, on the other hand, indicates a common optimum coverage overlap index for optimum downtilt angles. Naturally, the expected SHO probability with optimum downtilt angle depends on the SHO window setting. Nevertheless, the reduction of SHO connections is one reason for improved system capacity in the downlink due to downtilt.

12.3 Automated Optimisation

In this section, we present a radio network automated optimisation algorithm for finding the best settings of the antenna tilt and common pilot channel power of the base stations. This algorithm is a parameter method, which improves on topology optimisation discussed in Section 12.2. This method allows an operator to deal with 3G network design complexity, which is beyond the reach of a manual approach (see Figure 12.11). All the parameters as shown

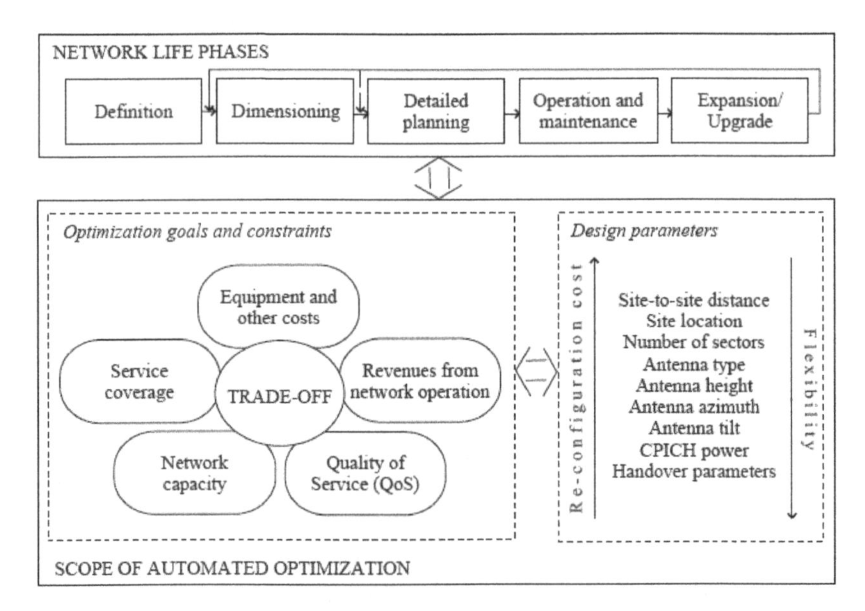

Fig. 12.11 Network planning and optimization.

in Figure 12.11 have strong influence on the interference in the system and therefore on the network capacity.

CPICH Power

The CPICH is used by mobile phones to obtain initial system synchronisation and to aid the channel estimation for the dedicated channel. After turning on the power and while roaming in the network, a mobile phone determines its serving cell by choosing the best CPICH signal. Thus, CPICH power determines the cell coverage area. Increasing or decreasing the CPICH power will enlarge or shrink the cell coverage area. Therefore, by appropriately adjusting the CPICH power to the base stations, the number of users per cell can be balanced among neighbouring cells, which reduces the inter-cell interference, stabilise network operation and facilitates radio resource management [27].

On the other hand, there are some constraints on setting the CPICH power: values too high will create interference called 'pilot pollution' to the neighbouring cells, which decreases the network capacity. Setting CPICH power too low will cause uncovered areas between cells. In an uncovered area, the CPICH power becomes too weak for the mobile phone to decode the signal, rendering network access impossible.

Antenna Tilt

Antenna tilt is defined as the elevation angle of the main beam of the antenna relative to the azimuth plane. Antenna downtilt is often used in mobile wireless systems, particularly in the UMTS network, where traffic in all cells is simultaneously supported using the same carrier frequency. The desired effect is a reduction of the other-to-own-cell interference ratio i, which is defined in [28] as,

$$i = \frac{I_{oth}}{I_{own}} \tag{12.4}$$

In equation (12.4), I_{oth} denotes the inter-cell interference and I_{own} is the intra-cell interference. By downtilting the antenna, the other-to-own-cell interference ratio i can be reduced: The antenna main beam delivers less power towards the neighbouring base stations, and therefore most of the radiated power goes to the area that is intended to be served by this particular base station [9]. Additionally, an antenna tilt adjustment also affects the cell coverage area, which limits the tilt to reasonable values.

Influence of CPICH Power and Antenna Tilt

Adjusting the CPICH power and antenna tilt can increase the network capacity by

(i) Reducing inter-cell interference and pilot pollution
(ii) Optimizing base station transmit power resources
(iii) Load sharing and balancing between cells, and
(iv) Optimizing SHO areas

Optimisation Technique

In this section, we describe a rule based optimisation algorithm developed in [29]. The rule-base optimisation algorithm is an extension of the approach introduced in [30]. The optimisation of the base station parameters (i.e., CPICH power and tilt), begins with an initial evaluation of the network. After analyzing the results of the initial evaluation, the iterative optimisation process is stated. Each loop includes two steps. In the first step, the parameters are changed according to a rule-based optimisation technique, described in this section. This new technique simultaneously changes the CPICH power and antenna tilt. In contrast to [30], an increase of CPICH power and antenna uptilting is also possible. After changing the parameters, in the second step, the network is evaluated. Then, the next iteration of the optimisation loop is started and the parameters are changed again. Figure 12.12 shows a flowchart of the optimisation process.

Grade of Service

During the optimisation process, an indicator is used to characterize the number of users provided with a service. This value is called Grade of Service (GoS) and describes the ratio of served users over all existing users. We define the GoS as

$$GoS = \frac{served\_users}{existing\_users} \tag{12.5}$$

In equation (12.5), served users denotes the total number of served users in a defined area (e.g., the whole simulation area), and existing users is the total number of simulated users in the same area. During the optimisation process, the GoS increases from its initial value of 95% until it has reached 100% [29]. Then all users are served and the optimisation algorithm cannot

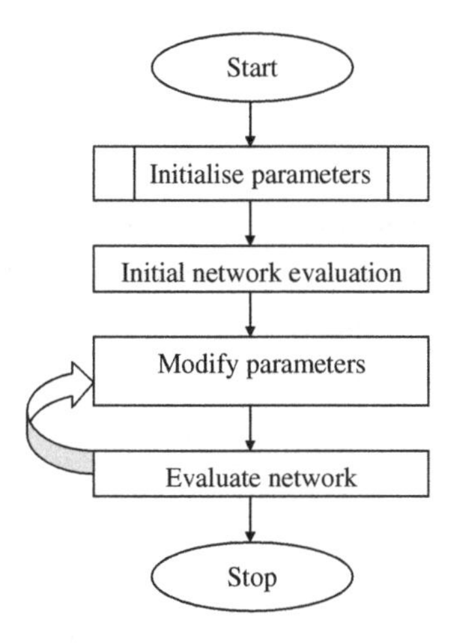

Fig. 12.12 Structure of optimisation process.

proceed any further. Now, however, the network can accept more users. Thus, the proposed optimisation algorithm applies the following approach: When the GoS reaches a threshold value of 97%, new users are added to the network until the initially defined GoS of 95% is reached again. In the following, this function is referred to as add users.

For the evaluation of the network, a fitness function has to be defined. The fitness function represents the optimisation goal. In this section, we desire to maximise the capacity of the network; therefore we consider the number of served users as the goal of the optimisation.

Quality Factor

To describe the quality of a cell in the network, a performance indicator denoted as quality factor (QF) is introduced. The QF indicates whether a cell is heavily loaded or not, and it is determined by the following three values: f_{load}, f_{pwr}, and f_{OVSP}. The parameter, f_{load} is a measure of the uplink cell loading condition and is defined as

$$f_{load} = \frac{\eta}{\eta_{thres}} \tag{12.6}$$

In Equation (12.6), η_{thres} denotes the planned uplink cell load, and η is the actual cell loading. The second value, f_{pwr} is a measure of base station (BS) transmit power utilisation in the downlink and is defined as

$$f_{pwr} = \frac{P_{transmit}}{P_{\max}} \tag{12.7}$$

In Equation (12.8), P_{max} denotes the maximum transmit power of a cell; while $P_{transmit}$ represents the current total transmit power of that cell. The third value, f_{OVSF}, is a measure of the orthogonal variable spreading factor (OVSF) code utilisation in the downlink and is defined as

$$f_{OVSF} = \frac{ovsf_{used}}{ovsf_{\mathrm{lim}it}} \tag{12.8}$$

In Equation (12.9), $ovsf_{limit}$ is the maximum available number of OVSF codes, which is 512; $ovsf_{used}$ is the number of used OVSF codes. With the three values, f_{load}, f_{pwr} and f_{OVSF}, the *QF* is defined as

$$QF = 1 - \max(f_{load}, f_{pwr}, f_{OVSF}) \tag{12.9}$$

The range of the *QF* is between zero and unity. A low *QF* describes a heavily loaded cell, and a high *QF* describes a weakly loaded cell. Therefore, by using the *QF* as the performance indicator, CPICH power and antenna tilt settings can be adjusted adaptively according to the loading condition in the uplink and downlink of a cell.

Changing CPICH Power and Antenna Tilt

For each iteration of the optimisation loop, illustrated in Figure 12.13, the *QF* is computed for each cell, while the CPICH power and antenna tilt are changed according to the developed rules, as shown in Table 12.7. The rules are designed in such a way that a highly loaded cell ($QF < 0.5$) having outage users is required to shrink its coverage area by decreasing the CPICH power and increasing the antenna downtilt. Conversely, if the cell has a low user density ($QF > 0.7$), it has to expand its coverage area to cover more users by increasing the CPICH power and antenna uptilt. With this approach, load balancing within the cells in a network can be achieved, resulting in higher network capacity.

CPICH power and antenna tilt modifications are controlled by a set of rules with different step size and limitation settings, as shown in Table 12.8. One rule consists of one instruction for the CPICH power and one instruction for antenna tilt.

Table 12.7. CPICH Power and Antenna Tilt Adjustment.

QF	CPICH power and antenna tilt change
$Qf < 0.5$	Decrease CPICH power and tilt antenna down
$0.5 \leq QF \leq 0.7$	No change
$QF > 0.7$	Increase CPICH power and tilt antenna up

Table 12.8. An example of CPICH power and antenna tilt step size and limitation settings.

Rule	Parameter	Stepsize	Limit	Iteration
0	CPICH	3dB	25 dBm	50
	Tilt	1.5°	4°	
1	CPICH	0.5dB	10 dBm	50
	Tilt	0.25°	7°	

In Table 12.8, parameter specifies the modified parameter; stepsize denotes the maximum allowed adjustment of CPICH power and antenna tilt settings per iteration; limit describes the lower or upper limit of the parameter, and iteration specifies how often the rule can be applied at most. When the *QF* of a cell is greater than 0.7, both its CPICH power and antenna uptilt will be increased by

$$stepsize. \left| \frac{QF - 0.7}{0.3} \right| \tag{12.10}$$

On the other hand, if the cell's *QF* is less than 0.5, the CPICH power will be reduced, and the antenna downtilt will be increased by

$$stepsize.(1 - QF) \tag{12.11}$$

Consequently, the adjustment of CPICH power and antenna tilt depend on the cell's actual *QF*. With this strategy, CPICH power and antenna tilt are adjusted adaptively according to the loading condition of a cell. The modification of antenna downtilt and CPICH power has to be limited (lower limit for CPICH power and upper limit for antenna downtilt) in each rule by the parameter limit in order to avoid changes that are too large in a particular cell, which can be seen in Table 12.7. There are no limitations set for the maximum CPICH power and maximum antenna uptilt, so a larger service coverage area is allowed in regions where the user density is low. Furthermore, the developed algorithm is biased toward reducing the initial CPICH power and increasing the antenna downtilt. When the optimisation process is launched, the algorithm starts with the first rule of the used rule set (e.g., from Table 12.8). For each rule, several iterations are performed. According to Table 12.8, we can see that in this case the algorithm proceeds to rule number 2 after the first 50 iterations. If the GoS after one of the iterations within any one rule is lower than 95% and lower than the GoS from the previous iteration, the new result is rejected. In this case, the same iteration is repeated, but only in two-thirds of the previously modified cells (priority according *QF*) are the CPICH power and tilt settings changed (Cell Reduction). The algorithm terminates when either all rules of the rule set have been processed or if the *QF* in each cell is between 0.5 and 0.7, which can be seen in Table 12.7. This means the network is balanced, or that the algorithm cannot process further due to the limits for CPICH power and tilt. Figure 12.13 shows a detailed flowchart of the optimisation loop.

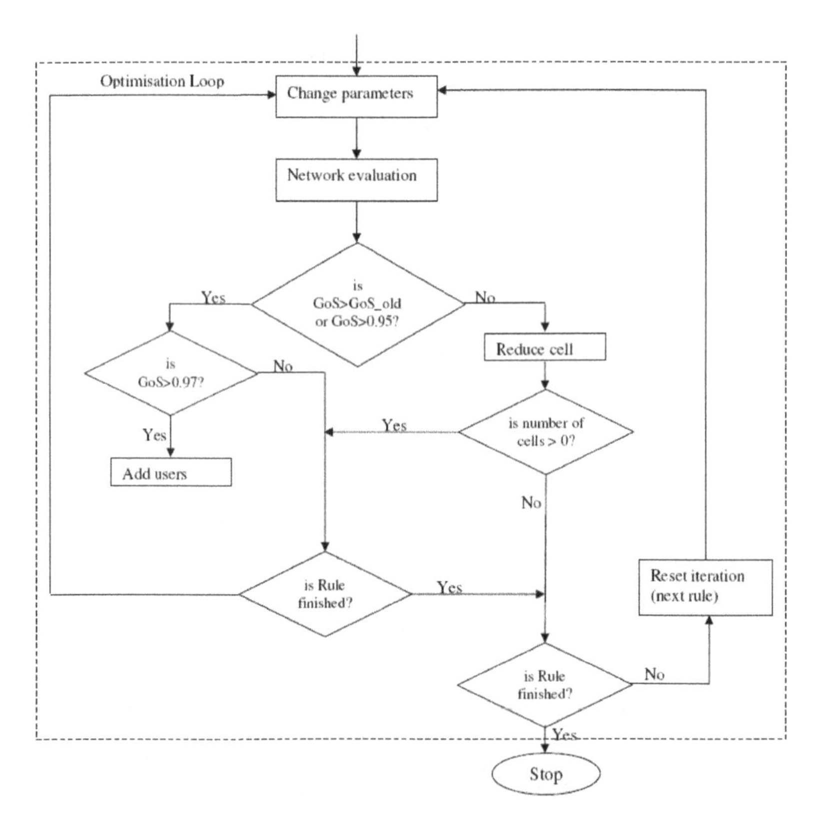

Fig. 12.13 Detailed flowchart of the optimisation tool.

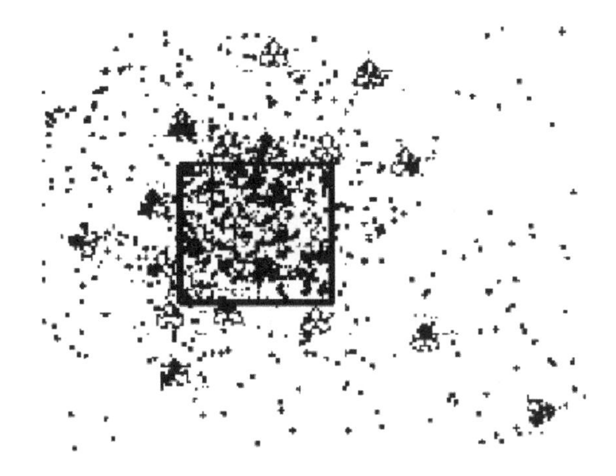

Fig. 12.14 Base station locations and user distribution of the network.[29]

Optimisation Results

In the network scenario, 25 base stations equipped with 3-sector antennas is used, thus comprising 75 cells. The distribution of the mobile terminals is assumed according to the population density, with a service mix of 40% speech users and 60% two-way 64 kb/s data users. The initial number of users in the whole network is 1057. The distribution of the base stations as well as the distribution of the users is shown in Figure 12.14.

The arrows in the figure represent the sectors of a base station, and the dots symbolize the users in the system. The area inside the rectangle is defined as the optimisation area. This is the area of interest, and the GoS is evaluated over this part. The initial value of the GoS in this optimisation area is 95.11%. The most important parameters used in the simulation scenario are introduced in Table 12.9.

Table 12.10 shows the results of the optimisation. Before and after optimisation, 50 snapshots with different user distributions are simulated. The optimisation process itself is performed with one fixed users distribution. Before optimisation, the mean number of served users in the optimisation area is 511.

After the optimisation, with the rule based approach [29], 831 users are served on the average. This increase in served users leads to a capacity gain of 62.6%. It is noticeable that the standard deviation of served users with respect to different user distributions after the optimisation is much higher than before. This means that the optimised network is less stable regarding the

Table 12.9. Simulation parameters.

Number of sites	25
Number of sectors per site	3
Number of sites in optimization area	9
Total number of initial users	1057
Service mix	40% 12.2 kb/s speech user 60% 64 kb/s data user
Activity factor	0.5 (speech); 1.0 (data)
Max. BS TX power	43 dBm
Max. MS TX power	21 dBm
Antenna height	30 m
BS antennas	65°/16 dBi
MS antennas	Omni/0 dBi
Active set size	2
Active set window	3 dB
Initial CPICH power	33 dBm
Initial tilt	0°

Table 12.10. Optimisation results.

Simulation	Before optimization	After optimization
Mean number of served user in target area	511	831
Standard deviation (number of served users)	7.54(1.48%)	31(3.7)
Capacity gain (%)		62.6

user distribution. A possible approach for reducing this dependence would be to use different user distributions during the optimisation process.

12.4 Network Optimisation

Radio network optimisation amounts to finding different techniques for network maintenance, load balancing, monitoring and improving the link budget by adjusting different performance parameters. In Figure 12.15, a simplified WCDMA radio access network optimisation procedure is described. An operator's master/business plan sets the framework for both short-term and long-term performance criteria in terms of planned service area, call blocking and service mix used in dimensioning, soft handover overhead, etc. The process flow diagram in Figure 12.15 can be divided into three different areas namely, preparations, measurements and optimisation. During the process flow, it is important to support rollback functionality to ensure a restore to previous configuration if needed.

The starting point for the preparation phase is related to short-term and long-term planning of new services, capacity expansion needs, service area

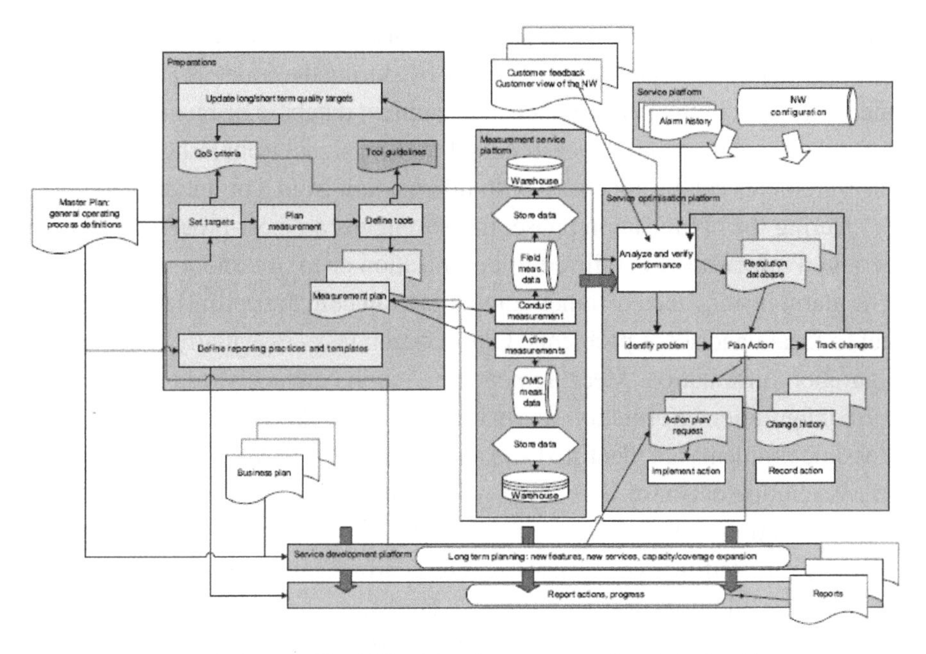

Fig. 12.15 A simplified WCDMA radio access network optimisation [8].

expansions and additional tasks related to new features in network elements or network element upgrades. During the preparations phase the performance targets are defined for short-term and long-term actions.

KPI targets are defined for capacity, coverage and quality related measures. Once these measures are defined, the method of information collection is chosen. Two main sources of measurements can be identified: management system statistics and field measurement tool. For management system measurements the following are defined:

- resources and their availability
- list of KPIs to be followed
- measurement schedules (start and stop time)
- summarisation level
- scope (the objects to be followed)
- reporting.

In the case of the field measurement tool, the measured indicators and their post-processing, measurement routes, services to be used (voice, CS/PS data) and number of calls to be generated are addressed. During the measurement phase, measurements are collected either by storing the counters and other raw measurements in the performance management database or by collecting the information with the field measurement tool. In signal troubleshooting cases, measurements can be appended with information from a protocol analyser.

During the optimisation phase, measurements are post-processed and the acquired information is utilised in optimisation or troubleshooting activities. The alarm history and customer complaints can help the optimisation personnel to locate and solve problems. Optimisation tasks essentially require configuration information. Measurement data/reports are analysed together with the configuration information, alarm history and feedback from customer care. Possible problems are identified and an action plan is derived with the support of a resolution database. Two main types of resolutions can be distinguished: hard resolutions and soft resolutions. Hard resolutions are associated with hardware-related changes such as antenna bearing, antenna type or antenna tilt change, site additions, network element hardware changes, etc. Soft changes are parameter changes that can be done centrally using management system capabilities. Also cost and benefit considerations are made when deciding what type of change to make and in which order to perform the changes.

After the planned action has been implemented, the measurement results are re-analysed to observe whether the action taken positively changed in the performance metrics. Service optimisation can be considered a never-ending task, aimed at maximising the amount of traffic without sacrificing quality.

Measurement Applications in Network Elements and in the Network Management System

In this section, the measurement support in the Radio Network Controller (RNC) and network-wide measurement support at the network management service level are introduced.

Real Time Monitoring in the Radio Network Controller

In addition to earlier described and presented measurements — which are processed with a delay and thus the results cannot be utilised immediately — there should also be real time tools providing monitoring possibilities, focus mostly on the radio network. The tool presented here is online monitoring, in which the user selects a specific monitored item — e.g., monitoring type — and according to parameters — e.g., cell-by-cell — tells from which part of the

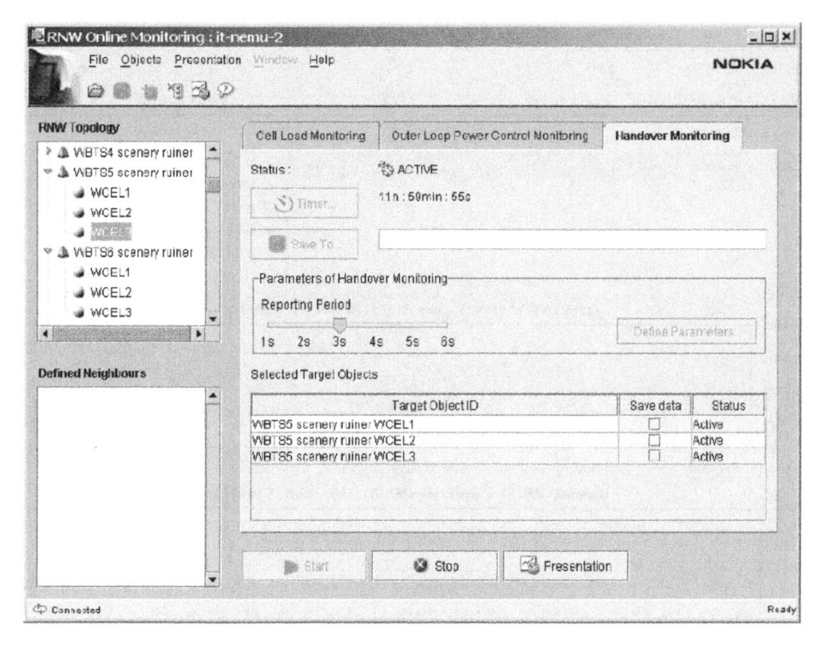

Fig. 12.16 Cell-level online monitoring administration.

radio network the detailed information should be obtained. In Figure 12.16, the administration window is shown.

Two different screenshots are chosen as examples:

(i) The monitoring of load behaviour in a cell.
(ii) The monitoring of handovers in a cell.

The result of (i) can is presented in Figure 12.17 and (ii) in Figure 12.18, respectively.

The first screenshot shows the load behaviour in the uplink of a selected cell in relation to basic admission control parameters. Uplink total interference is presented as PrxTotal and the own-cell real time user and non-real time user loads are seen as L_RT and L_NRT. The admission control parameters are presented as PrxNoise (noise floor), PrxTarget (planned uplink interference) and PrxOffset [8].

The second screenshot shows the distribution of different handover types in a cell and the relation of successful and unsuccessful handovers per handover type. There can also be other possibilities to be monitored in real time.

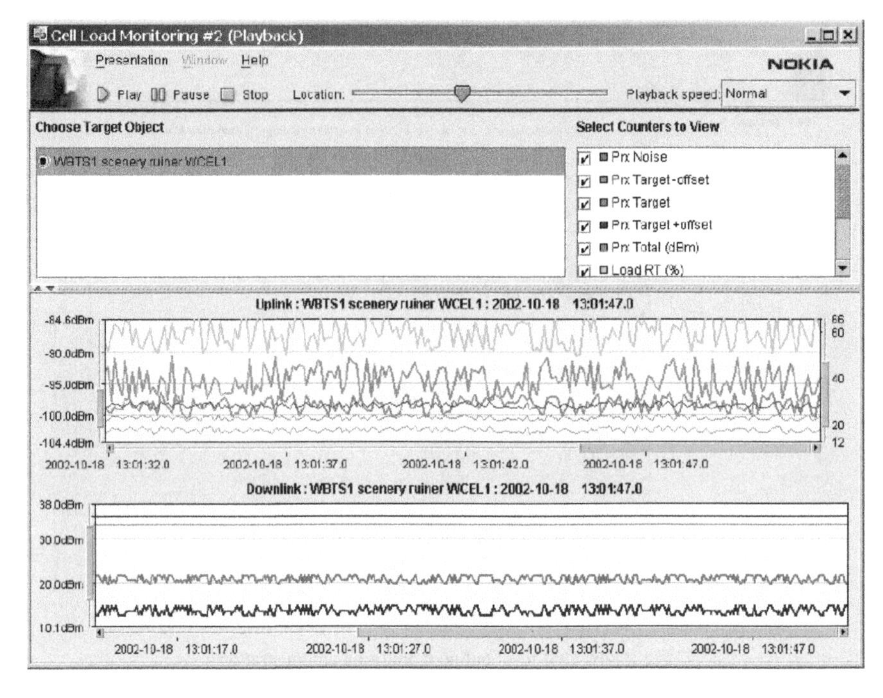

Fig. 12.17 Cell-level online monitoring example for load behaviour monitoring.

Mobile Tracing Functionality

Measurements and online monitoring reveal the service usage, traffic amounts, radio link characteristics and all kinds of errors detected in the radio network. But for locating problems for individual users/mobiles, the operator must also have a means to pinpoint a specific user (IMSI) or a specific mobile (IMEI) from the network. This is done by the tracing functionality. Trace is a network-wide tool used to collect all kinds of information for a user/mobile from all the network elements (BTS, RNC, MSC, etc.) and all the interfaces (RRC, Iub, Iur, etc.). Figure 12.18 depicts the trace activation mechanism.

Trace Activation

The activation of trace is not possible from any part of the network. Only the HLR and the Visitor Location Registers (VLRs) know if and when a certain UE is registered to the network. Therefore, the real activation of UE trace is done when a UE is registered to the network and there is a trace activation set 'on' in the HLR or a VLR.

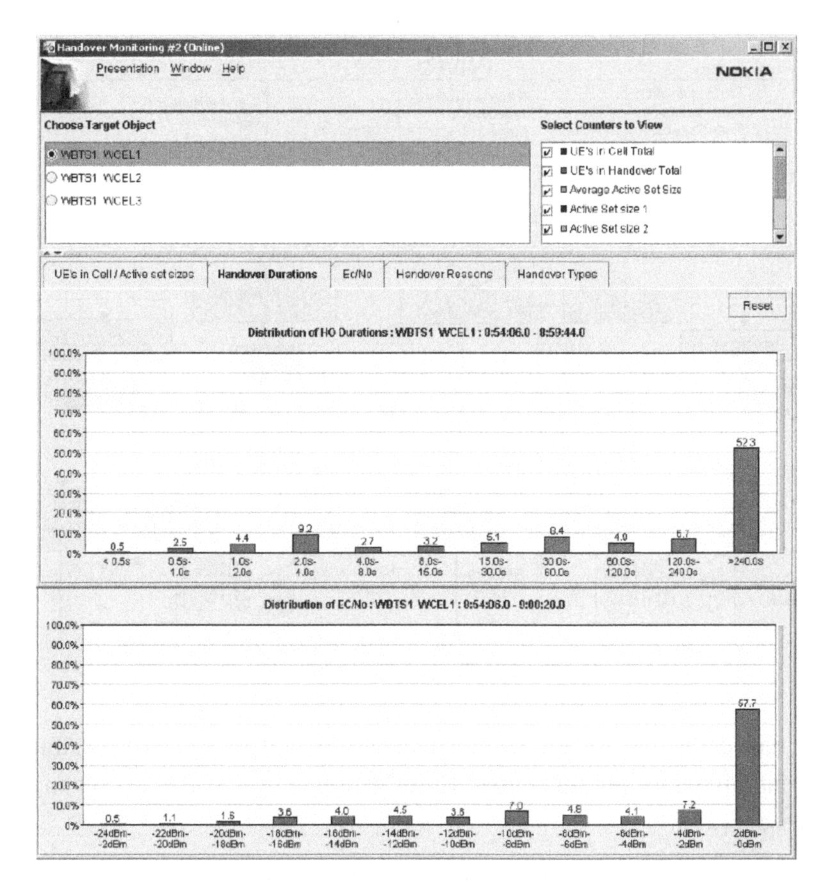

Fig. 12.18 Cell-level online monitoring example for handover. Handover duration and $E_c = N_0$.

Reference Numbers

Figure 12.19 shows that — in contradiction to non-real time measurements and online monitoring — trace data can come from any part of the UMTS (or GSM) network. In order to handle the data coming from different domains (NSS, PaCo, RAN) the system must use reference numbers to make it possible for the OS to gather data related to the same IMSI (International Mobile Subscriber's identity) or IMEI (International Mobile station Equipment Identity) that are sent from the network to the network management system.

Trace Records

The data are sent from the network to the network management system in trace records. A trace record contains the collected data for a UE. The contents of the

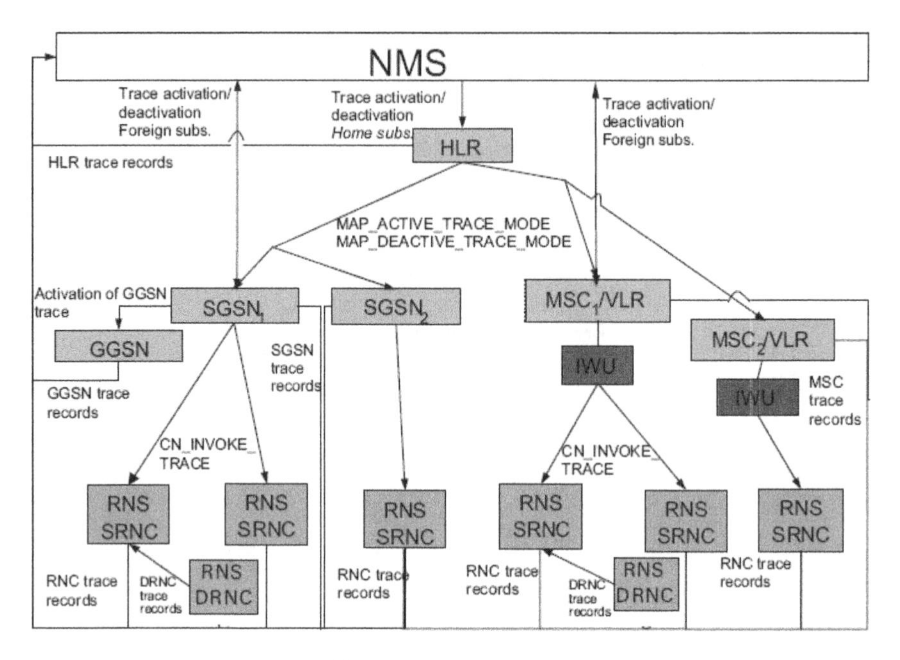

Fig. 12.19 The activation of trace [8].

trace data depend on trace activation and pre-defined trace record possibilities. Currently the trace record content is not specified by 3GPP but is planned for Release 6 [22].

Reporting Tools in the Network Management System

Generally, the differentiating factor between the element management and network management function is related to data amounts. The reporting functions at the network management system level contain information from a multitude of network elements (like RNCs) network-wide. The requirement of being network-wide can limit the usability of certain data sources, because there are numerous data sources that do not provide naturally network-wide data — e.g., field measurement tools. In addition to network-wide highly summarised data, the network management system supports drill-in functionality in order to zoom down to the needed details — for instance, in troubleshooting cases. Summarisation can be done in terms of elements/objects, time or the measurement content. Element-level summarisation provides averaged reports over network elements (like RNC-level KPI reporting). Time axis summarisation provides, for example, daily average or busy hour KPIs instead of each value collected separately from the network elements. Measurement content summarisation is related to the grouping of measurements. In high-level reports, for example, the handover performance figure is presented as one value. In the case of poor handover performance, support is offered to zoom into the details of different handover types.

The power of the network management system lies in the fact that large amounts of data are available and that different views of these data can be provided. This is depicted in Figure 12.20. If the scope is the whole network (high number of objects), then it is obvious that the detail level of data in the report is small. Users should be supported so that they can easily find the relevant area and focus. In high-detail views, the number of elements to be handled is typically lower but a different kind of data is available as well as time series or distributions.

If the level of detail is very high (high number of counters, KPIs, time series, etc.) then the number of objects should be really low. The reasons are

(i) the user cannot abstract too much information at a time and
(ii) the data interfaces have a limited performance.

For practical reasons the area (upper right in Figure 12.20) where we have all the objects within a region with best possible measurement detail level in reports is not feasible and thus not likely. Such reporting would cause high performance requirements of the interfaces from network elements to network management systems.

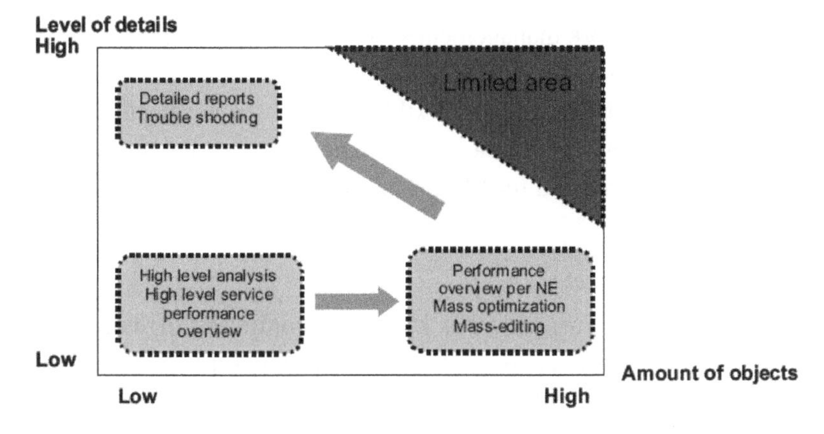

Fig. 12.20 Network management system level reporting.

In high-level analysis, the data are heavily summarised and there are few objects visualised. If the number of objects is high, the data are summarised and the number of different types of data to be visualised is low. If the number of elements is low, then several key performance indicators, parameters, time series and distributions can be used. Arrows show the typical flow from one use case to another.

The support provided by the network management system level reporting tools is as follows:

(i) 'Drill-in' functionality to support the analysis of data by moving to a higher level of detail — e.g., through Web links in the result tables. This is used in data 'crunching' when moving from higher to lower object levels or from high to more detailed data summarisation levels, as depicted in Figure 12.20.
(ii) Automatic creation of Top-N[1] lists into working sets from output-scheduled reports.
(iii) Support for area/grouped selection of objects by the use of working sets. This is used in optimisation/troubleshooting to focus on a specific part of the network or a set of objects with a similar performance — like Top-N lists.
(iv) Detailed reports at KPI level including trend graph and breakdown of KPI formulas showing the values of individual counters (used in optimisation/troubleshooting to understand sudden changes in the KPI trend).

Examples of network management system level reporting application output are shown in Figure 12.21 and 12.22.

Reporting solutions should be flexible towards scope selection. The size of the working set can be from a single element to a whole cluster. Another level of reporting is needed for SLA support and is beyond the scope of this chapter.

Active Service Monitoring

Traditional network management is based on monitoring of individual network elements. This gives a view of the network and service status, but does not always guarantee that the end-user service itself is working and what its quality is like. Active and passive measurements at the conceptual level are

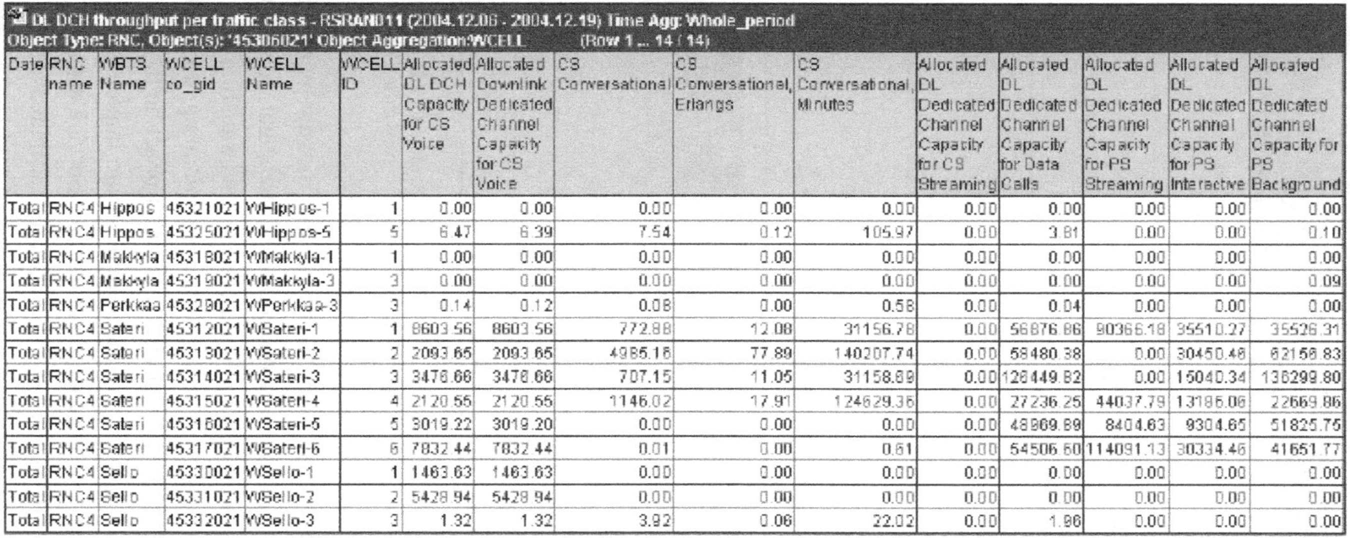

DL DCH throughput per traffic class - RSRAN011 (2004.12.06 - 2004.12.19) Time Agg: Whole_period
Object Type: RNC, Object(s): '45306021' Object Aggregation:WCELL (Row 1 ... 14 / 14)

Date	RNC name	WBTS Name	WCELL co_gid	WCELL Name	WCELL ID	Allocated DL DCH Capacity for CS Voice	Allocated Downlink Dedicated Channel Capacity for CS Voice	CS Conversational	CS Conversational, Erlangs	CS Conversational, Minutes	Allocated DL Dedicated Channel Capacity for CS Streaming	Allocated DL Dedicated Channel Capacity for Data Calls	Allocated DL Dedicated Channel Capacity for PS Streaming	Allocated DL Dedicated Channel Capacity for PS Interactive	Allocated DL Dedicated Channel Capacity for PS Background
Total	RNC4	Hippos	45321021	WHippos-1	1	0.00	0.00	0.00	0.00	0.00	0.00	0.00	0.00	0.00	0.00
Total	RNC4	Hippos	45325021	WHippos-5	5	6.47	6.39	7.54	0.12	105.97	0.00	3.81	0.00	0.00	0.10
Total	RNC4	Makkyla	45318021	WMakkyla-1	1	0.00	0.00	0.00	0.00	0.00	0.00	0.00	0.00	0.00	0.00
Total	RNC4	Makkyla	45319021	WMakkyla-3	3	0.00	0.00	0.00	0.00	0.00	0.00	0.00	0.00	0.00	0.09
Total	RNC4	Perkkaa	45328021	WPerkkaa-3	3	0.14	0.12	0.08	0.00	0.58	0.00	0.04	0.00	0.00	0.00
Total	RNC4	Sateri	45312021	WSateri-1	1	8603.56	8603.56	772.88	12.08	31156.78	0.00	56876.86	90366.18	35510.27	35526.31
Total	RNC4	Sateri	45313021	WSateri-2	2	2093.65	2093.65	4985.16	77.89	140207.74	0.00	58480.38	0.00	30450.46	62156.83
Total	RNC4	Sateri	45314021	WSateri-3	3	3478.66	3478.66	707.15	11.05	31158.89	0.00	128449.82	0.00	15040.34	136299.80
Total	RNC4	Sateri	45315021	WSateri-4	4	2120.55	2120.55	1146.02	17.91	124629.36	0.00	27236.25	44037.79	13186.06	22669.86
Total	RNC4	Sateri	45316021	WSateri-5	5	3019.22	3019.20	0.00	0.00	0.00	0.00	48969.89	8404.63	9304.65	51825.75
Total	RNC4	Sateri	45317021	WSateri-6	6	7832.44	7832.44	0.01	0.00	0.61	0.00	54506.60	114091.13	30334.46	41651.77
Total	RNC4	Sello	45330021	WSello-1	1	1463.63	1463.63	0.00	0.00	0.00	0.00	0.00	0.00	0.00	0.00
Total	RNC4	Sello	45331021	WSello-2	2	5428.94	5428.94	0.00	0.00	0.00	0.00	0.00	0.00	0.00	0.00
Total	RNC4	Sello	45332021	WSello-3	3	1.32	1.32	3.92	0.06	22.02	0.00	1.96	0.00	0.00	0.00

Fig. 12.21 An example report generated by a network management system application. Key performance indicators are in columns, elements in rows.

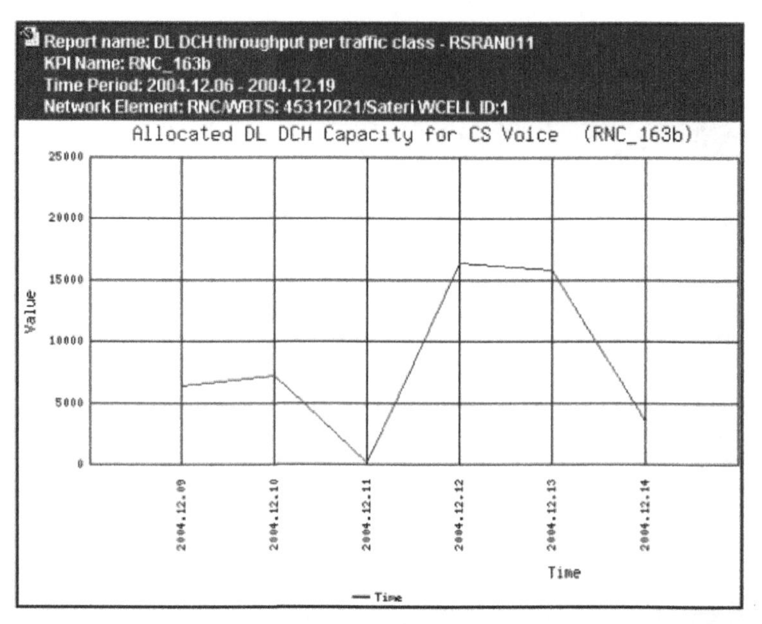

Report name: DL DCH throughput per traffic class - RSRAN011
KPI Name: RNC_163b
Time Period: 2004.12.06 - 2004.12.19
Network Element: RNC/WBTS: 45312021/Sateri WCELL ID:1 (Row 1 ... 6 / 6)

Date	RNC_163b	DUR_FOR_AMR_4_75_DL_IN_SRNC	DUR_FOR_AMR_12_2_DL_IN_SRNC	DURA_FO
2004.12.09	6221.93	0	1285184557	
2004.12.10	7118.02	0	2520479494	
2004.12.11	0.00	0	0	
2004.12.12	16235.68	0	1916342608	
2004.12.13	15699.25	0	4169309273	
2004.12.14	3474.75	0	1025337453	

Fig. 12.22 An example of drill-in into a specific key performance indicator and trend analysis. Starting point is the report in Figure 12.21. The first column of the table is opened to time series.

given in Figure 12.23. Typically these passive measurements, like counters and gauges from network elements, provide a statistical view of a component in the whole end-to-end chain. The network load situation can be analysed for business and resource (hardware/software) planning, and problems in the network may be found which currently has no impact on the service, but will have if not fixed in time. Active measurements are implemented with probes performing regular tests at scheduled intervals. Service usage is simulated as if an end-user uses the service. The probes are distributed around the network to gather results from all its parts. This method provides 24/7 statistics. With

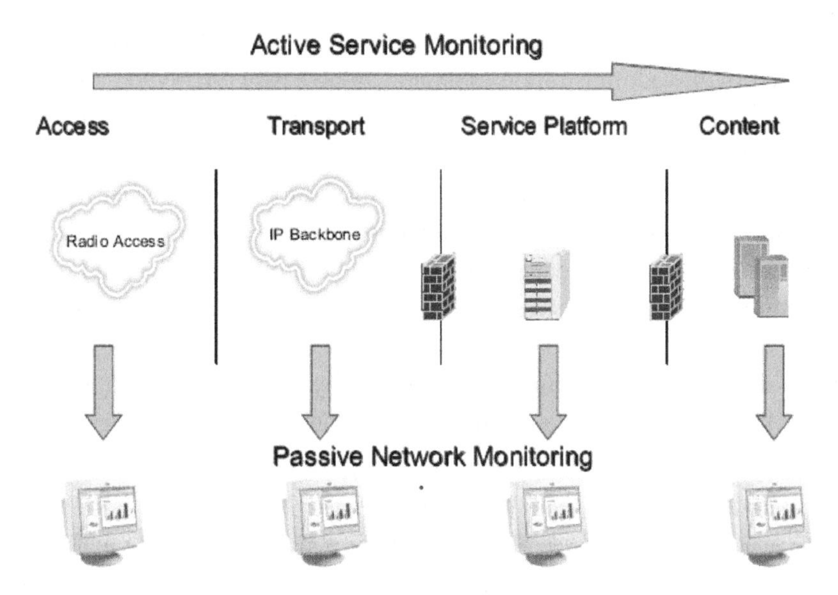

Fig. 12.23 Active vs. passive network monitoring.

consistent test transactions performed at regular intervals, fault situations are immediately detected. Comprehensive statistical data are obtained as well, offering a reliable analysis of historical trends. Utilising active probing also allows proactive fault detection.

Active measurements generate additional traffic in the network. However, with a well-planned configuration of QoS data collection, traffic can be minimised, and in normal conditions the volume of the test traffic is likely to be very small compared with the actual end-user traffic. The amount of traffic generated using active measurements has an effect on the coverage of the solutions. Thus it needs to be planned carefully, taking into consideration the use case, all of which begs the question: Is it troubleshooting or general performance monitoring? The active service test can be seen as a simple, straightforward way to verify the service in the same way the customer will experience it. This concept provides an easy way to monitor multi-vendor networks, as well as parts of the network that do not belong to the operator such as transmission lines, application server and IP backbone. Passive network and active service monitoring should not be seen in opposition but as complementary methods to get the most out of the network. Monitoring services using active monitoring is perceived as cost-effective because — with only a few or even a single probe — a service can be fully monitored end-to-end, probes can be centrally managed by the administration tool and can be easily distributed around the network. Measurements can be combined with other network management system level performance management applications. Active measurements provide a comprehensive solution for end-user service monitoring in mobile operator and service provider environments.

Probe(s) collect QoS data from verified (monitored) services and forward the resulting data to the network management system, where monitoring, analysis and reporting products utilise the data, as depicted in Figure 12.24. All probes are centrally configured and managed by a configurator running in the network management system. The probe verifies the service at regular intervals by simulating end-user behaviour. In Figure 12.24, the service is verified end to end from the mobile to the application server — e.g., a WWW or MMS server. Another possible scenario is verification of the service beginning from the Gb, Gn or Gi interfaces towards the application server. In case of Gb or Gn the BSS (Base Station System) — respectively, the SGSN — is emulated. This allows an isolated view of how the radio access network, packet core network or service platform behave for a certain service. Example use cases are given in Figure 12.25. In some cases two probes are involved in a single service verification. This is the case for end-user services where an item is sent

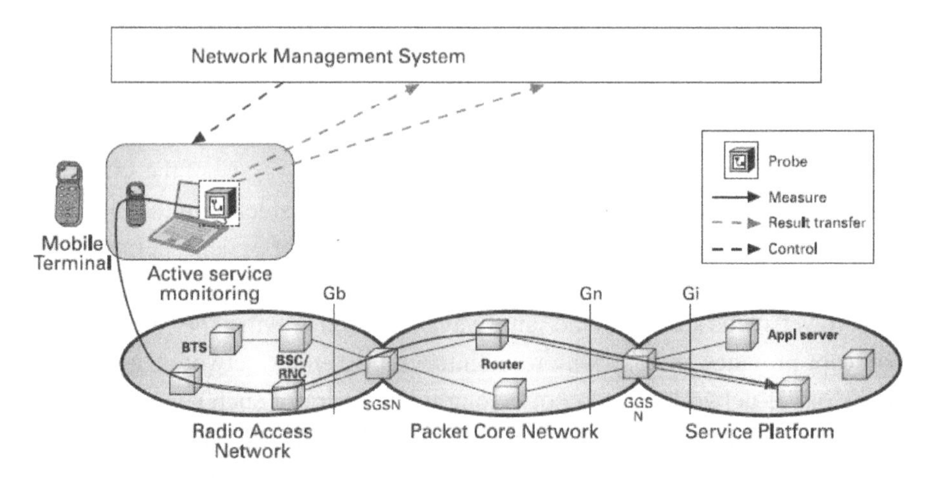

Fig. 12.24 Probe(s) collect quality of service data from verified (monitored) services and forward the resulting data to the network management system, where monitoring, analysis and reporting products utilise the data.

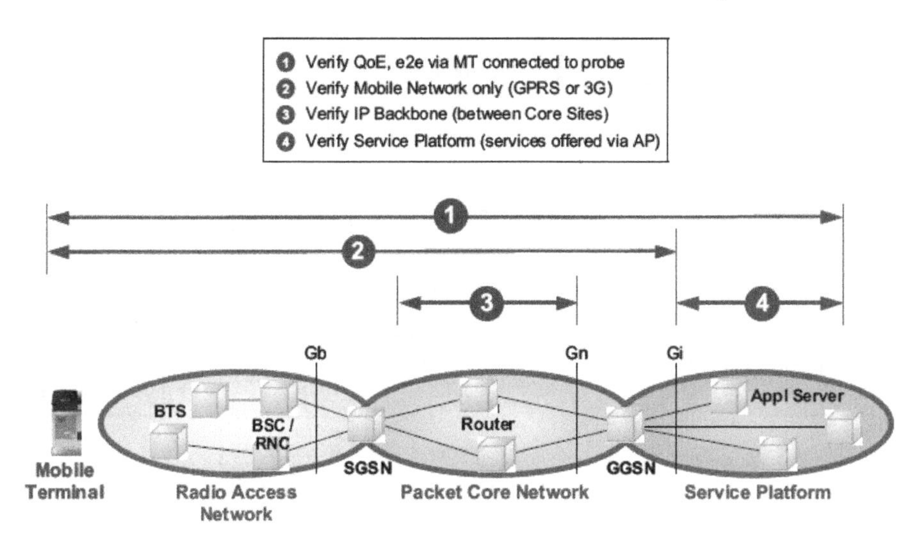

Fig. 12.25 Example use cases.

from one end-user to another, like MMS and SMS. Usually the probes will produce performance management and fault management data for each service verification. Fault management data are also known as 'alarms'. Alarms are generated based on configurable thresholds per probe. This information is sent to the network management system where several applications can use these data to conclude both network and service performance [8].

Optimisation Using Operations System Tools

One of the aspects of this chapter is automation and operational efficiency. Network management systems with workflow support and the availability of actual data from the network aim at OPEX-efficient optimisation solutions. Network management system level statistical optimisation aids the operator to control, visualise, analyse and optimise the mobile network using both performance data and configuration data on a network-wide basis.

As depicted in Figure 12.26, automation in connection with statistical optimisation includes automatic access to performance, fault and configuration data, functions to support network performance analysis and reporting, automated workflow support in order to move from one process phase to another. Data availability enables effective diagnostics support for decision making in troubleshooting cases. The most visible differences between the extensive optimisation solution and service assurance tools is the utilisation

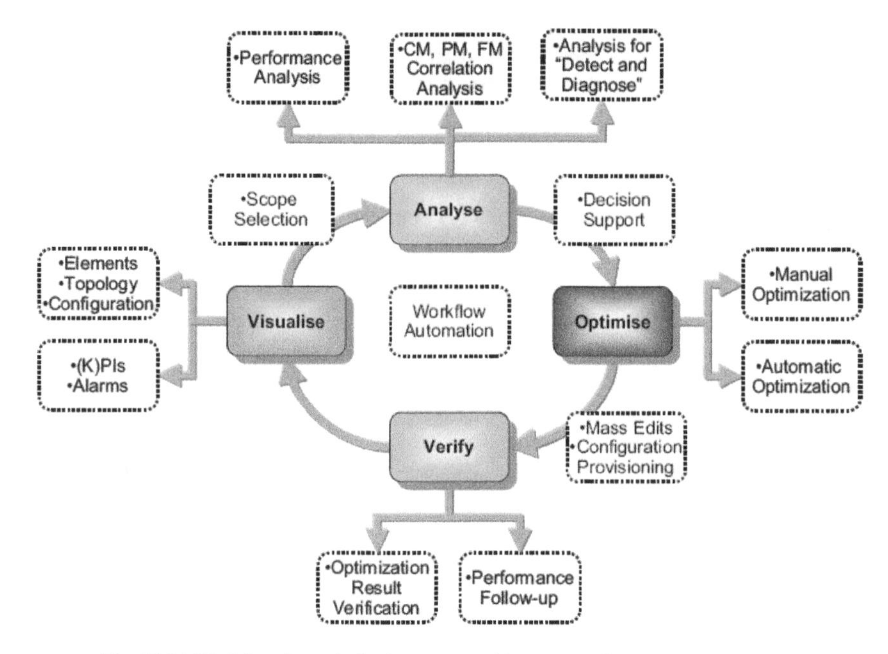

Fig. 12.26 Workflow for optimisation supported by a network management system.

of configuration data in the process and the possibility to react and change the configuration when needed. The optimisation solution utilises the performance management and configuration management issues discussed in earlier sections of this chapter.

KPI, alarm and configuration visualisation can happen in browser, GIS (map) or navigator view. The GIS functionality brings geographical aspects to the management system and enables measured KPI visualisation on cell dominance or cell icon, as well as visual verification of adjacency definitions for location and routing areas. This visualisation function can be used in the scope selection of the optimisation case. For example, the KPIs are sorted according to the performance level. Objects (or network elements) whose performance is below a set target are selected for further analysis and optimisation.

The analysis functionality contains further processing of the measurements and correlating the configuration with the performance, among other things. The possibility to visualise configuration management and related performance management simultaneously aids in troubleshooting activities. Often differences in performance can be explained simply by an anomaly in the configuration settings.

The optimisation tasks can proceed once the situation in the network is analysed. Based on the analysis results, corrective actions can be taken. These actions can be either reactive or proactive, anticipating the changes required in order to guarantee stable quality in the network for the future. Optimisation can be done for individual objects and network elements or on a network-wide basis. Workflow supports the provisioning of the changes. In some cases, the optimisation results are verified prior to provisioning. This is the case especially with automated optimisation. The proposed solution is further verified and analysed in order to gain confidence on the proposed changes. Another function used for verification is performance follow-up after the change has been made. An example realisation of an optimisation product in the network management system is given in Figure 12.27.

Field Measurement Tool

(i) The optimisation process of a network is easily understood as something conducted purely in a centralised manner — e.g., by utilising network management systems that allow control of the network by,

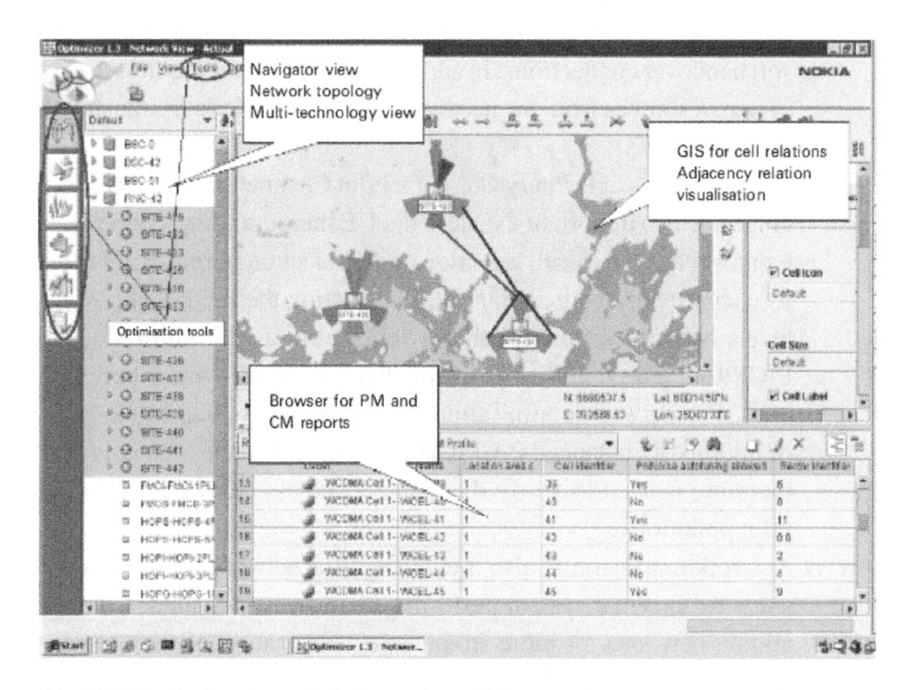

Fig. 12.27 Realisation of an optimisation product utilising network management system level data.

say, parameter adjustment. In many cases, however, a network planner and optimiser must also study the network from a customer's point of view — i.e., conduct field measurements with a proper field measurement tool. In WCDMA, such a need may arise in, for example, the following cases:

(ii) During a network launch phase, a set of field measurements to verify the basic performance of the network is required. In addition, the coverage areas of various services and bit rates need to be verified. Such tests should be conducted in both loaded and unloaded network conditions.

(iii) In general, propagation maps used as input for planning tools in the network planning phase often contain inaccuracies, and need to be verified by means of field measurements to ensure proper network coverage in critical areas.

(iv) Knowledge of soft handover areas is important, and is related to cell coverage measurements in general. Although soft handover is a desirable feature of WCDMA that adds reliability to the network, excessive soft handover (probability >35–40%) may decrease the downlink capacity due to additional downlink interference from soft handover connections. In addition, soft handover requires more system resources from the network (baseband in Node B, transmission and RNC).

(v) Measurements of Primary Common Pilot Channel (P-CPICH) coverage are also important, because the UE measures this channel for handover as well as cell selection and reselection purposes. Therefore, cell loads can be balanced by adjusting the P-CPICH power levels between different cells, as UEs from a cell with reduced P-CPICH power are more easily handed over to neighbouring cells. For network optimisation, this challenge is due to the risk of improperly balanced P-CPICH powers. Field measurements are therefore needed to verify this balance in selected areas to avoid pilot pollution.

(vi) Network expansion is also a challenge for network optimisation, since the original system performance must be maintained when adding new sites or more frequencies. Estimation of the new or

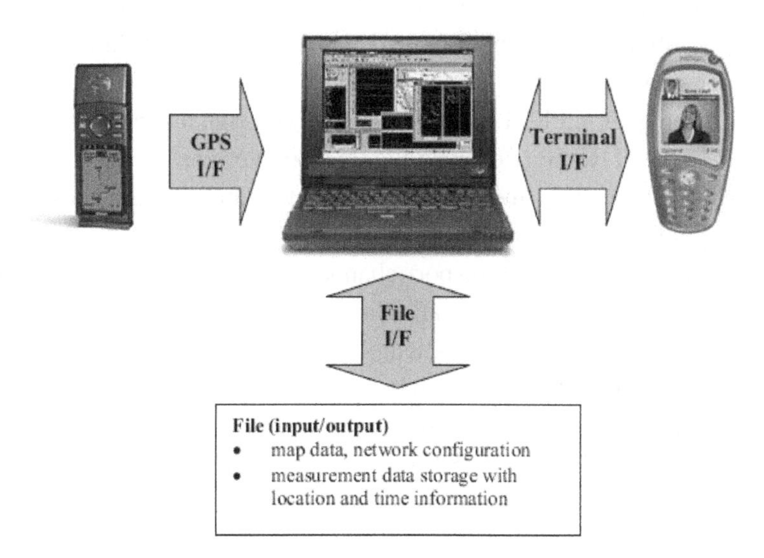

Fig. 12.28 Typical field measurement tool composition with the main functional modules and interfaces.

changed coverage areas and QoS before the integration of the new site is also possible in this case only by conducting field measurements.

Typical modules of a field measurement tool for WCDMA networks and the various interfaces are shown in Figure 12.28.

Parameters in Field Measurement Tool

To conduct optimisation tasks in WCDMA networks properly, the field measurement tool employed should support a number of parameters describing the network performance. In addition to direct quality-related parameters, such as number of call attempts, other parameters of a more supporting nature should be monitored by the tool. The following list summarises some of the most relevant parameters to be measured from a WCDMA network using a field measurement tool. As mentioned, however, not all of these parameters are directly linked to optimisation tasks. These supporting parameters can be used, for example, to investigate on a more detailed level of a detected problem in the network, when other indicators are not providing enough information:

(i) General information. Service type (voice, data), Mobile Country Code (MCC), Mobile Network Code (MNC), downlink primary scrambling code number for active and neighbour set, carrier number, cell ID.
(ii) Coverage. P-CPICH Rx E_c/I_0, UTRA cell signal strength (RSSI), P-CCPCH RSCP (Received Signal Code Power).
(iii) Signalling. L2 messages (uplink/downlink), L3 messages (uplink/downlink).
(iv) Quality. Downlink transport channel Block Error Rate (BLER), number of call attempts, call setup success ratio, call success ratio, uplink/downlink coding scheme.
(v) Handovers (soft/hard). Number of handover attempts, handover success ratio, handover type, active set list, neighbour list, E_c/I_0 values of individual RAKE fingers.
(vi) PC. UE transmit power per call, Dedicated Physical Channel (DPCH) SIR.

Field Measurement Sub-process of the Network Optimisation Process

In general, network optimisation is the part of the WCDMA network planning process that enables the availability of various network services and provides a defined service quality and performance. During the network launch period, however, troubleshooting is the main focus of pre-launch optimisation. This concentrates on locating problem areas and fixing them accordingly. Network expansion and/or traffic growth is another period in the network life cycle when field measurements are a vital part of the optimisation process.

Field measurements are naturally just a part of the whole network optimisation process, where network management system data and field measurements together are employed to determine the KPIs, and hence the QoS of the network. The role of the field measurement tool is important during pre-operational mode optimisation, when there is only a limited amount of performance data available from the network elements, for the simple reason that the network carries little commercial traffic.

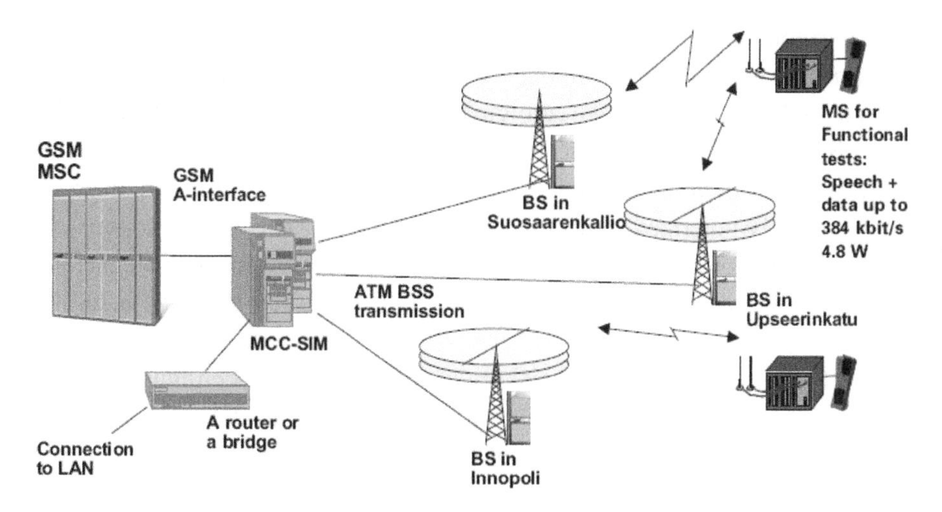

Fig. 12.29 WCDMA experimental systems, located in Espoo, Finland. [8]

Espoo (Finland) Case Study [8]

We present a single use case for the field measurement tool — i.e., evaluation of diversity gain in the WCDMA network. The basic layout for the experimental WCDMA system used in the measurements is shown in Figure 12.29.

In this case study the effect of uplink diversity on UE transmitted power level is demonstrated with field measurements. The propagation environment in the area consists of only suburban. Rural areas can be found within the cell edges. Two routes were defined for the measurement campaigns, allowing various sub-types of these propagation environments to be investigated:

(i) Route A: This route goes through a semi-urban area (four-storey and five-storey buildings rather closely spaced) along a local main street. Towards the end of the route, the buildings are clearly more widely spread. The maximum speed along this route is 50 km/h, although there are several traffic lights in the early part of the route, resulting in highly variable measurement — vehicle speeds along the route (from 0 to 50 km/h).

(ii) Route B: This route goes north along a regional highway as far as the limit of coverage. The speed limit is 70–80 km/h, with a few traffic lights. Route B is a rather open area and thus contains fewer multi-paths than route A.

Two antenna configurations were investigated by measuring them over the same routes with otherwise similar settings in the network elements (UE, Node B and MCC-SIM). These two configurations included case VV with vertically polarised Rx and diversity receive branches, vertically polarised transmitter and case V with vertically polarised receive branch, but no receive diversity and vertically polarized transmitter. Both routes A and B are approximately aligned with Node B antenna directions. Example results for the two antenna configurations are shown in Figure 12.30(a)–(b) for route A (route B results are not separately presented).

In Figure 12.30, three identical drives for both antenna configurations are shown. The two Node B antenna configuration cases are VV and V — i.e., V-polarised receiver and receive diversity branches, and V-polarised receive branch and no uplink diversity, respectively. Note that 0dBu corresponds to −115 dBm in the received signal strength indicator graphs.

The results of this limited comparison show that the antenna gain with and without uplink diversity is roughly 5 dB on route A and 2.7 dB on route B. The difference in the gains between routes A and B is due to different propagation channel characteristics. However, the results clearly emphasise the importance and benefit of uplink diversity schemes in adding network capacity in WCDMA systems.

12.5 Summary of Chapter

Detailed optimisation of CDMA-based networks is very necessary for high system capacity with minimum network deployment costs. This chapter has addressed different radio network optimisation and their impact on network coverage and system capacity. This approach enabled us to report on important algorithms and strategies for optimising the performance of CDMA-based networks configuration and study the results in realistic planning scenarios. From a practical viewpoint, the results indicate that network optimisation is a crucial step in radio access development to fully support the offered possibilities in the true cellular world.

Problems

Problem 12.1

Explain the 3G CDMA-based network optimisation concept. How is this concept related to network planning?

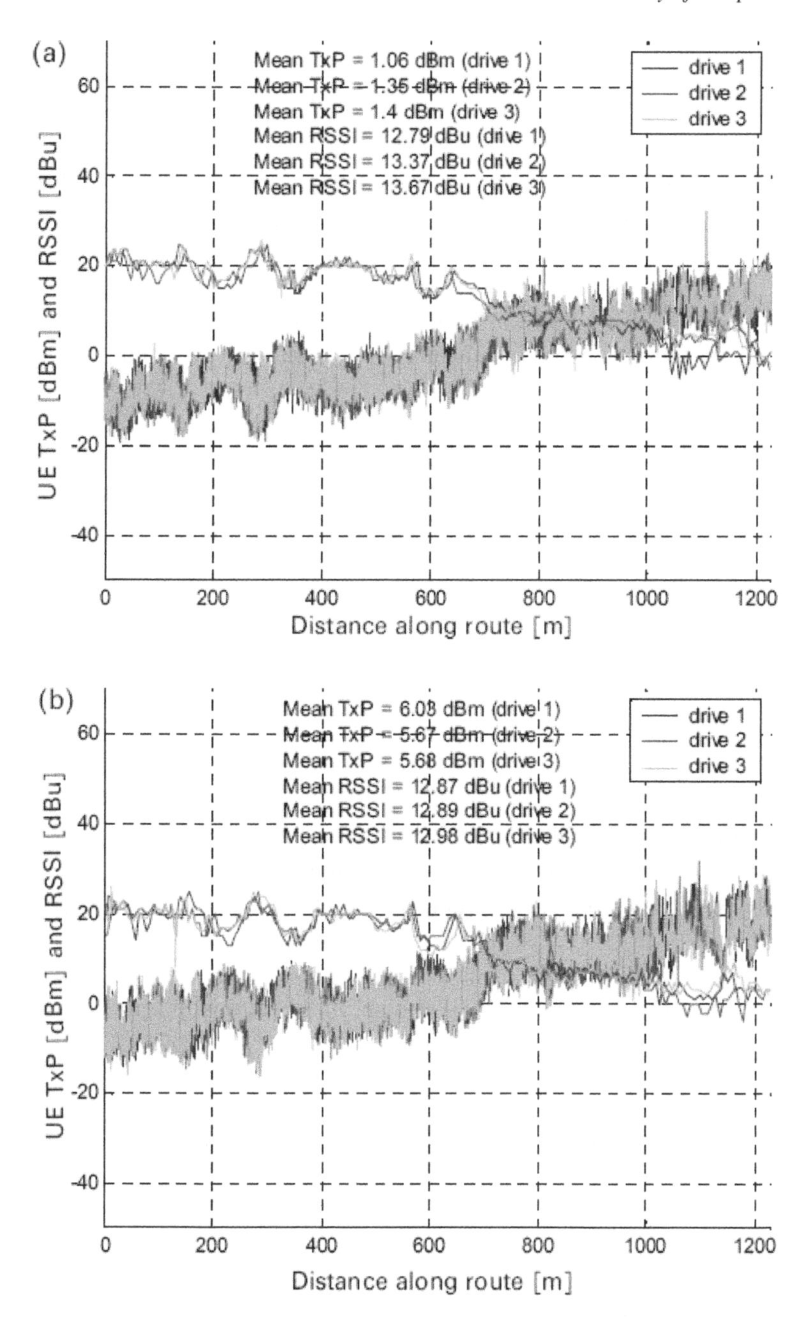

Fig. 12.30 User equipment transmitter power [dBm] and received signal strength indicator [dBu] as a function of distance along route A, both with and without uplink diversity (shown in plots (a) and (b), respectively).

Problem 12.2

Given a network configuration having user distribution with known traffic demand, and assuming that it is known which user is to be served by which cell. Discuss how you will minimise the number of cells in which at least one user cannot be served due to limited cell power capacity.

Problem 12.3

Compare and contrast topology optimisation, automated optimisation and network optimisation. List their advantages for 3G network configuration and transmission.

Problem 12.4

Using appropriate diagrammatic illustrations, explain pre-planning optimization, detail planning optimization and post planning optimisation.

Problem 12.5

(i) Why is antenna tilt important for CDMA network optimization?
(ii) Discuss two methods of implementing antenna tilt and the difference between the two methods.

Problem 12.6

Assuming an antenna has $6\lambda/2$ dipoles on top of each other, so that narrow vertical beam can be found. What is the antenna gain (in dBi's) of an ideal panel antenna when the horizontal beam width is 65 degrees (3-sector site)?

Problem 12.7

Produce a norminal cell planning optimisation including optimisation plan for coverage, capacity and low interference.

Problem 12.8

Relate network optimisation with resource management in 3G CDMA networks.

References

[1] J. Niemela, Aspects of Radio Network Topology Planning in Cellular WCDMA, P.hD. Thesis, Tampere University of Technology, 2006.
[2] J. Lempiainen and M. Manninen, Radio Interface System Planning for GSM/GPRS/UMTS. Kluwer Academic Publishers, 2001.

[3] J. Itkonen, B. P. Tuzson, and J. Lempiainen, "A novel network layout for CDMA cellular networks with optimal base station antenna height and downtilt," in Proc. 63rd IEEE Vehicular Technology Conference, 2006.

[4] W. C. Y. Lee, Mobile Communication Engineering. McGraw-Hill, 1997.

[5] E. Benner and A. B. Sesay, "Effects of antenna height, antenna gain, and pattern downtilting for cellular mobile radio," IEEE Trans. Vehicular Technology, 45(2), pp. 217–224, May 1996.

[6] J. Niemela, T. Isotalo, and J. Lempiainen, "Optimum Antenna Downtilt Angles for Macrocellular WCDMA Network," in EURASIP Journal on Wireless Communications and Networking, 5, pp. 816–827, December 2005.

[7] J. Lempiainen and M. Manninen, Eds., UMTS Radio Network Planning, Optimisation and QoS Management. Kluwer Academic Publishers, 2003.

[8] J. Laiho, A. Wacker, and T. Novosad, Eds., Radio Network Planning and Optimisation for UMTS. JohnWiley & Sons Ltd, 2002.

[9] C. Braithwaise and M. Scott, Eds., UMTS Network Planning and Development: Design and Implementation of the 3G CDMA Infrastructure. Elseview News, 2004.

[10] T. X. Brown, "Cellular performance bounds via shotgun cellular systems," IEEE Journal on Selected Areas in Communications, 18(11), pp. 2443–2455, 2000.

[11] W. C. Y. Lee, Mobile Communication Design Fundamentals. John Wiley & Sons, Inc., 1993.

[12] M. J. Nawrocki and T. W. Wieckowski, "Optimal site and antenna location for UMTS — output results of 3G network simulation software," in 14th International Conference on Microwaves, vol. 3, pp. 890–893, May 2002.

[13] J. Niemela and J. Lempiainen, "Impact of Base Station Locations and Antenna Orientations on UMTS Radio Network Capacity and Coverage Evolution," in Proc. IEEE 6th International Symposium onWireless Personal Multimedia and Communications, vol. 2, pp. 82–86, October 2003.

[14] J. Niemela and J. Lempiainen, "Impact of the Base Station Antenna Beamwidth on Capacity in WCDMA Cellular Networks," in Proc. IEEE 57th Semiannual Vehicular Technology Conference, vol. 1, pp. 80–84, April 2003.

[15] B. C. V. Johansson and S. Stefansson, "Optimizing antenna parameters for sectorised W-CDMA networks," in Proc. IEEE 52nd Vehicular Technology Conference, vol. 4, 2000, pp. 1524–1531.

[16] J. Niemela, T. Isotalo, J. Borkowski, and J. Lempiainen, "Sensitivity of Optimum Downtilt Angle for Geographical Traffic Load Distribution in WCDMA," in Proc. IEEE 62nd Semiannual Vehicular Technology Conference, vol. 2, pp. 1202–1206, September 2005.

[17] Y. K. H. Cho and D. K. Sung, "Protection against cochannel interference from neighboring cells using downtilting of antenna beams," in Proc. IEEE 53rd Vehicular Technology Conference, vol. 3, pp. 1553–1557, 2001.

[18] J. Wu, J. Chung, and C. Wen, "Hot-spot traffic relief with a tilted antenna in CDMA cellular networks," IEEE Trans. Vehicular Technology, vol. 47, pp. 1–9, February 1998.

[19] D. H. Kim, D. D. Lee, H. J. Kim, and K. C. Whang, "Capacity analysis of macro/microcellular CDMA with power ratio control and tilted antennas," IEEE Trans. Vehicular Technology, vol. 49, pp. 34–42, January 2000.

[20] A. Wacker, K. Sipil¨a, and A. Kuurne, "Automated and remotely optimization of antenna subsystem based on radio network performance," in Proc. IEEE 5th Symposium on Wireless Personal Multimedia Communications, vol. 2, pp. 752–756, 2002.

[21] 3GPP Technical Specification, UTRAN Iuant interface: Remore Electrical Tilting (RET) antennas application part (RETAP) signalling, 3GPP TS 25.463, version 6.0.0, Release 6.

[22] M. Garcia-Lozano and S. Ruiz, "Effects of downtilting on RRM parameters," in Proc. IEEE 15th Personal, Indoor, and Mobile Radio Communications, vol. 3, pp. 2166–2170, 2004.

[23] M. Garcia-Lozano and S. Ruiz, "UMTS optimum cell load balancing for inhomogenous traffic patterns," in Proc. IEEE 60th Vehicular Technology Conference, vol. 2, pp. 909–913, September 2004.

[24] M. Pettersen, L. E. Braten, and A. G. Spilling, "Automatic antenna tilt control for capacity enhancement in UMTS FDD," in Proc. IEEE 60th Vehicular Technology Conference, vol. 1, pp. 280–284, September 2004.

[25] J. Niemela, J. Borkowski, and J. Lempiainen, "Verification Measurements of Mechanical Downtilt in WCDMA," in Proc. IEE 6th International Conference on 3G and Beyond, pp. 325–329, November 2005.

[26] K. Valkeakahti, A. Hoglund, J. Parkkienen, and A. Hamalainen, "WCDMA Common Pilot Power Control for Load and Coverage Balancing", in Proc. 13th IEEE Int'l Symp, Personal, indoor and Mobile Radio Communications, Vol. 3, pp. 1412–1416, 2002.

[27] J. Laiho, A. Wacker, and T. Novosad, Radio Network Planning Optimization for UMTS, John Wiley and Sons, 2002.

[28] A. Gerdenitsch, S. Jakl, Y. Chong, and M. Toeltsch, "A Rule-Based Algorithm for Common Pilot Channel and Antenna Tilt Optimization in UMTS FDD Networks", ETRI Jornal, 26(5), pp. 437–442, 2004.

[29] A. Gerdenitsch, S. Jakl, M. Toeltsch, and T. Neubauer "Intelligent Algorithms for System Capacity Optimization of UTMS FDD Networks", Proc. 4th Int'l Conf. 3G Mobile Communication Technologies, London, June 25–27, 2003.

13

Narrowband Interference Suppression in Wireless OFDM Systems

Georgi Iliev, Zlatka Nikolova, Miglen Ovtcharov
and Vladimir Poulkov

University of Sofia, Bulgaria

Signal distortions in communication systems occur between the transmitter and the receiver; these distortions normally cause bit errors at the receiver. In addition interference by other signals may add to the deterioration in performance of the communication link. In order to achieve reliable communication, the effects of the communication channel distortion and interfering signals must be reduced using different techniques. The aim of this chapter is to introduce the fundamentals of Orthogonal Frequency Division Multiplexing (OFDM) and Orthogonal Frequency Division Multiple Access (OFDMA), to review and examine the effects of interference in a digital data communication link and to explore methods for mitigating or compensating for these effects.

13.1 OFDM Principles

13.1.1 Basics of OFDM

OFDM is quite similar to the well-known and often-used technique of Frequency Division Multiplexing (FDM) [1]. OFDM uses the principles of FDM to allow several users to send messages over a single radio link. However, OFDM differs from FDM in a number of ways. In conventional broadcasting, each radio station transmits at a different frequency, using FDM to maintain the separation of the stations; there is no coordination or synchronization between these stations. Within an OFDM transmission, the information signals from numerous stations are combined into a single multiplexed stream of data. Afterwards, this data is transmitted using an OFDM ensemble made up of a dense packing of many subcarriers. All the subcarriers within the OFDM signal are time- and frequency-synchronized with each other, thus permitting the interference between subcarriers to be carefully controlled. The multiple subcarriers overlap in the frequency domain but do not cause Inter-Carrier Interference (ICI), due to the orthogonal nature of the modulation. As a rule, to prevent interference when using FDM, the transmitted signals need a frequency guard-band between channels, which results in a lowering of the overall spectral efficiency. In contrast, OFDM allows a narrower-frequency guard-band, thus improving the spectral efficiency.

OFDM achieves orthogonality in the frequency domain by mapping each of the separate information signals onto different subcarriers. The baseband frequency of a subcarrier is chosen to be an integer multiple of the fundamental frequency of the OFDM symbol. As a result, all subcarriers have a whole number of cycles per symbol and are orthogonal to each other.

Figure 13.1 shows the construction of an OFDM signal with four subcarriers. Their time waveforms are depicted on 1b, 2b, 3b and 4b, while 1a, 2a, 3a and 4a show the Fast Fourier Transform (FFT) responses. Individual subcarriers have 1, 2, 3, and 4 cycles per symbol and zero phases. The result of the summation of the all four subcarriers is illustrated on 5a and 5b (Figure 13.1).

A set of functions are reciprocally orthogonal if they meet the conditions in (13.1). If any two different functions within the set are multiplied, and integrated over a symbol period, the result is zero [2].

$$\int s_i(t)s_j(t)dt = \begin{cases} C, & i = j \\ 0, & i \neq j. \end{cases} \tag{13.1}$$

Equation (13.2) represents a set of orthogonal sinusoids, which correspond to the subcarriers for a non-modulated real OFDM signal.

$$s_k(t) = \begin{cases} \sin(2\pi k f_a t), & \text{for } 0 < t < T,\ k = 1, 2, \ldots, M \\ 0, & \text{otherwise} \end{cases}, \tag{13.2}$$

where f_a is the carrier spacing, M is the number of carriers, T is the symbol period. Since the highest frequency component is Mf_a, the transmission

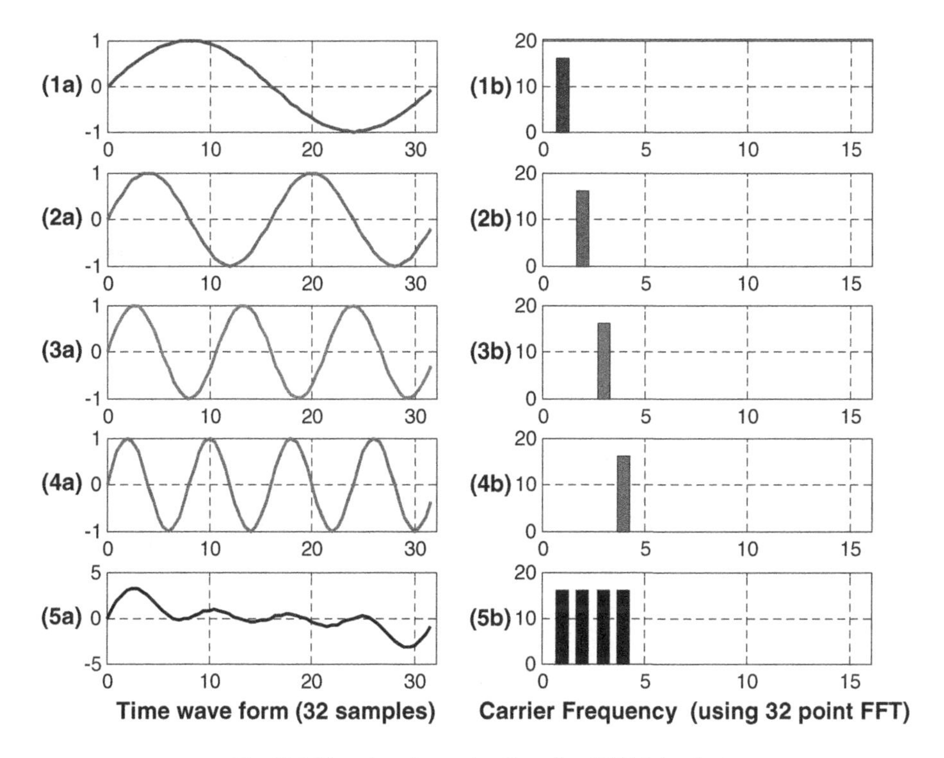

Fig. 13.1 Time domain construction of an OFDM signal.

bandwidth (BW) is also Mf_a. These subcarriers are orthogonal to each other, i.e., the result of the multiplication and integration of any two subcarriers waveforms over the symbol period is zero.

Another approach to observing the OFDM signals' orthogonality is via their spectrums. In the frequency domain each OFDM subcarrier has a $\sin(x)/x$ frequency response, as shown in Figure 13.2. The orthogonal nature of the transmission is due to the fact that each subcarrier peak coincides with the nulls of all other subcarriers. After a Discrete Fourier Transform (DFT) is applied, a discrete spectrum is achieved and its samples are depicted in Figure 13.2(a) by symbol 'o'. When the DFT is time-synchronized, the frequency samples of the DFT correspond only to the peaks of the subcarriers. Thus, the overlapping frequency region between subcarriers does not affect the receiver. The measured peaks correspond to the nulls for all other subcarriers, providing the orthogonality of the subcarriers.

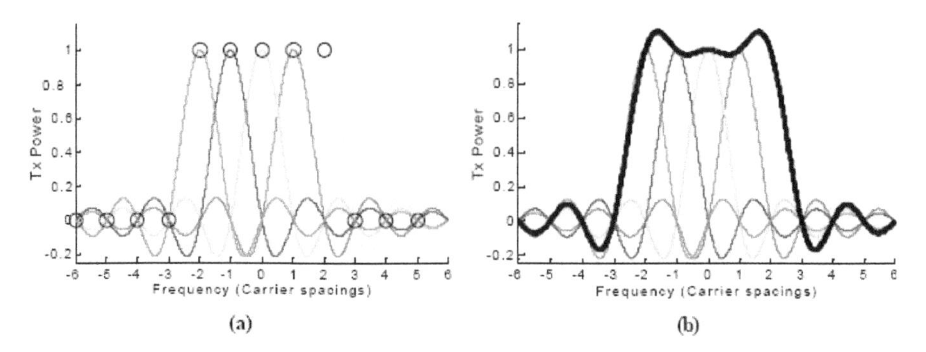

Fig. 13.2 Frequency response of the subcarriers in a 5 tone OFDM signal.

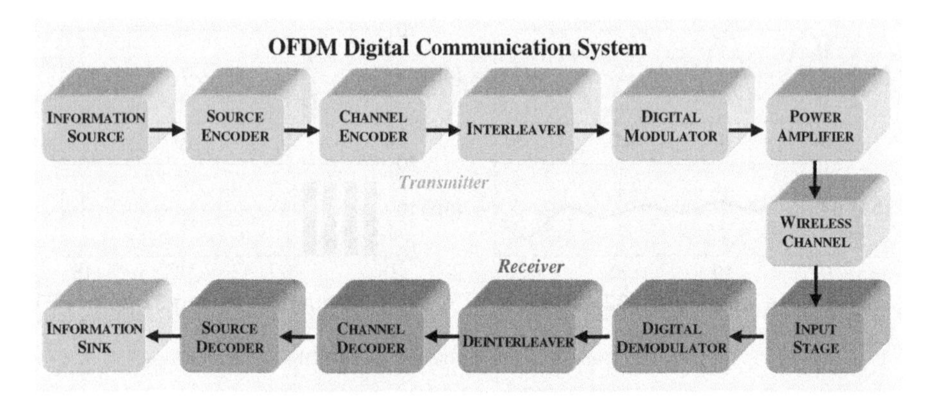

Fig. 13.3 Block diagram of a digital communication system (physical layer).

Figure 13.2(a) shows the spectrums of the carriers, and the discrete frequency samples seen by an OFDM receiver. The thick black line on Figure 13.2(b) shows the overall combined response of all five subcarriers.

13.1.2 Overview of OFDM Wireless Communication System

As Figure 13.3 indicates, an OFDM wireless communication system can consist of three major parts: *transmitter*, *channel*, and *receiver* [2, 3, 8].

Transmitter

The transmitter takes the information data supplied by the information source, i.e., the information to be transmitted, and processes it in several steps to ensure reliable communication. If the source is analog, the information is first digitized by an Analog to Digital Converter (ADC). Generally the transmitter consists of a source encoder, a channel encoder, an interleaver, and a modulator.

Source Encoder

The purpose of the source encoder is to remove as much redundancy as possible from the (digitized) output of the information source. Source encoding is also termed *data compression*. The sequence of binary digits from the source encoder is called the *information sequence*.

Channel Encoder

The channel encoder introduces controlled redundancy into the binary information sequence by applying error-correcting codes. This redundancy is used by the channel decoder in the receiver to overcome the effects of noise and interference during the transmission of the signal through the channel. The two most important error-correcting coding techniques are block coding and convolutional coding.

Interleaver

If errors occur in bursts, an effective method for improving error correction is to use interleaving. Interleaving is a controlled reordering of bits to avoid bit errors occurring in bursts at the decoder. By rearranging the coded data so that errors are separated by a large rather than a small number of bits, errors can be efficiently corrected using a random-error-correcting code. The two major interleaving techniques are block-interleaving and convolutional-interleaving.

Digital Modulator

The primary purpose of the digital modulator is to map the binary information sequence into signal waveforms [4, 5]. If the information sequence is to be transmitted one bit at a time, the modulator may map each "0" into a waveform $s_0(t)$ and each "1" into a waveform $s_1(t)$. This is called binary modulation. Alternatively, the modulator may transmit l information bits at a time by using $M = 2^l$ different waveforms $s_i(t)$, $i = 0, 1, \ldots, M - 1$, one waveform for each of the 2^l possible l-bit sequences. This is called M-ary modulation ($M > 2$), and the sequences of l bits are referred to as M symbols. A block diagram of an OFDM digital modulator is shown in Figure 13.4.

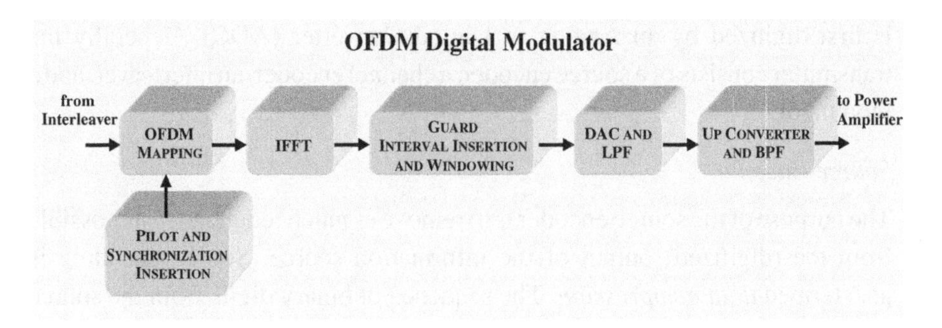

Fig. 13.4 Block diagram of a digital modulator.

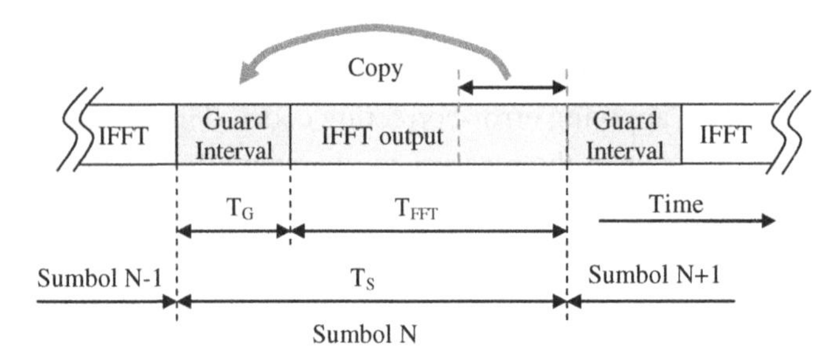

Fig. 13.5 Addition of a guard interval to an OFDM signal.

Guard Interval

The effect of Inter-Symbol Interference (ISI) on an OFDM signal can be improved by the addition of a guard interval to the start of each symbol. This guard interval is a cyclic copy that extends the length of the symbol. Figure 13.5 shows the insertion of a guard interval. The total length of the symbol is:

$$T_S = T_G + T_{FFT}, \tag{13.3}$$

where T_S is the total length of the symbol in samples, T_G is the length of the guard interval in samples, and T_{FFT} is the size of the Inverse Fast Fourier Transform (IFFT) used to generate the OFDM signal.

Power Amplifier

This is the output stage of an OFDM system. The Radio Frequency (RF) signal is amplified, shaped by the bandpass (BP) channel filter and supplied to the transmit antenna system. Two of the most crucial functions, determining the Quality of Service (QoS) of wireless systems are implemented here: transmitted power control and antenna beam forming.

Wireless Channel

The wireless channel introduces unintended distortion to the transmitted signal by, for example, adding noise and introducing fading, delayed reflections, and interference [5–7].

Receiver

The task of the receiver is significantly more difficult than that of the transmitter. This is largely due to the need for synchronizing the timing, phase and frequency of the distorted received signal. Basically, a receiver consists of: input stage, demodulator, deinterleaver, channel decoder and source decoder.

Input Stage

The input stage incorporates several functions, such as: receive antenna gain control, adaptive notch filter for Narrowband Interference (NBI) suppression, low noise amplifier, channel estimation and equalization and automatic gain control.

Digital Demodulator

The digital demodulator is usually the most complex part of a communication system. Here, the timing, phase and frequency of the received signal is detected and then tracked to ensure synchronization [3, 4, 6, 7]. The demodulator processes the channel-corrupted transmitted waveform and converts each waveform to a single number that represents an estimate of the transmitted data symbol. The demodulator consists of several parts. Usually, the received signal is fed through a matched filter or, alternatively, a correlator. The output from the filter/correlator is sampled at symbol rate, and passed on to a detector, which makes a decision as to which symbol was transmitted. The detector can employ soft decision or hard decision detection. A block diagram of a demodulator is shown in Figure 13.6.

Deinterleaver

At the receiver, a deinterleaver is employed to undo the effect of the interleaver in the transmitter [1, 2]. In order to ensure proper deinterleaving, the interleaver and deinterleaver must be synchronized. In the case of block interleaving, this means that bits interleaved in the same frame must also be deinterleaved in the same frame. Convolutional interleavers and deinterleavers are finite state machines, and must be in the same initial state.

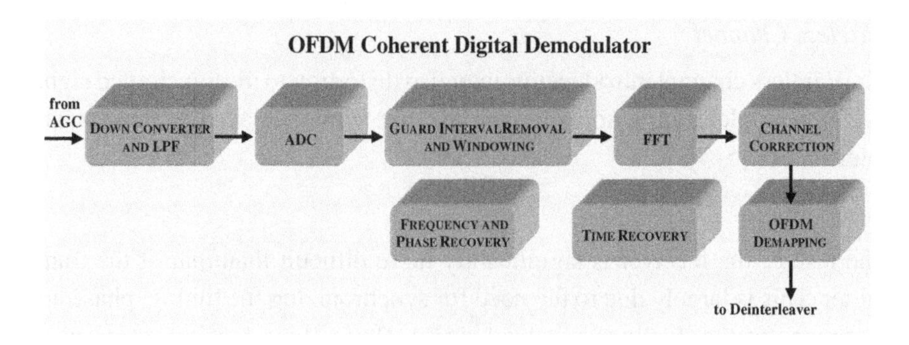

Fig. 13.6 Block diagram of a digital coherent demodulator.

Channel Decoder

The channel decoder attempts to reconstruct the original information sequence from knowledge of the code used by the channel encoder and the redundancy in the received data [1, 2]. As with deinterleaving, decoding requires synchronization. The block decoder must know where in the received bit sequence a code word begins, and the convolutional decoder must know what state to start in.

Source Decoder

From the knowledge of the source encoding method used, the source decoder attempts to reconstruct the original signal of the source. The result, an estimation of the transmitted information, is then finally passed to the information sink.

13.1.3 Wireless Fading Channel

Small-scale fading, or simply fading, is a random variation of the received signal's envelope, phase, and frequency, and is a result of the summing of signals from several paths (direct, reflected, scattered, etc.) at the receiver antenna [8, 9]. A wireless channel is a good example of a fading channel. The three most important fading effects caused by this multipath propagation are: rapid changes in signal strength over a small travel distance or time interval, random frequency modulation due to varying Doppler shifts in different multipath signals, and time dispersion (echoes) caused by multipath propagation delays. Examples of short term fading and frequency selective fading are shown in Figures 13.7 and 13.8.

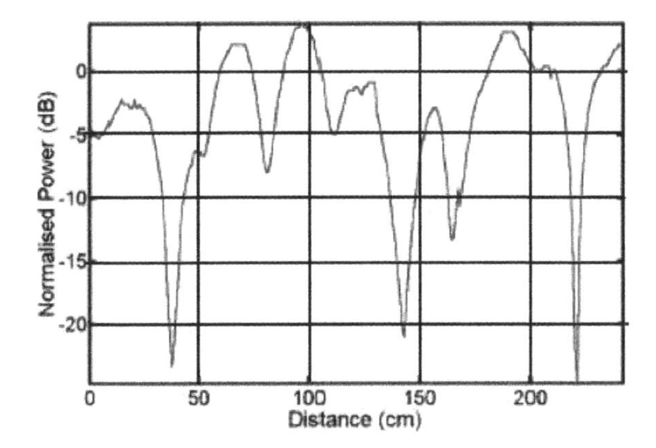

Fig. 13.7 Plot of Short-Term Fading with distance.

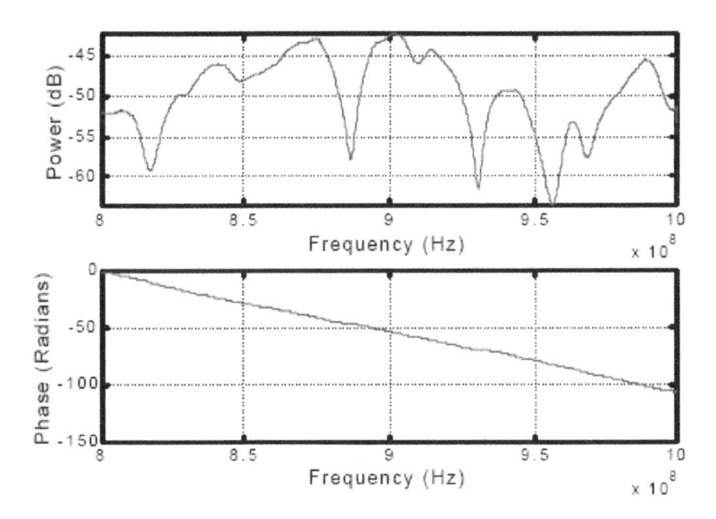

Fig. 13.8 Plot of frequency selective fading.

To better characterize the fading signal, a number of statistical measures are used [4, 7]. The time and frequency dispersive properties of the channel are described by its rms delay spread and coherence BW, and its coherence time and Doppler spread, respectively. Although analytical expressions do not exist for the general case, probability distributions that describe these variations are known for a few well-defined situations. Among these, the most frequently used are the Rayleigh and Rician fading models. An example of time varying impulse response of a fading channel is given in Figure 13.7.

Fading Channel Parameters

The time dispersive properties of the channel, i.e., its root mean square (rms) delay spread and coherence BW, are derived from its power delay profile. The power delay profile defines the delays and relative powers of different multipath signals, and is heavily dependent on the particular surroundings. The coherence BW is a statistical measure of the range of frequencies over which the channel can be considered "flat" i.e., a channel that passes all spectral components with approximately equal gain and linear phases.

If the coherence BW is defined as the bandwidth over which the frequency correlation function is above 0.5 (which is sufficient for all practical purposes), the coherence BW is approximately [8, 9]:

$$B_C \approx \frac{1}{5\sigma_\tau}, \tag{13.4}$$

where σ_τ is the rms delay spread defined as:

$$\sigma_\tau = \sqrt{\bar{\tau}^2 - (\bar{\tau})^2}, \tag{13.5}$$

with

$$\bar{\tau} = \frac{\sum_j P(\tau_j)\tau_j}{\sum_j P(\tau_j)}, \tag{13.6}$$

and

$$\bar{\tau}^2 = \frac{\sum_j P(\tau_j)\tau_j}{\sum_j P(\tau_j)} \tag{13.7}$$

Here, $P(\tau_j)$ refers to the received signal power at delay τ_j from the first received signal component at $\tau_0 = 0$.

The Doppler frequency spread B_D is defined as the range of frequencies over which the received Doppler spectrum is essentially non-zero. When a pure sinusoidal tone of frequency f_c is transmitted, the received signal spectrum, the Doppler spectrum, will have components in the range from $f_c - f_d$ to $f_c + f_d$, where f_d is the Doppler shift. The coherence time T_C is a statistical measure of the time duration over which the channel impulse response is essentially invariant, and is inversely proportional to the Doppler spread B_D:

$$T_C = \frac{3}{4\sqrt{\pi}B_D}. \tag{13.8}$$

Fading Channel Types

Depending on the relation between the signal parameters (such as BW and symbol period) and the channel parameters (such as rms delay spread and Doppler spread), transmitted signals will undergo different types of fading. The multipath time delay spread determines whether the channel fading is flat or frequency-selective, while the Doppler spread determines whether it is fast or slow [6, 9].

The channel is classified as a flat fading channel if the rms delay spread is smaller than the symbol period. The Rayleigh and Rician fading models both describe flat fading channels. If the rms delay spread is larger than the symbol period, the channel exhibits frequency selective fading. As a rule of thumb, the channel is considered frequency-selective if $\sigma_\tau \geq T_S/10$. Models for such channels are difficult to construct and are usually modeled as linear filters.

In a fast fading channel, the Doppler spread is large and the coherence time is smaller than the symbol period. The channel variations are faster than those of the baseband signal, causing distortion at the receiver. If the Doppler spread is so small that the coherence time is much larger than the symbol period, the channel variations are slower than those of the baseband signal, and the channel is classified as a slow fading channel. In such a channel, the effects on signal amplitude and phase can be considered constant during one or several modulation symbol periods. Moreover, in a slow fading channel, the effects of Doppler spread are negligible at the receiver.

The two most commonly used models of fading channels are the Rayleigh and the Rician fading channel models. Both describe slow, flat fading channels. Also they are often essential parts of models of frequency-selective channels.

13.2 OFDM Implementation in Multiple Access Schemes

Besides as a modulation scheme OFDM could be used as part of a multiple access technique. "Multiple access" refers to sharing a communication resource by different users. Multiple access schemes attempt to distribute the resources in such a way in order to provide orthogonal, i.e., non-interfering, communication channels for each user. The basic ways to distribute a communication resource are Frequency Division (FD), Time Division (TD) and Code Division (CD). In Frequency Division Multiple Access (FDMA), each user receives a unique carrier frequency and bandwidth. In Time Division Multiple

Access (TDMA), each user is given a unique time slot, either on demand or in a fixed rotation. Orthogonal Code Division Multiple Access (CDMA) systems allow each user to share the bandwidth and time slots with many other users, and rely on orthogonal binary codes to separate the users [10–12].

Frequency division multiple access (FDMA) can be implemented in OFDM systems by assigning different users their own sets of subcarriers, as shown in Figure 13.9. There are a number of ways this allocation can be performed. The simplest method is a fixed allocation of subcarriers to each user, as shown on the left of Figure 13.9. An improvement of the fixed allocation scheme is dynamic subcarrier allocation based upon the channel state conditions. For example, due to frequency selective fading one user may have relatively good channels on some subcarriers, while another user might have good channels on other subcarriers. It could be mutually beneficial for these users to swap the fixed allocations given above based on a dynamic allocation scheme.

Besides FDMA, multiple users can also be accommodated with TDMA, where they are provided with a fixed or dynamic assignment in time. Fixed TDMA is shown on the right of Figure 13.9. Such a TDMA scheme is appropriate for constant data-rate (i.e., circuit switched) applications like voice and streaming video. Dynamic assignment schemes (such like packet based systems employ more sophisticated scheduling algorithms based on queue-lengths, channel conditions, and delay constraints to achieve much better performance than fixed TDMA.

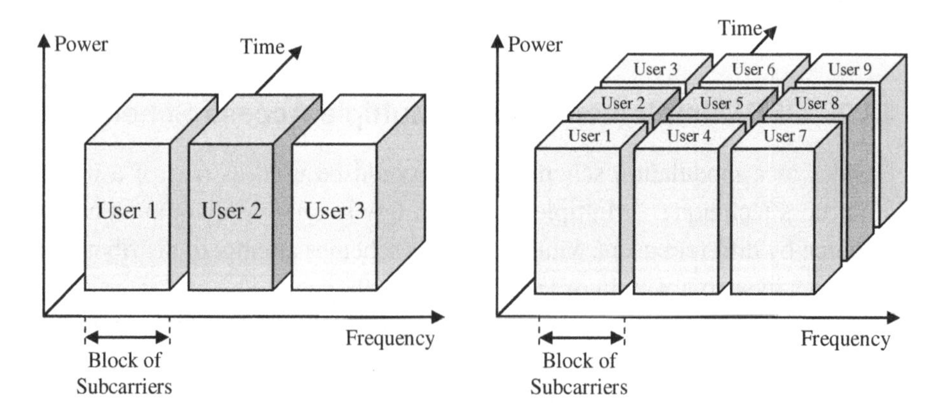

Fig. 13.9 FDMA (left) and FDMA with TDMA (right).

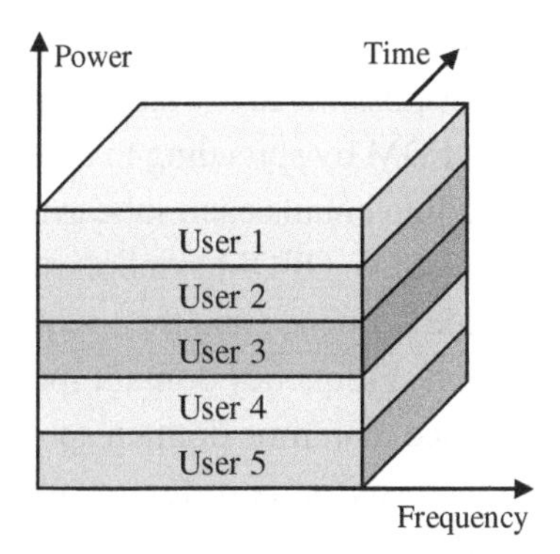

Fig. 13.10 User assignment in CDMA.

In OFDM-TDMA, a particular user is given all the sub-carriers of the system for any specific OFDM symbol duration. Thus, the users are separated via time slots. All symbols allocated to all users are combined to form an OFDM-TDMA frame. The number of OFDM symbols per frame can be varied based on each of the users' requirement.

In Code Division Multiple Access (CDMA), each user is assigned a unique code sequence (spreading code) it uses to encode its information bearing signal. In CDMA the users share time and frequency slots but employ different codes that allow the users to be separated by the receiver as illustrated in Figure 13.10. The receiver, knowing the code sequences of the user, decodes a received signal after reception and recovers the original data. This is possible because the cross-correlations between the code of the desired user and the codes of the other users are small. Because the bandwidth of the code signal is chosen to be much larger than the bandwidth of the information bearing signal, the encoding process enlarges (spreads) the spectrum of the signal and is therefore also known as spread spectrum modulation. A channel in a CDMA system is then defined by the spreading code. Ideally, the codes among terminals should be orthogonal so that the receiver can detect the signal addressed to it in the presence of interference from other users.

CDMA is the dominant multiple access technique for present cellular systems, but is not particularly appropriate for high-speed data since the entire premise of CDMA is that a bandwidth much larger than the data rate is used to suppress the interference. In wireless broadband networks the data rates already are very large, so spreading the spectrum further is not viable.

OFDM and CDMA can be combined to create a Multi-Carrier CDMA (MC-CDMA) waveform. It is possible to use spread spectrum signaling and to separate users by codes in OFDM by spreading in either the time or frequency domain. Time domain spreading entails each subcarrier transmitting the same data symbol on several consecutive OFDM symbols, that is, the data symbol is multiplied by a length N code sequence and then sent on a specific subcarrier for the next N OFDM symbols. Frequency domain spreading, which generally has slightly better performance than time domain spreading, entails each data symbol being sent simultaneously on N different subcarriers. In such a case the advantage is that OFDM provides a simple method to overcome the Inter-Symbol Interference (ISI) effect of the multi-path frequency selective wireless channels, while CDMA provides the frequency diversity and the multi-user access scheme. Multi-Carrier CDMA has gained much attention, because the signal can be easily transmitted and received using the FFT device without increasing the transmitter and receiver complexities and is potentially robust to channel frequency selectivity with good frequency use efficiency.

OFDM could be used both as a modulation scheme and as part of the multiple access technique, by applying a spreading code in the frequency domain which as a result gives orthogonal frequency division multiple access (OFDMA). In OFDMA, multiple access is realized by providing each user with a fraction of the available number of subcarriers. The available sub-carriers are distributed among all the users for transmission at any time instant. The subcarrier assignment is made for the user lifetime, or at least for a considerable time frame. In this way, OFDMA could be considered as an ordinary frequency division multiple access (FDMA); however, OFDMA avoids the relatively large guard bands that are necessary in FDMA to separate different users. The scheme was first proposed for CATV systems, and later adopted for wireless communication systems. OFDMA can support a number of identical downstreams, or different user data rates, (e.g., assigning a different number of sub-carriers to each user). Based on the sub-channel condition, different baseband modulation schemes can be used for the individual sub-channels, e.g., QPSK, 16-QAM and 64-QAM etc. This is referred to as adaptive subcarrier, bit, and power allocation or Quality of Service (QoS) allocation. Very often a mixture of OFDMA and TDMA is applied, arranging the transmission in such a way so that different users transmit in different timeslots, which may contain one or several OFDM symbols. Also frequency hopping (one form of spread spectrum realization) could be employed to provide security and resilence to inter-cell interference.

13.2.1 Advantages of OFDMA

OFDM and in particular, its IFFT/FFT implementation gave FDMA a new life as a broadband multiple access scheme. The use of IFFT/FFT allowed terminals to arbitrarily combine multiple frequencies (subcarriers) at baseband, leading to orthogonal frequency division multiple access (OFDMA). As stated above OFDMA could be viewed as a hybrid of FDMA and TDMA: users are dynamically assigned subcarriers (FDMA) in different time slots (TDMA). The advantages of OFDMA start with the advantages of single user OFDM in terms of robust multipath suppression and frequency diversity. In addition, OFDMA is a flexible multiple access technique that can accommodate many users with widely varying applications, data rates, and quality of service (QoS) requirements. Because the multiple access is performed in the digital domain (before the IFFT operation), dynamic and efficient bandwidth allocation is possible. This allows sophisticated time and frequency domain scheduling algorithms to be integrated in order to best serve the users.

In comparison of the three major types of systems for OFDM multiple access, i.e., OFDM-TDMA, OFDM-CDMA (MC-CDMA or MC DS-CDMA) and OFDMA, it must be noted that OFDMA is fundamentally advantageous over OFDM-TDMA and OFDM-CDMA when it comes to real system applications.

The first major advantage is the *better data rate granularity* based on both time and frequency domain assignment. Early broadband systems utilize OFDM-TDMA to offer a straightforward way of multiple accessing, where each user uses a small number of OFDM symbols in a time slot and multiple users share the radio channel through TDMA. The method has two obvious shortcomings: first, every time a user utilizes the channel, it has to burst its data over the entire bandwidth, leading to a high peak power and therefore low RF efficiency; second, when the number of sharing users is large, the TDMA access delay can be excessive. OFDMA is a much more flexible and powerful way to achieve multiple access with an OFDM modem. In OFDMA, the multiple access is not only supported in the time domain, but also in the frequency domain, just like traditional FDMA minus the guard-band overhead.

As a result, an OFDMA system can support more users with much less delay. The finer data granularity in OFDMA is an advantage with multi-media applications with diverse QoS requirements.

Another important advantage is the *smaller link budget* for low rate users. Since TDMA user must burst its data over the entire bandwidth during the allocated time slots, the instantaneous transmission power (dictated by the peak rate) is the same for all users, regardless of their actual data rates. This inevitably creates a link budget deficit that handicaps the low rate users. Unlike TDMA, an OFDMA system can accommodate a low-rate user by allocating only a small portion of his band, proportional to the requested data rate.

Concerning practical realization, advantage of the OFDMA system is the *receiver simplicity* with multi-user interference-free detection. OFDMA has the merit of easy decoding at the receiver side, as it eliminates the intra-cell interference avoiding CDMA type of multi-user detection. This is not the case in MC-CDMA, even if the codes are designed to be orthogonal. The signals of different users can only be detected jointly since the code orthogonality is destroyed by the frequency selective fading. In MC-CDMA users' channel characteristics and responses must be estimated using complex jointly estimation algorithms, which is not the case with OFDMA. Furthermore, OFDMA is the least sensitive multiple access scheme to system imperfections. Due to these features OFDMA has been adopted in several modern wireless systems, e.g., IEEE 802.16a-e and IEEE 802.20.

A significant advantage of OFDMA relative to OFDM is its potential to *reduce the transmit power* and to also *relax the peak-to-average power ratio* (PAPR) problem. The PAPR problem is particularly acute in the uplink, where power efficiency and cost of the power amplifier are extremely sensitive issues. By splitting the entire bandwidth among many users in the cell, each user utilizes only a small subset of subcarriers. Therefore, each user transmits with a lower PAPR (PAPR increases with the number of subcarriers), and also with far lower total power than if it had to transmit over the entire bandwidth. Lower data rates and bursty data are handled much more efficiently in OFDMA than in single user OFDM, or with TDMA or CSMA, since rather than having to blast at high power over the entire bandwidth, OFDMA allows the same data rate to be sent over a longer period of time using the same total power [13, 14].

Another important advantage of OFDMA is in relation with the *multiuser diversity capability*. Multiuser diversity gain arises from the fact that in a wireless system with many users, the achievable data rate of a given resource unit varies from one user to another. Such fluctuations allow the overall system performance to be maximized by allocating each radio resource unit to the user that can best exploit it. OFDMA allows different users to transmit over different portions of the channel spectrum (traffic channel). Since different users perceive different channel qualities, a deep faded channel for one user may still be favorable to another. Therefore, through judicious channel allocation, the system can potentially outperform interference - averaging techniques by a factor of up to 3 in spectrum efficiency [15, 16].

One of the major problems associated with transmitting information from subscriber premises is the *Narrow-Band Interference* (NBI). In OFDM-based systems, the effect can be loss of sub-carriers and degraded BER performances. NBI affects OFDM-TDMA much more severely than OFDMA systems. As only one user occupies the whole bandwidth, if the particular user is subdued with high Signal to Interference Noise Ratio (SINR), then the whole system collapses. A low interference level jams a few carriers in OFDMA and slightly reduces the system capacity, whereas an OFDM-TDMA system can still operate at its full capacity. However, as the interference level exceeds a certain threshold OFDM-TDMA entirely breaks down, whereas OFDMA only loses a small percentage of its total capacity. NBI affects certain sub-carriers and only the users who are allocated those sub-carriers are affected by increasing NBI [17].

13.3 NBI Suppression Methods

13.3.1 An Overview

In this section we will focus on an NBI which is an in-band signal in an OFDM system. This type of signal interference can be found in the new unlicensed frequency bands, e.g., the Industrial Scientific Medical (ISM) band, coming from systems such as Bluetooth or microwave ovens which interfere with OFDM based Wireless Local Area Networks (WLAN), like Hiperlan II. Other examples of NBI are strong Radio Frequency Interference (RFI) from short-wave radio, Citizen's Band (CB) radio and amateur "ham" radio which interfere with Hybrid Fiber Coaxial (HFC) networks and Digital Subscriber Lines (DSL). Ineffective cable shielding of a network may also cause the

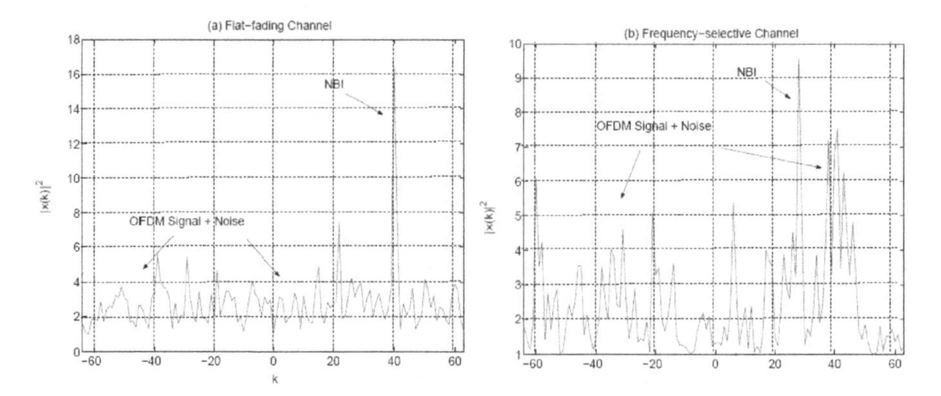

Fig. 13.11 The magnitude-squared of FFT bins: $|X[k]|^2$.

ingress from external electrical devices in the homes, such as TVs and computers.

Let's consider a strong NBI signal which resides within the same frequency band as a wideband OFDM signal. In this case, severe Signal to Noise Ratio (SNR) degradation is likely to occur across all OFDM subcarriers due to spectral leakage of the NBI signal from block processing in the OFDM receiver. In Figure 13.11, the magnitude-squared of the received signal samples $|X[k]|^2$ is plotted. For flat-fading channels it is straightforward to ascertain the fact that the highest peak corresponds to the subcarrier affected by NBI, as shown in Figure 13.11(a). Figure 13.11(b) shows that in frequency-selective channels the OFDM signal presents a large dynamic range, and some signal peaks can have values close to the peak induced by NBI. This observation indicates that the detection of NBI becomes difficult in frequency-selective channels, especially if the Signal to Interference Ratio (SIR) is high [18].

13.3.2 Frequency Excision Method

In this approach, an FFT based frequency-domain excision method is used to remove NBI [19]. The DFT output of each block of N_{FFT} samples, $r_{m,n}$ is given by:

$$r_{m,n} = \sum_{k=1}^{N_{FFT}} r_{m,k} e^{-j2\pi kn/N_{FFT}}, \quad k = 1, \ldots, N_{FFT} \tag{13.9}$$

where $r_{m,n}$ is the digital complex baseband signal at the demodulator output.

In the frequency domain, the NBI manifests itself as a peak in the spectra. By comparing the magnitude of each frequency bin to a threshold, and limiting those bins within the threshold, interferences can be excised. The effectiveness of the FFT based method depends on the selection of the threshold. The following method is used for the determination of the threshold. The mean value of the logarithm amplitude of the frequency bins and its variance are computed [19]:

$$T_{\text{mean}} = \sum_{n=1}^{N_{FFT}} \frac{10\log_{10}|r_{m,n}|}{N_{FFT}} \tag{13.10}$$

$$T_{\text{var}} = \frac{1}{N_{FFT}}\left[\sum_{n=1}^{N_{FFT}} (10\log_{10}|r_{m,n}|)^2 - \frac{1}{N_{FFT}}\left(\sum_{n=1}^{N_{FFT}} (10\log_{10}|r_{m,n}|)\right)^2\right]. \tag{13.11}$$

The threshold is determined according to the mean value and variance and is given by [19]:

$$T_{\text{excision}} = T_{\text{mean}} - \alpha T_{\text{var}}^{1/2}. \tag{13.12}$$

The scale factor α in the above equation is adjusted to maintain the threshold at some value of the noise floor. Each frequency bin is compared to the threshold and, if it exceeds the threshold, its value is held at the threshold. After applying the IFFT, the signal is much less contaminated with narrow band interferences.

13.3.3 Frequency Identification and Cancellation Method

The complex baseband signal at the input of the receiver can be expressed as [20]:

$$r(t) = s_t(t) * h(t) + n(t) + i(t), \tag{13.13}$$

where $r(t)$ is the complex baseband signal at the input of the receiver, $s(t)$ is the complex output signal of the transmitter, $h(t)$ is the complex impulse response of the channel, $n(t)$ is the complex Additive White Gaussian Noise (AWGN) and $i(t)$ is the complex single tone NBI. In order to approach the problem of the NBI identification, firstly the frequency position of the interference tone has to be estimated, via an FFT-based algorithm. After sampling the received signal

$r(t)$ to obtain $r(k)$, the FFT must be applied to this sequence via a Goertzel algorithm or using a butterfly lattice. It is important to note that appropriate setting of the sampling time T is a fundamental step; therefore, to increase the frequency resolution, an 8 times oversampling in a frequency domain is proposed.

Firstly, all power spectrum properties of all signals constituting $r(t)$ are known. In fact, the power spectrum of an OFDM is strictly defined. Furthermore, the narrow band interfering signal can be modeled as a complex sinusoidal tone [20]:

$$i_n(t) = A_n e^{j(\omega_n t + \varphi_n)} = a_n \cos(\omega_n t) + j b_n \sin(\omega_n t). \tag{13.14}$$

Secondly, it is clear that frequency domain processing is an appealing approach to estimate frequency, because the spectral properties of $r(t)$ are known.

The frequency cancelation method is implemented as a two-phase algorithm. At the beginning of the first phase, the complex NBI frequency is estimated by finding the maximum amplitude in the oversampled signal spectrum [20]:

$$\omega_n = \arg \omega \in L_\infty \max(P_R(\omega)). \tag{13.15}$$

After that, the amplitude and phase estimation is done. The received sampled signal $r(k)$ can be expressed as:

$$r(k) = s(k) + n(k) + in(k) \tag{13.16}$$

and the interference signal defined as:

$$i_n(k) = a_n \cos(\omega_n k/T) + j b_n \sin(\omega_n k/T). \tag{13.17}$$

A matrix form to represent the sampled interference can be used. In particular, it can be rewritten in matrix form as [20]:

$$[I]_n = [M].[X]. \tag{13.18}$$

The matrix $[M]$ is defined as:

$$[M] = \begin{bmatrix} \cos(\omega_n k_1/T) & j\sin(\omega_n k_1/T) \\ \cos(\omega_n k_2/T) & j\sin(\omega_n k_2/T) \\ \cdots\cdots & \cdots\cdots \\ \cos(\omega_n k_N/T) & j\sin(\omega_n k_N/T) \end{bmatrix}. \tag{13.19}$$

and vector $[X]$ gathers the coefficients a_n and b_n:

$$[X] = \begin{bmatrix} a_n \\ b_n \end{bmatrix}. \tag{13.20}$$

Applying the Maximum Likelihood (ML) algorithm, the solution is given by [22]:

$$[X] = ([M]^T[M])^{-1}[M]^T[R]. \tag{13.21}$$

Where the input signal vector $[R]$ is defined as:

$$[R] = \begin{bmatrix} r(k_1) \\ r(k_2) \\ \dots \\ r(k_N) \end{bmatrix}. \tag{13.22}$$

Thus the information about the complex amplitude of the NBI tone is derived.

The second phase of the algorithm uses the estimation results from the first phase as the initial conditions of the Normalized Least Squares (NLS) optimization procedure [20]:

$$f(\omega, A, \varphi) = \sum_{i=1}^{N} \left| r(t) - \sum_{m=1}^{M} A_m e^{j(\omega_m t + \varphi_m)} \right|^2. \tag{13.23}$$

Here ω_m, is the frequency of the m-th interfering tone and A_m and φ_m are the amplitude and phase respectively of the m-th interfering sinusoidal tone and M is the total number of interfering tones. The sum from 1 to N is referred to the N samples representing the sampled monocycle.

13.3.4 Nonlinear Filtering Methods

Nonlinear ACM Filter for Autoregressive Interference

The analysis begins by modeling the narrow band interference $\{i_k\}$ as a Gaussian autoregressive process of order p, i.e., assuming a model of the form [21, 22]:

$$i_k = \sum_{i=0}^{p} \phi_i i_{k-i} + e_k, \tag{13.24}$$

where $\{e_k\}$ is a white Gaussian process, and where the autoregressive parameters $\Phi_1, \Phi_2, \ldots, \Phi_p$ are known to the receiver.

Under this model, the received signal has a state space representation as follows [21]:

$$x_k = \Phi x_k - 1 + w_k, \tag{13.25}$$

$$z_k = H x_k + s_k + n_k, \tag{13.26}$$

where

$$x_k = [i_k \quad i_{k-1} \cdots i_{k-p+1}]^T, H_k = [1 \quad 0 \cdots 0]^T, w_k = [e_k \quad 0 \cdots 0]^T,$$

$$\Phi = \begin{bmatrix} \phi_1 & \phi_2 & \cdots & \phi_{p-1} & \phi_p \\ 1 & 0 & \cdots & 0 & 0 \\ \cdots & \cdots & \cdots & \cdots & \cdots \\ 0 & 0 & \cdots & 1 & 0 \end{bmatrix}.$$

The first component of the state vector x_k is the interference i_k. Hence, by estimating the state, an estimate of the interference which can be subtracted from the received signal to reject the interference, can be obtained.

In [23], Masreliez has developed an Approximate Conditional Mean (ACM) filter for estimating the state of a linear system with Gaussian state noise and non-Gaussian measurement noise. The nature of the nonlinearity is determined by the probability density of the observation noise. Under this assumption, the filtered estimate and its conditional covariance P_k are $\widehat{x}_k$ obtained recursively through the update equations:

$$P_k = M_k - M_k H^T G_k(z_k) H M_k, \tag{13.27}$$

$$\widehat{x}_k = \bar{x}_k + M_k H^T g_k(z_k), \tag{13.28}$$

$$M_{k+1} = \Phi P_k \Phi^T + Q_k, \tag{13.29}$$

$$\bar{x}_{k+1} = \Phi\, \widehat{x}_k, \tag{13.30}$$

where

$$g_k(z_k) = -\left[\frac{\partial p(z_k|Z^{k-1})}{\partial z_k}\right] \cdot [p(z_k|Z^{k-1})]^{-1},$$

$$G_k(z_k) = \frac{\partial g_k(z_k)}{\partial z_k}, \quad Q_k = E\{w_k w_k^T\}. \tag{13.31}$$

The ACM filter is thus seen to have a structure similar to that of the standard Kalman-Bucy filter. The time-update equations (13.28)–(13.29) are identical to those in the Kalman-Bucy filter. The measurement update of (13.31) involves correcting the predicted value by a nonlinear function of the prediction residual $z_k - Hx_k$. The nature of the nonlinearity is determined by the probability density of the observation noise.

Adaptive Nonlinear Filter Based on an LMS Algorithm

It has been shown in [24] that better interference rejection can be obtained by using a two-sided interpolation filter. For such a two-sided filter, shown in Figure 13.12, the following Widrow Least Mean Squared (LMS) algorithm equations hold [22], [25]:

$$X_k = [z_{k+N}, z_{k+N-1}, \ldots, z_{k+1}, z_{k-1}, \ldots, z_{k-N}]^T, \tag{13.32}$$

$$\theta_k = [a_{-N}(k), a_{-N+1}(k), \ldots, a_{-1}(k), a_1(k), \ldots, a_N(k)]^T, \tag{13.33}$$

where $2N + 1$ is the length of the data window.

The LMS algorithm can be normalized as follows [25]:

$$\theta_k = \theta_{k-1} + \frac{\mu_o}{r_k}\xi_k X_k, \quad r_k = r_{k-1} + \mu_o[|X_k|^2 - r_{k-1}]. \tag{13.34}$$

The same approach as (13.24)–(13.31) is used to modify the adaptive linear prediction filter:

$$\hat{z}_k = \sum_{i=0}^{L} a_i(k-1)[\hat{z}_{k-i} + \xi_{k-i}], \tag{13.35}$$

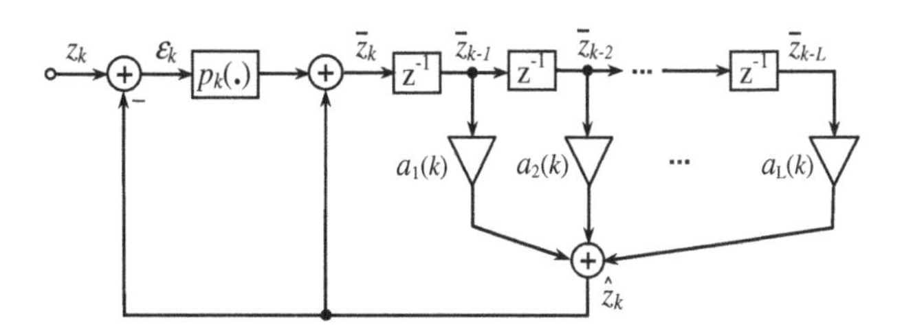

Fig. 13.12 Adaptive nonlinear filter based on LMS algorithm.

If the variance of the Gaussian random variable is σ_k^2 then the nonlinear transformation appearing in the ACM filter can be written as [25]:

$$\rho_k(\xi_k) = \xi_k - \tanh\left(\frac{\xi_k}{\sigma_k^2}\right). \tag{13.36}$$

By transforming the prediction error in (13.35) using the above nonlinearity, we get a nonlinear transversal filter for the prediction of z_k [25]:

$$\widehat{z}_k = \sum_{i=0}^{L} a_i(k-1)[\widehat{z}_{k-i} + \rho_{k-1}(\xi_{k-1})]. \tag{13.37}$$

In order to implement the filter of (13.37), an estimate of the parameter σ_k^2 and an algorithm for updating the tap weights must be obtained.

Nonlinear Gradient Algorithm

Finding an adaptation strategy that minimizes the squared prediction error is the subject of interest in this section. A popular approach to this problem is to use a gradient descent solution [26, 27]:

$$\theta_k = \theta_{k-1} + \mu\xi_k \frac{\partial z_k}{\partial \theta_{k-1}}. \tag{13.38}$$

For the nonlinear filter of (13.37), the LMS algorithm does not give a true gradient-based solution. Therefore, a modified gradient-based approach for (13.37) is considered [27].

Equation (13.37) can be rewritten as:

$$\widehat{z}_k = Y_k^T \theta_{k-1}, \tag{13.39}$$

where

$$Y_k = [\bar{z}_{k-1}, \bar{z}_{k-2}, \ldots, \bar{z}_{k-L}]^T, \quad \bar{z}_k = \widehat{z}_k + \rho_k(\xi_k) \tag{13.40}$$

Thus, the gradient algorithm (13.38) becomes, approximately [27]:

$$\theta_k = \theta_{k-1} + \mu\Psi_k\xi_k, \tag{13.41}$$

where

$$\Psi_k = Y_k + \sum_{i=0}^{L} a_i(k-1)x_{k-i}\Psi_{k-i}, \tag{13.42}$$

and

$$x_k = 1 - \frac{\partial \rho_k(\xi_k)}{\partial \xi_k} = \frac{1}{\sigma_k^2} \operatorname{sec} h^2 \left(\frac{\xi_k}{\sigma_k^2} \right). \tag{13.43}$$

From (13.42), it is seen that the vector Ψ_k is obtained by passing Y_k through an L-th order IIR filter whose coefficients are the current estimates obtained by the gradient algorithm [27]:

$$A(q^{-1}, k) = a_1(k)q^{-1} + a_2(k)q^{-2} + \cdots + a_L(k)q^{-L}, \tag{13.44}$$

At each stage, the stability of this polynomial must be checked and, if it is found to have any roots not within the unit circle, the parameter estimates must be projected onto that space where the polynomial is stable.

13.3.5 Complex Adaptive Narrowband Filtering

Compared with the desired wideband signal the NBI occupies a much narrower frequency band, but with a higher power spectral density. On the other hand, the wideband signal usually has autocorrelation properties quite similar to that of AWGN, so filtering in the frequency domain could be realized. The filtering is performed at the input of the OFDM demodulator. In order to do this, a complex variable filter section with independent tuning of the central frequency and the BW is used, which is then turned into adaptive to implement it in an OFDM receiver. A variable complex BP first-order realization named LS1 (Low Sensitivity) is used [28] — Figure 13.13.

The transfer functions (all of them are of BP type) of the LS1 section are [28]:

$$\begin{aligned} H_{RR}(z) &= H_{II}(z) = \hat{\beta} \frac{1 + 2\hat{\beta}\cos\theta z^{-1} + (2\hat{\beta} - 1)z^{-2}}{1 + 2(2\hat{\beta} - 1)\cos\theta z^{-1} + (2\hat{\beta} - 1)^2 z^{-2}}, \\ H_{RI}(z) &= -H_{IR}(z) = \hat{\beta} \frac{2(1 - \hat{\beta})\sin\theta z^{-1}}{1 + 2(2\hat{\beta} - 1)\cos\theta z^{-1} + (2\hat{\beta} - 1)^2 z^{-2}}, \end{aligned} \tag{13.45}$$

where the composed multiplier $\hat{\beta}$ is $\hat{\beta} = \beta + 2\beta(\beta - 1)$.

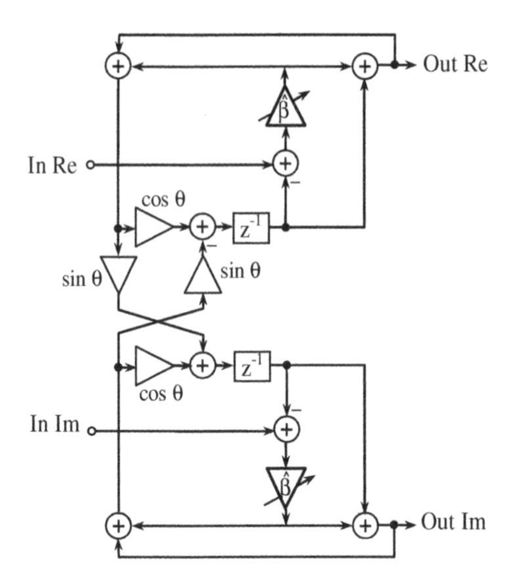

Fig. 13.13 Variable complex-coefficient first-order BP LS1 filter section.

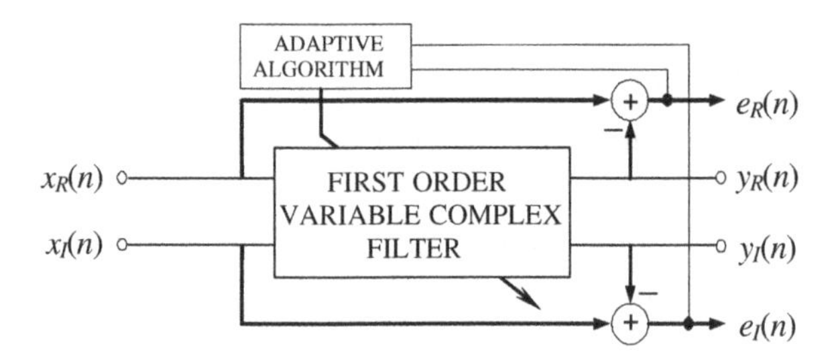

Fig. 13.14 Block-diagram of a BP/BS adaptive complex filter section.

The BW can be tuned by trimming the single coefficient β, whereas θ controls the central frequency ω_0.

This implementation has two very important advantages: firstly, an extremely low passband sensitivity increases the resistance to quantization effects; secondly, the central frequency and filter bandwidth can be independently controlled over a wide frequency range.

In Figure 13.14 the adaptive complex notch/bandpass narrowband system based on the LS1 variable complex filter section is shown.

In the following we consider the input/output relations for corresponding BP/BS filters (13.46)–(13.53). For the BP filter we have the following real output:

$$y_R(n) = y_{R1}(n) + y_{R2}(n), \tag{13.46}$$

$$y_{R1}(n) = -2(2\beta - 1)\cos\theta(n)y_{R1}(n-1) - (2\beta - 1)^2 y_{R1}(n-2) + 2\beta x_R(n) + 4\beta^2\cos\theta(n)x_R(n-1) + 2\beta(2\beta - 1)x_R(n-2), \tag{13.47}$$

$$y_{R2}(n) = -2(2\beta - 1)\cos\theta(n)y_{R2}(n-1) - (2\beta - 1)^2 y_{R2}(n-2) - 4\beta(1-\beta)\sin\theta(n)x_I(n-1), \tag{13.48}$$

where y_R is the real output and x_R is the real input.

The imaginary output is given by the following equation:

$$y_I(n) = y_{I1}(n) + y_{I2}(n), \tag{13.49}$$

$$y_{I1}(n) = -2(2\beta - 1)\cos\theta(n)y_{I1}(n-1) - (2\beta - 1)^2 y_{I1}(n-2) + 4\beta(1-\beta)\sin\theta(n)x_R(n-1), \tag{13.50}$$

and

$$y_{I2}(n) = -2(2\beta - 1)\cos\theta(n)y_{I2}(n-1) - (2\beta - 1)^2 y_{I2}(n-2) + 2\beta x_I(n) + 4\beta^2\cos\theta(n)x_I(n-1) + 2\beta(2\beta - 1)x_I(n-2), \tag{13.51}$$

where y_I is the imaginary output and x_I is the imaginary input.

For the bandstop (BS) filter we have a real output:

$$e_R(n) = x_R(n) - y_R(n), \tag{13.52}$$

and imaginary output:

$$e_I(n) = x_I(n) - y_I(n). \tag{13.53}$$

The cost function is the power of the BS filter output signal:

$$[e(n)e^*(n)], \tag{13.54}$$

where

$$e(n) = e_R(n) + je_I(n). \tag{13.55}$$

We apply the LMS algorithm to update the filter coefficient responsible for the central frequency as follows:

$$\theta(n+1) = \theta(n) + \mu Re[e(n)y'^{*}(n)]. \tag{13.56}$$

where μ is the step size controlling the speed of convergence, (*) denotes complex-conjugate, $y'(n)$ is the derivative of $y(n) = y_R(n) + jy_I(n)$ with respect to the coefficient subject of adaptation.

The real part of $y'(n)$ is:

$$\begin{aligned} y'_R(n) &= 2(2\beta - 1)\sin\theta(n)y_{R1}(n-1) - 4\beta^2\sin\theta(n)x_R(n-1) \\ &\quad + 2(2\beta - 1)\sin\theta(n)y_{R2}(n-1) - 4\beta(1-\beta)\cos\theta(n)x_I(n-1). \end{aligned} \tag{13.57}$$

The imaginary part of $y'(n)$ is:

$$\begin{aligned} y'_I(n) &= 2(2\beta - 1)\sin\theta(n)y_{I1}(n-1) + 4\beta(1-\beta)\cos\theta(n)x_R(n-1) \\ &\quad + 2(2\beta - 1)\sin\theta(n)y_{I2}(n-1) - 4\beta^2\sin\theta(n)x_I(n-1). \end{aligned} \tag{13.58}$$

In order to ensure the stability of the adaptive algorithm the range of the step size μ should be set according to [29]:

$$0 < \mu < \frac{K}{L\sigma^2}. \tag{13.59}$$

In this case L is the filter order, σ^2 is the power of the signal $y'(n)$ and K is a constant depending on the statistical characteristics of the input signal. In most practical situations, K is approximately equal to 0.1.

13.3.6 Comparison of Narrowband Interference Suppression Methods for OFDM Systems

13.3.6.1 Simulation Model

In order to evaluate the performance of the NBI suppression methods. simulations in relation to the baseband are conducted, assuming a standard OFDM receiver. The channel encoder is implemented as a convolutional encoder. In the simulation, the code rate $R_C = 1/2$ is selected. In the receiver, a Viterbi hard threshold convolutional decoder is used. The simulation employs a block interleaver-deinterleaver which randomly selects a permutation table, using the initial state input which is provided.

The digital modulator is implemented as a 256-point IFFT. The OFDM symbol consists of 128 data bins and 2 pilot tones. Each piece of OFDM data can use different modulation formats. In the experiments Grey encoded 64-QAM modulation format is used. After the IFFT process, the prefix and suffix guard intervals are added.

The output signal s(t) in the transmitter is a complex OFDM symbol starting at time $t = t_s = kT_s$:

$$s_k(t) = \begin{cases} w(t-t_s)\sum_{i=N_s/2}^{N_s/2-1} di + N_s(k+1/2)e^{j2\pi\left(f_c - \frac{i+0.5}{T_s}\right)(t-t_s-T_{prefix})}, \\ \qquad t_s \leq t < t_s + T_s(1+\beta) \\ 0, \qquad t < t_s \wedge t > t_s + T_s(1+\beta) \end{cases} \tag{13.60}$$

In order to minimize the spectrum leakage and limit the frequency bandwidth, windowing $w(t)$ is applied to the individual OFDM symbols. A commonly used window type is Raised Cosine Window, defined as:

$$w(t) = \begin{cases} 0.5 + 0.5\cos\left(\pi + \dfrac{t\pi}{\beta T_s}\right), & 0 \leq t \leq \beta T_s \\ 1, & \beta T_s \leq t \leq T_s \\ 0.5 + 0.5\cos\left((t-T_s)\pi + \dfrac{t\pi}{\beta T_s}\right), & T_s \leq t \leq (1+\beta)T_s \end{cases} \tag{13.61}$$

In the simulations a roll-off factor $\beta = 0.025$ is used.

For the wireless channel a multi-ray model with direct and delayed (reflected) components is used. The delayed components are subject to fading, while the direct one is not. In order to preserve total signal energy, the direct and delayed signal components are scaled by the square roots of $K/(K+1)$ and $1/(K+1)$, respectively. The delay τ is the difference between the propagation time of the delayed component and that of the direct one. To simplify simulations, a complex baseband representation of the system is used [62, 63]. Moreover, to keep simulation memory and computational loads to a minimum, it is desirable to sample at twice the modulation symbol rate. This requires delay τ to be a multiple k of the symbol period T_s. With $\tau = kT_s$, the discrete equivalence of the wireless channel simulation model, can be written:

$$r_i = y_i + n_i + i_i, \tag{13.62}$$

where r_i is the complex baseband signal at the receiver side, s_i is the transmitted symbol, y_i is the fading sample, n_i is the complex noise sample and i_i is the complex NBI sample.

The fading channel is represented by a FIR filter, where the subscript i indicates that the sample is taken at time $t = iT_s$ and with tap weights given by h_k:

$$y(i) = \sum_{j=0}^{N-1} s(i-j)h(j), \tag{13.63}$$

and

$$h(j) = \sum_{k} \mathrm{sinc}\left(\frac{\tau_k}{T} - j\right) g_k. \tag{13.64}$$

Here, N is the number of major paths, $\{\tau_k\}$ is the set of path delays, T is the input sample period, $\{g_k\}$ is the set of complex path gains, which are not correlated with each other. To generate a particular path gain g_k, the model performs the following steps. First, the AWGN is generated. Then the AWGN is passed through a filter, whose transfer function corresponds to the Jakes Doppler spectrum, and the output values are interpolated so that the sample period is consistent with that of the signal. Finally, the filter is adjusted accordingly to obtain the correct average path gain.

The excision method is applied to the OFDM signal with an NBI at the input of the demodulator. The signal is converted into the frequency domain by FFT and the noise peaks in the spectra of the signal are limited to the determined threshold. After this, the signal is converted back in the time domain and applied to the input of the demodulator. It should be noted that, for more precise frequency excision, FFT of higher order than the one in the demodulator is applied.

For the realization of the suppression method, the adaptive notch filter is connected at the receiver's input. An adaptation algorithm tunes the filter in such a way that its central frequency and bandwidth match that of the NBI signal spectrum. In the simulations, the central frequency of the notch filter is chosen in such a way that it is equal to the NBI central frequency, while its bandwidth is equal to 20% of the bandwidth between two adjacent OFDM subcarriers.

In the OFDM demodulator the guard prefix and suffix intervals are removed and 256-point FFT is applied. The pilot tones are removed and a respective channel equalization of the OFDM symbol is performed. Finally, corresponding 64-QAM demodulation is carried out.

13.3.6.2 Simulation Results

Using this general simulation model different experiments are performed, estimating the bit error ratio (BER) as a function of the Signal to Interference Ratio (SIR). The NBI is modeled as a single tone, the frequency of which is located in the middle of two adjacent OFDM subcarriers. Three types of channels are considered: AWGN, Rayleigh and Rician. The Rayleigh and Rician channels are subject to strong fading and additionally background AWGN is applied, so that the signal to AWGN ratio at the input of the OFDM receiver is 20 dB. In Figure 13.15, a standard Gaussian channel is considered. The SIR is varied from −20 dB to 10 dB. It can be seen that for high NBI, where the SIR is

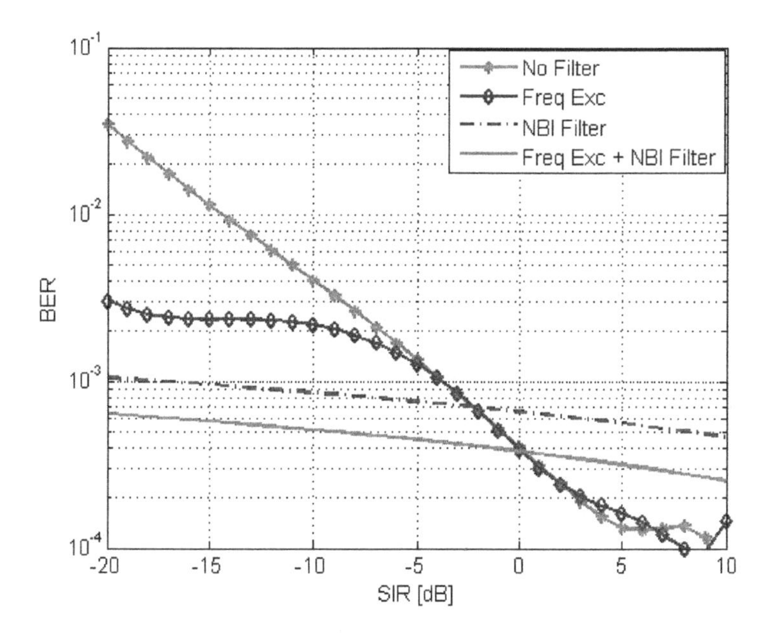

Fig. 13.15 BER as a function of SIR for AWGN channel.

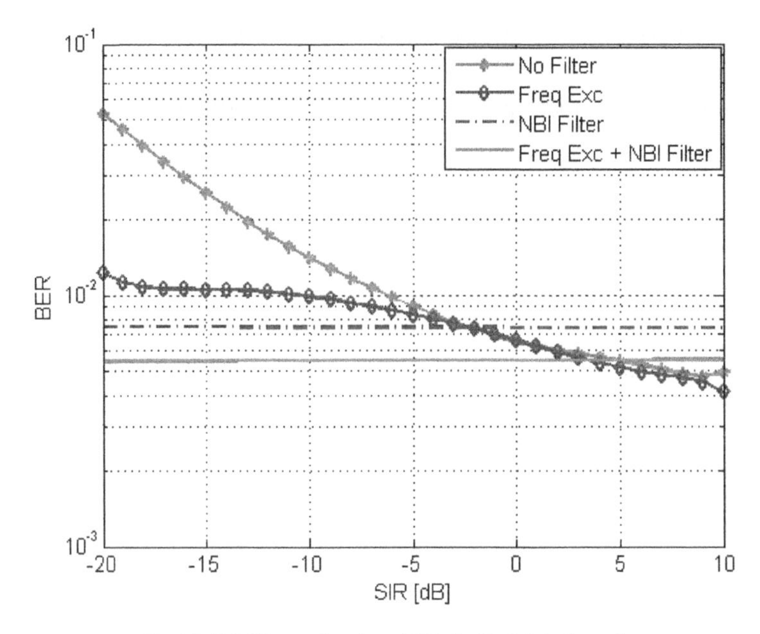

Fig. 13.16 BER as a function of SIR for Rayleigh channel.

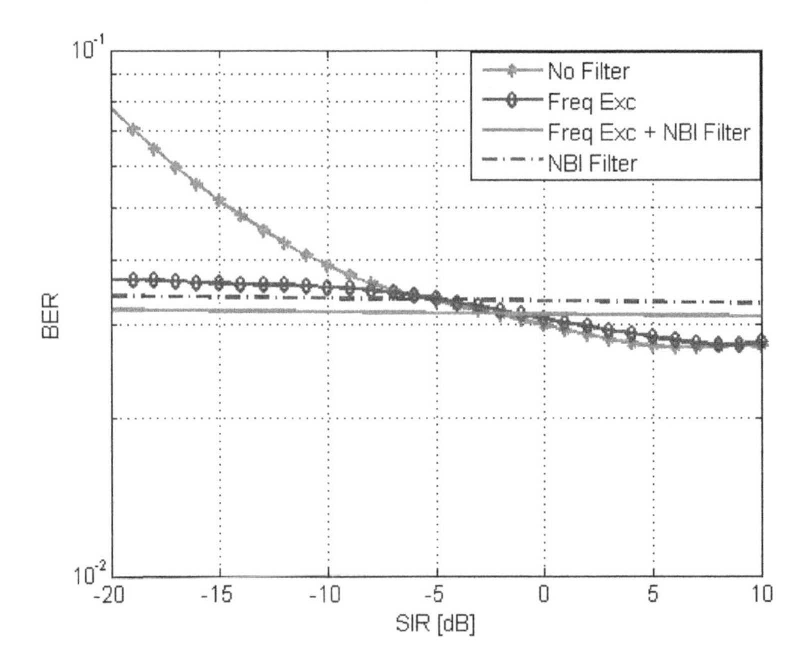

Fig. 13.17 BER as a function of SIR for Rician channel.

less than 0 dB, both suppression methods lead to a significant improvement in performance. The filtering scheme gives a better performance than the frequency excision method. This could be explained by the NBI spectral leakage effect caused by DFT demodulation at the OFDM receiver, when many subcarriers near the interference frequency suffer degradation. Thus, filtering out the NBI before demodulation is better than frequency excision. Additional improvement is attained by employing both methods together. In the case of the Rayleigh and Rician types of channels Figures 13.16 and 13.17, it could be seen that similar results are obtained. It should be noted that the filtering scheme leads to a degradation of the overall performance when SIR > 0, which is due to the amplitude and phase distortion of the filter. The degradation could be reduced by the implementation of a higher-order notch filter or avoided by simply switching off the filter, when SIR > 0.

13.4 Summary of Chapter

A brief overview of OFDM and OFDMA principles is presented in this chapter. Special attention is given to several frequently used NBI suppression methods. An example for NBI suppression in AWGN, Rayleigh and Rician channels is presented illustrating the benefits of the combined application of different NBI suppression approaches.

Problems

Problem 13.1

Describe the main differences between FDM and OFDM. List the advantages of OFDM compared to FDM.

Problem 13.2

Define the orthogonality of OFDM in the time domain and in the frequency domain.

Problem 13.3

List the major parts of an OFDM communication system. Describe the basic functions of each part.

Problem 13.4

Assume an OFDM communication system that uses FFT with size 1024. If the total length of the OFDM symbol in samples is 1280, calculate the length of the guard interval.

Problem 13.5

List the most important fading effects caused by multipath propagation.

Problem 13.6

Define the terms "flat fading channel", "frequency selective fading channel", "fast fading channel" and "slow fading channel.

Problem 13.7

Compare the basic OFDM multiple access schemes.

Problem 13.8

Assume that a strong NBI signal is presented within the same frequency band as a wideband OFDM signal. Explain the influence of NBI signal in the case of flat fading channel and in the case of frequency selective fading channel.

Problem 13.9

Explain the essence of the frequency excision method for NBI suppression. Define the choice of the scale factor α.

Problem 13.10

Explain the two phases of the frequency cancellation method. Define the role of oversampling.

Problem 13.11

Assume that a complex adaptive narrowband filter is used for NBI suppression. If the filter order is 6 and the normalized signal power is 0.5, define the range of the step size μ.

References

[1] R. van Nee, R. Prasad, OFDM for Wireless Multimedia Communications, Artech House, 2000.
[2] H. Shulze, C. Luders, Theory and Applications of OFDM and CDMA, Wiley, 2005.
[3] H. Lui, G. Li, OFDM-Based Broadband Wireless Networks, Wiley, 2005.
[4] S. Sampei, Applications of Digital Wireless Technologies to Global Wireless Communications, Prentice Hall, 1997.

[5] M. Jeruchim, P. Balaban, and K. Shanmugan, Simulation of Communication Systems, Kluwer Academic, 2000.

[6] W. Jakes, ed. Microwave Mobile Communications, IEEE Press, 1974.

[7] W. Lee, William, Mobile Communications Design Fundamentals, Wiley, 1993.

[8] W. Zou, Y. Wu, "COFDM: An Overview," IEEE Transactions on Broadcasting, 41(1), March 1995.

[9] M. Chryssomallis, "Simulation of Mobile Fading Channels," IEEE Antenna's and Propagation Magazine, 44(6), December 2002.

[10] R. Prasad, OFDM for Wireless Communications Systems, Artech House, 2004.

[11] A. Goldsmith, Wireless Communications. Cambridge University Press, 2005.

[12] P. Jung, P. Baier, and A. Steil. "Advantages of CDMA and spread spectrum techniques over FDMA and TDMA in cellular mobile radio applications. IEEE Trans. on Vehicular Technology, pp. 357–364, August 1993.

[13] S. Hara and R. Prasad. "Overview of multicarrier CDMA," IEEE Communications Magazine, Vol. 35, pp. 126–33, December 1997.

[14] X. Gui and T. S. Ng. Performance of asynchronous orthogonal multicarrier system in a frequency selective fading channel. IEEE Trans. on Communications, Vol. 47, pp. 1084–1091, July 1999.

[15] H. Liu and G. Li. OFDM — Based Broadband Wireless Networks. Design and Optimization. Wiley, 2005.

[16] WiMAX Forum website, MobileWiMAX — Part I: A Technical Overview and Performance Evaluation, Online: available http://www.wimaxforum.org.

[17] H. Sari et al., "An Analysis of Orthogonal Frequency-Division Multiple Access," in GLOBECOM, vol. 3, pp. 1635–1639, November 1997.

[18] A. Giorgetti, M. Chiani, M. Win, "The Effect of Narrowband Interference on Wideband Wireless Communication Systems," IEEE Trans. Communications, 53(12), pp. 2139–2149, December 2005.

[19] J.-C. Juang, C.-L. Chang, Y.-L. Tsai, "An Interference Mitigation Approach Against Pseudolites," 2004 International Symposium on GNSS/GPS, Sydney, Australia, 6–8 December 2004.

[20] E. Baccareli, M. Baggi, L. Tagilione, "A Novel Approach to In-Band Interference Mitigation in Ultra Wide Band Radio Systems," IEEE Conference on Ultra Wide Band Systems and Technologies, 2002.

[21] F. M. Hsu and A. A. Giordano, "Digital whitening techniques for improving spread-spectrum communications performance in the presence of narrow-band jamming and interference," IEEE Trans. on Communications, COM-26, pp. 209–216, February 1978.

[22] R. A. Iltis and L. B. Milstein, "An approximate statistical analysis of the Widrow LMS algorithm with application to narrow-band interference rejection," IEEE Trans. On Communications, COM-33, pp. 121–130, February 1985.

[23] C. J. Masreliez, "Approximate non-Gaussian filtering with linear state and observation relations," IEEE Trans. Automatic Control, AC–20, pp. 107–110, February 1975.

[24] S. Sampei, Applications of Digital Wireless Technologies to Global Wireless Communications, Prentice Hall, 1997.

[25] C. R. Johnson, "Adaptive IIR filtering: Current results and open issues," IEEE Trans. on Information Theory, IT-30, pp. 237–250, March 1984.

[26] B. Friedlander, "System identification techniques for adaptive signal processing," IEEE Trans. Acoustic, Speech, Signal Processing, ASSP-30, pp. 240–246, April 1982.

[27] R. Vijayanan, H. Vincent Poor, "Nonlinear Techniques for Interference Suppression in Spread-Spectrum Systems," IEEE Trans. on Communications, 38(1), July 1990.

[28] G. Iliev, Z. Nikolova, G. Stoyanov, K. Egiazarian, "Efficient Design of Adaptive Complex Narrowband IIR Filters," Proceedings of XII European Signal Processing Conference, EUSIPCO 2004, Vienna, Austria, pp. 1597–1600, 6–10 September 2004.

[29] S. Douglas, "Adaptive filtering," in Digital Signal Processing Handbook, D. Williams and V. Madisetti, Eds., CRC Press, pp. 451–619, 1999.

14

Synchronization

Ramin Vali, Stevan Berber and Sing Kiong Nguang

The University of Auckland, New Zealand

In this chapter we study the problem of synchronization between transmitter and receiver in the code division multiple access (CDMA) system. In this system it is important to acquire the synchronization of waveforms in both time and frequency. Lack of synchronization will result in complete system failure since insufficient received signal energy will reach the demodulator to detect the data [1, 2]. Acquiring and maintaining time synchronization is difficult because of various channel impediments and uncertainties [1–3]. Moreover, the synchronization should be achieved using minimum hardware, power and time. Broadly speaking synchronization in CDMA systems is achieved in two stages. The first stage is rough or coarse synchronization. This stage is called *acquisition* and involves reducing the time difference between the transmitter and receiver to within a chip duration. A chip is a single fraction of the CDMA spreading code that is each bit of the message is spread by a certain number of chips [4]. Once the acquisition has been achieved the second stage, *fine synchronization*, also known as *tracking* will begin. In this chapter we will focus on the most popular and widely used acquisition and tracking techniques for time synchronization [1, 3].

The rest of the chapter is organised as follows. First, a brief introduction to the classical CDMA system is given, then the acquisition phase of the synchronization block is explained in detail and its performance is analysed using different spreading codes. The next section presents the tracking phase analyses in detail followed by an explanation on the effect of sampling on signal tracking. The dependability of the system performance on synchronization is shown in the last section.

14.1 Introduction

14.1.1 Overview of a CDMA System

In this section, a very brief description of the analysed CDMA system will be presented. The aim from this system description is to put the synchronization into its proper context. Figure 14.1 will give insight into the workings of the CDMA system, note that the synchronization block is not shown in detail and will be dealt with separately.

The CDMA system that we are concerned with uses different sets of spreading codes ($x_t^{(g)}$) to modulate users' data ($\gamma_i^{(g)}$ where $g \in \{1, 2, 3, \ldots, N\}$). The result of spreading is then added together and sent across the communication channel after filtering and up conversion [4].

The channel impairments present are slow flat fading denoted by α_t, which is a random variable with a Rayleigh distribution and mode of b, which is chosen so that $E[\alpha_t^2] = 1$. Additive white Gaussian noise (AWGN) denoted by $\xi_t(t)$ is also present which has a zero mean and power spectral density of $N_0/2$.

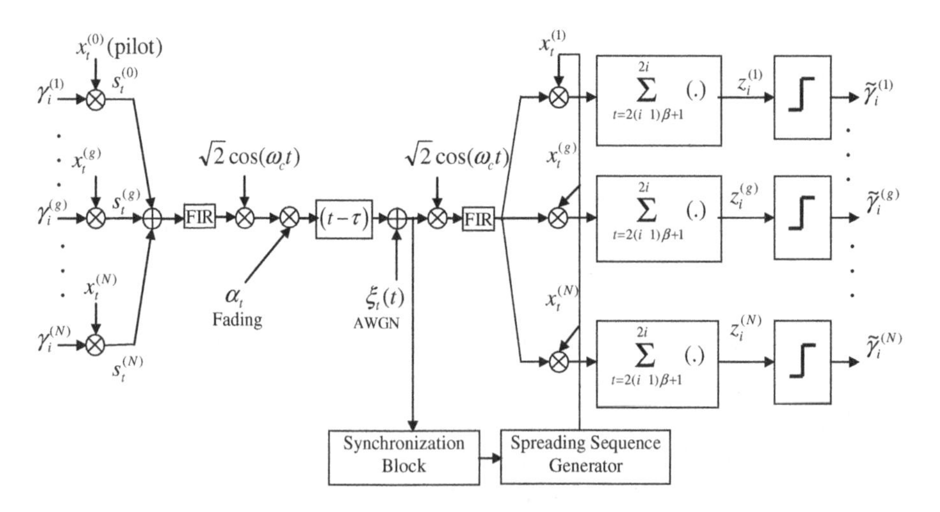

Fig. 14.1 A CDMA system complete with channel impairments and sequence generator.

The receiver has to de-spread each user's message correctly. In order to achieve this, the receiver has to have an exact copy of the spreading sequence used for each user in the transmitter side. Since the spreading sequences are *orthogonal* the receiver is able to de-spread each user's message [4]. This is done by multiplying the down converted received sequence with each of the spreading sequences and accumulating the results. These accumulated values are denoted by $z_i^{(g)}$ (where $g \in \{1, 2, 3, \ldots, N\}$). These values then go through the threshold detectors for each user, where the transmitted bits values are estimated. These estimations are denoted by $\tilde{\gamma}_i^{(g)}$ (where $g \in \{1, 2, 3, \ldots, N\}$).

The spreading sequences can be categorised into two broad groups. The first are binary spreading sequences which, as their name suggests, can assume only two fixed values. The classical CDMA systems use spreading sequences that are in this category such as *Walsh functions* and *Pseudo Random Binary Sequences* (PRBS) [4]. The second category are the non-binary sequences in which the sequences do not have any fixed value. Two types of non-binary sequences that have been investigated recently are chaotic sequences and noise-like sequences. In theory, chaotic and noise-like sequences have infinite periods and look like channel noise in time and frequency domain [5, 6]. This means that the signals will be hidden from interception by a large degree which gives more security to the systems that use these sequences [6, 7]. We will discuss this issue later in the chapter when we compare the binary and non-binary spreading sequences.

14.2 Acquisition

14.2.1 Theoretical Analysis

We mentioned that the receiver has to have an exact copy of the spreading sequence in order to de-spread the message correctly. However, looking at Figure 14.1, we see that there is a relative time delay of τ between the received and the locally generated sequences, caused by propagation delays, clock drift and other non-linear behaviour in the wireless channel. Any sequence misalignment causes the demodulator output to fall and increases the probability of bit error. In order to estimate this delay and compensate for it on the receiver spreading sequence generator, we need the acquisition phase. The acquisition phase estimates the relative time delay to within a chip duration (T_c).

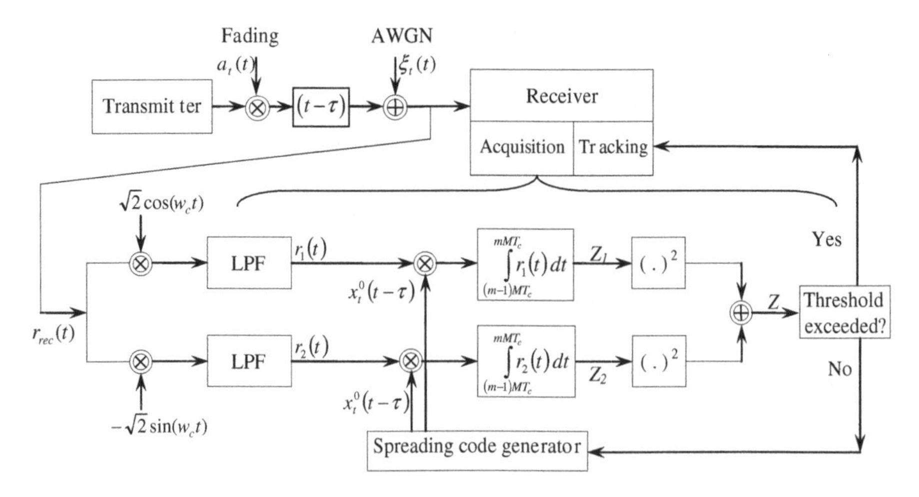

Fig. 14.2 Acquisition phase of the synchronization block.

Figure 14.2 shows the internal workings of the acquisition phase which we will go through in detail.

The presence of message impedes sequence synchronization, so a separate channel is used to transmit an unmodulated *pilot signal* [4]. The pilot signal has a fixed period and is repeated, which allows the system multiple attempts to gain synchronization. The period of the pilot sequence has no impact on the system, except the range of possible delays that can be synchronized [7]. Walsh or PRBS pilot sequence periods are typically a power of two. In a CDMA system, users begin transmission at the start of the pilot sequence. Therefore, once the pilot is synchronized, and the relative offset estimated, user sequences can be aligned in time. Non-coherent code acquisition must precede carrier synchronization as the signal energy is spread over a wide spectral band [1, 2]. Synchronization is attained by advancing the locally generated pilot one chip and computing its correlation with the received signal. The result is processed using an appropriate decision-rule and search strategy [1, 2]. This way of achieving synchronization is called the *serial search algorithm* which is one of the most commonly used CDMA synchronization techniques. The serial-search acquisition system attempts to align the pilot sequence by shifting the locally generated sequence in discrete steps. Other strategies, such as parallel or windowed serial searches, are generally more computationally expensive [1, 4]. We will now present the theoretical analysis of the serial search algorithm.

We consider the case were no fading is present to begin with. Looking at Figure 14.1 we can see that the received signal is

$$
\begin{aligned}
r_{rec}(t) &= \left(\sum_{g=0}^{N} \gamma_i^g x_t^g\right) \sqrt{2} A \cos(\omega_c t + \phi) + \xi_t(t) \\
&= \left(\sum_{g=0}^{N} \gamma_i^g x_t^g\right) \sqrt{2} A [\cos(\omega_c t)\cos(\phi) + \sin(\omega_c t)\sin(\phi)] \\
&\quad + \sqrt{2}\xi_t^I(t)\cos\omega_c t - \sqrt{2}\xi_t^Q(t)\sin\omega_c t,
\end{aligned}
\tag{14.1}
$$

where ϕ is the carrier phase difference between the transmitter and receiver, $\xi_t^I(t)$ is the in-phase and $\xi_t^Q(t)$is the quadrature noise component, ω_c is the carrier frequency and A is the pilot amplitude. Note that g has started from 0. This is because we have assumed the unmodulated pilot sequence to be x_t^0. Now that we have the received signal, we will focus on Figure 14.2.

The received signal is down converted so for the upper branch we have

$$
\begin{aligned}
r_1(t) &= \left(\sum_{g=0}^{N} \gamma_i^g x_t^g\right) A \cos(\phi) + \xi_t^I(t) \\
&= \gamma_t^0 x_t^0 A \cos(\phi) + \left(\sum_{g=1}^{N} \gamma_i^g x_t^g\right) A \cos(\phi) + \xi_t^I(t)
\end{aligned}
\tag{14.2}
$$

As the pilot is unmodulated so its data stream is always 1. Therefore we have

$$
r_1(t) = x_t^0(t) A \cos(\phi) + \xi_1(t) \tag{14.3}
$$

where the noise term ($\xi_1(t)$) includes both the AWGN and the inter-user interference.

Similarly we can find $r_2(t)$ to be

$$
r_2(t) = x_t^0(t) A \sin(\phi) + \xi_2(t). \tag{14.4}
$$

Now the pilot correlation process can begin. Both $r_1(t)$ and $r_2(t)$ will be correlated with the locally generated pilot sequence which has m-bits containing

M chips. This locally generated pilot is the time shifted version of the received signal. The correlation result for $r_1(t)$ is

$$\begin{aligned}
Z_1 &= \int_{t=(m-1)T_c}^{t=mT=mMT_c} r_1(t)x_t^0(t-\tau)dt \\
&= \int_{t=(m-1)T}^{t=mT} \left[Ax_t^0(t)\cos(\phi) + \xi_1(t)\right]x_t^0(t-\tau)dt \\
&= A\cos(\phi)\int_{t=(m-1)T}^{mT} x_t^0(t)x_t^0(t-\tau)dt + \int_{t=(m-1)T}^{mT} \xi_1(t)x_t^0(t-\tau)dt \\
&= A\cos(\phi)TR(\tau) + N_1, \qquad (14.5)
\end{aligned}$$

where $R(\tau)$ is the measure of correlation between the received and locally generated pilot sequences and T is the overall correlation period.

Both Z_1 and Z_2 branch outputs are squared and summed to form a decision variable which is compared with a certain correlation threshold. If this decision variable is greater than the threshold, synchronization is declared and the tracking phase will begin. Otherwise, the locally generated pilot sequence will be shifted in time by one chip duration (one time delay unit) and the correlation will be repeated.

Since $A = \sqrt{2E_c/T_c}$ and the number of chips in the interval T is $M = T/T_c$ we will have

$$Z_1 = T\sqrt{\frac{2E_c}{T_c}}\cos(\phi)R(\tau) + N_1 \qquad (14.6)$$

and similarly for Z_2 we will have

$$Z_2 = T\sqrt{\frac{2E_c}{T_c}}\sin(\phi)R(\tau) + N_2, \qquad (14.7)$$

where E_c is the energy per chip, N_1 and N_2 are zero-mean Gaussian random variables with the variance $N_0'T/2$, expressed in these forms $N_1 \Rightarrow G(0, N_0'T/2)$ and $N_2 \Rightarrow G(0, N_0'T/2)$.

Given the above we can rewrite Z_1 as

$$Z_1 \Rightarrow G\left(T\sqrt{\frac{2E_c}{T_c}}\cos(\phi)R(\tau),\ N_0'T/2\right)$$

$$= \sqrt{N_0' T/2} \cdot G\left(2\sqrt{\frac{TE_c}{T_c N_0'}}\cos(\phi)R(\tau),\, 1\right)$$

$$= \sqrt{N_0' T/2} \cdot G\left(2\sqrt{\frac{ME_c}{N_0'}}\cos(\phi)R(\tau),\, 1\right), \tag{14.8}$$

similarly

$$Z_2 \Rightarrow \sqrt{N_0' T/2} \cdot G\left(2\sqrt{\frac{ME_c}{N_0'}}\sin(\phi)R(\tau),\, 1\right). \tag{14.9}$$

Now we are at a position to formally define the decision variable which is compared with the present threshold for every correlation result and determines the synchronization status. The decision variable is $Z = Z_1^2 + Z_2^2$. This variable is a non-central chi-squared distribution with two degrees of freedom, multiplied by a factor of $\sigma^2 = N_0' T/2$. The non-centrality parameter is

$$\lambda = \left[2\sqrt{\frac{ME_c}{N_0'}}\sin(\phi)R(\tau)\right]^2 + \left[2\sqrt{\frac{ME_c}{N_0'}}\cos(\phi)R(\tau)\right]^2$$

$$= 4\frac{ME_c}{N_0'}R^2(\tau) = 4MR^2(\tau) \cdot \frac{E_c}{N_0'}. \tag{14.10}$$

Given the above, we can write down the probability density function (PDF) for Z as

$$p_Z(z) = \begin{cases} \frac{1}{2\sigma^2} e^{-\frac{1}{2}(\lambda + z/\sigma^2)} I_0\left(\sqrt{\frac{\lambda z}{\sigma^2}}\right) & z > 0 \\ 0 & \text{otherwise} \end{cases}. \tag{14.11}$$

Now that we have the PDF of Z we can evaluate the performance of the acquisition phase. The decision about the synchronization being achieved entails two hypothesis tests. The first one (H_1) is that the locally generated sequences are synchronized to the received sequences to within one chip duration. The second hypothesis (H_0) is that the received signals are not synchronized with the locally generated ones.

These two hypotheses can be formally expressed as

$$H_0 : |\tau| > T_c \Rightarrow R(\tau) \cong 0, \quad N_0' > N_0 \tag{14.12}$$

and

$$H_1 : |\tau| \leq T_c \Rightarrow R(\tau) > 0, \quad N_0' \cong N_0. \tag{14.13}$$

We can now find the conditional PDF of Z when it is modified based on each of the hypotheses, that is

$$p_Z(z|H_0) = \frac{1}{2\sigma^2} e^{-\frac{1}{2}(z/\sigma^2)} \tag{14.14}$$

and

$$p_Z(z|H_1) = \frac{1}{2\sigma^2} e^{-\frac{1}{2}(\lambda + z/\sigma^2)} I_0\left(\sqrt{\frac{\lambda z}{\sigma^2}}\right). \tag{14.15}$$

Note that the non-centrality parameter (λ) is zero for the first hypothesis and different from zero for the second.

Now we can define two probabilities which are going to be used regularly from now on. The first one is the *probability of failure* or *false alarm*, p_F, which is the probability that the sequences are not synchronized but the value of Z is higher than the threshold. This corresponds to declaring the achievement of synchronization where in fact synchronization has not been achieved. It is clear that the value of p_F has to be minimised. The second important probability is the *probability of detection*, p_D, which is the probability of Z value being larger than the threshold when in fact the sequences are synchronized. This corresponds to the probability of successful synchronization.

The single attempt, when only one correlation period is used for correlation i.e., ($m = 1$), probability of false alarm is

$$p_F(m = 1) = p(Z > z_T|H_0) = \int_{z_T}^{\infty} \frac{1}{2\sigma^2} e^{-\frac{1}{2}(z/\sigma^2)} dz = e^{-\frac{1}{2}(z_T/\sigma^2)}. \tag{14.16}$$

From this we can extract the threshold value denoted by z_T as

$$z_T = -2\sigma^2 \ln p_F(m = 1) = -N_0' T \ln p_F(m = 1). \tag{14.17}$$

For this value of threshold the single attempt probability of detection is

$$p_D(m = 1) = p(Z > z_T|H_1) = \int_{z_T}^{\infty} \frac{1}{2\sigma^2} e^{-\frac{1}{2}(\lambda + z/\sigma^2)} I_0\left(\sqrt{\frac{\lambda z}{\sigma^2}}\right) dz. \tag{14.18}$$

Table 14.1. Assumptions regarding the two hypotheses.

H_0	$R(\tau) \cong 0$	$N_0 < N_0'$
H_1	$R(\tau) \cong 1$	$N_0 \cong N_0'$

Performing the transform $z = x^2\sigma^2$ we will get

$$p_D(m=1) = p(Z > z_T|H_1) = \int_{z_T}^{\infty} \frac{1}{2\sigma^2} e^{-\frac{1}{2}(\lambda + z/\sigma^2)} I_0\left(\sqrt{\frac{\lambda z}{\sigma^2}}\right) dz$$

$$= \int_{\sqrt{z_T/\sigma^2}}^{\infty} x \cdot e^{-\frac{1}{2}(\lambda + x^2)} I_0(\sqrt{\lambda} x) dx. \quad (14.19)$$

Note that the above equation is the Marcum's Q-function, so

$$p_D(m=1) = Q_M(\sqrt{\lambda}, \sqrt{z_T/\sigma^2}). \quad (14.20)$$

Now we state our assumptions about the correlation values for the two hypotheses in the following table

Based on the assumptions for H_1 we can write λ as

$$\lambda = 4MR^2(\tau) \cdot \frac{E_c}{N_0}. \quad (14.21)$$

Therefore, inserting the above expression and also (1.17) into (1.20), we will have

$$p_D(m=1) = Q_M\left(2R(\tau)\sqrt{M \cdot \frac{E_c}{N_0}}, \sqrt{-2\frac{N_0'}{N_0} \ln p_F(m=1)}\right). \quad (14.22)$$

The upper bound on the probability of detection is found when synchronization is achieved, thus

$$p_D(m=1) \leq Q_M\left(2\sqrt{M \cdot \frac{E_c}{N_0}}, \sqrt{-2 \ln p_F(m=1)}\right)$$

$$\approx Q\left(\sqrt{-2 \ln p_F(m=1)} - 2\sqrt{M \cdot \frac{E_c}{N_0}}\right), \quad (14.23)$$

where Q is a Gaussian Q-function that accurately approximates the Marcum's Q-function Q_M. We can plot the Q_M function with M as a parameter. That

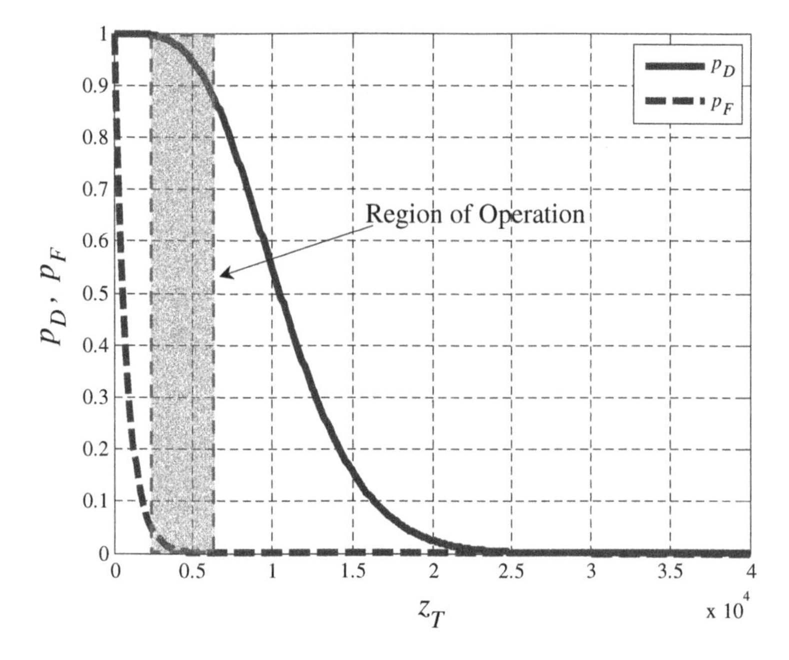

Fig. 14.3 Theoretical probabilities of detection and false alarm based on the threshold value.

means we can determine the number of reference sequence chips (M) needed to synchronize the receiver with a desired probability of detection and failure [4].

If p_D and p_F are to be plotted based on the value of z_T, we will have Figure 14.3. It is clear that the probability of false alarm should be minimized whilst the probability of detection is kept to its possible maximum. We therefore can select the region of operation as the dashed box on Figure 14.3.

If p_D and p_F are to be plotted against each other, then we will get the *receiver operating characteristics* or ROC for short. ROC plots are important because they are a very good way of comparing the performance of the acquisition phase for different signal to noise ratios. Figure 14.4 shows the ROC for our theoretical acquisition phase under different values of signal to noise ratio.

Note that a lower probability of detection is obtained for a fixed probability of failure as the signal to noise ratio decreases. Ideally, a curve on the upper left hand corner (where p_D is maximum and p_F is minimum) is desirable. This concludes our theoretical analysis of the acquisition phase.

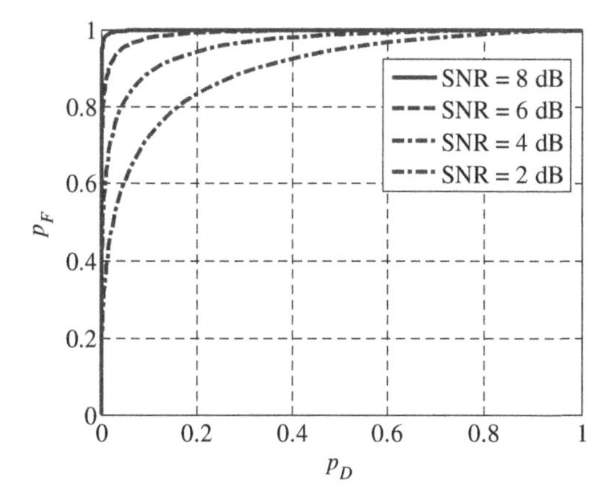

Fig. 14.4 ROC for different values of signal to noise ratio.

14.2.2 CDMA Acquisition Using Non-binary Sequences

In the beginning of this chapter, non-binary sequences were introduced as sequences which can have any value in the defined interval of possible values [6]. In this chapter we present results obtained from modelling CDMA systems that use two types of non-binary spreading codes, namely chaotic spreading codes and noise-like spreading codes.

We begin by a short description of chaotic sequences and presenting their merits as spreading codes for CDMA systems. We show how chaotic modulation changes the time and frequency domain representation of the message signal completely, hence adding an extra physical layer of security. Chaotic signals are non-periodic, bounded, deterministic signals that appear to be random-like in the time domain. They are derived from non-linear dynamical systems [8, 9]. These dynamical systems can be mapped into discrete iterative functions known as chaotic maps. In this chapter we use chaotic sequences derived from the *logistic map*. The logistic map is an iterative function where

$$x_{n+1} = 2(x_n)^2 - 1 \tag{14.24}$$

The chaotic sequences generated by the logistic map are bound between -1 and 1 [6].

An important characteristic of chaotic sequences is their hyper-sensitivity on their initial conditions. The chaotic sequences generated by slightly

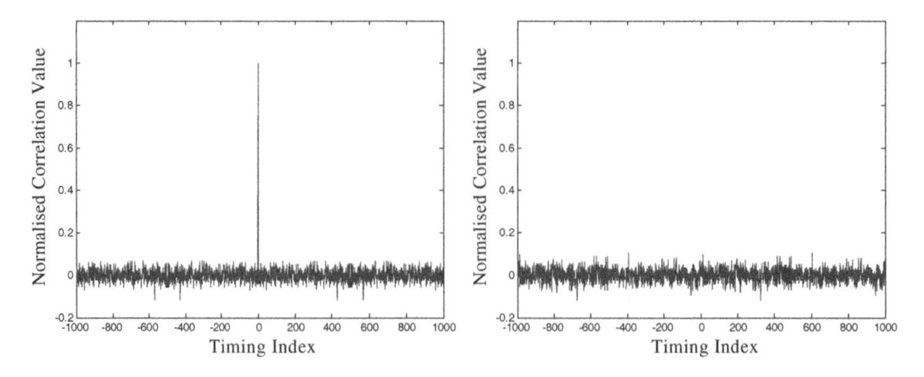

Fig. 14.5 (a) Auto-correlation of a chaotic sequence and (b) cross-correlation of two chaotic sequences with an initial condition difference of 0.0001.

different initial conditions are *orthogonal* as shown by the two figures below [6].

As can been seen, the auto-correlation function is impulse like and the cross-correlation function is nearly zero. By assigning each user to one of these orthogonal sequences, which can be easily obtained by chaotic maps, we can spread each user's message in the frequency domain, add it to the other users' messages and transmit it. Using the same orthogonal spreading codes the receiver can, knowing each user's unique code, de-spread the message for each user [6, 7, 9]. The normalised PDF of chaotic signals drawn from the logistic map is given in Figure 14.6. As can be seen from this figure, the chaotic values are bound between 1 and -1 with the highest probability belonging to these two numbers.

Now that chaotic sequences are introduced, we can present the results obtained by modelling an acquisition phase of a CDMA system using chaotic signals. In order to support the theoretical findings and simulation results, this section will also include the implementation results of the acquisition phase on a digital signal processor (DSP).

We begin by presenting the simulated receiver operation characteristics (ROC) curves for a CDMA system that uses chaotic spreading sequences. We change the pilot sequences correlation period (the M value in the theoretical analysis), and observe the effect on the ROC plot in Figure 14.7. Clearly, the reduction of correlation period M will result in a lower auto-correlation peak, which is more susceptible to be missed in the noise (8 dB in this case).

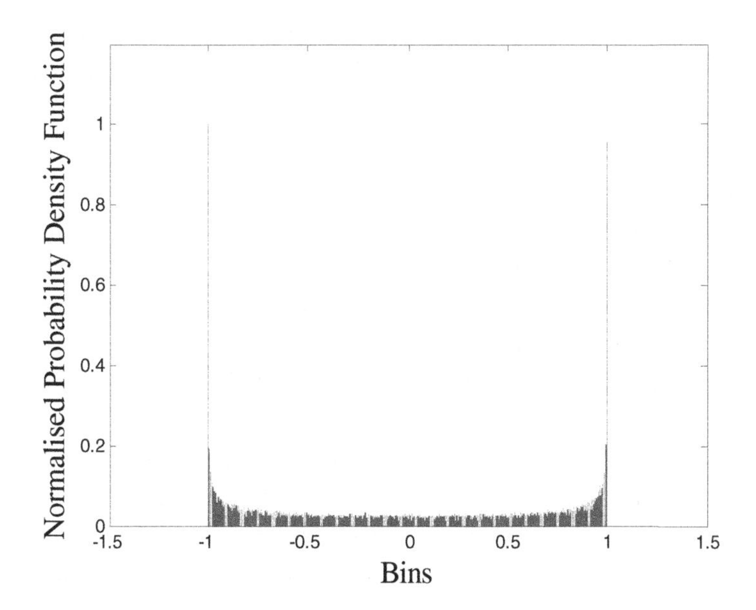

Fig. 14.6 PDF of chaotic values drawn from the logistic map.

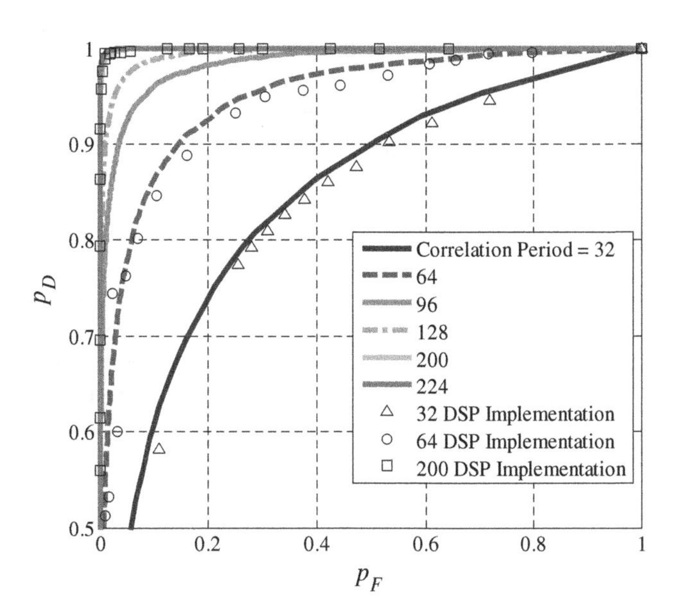

Fig. 14.7 ROC plots of simulated and implemented chaotic sequence acquisition at 8 dB.

As can be seen from the Figure 14.7, the DSP results follow the simulation results closely. Another factor which is important in using orthogonal spreading codes is the number of users that can be accommodated with the codes. Once the modulated messages of the users are added to the pilot, they act as interference. The larger the amount of interference, the worse performance for the chaotic pilot becomes.

The effect of the user interference on the chaotic sequence acquisition is shown in Figure 14.8. As can be seen from this figure, the performance worsens for an increased number of users. This is expected because the autocorrelation peak of the pilot signal will be affected by the increased number of users.

The results shown in Figures 14.9 and 14.10 are important and informative when we consider that the synchronization task has to be performed when the system is handling several users. Both figures are acquired using chaotic sequences. Figure 14.9 shows the trend of the z_T when system attempts to

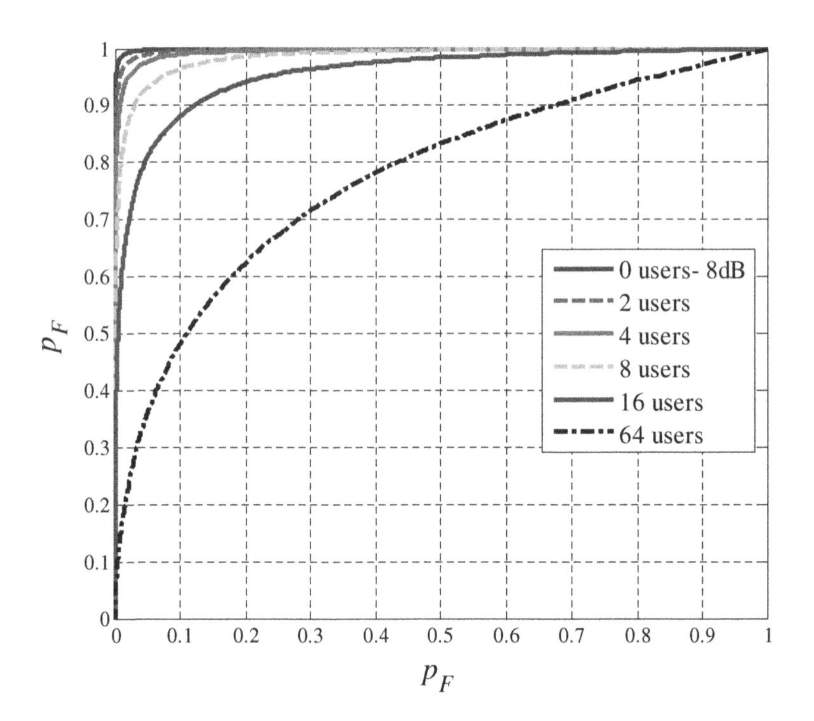

Fig. 14.8 ROC plots of simulated chaotic sequence acquisition at 8 dB for different user numbers.

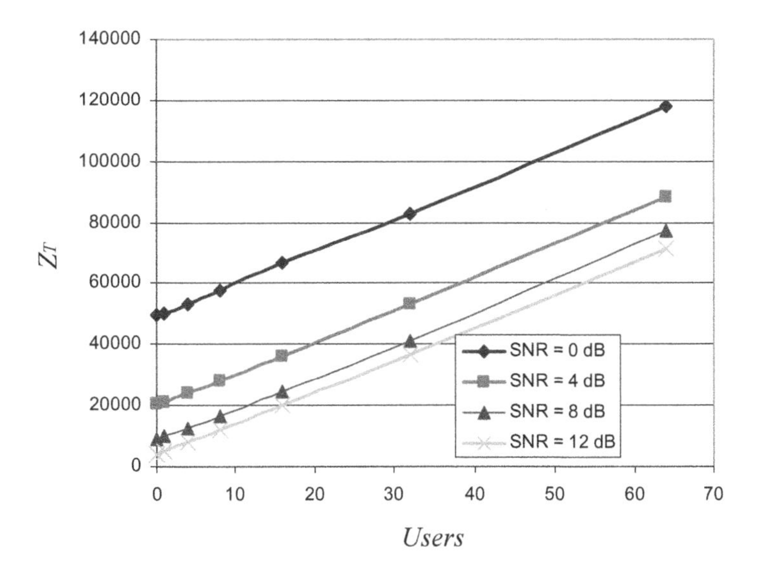

Fig. 14.9 Threshold value for different number of users at different noise levels.

synchronize with existing users. As can be clearly seen, for a fixed probability of false alarm ($p_F = 0.01\%$), the threshold value rises linearly as the number of users increase. The lines have the same slope for different signal to noise ratios.

Figure 14.10 shows the trend of the probability of detection when the number of users increases. As can be seen, there is an exponential-like drop in the probability of detection when the system is attempting to acquire the pilot signal with inter-user interference present. Again, in higher signal to noise ratios the probability of detection is high when there are few users, but in lower signal to noise ratios, the probability of acquiring the signal is not high even when the amount of inter user interference is not high.

Noise-like sequences are the second type of non-binary spreading sequences we will discuss in this chapter. Their orthogonality properties are similar to chaotic sequences i.e., their cross-correlation is close to zero. Unlike the chaotic sequences we have already discussed, the noise-like sequences have no bound [5]. These spreading sequences are designed to look like the channel noise with one difference and that is their predictability. Like AWGN present in the channel, noise-like sequences used in this chapter, are wideband and have a Gaussian PDF.

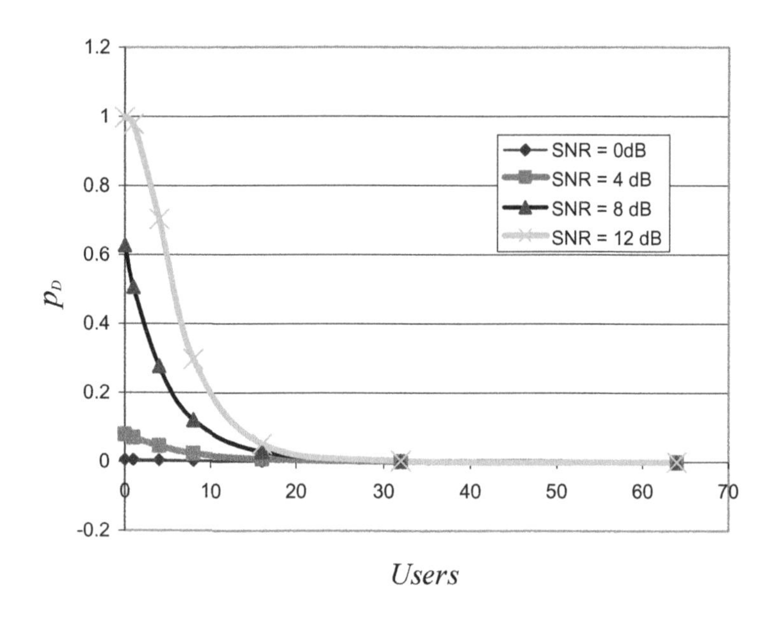

Fig. 14.10 Probability of detection for different number of users.

The Box-Muller method is one way to generate noise-like sequences in DSP technology. In this method two independent uniformly distributed random numbers, designated x_1 and x_2 are produced using linear feed back shift registers. In the next phase, the Gaussian distributed random variable is acquired using the equation [10]

$$x = (\sqrt{-\ln(x_1)} \times \sqrt{2}\cos(2\pi x_2)). \tag{14.25}$$

Figure 14.11 shows the PDF of the noise-like sequence used as the pilot for acquisition that is calculated using a sample of observations generated by a Gaussian random number generator.

Now we present the results obtained when the CDMA system and acquisition phase use noise-like sequences. We first look at the ROC graphs for the acquisition phase when the number of users changes (Figure 14.12). We can clearly see that more users reduce the probability of detection for a certain probability of false alarm. This is expected and is attributed to the increase in inter user interference.

In Figure 14.13, we have fixed the number of users to look at the acquisition performance when the signal to noise ratio changes. As expected, the

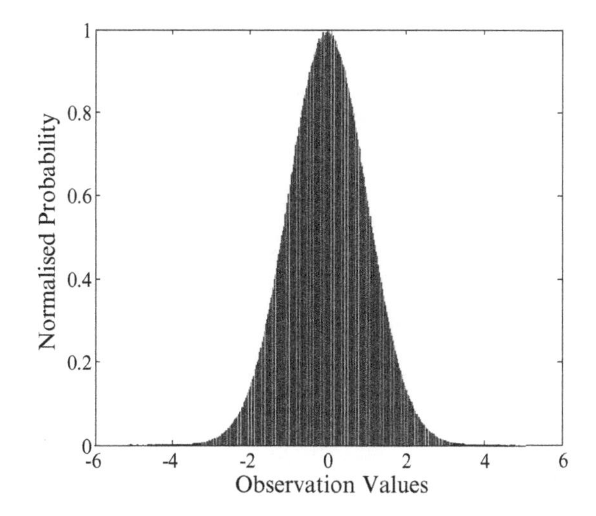

Fig. 14.11 The PDF of noise-like sequences used for acquisition.

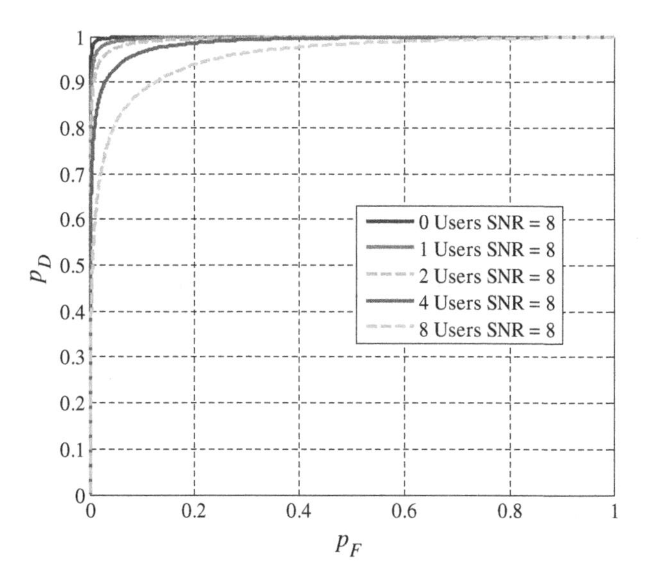

Fig. 14.12 ROC plots of simulated noise-like sequence acquisition at 8 dB for different user numbers.

system performance worsens. Figures 14.14 and 14.15, present the multiple user results that we have seen before for chaotic sequences.

Now we are at a position to compare the performance of chaotic sequences with the noise-like sequences and classical PRBS. As can be seen from

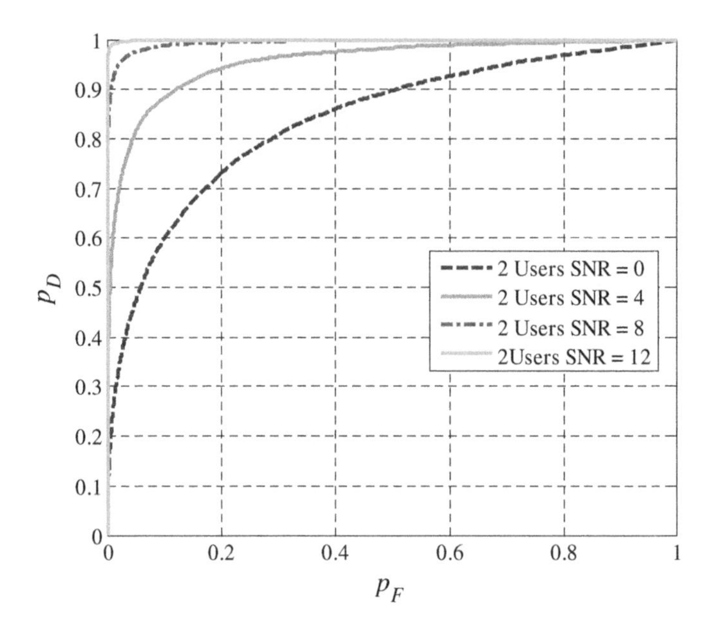

Fig. 14.13 ROC plots of simulated noise-like sequence acquisition for 2 users for different SNR.

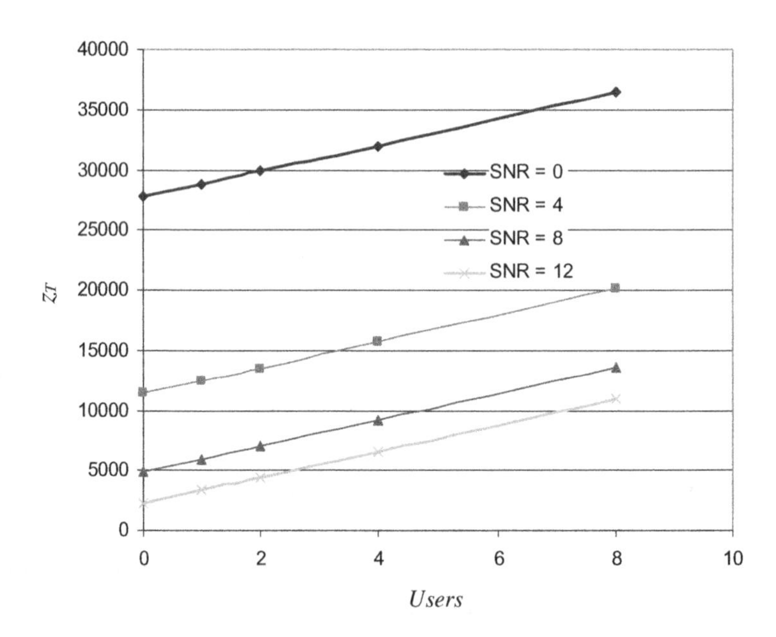

Fig. 14.14 Threshold value for different number of users at different noise levels.

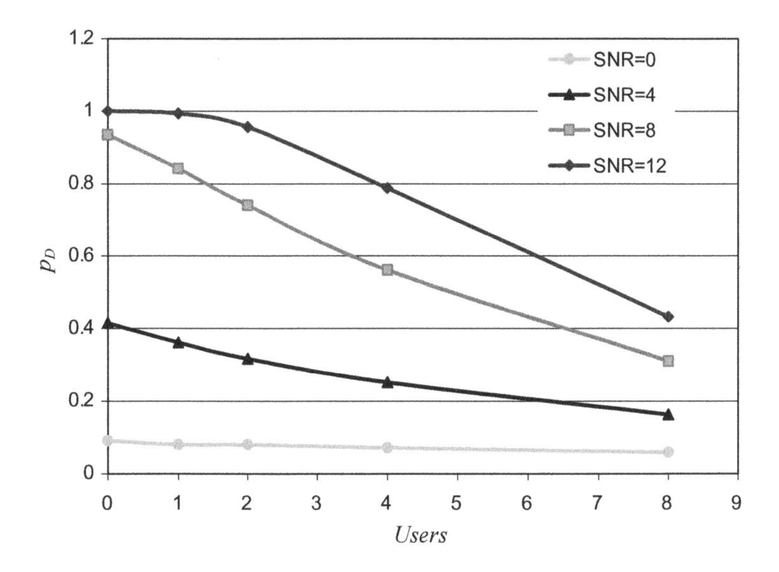

Fig. 14.15 Probability of detection for different number of users.

Figure 14.16, the performance of chaotic sequences is very close to the classical systems for various numbers of users. Noise-like sequences however, show a much lower probability of detection for a fixed probability of failure. This is attributed to the unbound nature of the noise-like sequences and well as their correlation properties.

The difference in performance is more evident in Figure 14.17, where the performance of chaotic and noise-like sequences is shown for two different signal to noise ratios and with 8 users.

14.2.3 Acquisition in Presence of Fading

At the beginning of this chapter we presented a system model that included a channel with the presence of fading. Fading is a phenomenon experienced when various reflected copies of the transmitted signal will get to the receiver with different time delays, which when added together and taken statistically will construct a signal power envelope of Rayleigh distribution [4]. In this chapter we examine the effects of slow, frequency flat Rayleigh fading. We therefore produce a set of Rayleigh distributed random samples and multiply it with the chips passing through the channel. For this, we have used the Jakes method as described in [10], which begins by generating two independent

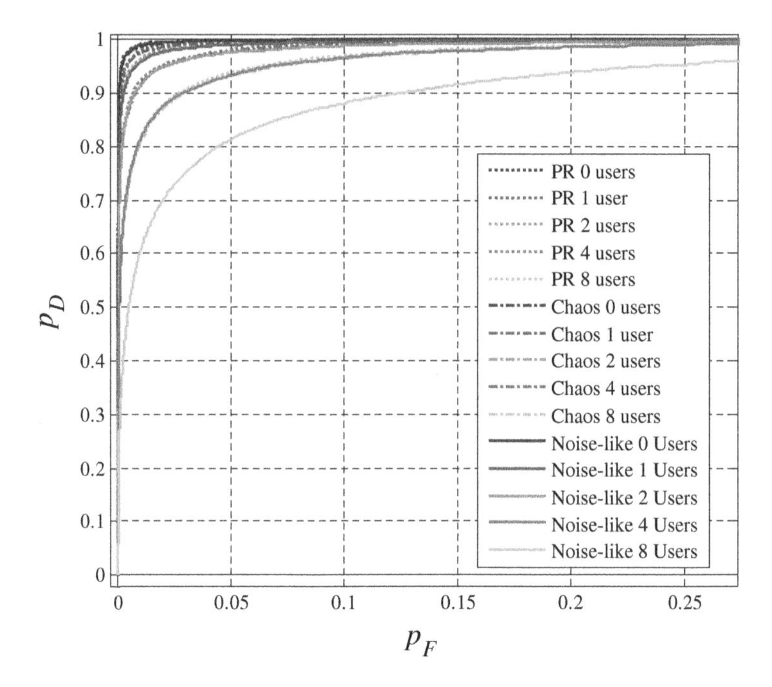

Fig. 14.16 ROC plots for comparison between chaotic, PR and noise-like sequences at 8 dB.

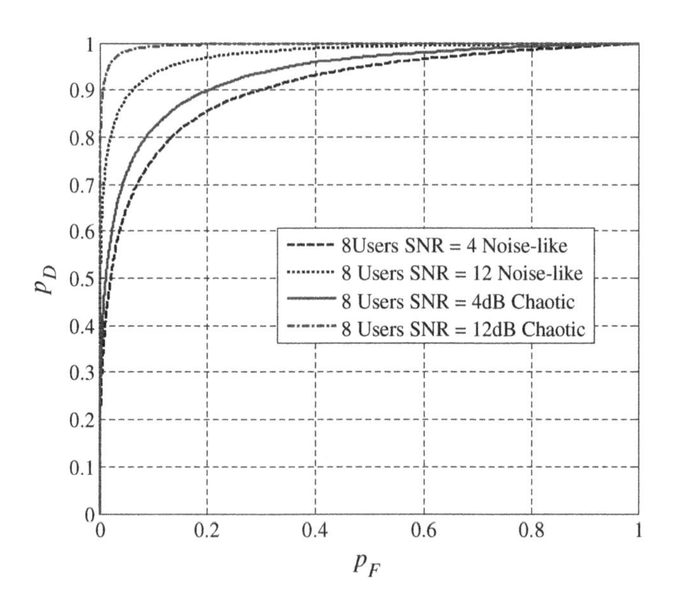

Fig. 14.17 Comparison of noise-like sequences acquisition performance with chaotic sequences.

Gaussian random variables and then filtering them separately. The peak of this filter is approximately 6 dB. The filtering will commence in time domain resulting in a set of non-deterministic fading samples. The reason for this filtering is to mimic the effects of isotropic scattering and an omnidirectional antenna considering the Doppler frequency shift of a moving receiver, as well as fulfilling the requirements about the fading coefficient being constant over a bit period [10].

It is important to point out however, that although the power spectral densities of the Gaussian samples are affected by the spectral shaping, the probability density functions of the processes are not affected resulting a Rayleigh distributed envelope. This fading generator was used to simulate real channel conditions when both the chaotic and noise-like spreading sequences were investigated.

As can be seen from Figure 14.18, the performance of the acquisition phase is adversely affected by the presence of slow flat fading in the channel. This is due to the effect of the fading channel on the auto-correlation peak of the pilot signal, and more importantly, the rising of the secondary peaks in the pilot auto-correlation function.

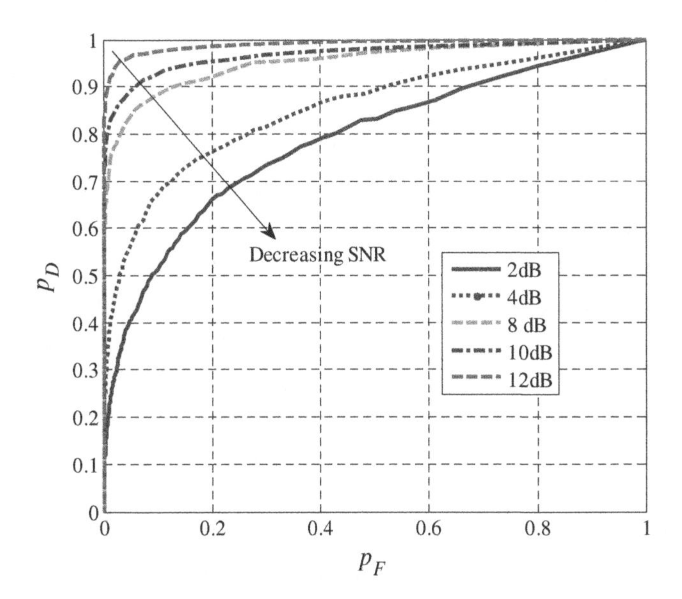

Fig. 14.18 ROC for a faded chaotic pilot (0 Users).

14.3 Tracking

This section presents the theoretical analysis and simulation results for the tracking phase of the synchronization block. Once the acquisition phase has finished, the synchronization block will automatically change to the tracking phase and the value of the estimated time delay will be passed on to the tracking circuit. We will denote this passed value as T_a [1]. The aim of the tracking phase is to finely align the timing between the transmitter and receiver to within a chip duration as well as maintaining synchronization. The synchronization needs maintenance because the transmitter and receiver are susceptible to clock drift and various other impediments that cause the eventual loss of synchronization [1, 3, 11].

The time difference between the transmitter and receiver changes by small random increments, called *timing jitter*. In order to get accurate timing estimation between the transmitter and receiver, the timing jitter needs to be accounted for [1, 3, 4].

Tracking phase involves continuous correlation of the incoming pilot signal with the locally generated pilot signal in the neighbourhood of the timing estimation provided by the acquisition phase. To achieve this, various techniques involving different loop topologies are used.

One of the most common ways is to use two correlators which are placed apart in time by a fraction of a chip duration T_c. The normalised time difference between the correlators is denoted by Δ.

In our analysis we chose $\Delta = 1$; this means that the sequence generator outputs two sequences. The first one is the sequence which is advanced from the acquisition timing estimate by one half of a chip duration ($T_c/2$) in time. The second one is delayed by the same amount. These two sequences are continually correlated by the incoming signal and the results are subtracted. The final output is an error signal which is used to control the speed at which the sequences are generated as shown in Figure 14.19. If the timing estimate coming from the acquisition is wrong by a fraction of T_c, the tracking phase will minimise this error and align the two signals. This configuration of correlators is known as a *Delay Lock Loop* (DLL) [7, 11–13].

The tracking loop can minimise the error in the acquisition estimation of the time delay between the transmitter and receiver provided that this error is

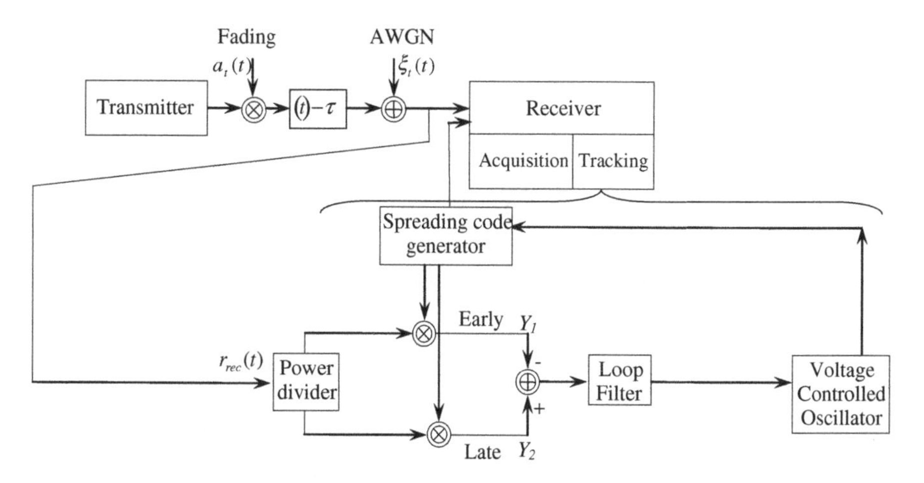

Fig. 14.19 Tracking phase of the synchronization block (DLL).

not more than one chip duration. The region in time for which the tracking loop can minimise the error is called the *pull in region*. If the acquisition estimation error is outside this pull in region, the tracking phase will be unable to perform its role and the acquisition phase will restart. We redefine the timing offset τ from the last section as T_d. This is to indicate that the acquisition phase has finished and now the delays we deal with are fractions of a chip rather than multiples of one. The pull in region is defined to be half of a chip duration therefore we have

$$-\frac{T_c}{2} \leq (T_a - T_d) \leq \frac{T_c}{2}. \tag{14.26}$$

The tracking phase tries to estimate the incoming time offset (T_d). We denote this estimate by $\hat{T}_d$ and introduce the variable, δ as the normalised difference between the incoming time offset and the estimated time offset. That is

$$\delta = \frac{T_d - \hat{T}_d}{T_c}. \tag{14.27}$$

The signal at Y_1 is the correlation of incoming and advanced spreading codes. The branch where this multiplication happens is called the early branch. The other branch (Y_2) where multiplication with the delayed code happens is called the late branch. For Y_1 and Y_2 we have

$$Y_1(t, T_d, \hat{T}_d) = \frac{A}{\sqrt{2}} x_t^0(t - T_d) x_t^0\left(t - \hat{T}_d + \frac{\Delta T_c}{2}\right) + \frac{1}{\sqrt{2}} x_t^0\left(t - \hat{T}_d + \frac{\Delta T_c}{2}\right) n(t) \tag{14.28}$$

$$Y_2(t, T_d, \hat{T}_d) = \frac{A}{\sqrt{2}} x_t^0(t - T_d) x_t^0\left(t - \hat{T}_d - \frac{\Delta T_c}{2}\right) + \frac{1}{\sqrt{2}} x_t^0\left(t - \hat{T}_d - \frac{\Delta T_c}{2}\right) n(t) \tag{14.29}$$

where A is the amplitude of the pilot signal and $n(t)$ is the AWGN noise present in the channel with power spectral density of $\frac{N_0}{2}$. The input to the loop filter is $\varepsilon(t, \delta)$ which can be rewritten as

$$\begin{aligned} \varepsilon(t, T_d, \hat{T}_d) &= Y_2(t, T_d, \hat{T}_d) - Y_1(t, T_d, \hat{T}_d) \\ &= \frac{A}{\sqrt{2}} x_t^0(t - T_d)\left[x_t^0\left(t - \hat{T}_d - \frac{\Delta T_c}{2}\right) - x_t^0\left(t - \hat{T}_d + \frac{\Delta T_c}{2}\right)\right] \\ &\quad + \frac{1}{\sqrt{2}} n(t)\left[x_t^0\left(t - \hat{T}_d - \frac{\Delta T_c}{2}\right) - x_t^0\left(t - \hat{T}_d + \frac{\Delta T_c}{2}\right)\right] \end{aligned} \tag{14.30}$$

The signal above consists of three parts, a dc part which is called the error signal and is used for code tracking. This parameter is zero when the transmitter and receiver are aligned exactly and its positive or negative value give an indication of the relative timing error between the transmitter and the receiver. This error signal is integrated and averaged to extract this dc part which is

$$D_\Delta(\delta) = R_c\left\{\left(\delta - \frac{\Delta}{2} T_c\right)\right\} - R_c\left\{\left(\delta + \frac{\Delta}{2} T_c\right)\right\} \tag{14.31}$$

where $\delta = \frac{T_d - \hat{T}_d}{T_d}$.

The second part is the time varying component of ε. This is called self noise and can be effectively ignored because the frequency of the self noise is far higher than the frequency of the tracking loop we are dealing with therefore it will be attenuated by the loop filter [1]. We will not discuss the self noise further in order to keep the focus on the important parts. The last part is the noise which is present in the correlation process.

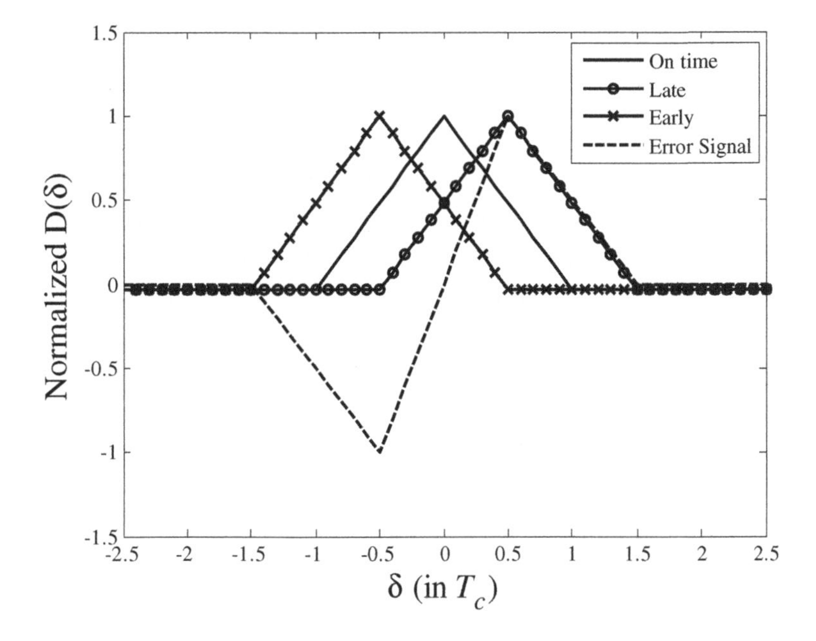

Fig. 14.20 Early, late and on time correlations for the DLL. The error signal is shown in dashed lines.

The on time auto-correlation function and the error signal are shown in Figure 14.20. The correlation function and the corresponding error function are related to a PRBS and the effects of noise and fading are excluded. The desired region of operation is between the positive and negative peaks of the error signal i.e., $-0.5 \leq \delta \leq 0.5$.

14.3.1 The Effect of Signal Sampling on Tracking

One important aspect of tracking phase of the synchronization algorithm is sampling of the signals. Since the tracking loop works in a time frame that is a fraction of a chip, the chip resolution becomes important. This means that instead of defining a spreading chip as one number, as it is sufficient for the analysis of the CDMA system characteristics, it is necessary to define it as a set of numbers with certain duration. This means that, the chip must have a finite width, which is represented by a number of samples, in order to modulate a carrier. This is possible in several ways, two of which are presented in this section.

The first one presented is the *Zero Order Hold* (ZOH) sampling, in which chip samples retain their value over the whole chip duration. This process, however, extends the bandwidth of the pilot and therefore is not desirable. The second way of sampling is interpolation, in which zeros are inserted between the chips and then the whole signal is low pass filtered. The latter method is called *interpolation* which preserves the bandwidth and is therefore desired.

To get a better insight, two types of spreading sequences were investigated. The PRBS, (the conventional sequences used in CDMA systems); and chaotic sequences used from the acquisition phase.

Figures 14.21 and 14.22 show the PRBS sampled with the ZOH method and the interpolation method respectively while Figures 14.23 and 14.24 show chaotic signals sampled with the two techniques mentioned above.

Once the sampling is performed, the resultant signal will be sent to the tracking phase described before. Depending on the sequence type and the sampling type the tracking loop result changes.

Figure 14.20 shows the tracking loop's correlation and error function when the PRBS are ZOH sampled. Figure 14.25 shows the same for interpolated

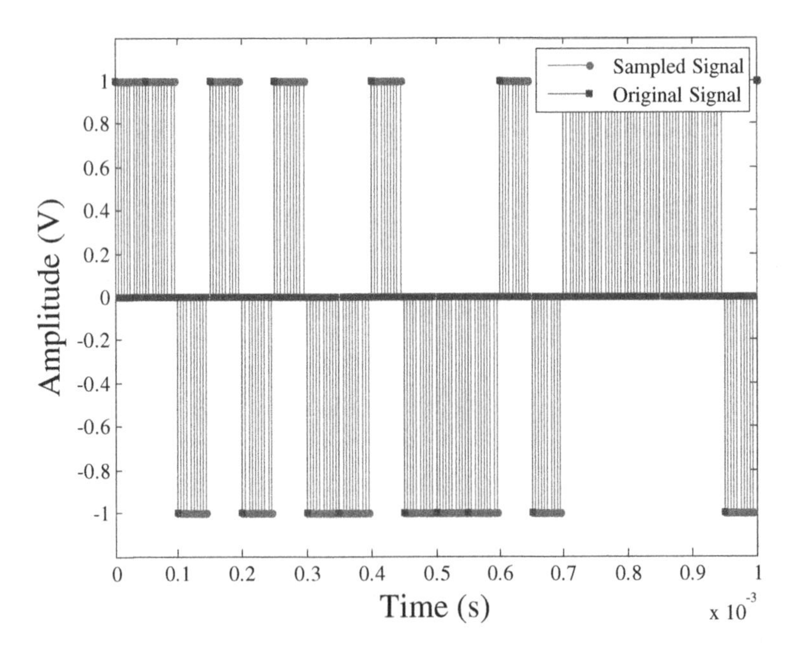

Fig. 14.21 ZOH sampled PR sequences.

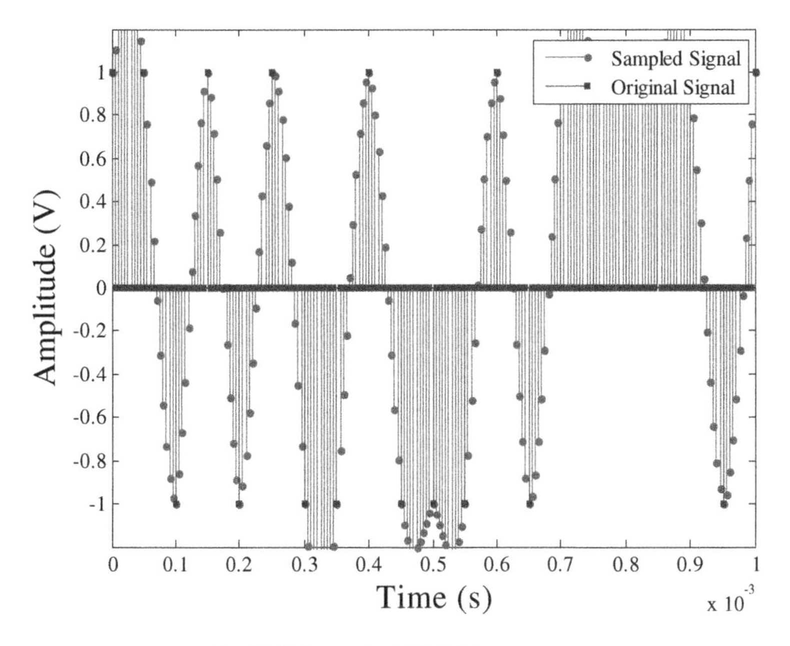

Fig. 14.22 Interpolated ZOH PR sequences.

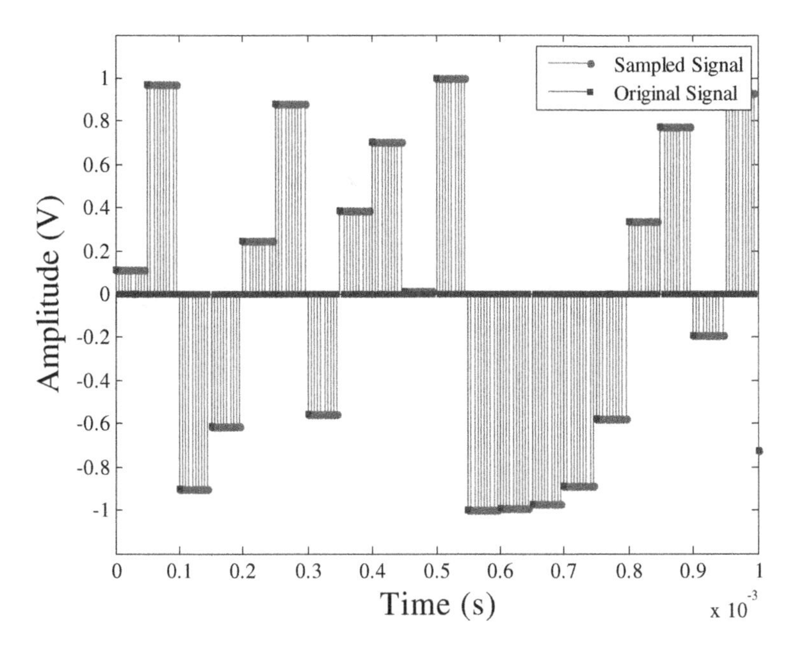

Fig. 14.23 ZOH sampled chaotic sequences.

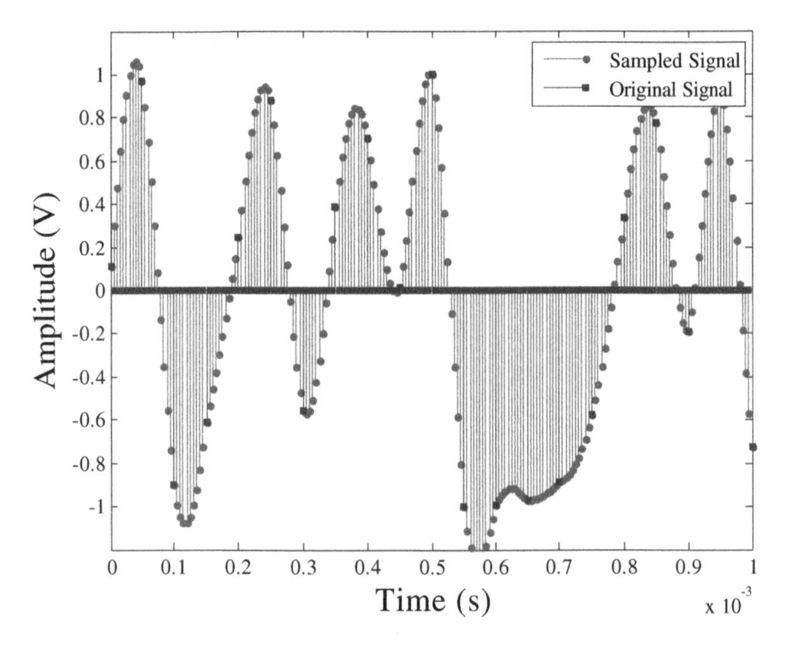

Fig. 14.24 Interpolated chaotic sequences.

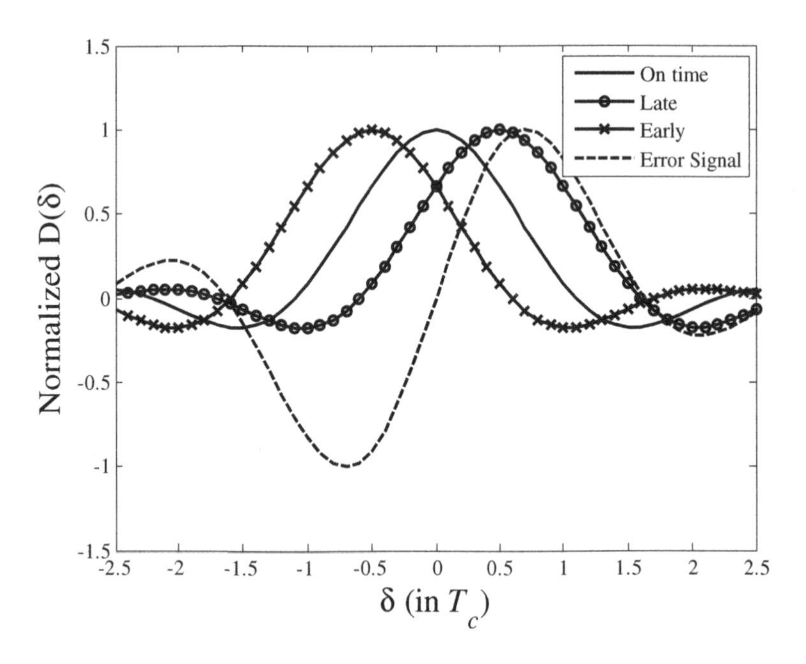

Fig. 14.25 DLL with interpolated PR sequences.

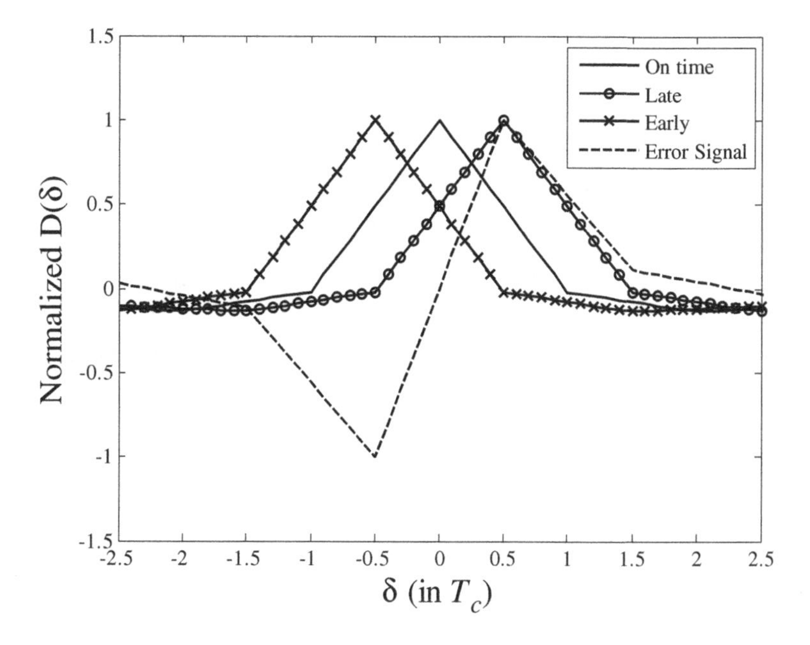

Fig. 14.26 DLL for ZOH sampled chaotic sequences.

PRBS. Figures 14.26 and 14.27 show the tracking loop correlation and error curves for ZOH sampled and interpolated chaotic sequences respectively.

Note that the useful part of the error signal is the linear section in the middle. As the delays get close to the non-linear region, the quality of tracking deteriorates. The amount of time it takes for the tracking phase to lose synchronization is called the *tracking slip time*.

Once the error signal is calculated, the random jitter in the incoming signal will be tracked using this error signal. The cumulative and non-cumulative values for T_d and $\hat{T}_d$ (T_d estimate) are shown in Figure 14.28. Note that $-\frac{T_c}{2} \leq (T_d) \leq \frac{T_c}{2}$ in the lower figure and that $\hat{T}_d$ is always one step behind T_d. The cumulative graph is useful for looking at the stability of the tracking loop once it executes many cycles.

The tracking results presented so far are based on the assumption of an ideal noiseless channel; however, a typical communication channel is impaired by a wide variety of factors, the most important of which is AWGN. The tracking loop performance was observed in the presence of noise. Figure 14.29 to Figure 14.32 show a representation of the tracking loop with different sampling

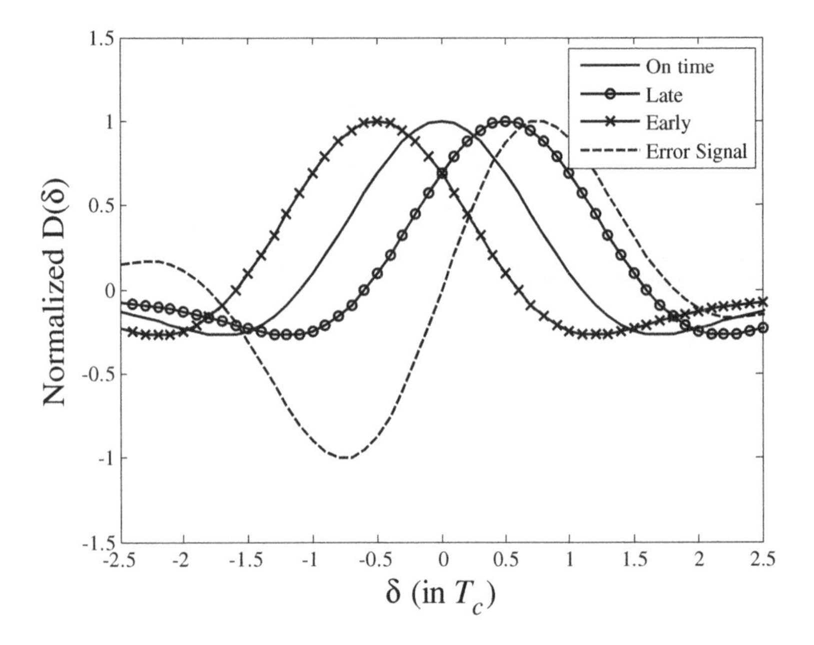

Fig. 14.27 DLL for interpolated chaotic sequences.

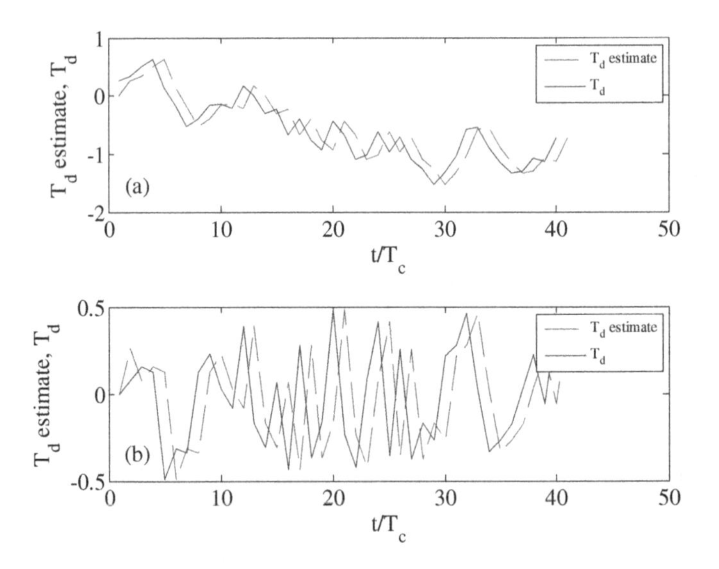

Fig. 14.28 (a) cumulative tracking performance (b) instantaneous tracking performance.

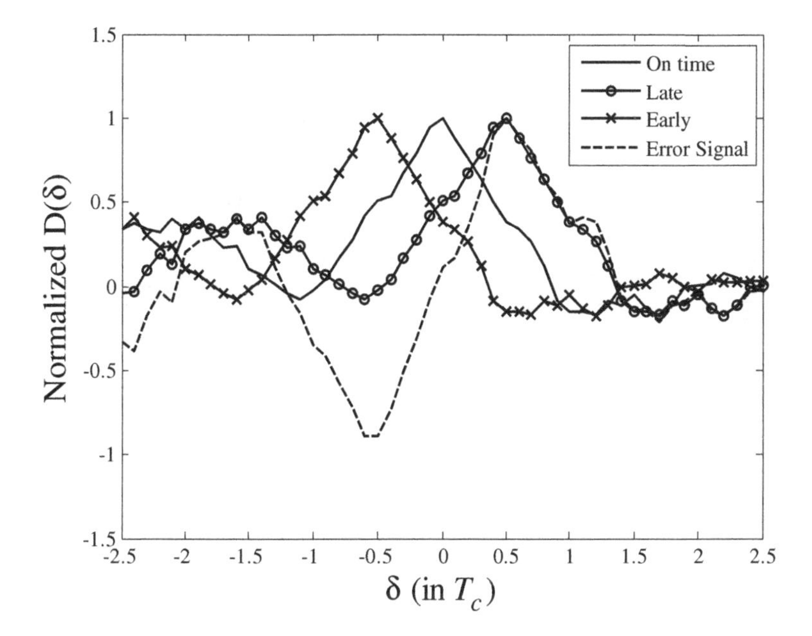

Fig. 14.29 ZOH sampled PR sequences in presence of noise.

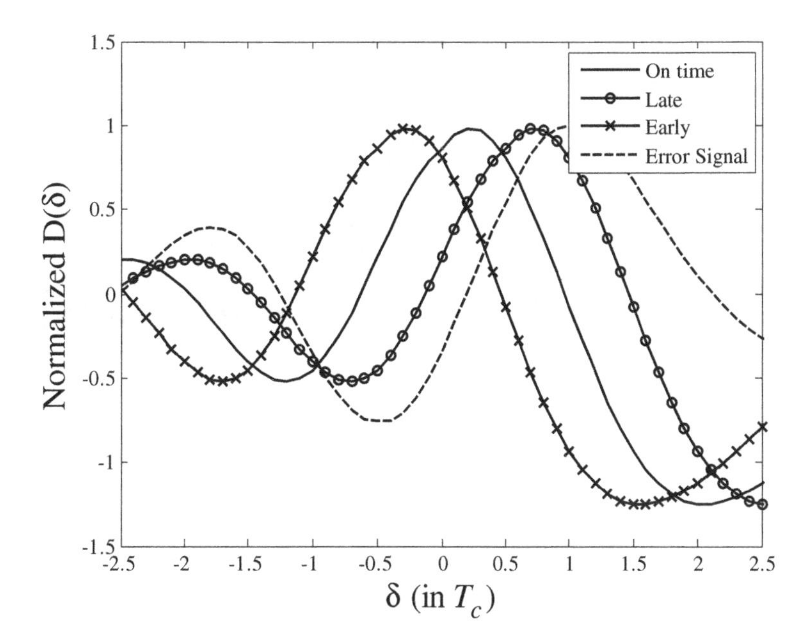

Fig. 14.30 Interpolated PR sequences in presence of noise.

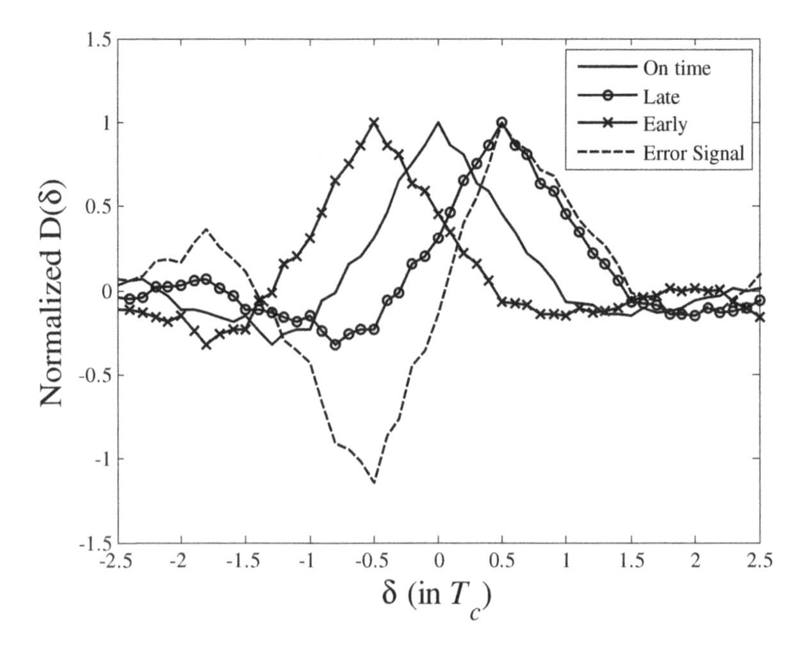

Fig. 14.31 ZOH sampled chaotic sequences in presence of noise.

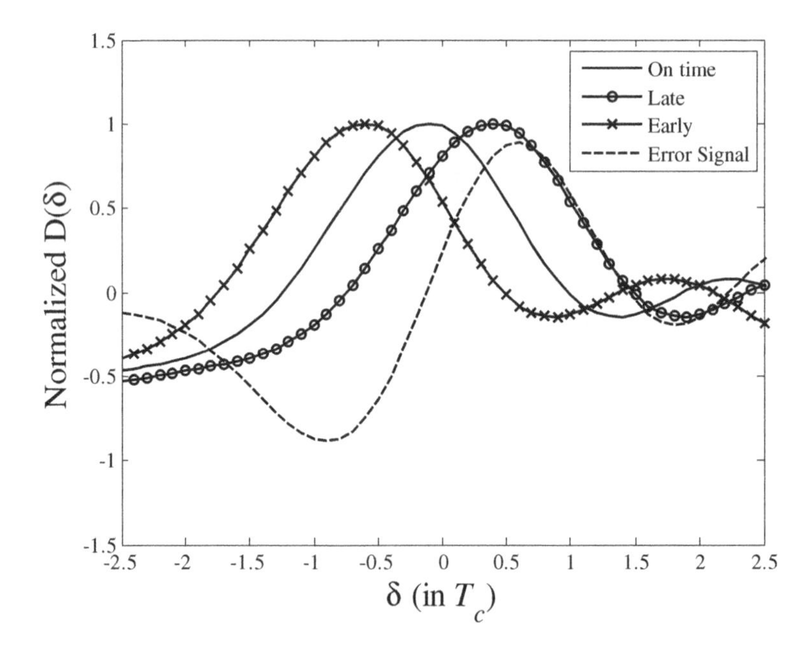

Fig. 14.32 Interpolated chaotic sequences in presence of noise.

methods and different spreading sequences in the presence of noise. As can be seen, the error signal loses its correct timing and linearity in the intended region of operation, which will introduce errors in the tracking process. The Signal to Noise Ratio (SNR) used for these simulations is 8 dB.

14.4 Overall System Performance

Now that we have covered the principles in acquisition and tracking, we focus on the overall system performance. This section shows the effect that timing error has on the overall system performance. We assume that the acquisition phase has been successful and the tracking phase has started. Although the tracking phase will track the timing jitter within the chip duration, some amount of error will be remaining. We presume that the amount of error in timing estimation that the tracking phase has will be a uniformly distributed random variable that is bounded by half the chip duration on each side i.e.,

$$-\frac{T_c}{2} \leq (T_d - \hat{T}_d) \leq \frac{T_c}{2}. \tag{14.32}$$

That is, the chips can be misaligned by the maximum of half a chip duration on either side, meaning that the tracking phase either over estimated or under estimated the time difference between the incoming sequence and the locally generated sequence. If the chips are misaligned by more than half a chip duration on either side, then the dispreading process will be incapable of extracting the correct user messages [1].

With this in mind we present the *Bit Error Rate* (BER) for a simulated chaotic multiuser system for different number of users in Figure 14.33. The theoretical predictions from [6] are given alongside the simulated curves. Note that in each case the simulated curves, which have the partial chip misalignment we discussed above, have a higher BER than their corresponding theoretical curves. As an example, we can concentrate on the single user case and see that the partially misaligned sequence needs a signal to noise ratio which is higher by 1 dB to get the same BER as the fully aligned system.

Note that the tracking loop discussed above is not the optimum for CDMA synchronization, other topologies of loops exist that exhibit a much better performance because of minimized delay estimation error. DLL was presented in this chapter for its learning value. The readers are encouraged to look for more powerful tracking loop topologies in [11, 14–19]. However, the performance

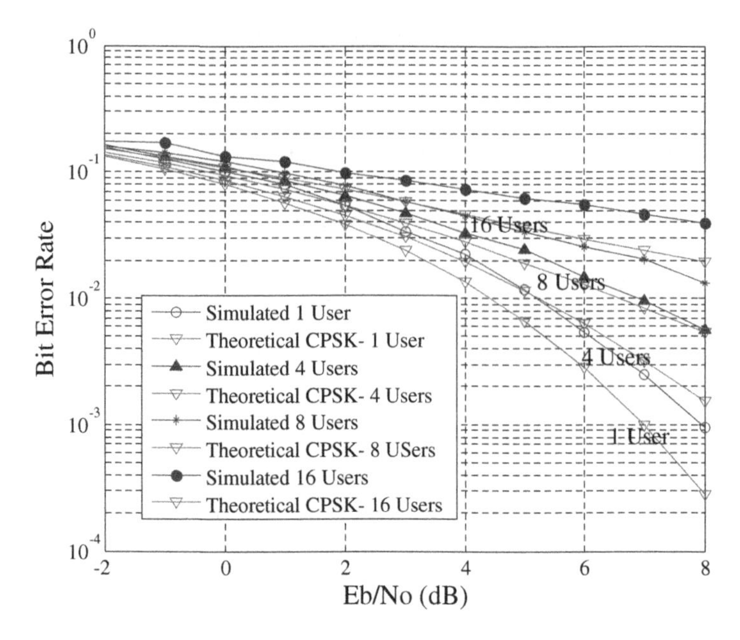

Fig. 14.33 BER curves for a multiuser chaotic phase shift keying with random jitter.

of these types of topologies is not investigated in presence of chaotic spreading sequences. This can be a way forward for an already active area of research into secure spread spectrum communication systems.

References

[1] R. L. Peterson, R. E. Zeimer, and D. E. Borth, *Introduction to Spread Spectrum Communicaion*. Englewood Cliffs, NJ Prentice Hall, 1995.

[2] L. Hanzo, L. Yang, E. Kuan, and K. Yen, *Single and Multi-Carrier DS-CDMA Multi-User Detection, Space-Time Spreading, Synchronisation and Starndards*. Chichester: IEEE Press Wiley, 2003.

[3] J. K. Holmes, *Spread Spectrum Systems for GNSS and Wireless Communications*London Artech House, 2007.

[4] J. S. Lee and L. E. Miller, *CDMA Systems Engineering Handbook*. London: Artech House, 1998.

[5] R. Vali and S. M. Berber, "Secure Communication in Asynchronous Noise Phase Shift Keying CDMA Systems," in *Spread Spectrum Techniques and Applications, 2008. ISSSTA '08. IEEE 10th International Symposium on*, pp. 528–533, 2008.

[6] G. S. Sandhu, G. S. Sandhu, and S. M. Berber, "Investigation on operations of a secure communication system based on the chaotic phase shift keying scheme," in *Information Technology and Applications, 2005. ICITA 2005. Third International Conference on*, vol. 2, pp. 584–587, 2005.

[7] B. Jovic, S. G., U. C., and B. S. M., "A robust sequence synchronization unit for multi-user DSCDMA," *Elsevere*, January 2007.
[8] L. M. Pecora and T. L. Carroll, "Synchronisation in Chaotic Systems," *Physical Review Letters*, 64(19), February 1990.
[9] G. Mazzini, R. Rovatti, and G. Setti, "Sequence synchronization in chaos-based DS-CDMA systems," vol. 4, pp. 485–488, 1998.
[10] M. Jeruchim and M. Jeruchim, "Techniques for Estimating the Bit Error Rate in the Simulation of Digital Communication Systems" *Selected Areas in Communications, IEEE Journal on*, 2, pp. 153–170, 1984.
[11] M. G. El-Tarhuni and A. U. H. Sheikh, "Performance analysis for an adaptive filter code-tracking technique in direct-sequence spread-spectrum systems," *Communications, IEEE Transactions on*, 46, pp. 1058–1064, 1998.
[12] J. J. Spilker and D. T. Magill, "The Delay-Lock Discriminator-An Optimum Tracking Device," *Proceedings of the IRE*, vol. 49, 1961.
[13] P. Hyung-Rae and P. Hyung-Rae, "Performance analysis of a decision-feedback coherent code tracking loop for pilot-symbol-aided DS/SS systems," *Vehicular Technology, IEEE Transactions on*, 55, pp. 1249–1258, 2006.
[14] P. Hyung-Rae, "Performance analysis of a decision-feedback coherent code tracking loop for pilot-symbol-aided DS/SS systems," *Vehicular Technology, IEEE Transactions on*, 55, pp. 1249–1258, 2006.
[15] H. Meyr, "Delay-Lock Tracking of Stochastic Signals," *Communications, IEEE Transactions on [legacy, pre - 1988]*, 24, pp. 331–339, 1976.
[16] E. Karami, "Tracking Performance of Least Squares MIMO Channel Estimation Algorithm," *Communications, IEEE Transactions on*, 55, pp. 2201–2209, 2007.
[17] P. M. Hopkins, "Double Dither Loop for Pseudonoise Code Tracking," *Aerospace and Electronic Systems, IEEE Transactions on*, AES-13, pp. 644–650, 1977.
[18] K. Kallman, K. Kallman, and G. Davis, "Jitter performance of a baseband sampled code tracking loop," *Communications, IEEE Transactions on*, 42, pp. 2919–2925, 1994.
[19] Y. Ma, "Recursive Code-Timing Estimation for DS-CDMA Signals in a Time-Varying Environment Through a Stochastic Optimization Approach," *Communications, IEEE Transactions on*, 55, pp. 665–669, 2007.

15

Topology Control for Effective Power Efficiency in Wireless Mesh Networks

Felix O. Aron, Anish Kurien and Yskandar Hamam

Tshwane University of Technology

15.1 Introduction

Today, wireless Local Area Networks (WLANs) are widely implemented in hospitals, banks and bank branches, schools, government offices, private offices and even in homes. Some of these WLANs are located in very far flung rural areas. In order to extend the network coverage to interconnect the distantly distributed WLANs, 802.11 wireless mesh networks (WMNs) have been proven to be a better alternative. WMNs have advantages such as low cost, easy and incremental deployment and fault tolerance [1]. Three types of WMN architectures exist [2]: client WMNs, infrastructure/backbone and Hybrid WMNs. A client WMN is more like a MANET. The infrastructure/backbone is a network of conventional wireless devices connecting only via wireless mesh routers and cannot communicate directly with each other. The hybrid architecture combines infrastructure and client meshing. It has more advantages and is deemed to be more appropriate for mesh network implementations [3]. The architecture consists of three distinct wireless network elements: a *Network Gateway* (a mesh router with gateway/bridge functionalities), *mesh access points* (mesh routers) and *mobile or stationary nodes* (mesh clients) [2, 4]. The Gateway and the mesh access points (MAPs) are however just a composition of mesh routers with varying degrees of strengths and can be categorized as backbone wireless mesh routers (WMRs). The front end is composed of wireless mesh clients (WMCs) as shown in Figure 15.1.

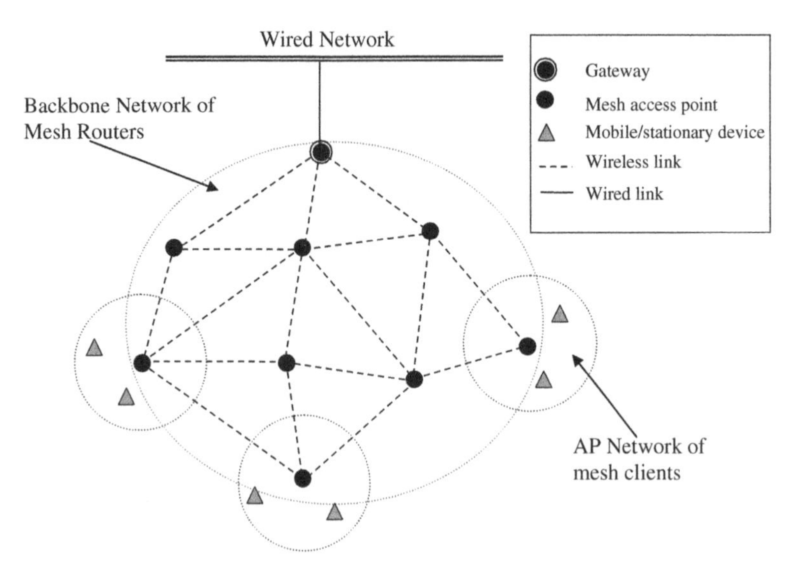

Fig. 15.1 WMN composed of nine mesh routers (mesh access points), with one of the routers acting as a gateway to the wired network.

However, even as the applications of WMNs continue to gain popularity in the networking and communication arena, several challenges are observed [2, 16]. One of the issues identified by the authors in [2] is power management in WMNs. The authors, however, assume that since WMRs are stationary, power is not a concern in WMNs. In this chapter, it is asserted that power efficiency is essential for the realization of the numerous benefits of WMNs. With specific considerations to rural area applications of WMNs, it is argued that mesh routers would be stationary but with power constraints. In rural areas, electrical mains power sources are limited and/or often not available. Mesh nodes (MNs) have thus to rely on exhaustible and renewable means of energy supply such as solar, battery or generators. Furthermore, mesh clients are definitely power constrained [2]. Additionally, power efficiency is essential for connectivity and interference control and for spectrum spatial reuse [6]. As a result, this chapter addresses the challenge of power efficiency by presenting a localized distributed power-efficient topology control algorithm for application of WMNs in rural areas. The approach taken in this work is to construct an overlay graph topology of the WMN with such desirable features as reduced physical node degree, increased throughput, increased network lifetime and maintenance of connectivity by varying the transmission power at each node. The main contribution of the chapter lies in the ability of the algorithm to balance energy efficiency and throughput in WMNs and without loss of connectivity.

This work is an improvement of the work that was carried out in [19]. In [19], an algorithm was presented with results showing an enhancement in the network lifetime. In this work, the algorithm is further optimized by an introduction of a constant β ranging from 1 to 4. Results show the ability of the algorithm to enhance throughput. Additionally, maintenance in connectivity is shown via topology graphs.

Figure 15.1 shows the WMN under consideration. It is assumed that a hybrid WMN is used to provide internet connectivity to the mobile or stationary devices in the AP networks. Hence, most traffic appears between the mesh access points (MAPs) and the gateway and between the MAPs themselves. Consequently, the simulations are based on the WMRs backbone network. The communication between the MAPs and the mobile or stationary devices is ignored. It is further assumed that there is no interference between the AP network and the backbone network. In a practical WMN, this can be achieved by having the MAPs fitted with two radio models [7]. This enables the AP networks to be assigned an independent channel different from that of the backbone mesh network. The WMN is therefore sufficiently described as a set of MNs trying to send packets between the MAPs and to the gateway.

15.2 Topology Control

The Topology control algorithms [10, 11, 13, 14] have largely been proven to be one of the ways of achieving energy efficiency in MWNs. Topology control can be defined as:

Definition 15.1. The technique of managing nodes' decisions concerning their transmission ranges so as to construct a network with the desired properties (e.g., connectivity) while reducing energy consumption at the node and/or increasing network capacity.

The authors in [20] state that this definition distinguishes topology control from other techniques of energy saving and/or network capacity enhancements in wireless networks. The distinguishing feature, as per their work, is that topology control involves setting the transmission power level of nodes with the aim of achieving a certain global network-wide property. Consequently, energy-efficient designs of wireless transceivers are not considered as a topology control technique, because it has a node perspective. Similarly, power control techniques that optimize the choice of the transmit power level for a single wireless transmission over a single hop or along several hops have a channel-wide perspective. Hence, they are not considered as topology control techniques.

The primary motivation to the study of the topology control problem is to maintain a minimum number of links between nodes while ensuring network connectivity [20]. The resultant overlay network topology, for instance as shown in Figure 15.2, should finally lead to: (15.1) energy saving at the transceivers, thereby increasing the lifetime of the network, (15.2) reduction of noise to other neighbours, thus decreasing channel contention and (15.3) an increase in spatial channel re-use, hence increasing the throughput of the network.

The topology control problems are subdivided into two main phases, the neighbour discovery and the network organization phase. In the neighbour discovery phase, the problem is to detect neighbouring nodes located within the transmission range. In the network organization phase, each node has to decide which communication links to establish with neighbouring nodes and which power management schemes to adopt e.g., sleep periods and transmission range adjustments.

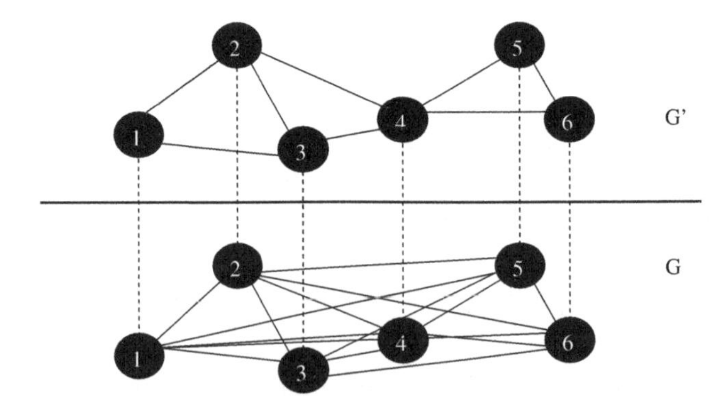

Fig. 15.2 Communication graph G and overlay graph G'.

15.2.1 The Graph Concept

In studying the topology control for wireless networks [20], it has often been assumed that a set V of nodes is given in some plane. The nodes could be static, mobile, or both. A wireless network is then represented by a graph $G = (V, E)$, where V is the set of nodes (vertices) and E is the set of links (edges). The routers, switches and client hosts are represented by nodes and the connections between them are represented by links. A node transfers information to another node in the form of data-packets if there is a link between them. If there is no direct link between the nodes, then a path in the network is the sequence of distinct nodes visited when transferring data-packets from one node to another in a multihop nature.

If nodes transmit at the maximum power threshold, then the resulting communication graph is referred to as the Maximum-power graph. The graph represents all possible communication links between the nodes. In cases where all the nodes have the same maximum transmission range, the resulting network is referred to as a *homogenous* network [20]. In such cases, the network is said to be composed of nodes with the same features, especially for those that concern the communication apparatus e.g., the same transceivers. If the network is composed of diverse nodes and with diverse wireless transceivers with diverse transmission ranges, then the network is said to be *heterogeneous* [15].

15.3 Related Work

Considerable work exists that addresses the problem of topology control in multihop wireless networks (MWNs). For instance, Rodoplu and Meng [8] described the first algorithm which is based on the concept of relay region. A node decides to relay through other nodes if less power will be consumed. The algorithm guarantees the preservation of minimum energy paths between every pair of nodes connected in the original graph. Based on the results of [8], Li and Halpern [9] proposed an improved protocol which is computationally simpler and better in performance with the resulting topology being a sub-network of the one generated by [8]. The work in [8, 9], however, implicitly assume that a long link consumes more power than a shorter link. This assumption is not practical for instance in heterogeneous networks according to [10]. In [11–13], the concept of local neighbourhood is introduced. This concept proposes that a logical topological view of a node in a network be constructed based only on its local information. In the work of Li et al. [11], a node builds its local minimum spanning tree (LMST) based only on its one hop neighbourhood information. It keeps only one hop nodes as neighbours in the final topology. The resulting topology has been shown to be connected and with node degree bounded by 6. However, [11] seems not to guarantee power efficiency. For instance, LMST with link removal could lead to too few neighbours per node which imply longer paths and thus higher energy consumption. Generally, all of the algorithms shown in [11–15] have not been applied to WMNs. Besides, they do not provide a clear guarantee for a balance between increase in capacity in the network and increase in the network lifetime in WMNs. It can not be assumed that the algorithms will automatically function in WMNs as the requirements on power efficiency and mobility are very different between WMNs and other MWNs [2].

15.4 Network Model

The network considered is modelled as a graph $G = (V, E)$. Here, $V = \{v_1, v_2, \ldots, v_n\}$ is the set of randomly distributed static MNs and E is the set of edges/links which are the wireless links found between MNs when nodes communicate at full transmission power, $P^{\max}$. Each node $u \in V$ has a unique $id(u_i) = i$, where $1 \le i \le N \quad (N = |V|$, number of nodes) and is specified by its location, $XY(u)$ i.e., its coordinates $(x(u), y(u))$ on a 2D plane at any instance. It is assumed that each node is equipped with an omni-directional antenna with an adjustable transmission power. For all $u \in V$, let $P_u(d_u)$ denote u's minimum transmission power needed to establish a communication link to another node that is located at d_u distance away from u. Let $P_u^{\max}$ denote the maximum transmission power for node u. A homogenous network is assumed in regard to their maximum transmission power ability and the maximum distance, D, needed for any two nodes $u, v \in V$ to communicate directly. Therefore, $\forall\ u \in V, P_u(D) = P_u^{\max}$. The topology is modelled as a weight directed graph $G' = (V, E')$ whereby for each edge $(u, v) \in E'$, node v must be within the transmission range of node u. The notation for the Euclidean distance between u and v is given by:

$$d(u, v) = \|XY(v) - XY(u)\|, \tag{15.1}$$

where $XY(.)$ denotes the coordinates of the physical location of the node. The set $E' = \{(u, v)|d(u, v) \leq D\}$. Clearly, the graph $G' \subseteq G$.

In [5], the authors investigated wireless signal fading models. They established that for there to be a successful reception of packets sent from node u to v, the transmitter power setting at node u must satisfy the following inequality equation:

$$\Pr(u) \geq k.(d(u, v))^{\sigma}. \tag{15.2}$$

Here, k is a constant whose value is dependent on the system parameters such as the antenna gains, the system path loss, the wavelength of the wireless signal and the receiving power threshold. The symbol $\sigma \in [2, 5]$ is a constant real number depending on the wireless transmission environment. Based on the work in [5], the weight of an edge $(u, v) \in E'$ is calculated using the weight function.

Definition 15.2 (Weight Function). An edge (u, v) has a weight given by the following expression:

$$w(u, v) = k.(d(u, v))^{\sigma} = k. \|XY(v) - XY(u)\|^{\sigma}. \tag{15.3}$$

It is assumed that every node $u \in V$ is fitted with a GPS receiver or uses some localization algorithm e.g., [17] such that each node has knowledge of its approximate location. Hence, $d(u, v)$ is easily obtained based on equation 15.1.

The following gives a list of other definitions to the terms used in the chapter.

Definition 15.3 (Accessible Neighbourhood Set)). The Accessible Neighbourhood Set, A_u^N, is defined as the set of all nodes that have a direct link with node u, when u transmits at maximum transmission power. The set is given by $A_u^N = \{v \in V|d(u, v) \leq D\}$.

Definition 15.4 (Logical Neighbour Set). A logical neighbour set of node u is given by NS_u^L. Node $v \in NS_u^L$ if and only if there exists an edge (u, v) in the topology generated by the algorithm and $NS_u^L = \{v \in V|u \to v\}$.

Definition 15.5 (Fully Connected). A network is fully connected if and only if $\forall\ u \in V$, there exists either a direct path or a multihop path from u to every other node $v \in V$ in the network.

Definition 15.6 (Relay Region). Given a node v, let the physical location of v be denoted by $Loc(v)$. The relay region of the transmit-relay node pair (u, v) is the physical region $RL_{u \to v}$ such that relaying through v to any other point in $RL_{u \to v}$ consumes less power than direct transmission to that point.

Definition 15.7(Network Lifetime). Given a set of nodes V and for all $v \in V$, v has an energy value $E(v)$, the lifetime of node v is $Lt_v = \{t \mid f_v(t) \leq E(v)\}$ until $E(v) = 0$, where $f_v(t)$ is the energy consumed by v. The network lifetime $Lt_v = Min_{v \in V}(Lt_v)$ i.e., the time taken till the first node goes off.

15.5 Proposed Algorithm

Taking an arbitrary node $u \in V$ in the network G and without loss of generality, a two phased design algorithm that executes in every node is presented. It is assumed that this original topology G is fully connected. Phase 1 constructs a Local Minimum Shortest-path Tree and in phase 2, unidirectional links are removed through link additions.

15.5.1 Phase 1 — Construction of Local Minimum Shortest Path-Tree (LM-SPT)

In this phase, each node gathers neighbour information and constructs an LM-SPT. The phase involves the following three stages.

15.5.1.1 Information collection and exchange

Each node in this step periodically broadcasts beacon '*Hello*' messages using maximum power $P_{tx}^{\max}$. The information exchanged here includes the node ID and the position in the 2D plane. This information is used to calculate the node to node distance, the link weights and the path weights. The link weight represents the power required for transmission along a link, and the path weight represents the sum of all minimum link weights of a path from source to destination. The result of this stage is the '*Accessible Neighbourhood Set*' A_u^N for each node $u \in V$. The '*Hello*' message is also sent by each node asynchronously and periodically giving each node's information about its neighbours.

15.5.1.2 Construction of a logical visible neighbourhood topology

Each node applies the concept of the relay region in order to gather the nodes in the set of '*Logical Visible Neighbourhood*'. The set $LVN(u,k) \subseteq A_u^N$ at $k = 1$. If a node $v \in A_u^N$ is in the relay region of another node $w \in A_u^N$, then node v is moved to the set of non neighbours called *NotNbr*. This is repeated for all the nodes $i \in A_u^N$. All nodes reachable via other nodes are moved out of the set A_u^N and the remaining set is called $LVN(u)$. Each node then applies the Dijkstra's algorithm independently from a source node to all the other nodes in Vin order to build its LM-SPT.

15.5.1.3 Computing the transmission power

Each node computes its minimal transmission power to cover only all of the nodes contained in the set $LVN(u, 1)$. It thus determines which node among the nodes is furthest. The node then adjusts its transmission power to reach this node and thus all other nodes in the set $LVN(u, 1)$ are covered. The set is given by $G' = (V', E')$. From the information on the location of the nodes, the inter node distance is calculated. The distance is applied in the propagation models to obtain the minimum transmission power. The free space model is used for short distances and the two ray ground reflection model is used for longer distances depending on the value of the Euclidean distance in relation to the cross over distance.

The cross over distance is calculated using the following expression

$$Cross\_Over\_dist = \frac{4\pi h_t h_r}{\lambda}, \tag{15.4}$$

where, $h_t h_r$, are the antenna heights of the transmitter and receiver respectively and λ, denotes the wavelength. If $d(u, v) < Cross\_over\_dist$, the Free Space model is used, otherwise, if $d(u, v) \geq Cross\_over\_dist$, then the Two-ray-ground model is used.

The free-space propagation model is given by the following expression:

$$P\min = \frac{RxThresh((4\pi d)^2 L)}{G_t G_r \lambda^2} \tag{15.5}$$

The two ray ground reflection model is given by the following expression:

$$P\min = \frac{RxThresh(d^4 L)}{G_t G_r h_t^2 h_r^2}, \tag{15.6}$$

where, G_t, G_r is the transmitter and receiver antenna gain respectively, L is path loss exponent. To achieve optimality, a value of β picked randomly in the range of 1 to 4 is used such that the optimal transmission power $P_{\min}^{opt} = \beta * P\min$. This ensures that the optimal value is slightly above the expected $P\min$ in order to protect an active link from interferences caused by parallel transmissions in the same time slot.

Figure 15.3 shows the algorithm that executes in each node $u \in V$ to compute the minimum transmission power. Let $p(u, v)$ be the minimum power required to transmit a data packet from node u to node v at any time instance. Also let initial power $P = P_{tx}^{\max}$ and $F(u, p)$ be the region that node u can reach if it broadcasts with power P. It is assumed that every node u knows its terrain and antennae characteristics and is able to compute the region $F(u, p)$.

The sets Accessible Neighbourhood (A_u^N), Not neighbour (*NotNbr*), and Logical Visible Neighbourhood (*LVN*(u)) of node u are initialized to *empty set*, ϕ. Node u broadcasts the '*Hello*' messages at full transmission power $P_{tx}^{\max}$ stating its position. It collects all the *ACKs* recording each nodes ID, and Location (*Loc*(v)) in the Accessible Neighbourhood set $A_u^N = \{v | Loc(v) \in F(u, P_{tx}^{\max}), v \neq u\}$ where *Loc*(v) is the location of the node v and $F(u, P_{tx}^{\max})$ is the region covered by node u at full transmission power.

For every $v \in A_u^N$, u computes the distance $d(u, v)$ and the power $p(u, v)$ and arranges them in ascending order. For every two nodes, $v, w \in A_u^N$, if v is in the relay region of node w i.e., $Loc(v) \in RL_{u \to w}$ then v is moved to the *NotNbr* set, otherwise it remains. If $Loc(w) \in RL_{u \to v}$, then node w is moved to the *NotNbr* set. Otherwise, it remains.

The Logical Visible Neighbourhood of node u i.e., *LVN*(u) is therefore given by the set A_u^N less *NotNbr* set of node u. $P\min = P(u)$ becomes the

Phase 1: *Calculating the minimal transmission power*
Input: *The set* $G = (V, E)$
Output: *Power assignment to node u*

1. $p = p_{tx}^{max} \leftarrow$ *initialize the maximum power*
2. $R_u^N = \phi \leftarrow$ *the reachable neighborhood set of node u at maximum power*
3. $NotNbr = \phi \leftarrow$ *the set of nodes that are considered none neighbors*
4. $LVN(u) = \phi \leftarrow$ *local Visible neighborhood set at reduced power*
    Begin
5. *Broadcast "Hello" message with full power p*
6. *Receive Acks and record the neighbors' id and their locations in the set* R_u^N
7. $R_u^N = \{v|Loc(v) \in F(u, p_{tx}^{max}), \quad v \neq u\}$
8. ***if*** $(R_u^N == \phi)$
9. ***Return,***
10. ***else***
11. $\forall v \in R_u^N$, *calculate distance* $d(u, v)$
12. *Sort distance in ascending order*
13. *Calculate required transmission power* $p(u, v), \quad \forall v \in R_u^N$ *using* $d(u, v)$ *and received power & propagation models.*
14. *Sort* R_u^N *by* $p(u, v)$, *in increasing order,* $\forall v \in R_u^N$
15. ***for*** *each* $v \in R_u^N$ *do*
16. ***for*** *each* $w \in R_u^N$ *do*
17. ***if*** $Loc(v) \in RL_{u \to w}$ ***then*** $NotNbr = NotNbr \cup \{v\}$
18. ***else if*** $Loc(w) \in RL_{u \to v}$ ***then*** $NotNbr = NotNbr \cup \{w\}$
19. $LVN(u) = R_u^N - NotNbr$
20. $p(u) = \max\{p(u, v)|v \in LVN(u), \quad F(u, p) \leq F(u, p_{tx}^{max})\}$

Fig. 15.3 Algorithm used to calculate the transmit power at node u.

maximum transmission power to reach the furthest node in the set $LVN(u)$. This value is multiplied by β to obtain a more optimal value.

15.5.2 Phase 2 — Removal of Unidirectional Links

Figure 15.4 depicts the algorithm executed to achieve bi-directionality. Bi-directional links are quite important for link level acknowledgements and for packet transmissions and retransmissions over the unreliable wireless medium. In phase two of the algorithm, unidirectional links generated in phase 1 are removed so as to obtain bi-directional edges using edge addition. The resulting topology is given by $G' = (V', E')$ where $E' = \{(u,v)|(u, v) \in E(G')$, and $V' = V$, and $(v,u) \in E(G')\}$. For every $v \in LVN(u)$, if there is an edge $u \to v$, then there must be an edge $v \to u$ and $u \in LVN(v)$.

Phase 2:*Conversion of Unidirectional links to bidirectional links.*
Input:*The set $LVN(u) \leftarrow$ 1 hop uni/bidirectional link neighbors of u.*
Output:*New $LVN(u) \leftarrow$ new $LVN(u)$ with bidirectional links only.*

Convert to bidirectional
*(*convert unidirectional links to bidirectional links*)*

1. $LVN(u) \leftarrow 1$ *hop unidirectional neighbors of node u*
2. ***if*** $LVN(u) \neq \phi$
3. ***for*** *each node k in* $LVN(u)$
4. ***if*** *edge* $u \rightarrow k$ *and* $u \notin LVN(k)$
5. ***then*** $p(k,u) = p(u,k)$
6. ***then*** *add edge* $k \rightarrow u$
7. ***then*** *add node u in* $LVN(k)$
8. *repeat 2–6 for all* $u \in V$
9. ***Return*** $LVN(u)$

Fig. 15.4 Algorithm used for removal of unidirectional.

Table 15.1. Simulation parameters.

Traffic pattern	CBR	RxThreshold	3.652e-10
Transmission protocol	UDP	CPThreshold	10 dB
Routing Protocol	OLSR	Packet Rate	4/s
Slot time	$20\,\mu s$	Data Rate	11 Mbps
Max Transmit Power	0.2818	Transmission bandwidth	2 MHZ
Max Distance	250 m	ACK frame	38bytes
h_t, h_r, G_t, G_r	1.0	Carrier Frequency	2.4 GHz
$CW_{\min}$	25	Basic rate	2 Mbps
$CW_{\max}$	1024	Packet payload	512 bytes

15.6 Simulation Results

In this section, a simulation study is carried out to evaluate the performance of the proposed algorithm. Table 15.1 depicts the parameters used.

The results of LM-SPT(Opt) are compared against the maximum transmission power scheme, and with those of other variations of Beta (β). A total of 10 sample networks that have been randomly generated are considered. Each network is set to have 10 to 100 nodes randomly distributed on a rectangular region of $1200 \times 1200\,m^2$. For each of the N nodes networks, N/2 CBR-traffic source-destination pairs are specified. The simulations are carried out in the NS-2 environment. The IEEE 802.11b based networks are considered.

The following metrics are of primary interest in this simulation study.

15.6.1 Average Node Degree

This is measured in order to evaluate the average connectivity, ψ, which is defined as follows

$$\psi = \frac{1}{N}\sum_{u=0}^{N-1} C_u, \tag{15.7}$$

where, $C_u = y/(N)$, y is the number of nodes reachable by node u and N is the total number of nodes in the network. The result is equivalent to summing up all the *mean connectivity* of every node in the entire network. The value of C_u should not be too large as this would imply that a node communicates even with very distant nodes and this increases interference and collision leading to reduction in throughput. It also wastes energy. On the other hand, it should not be made too small. Even though throughput would be greatly increased, too small a node degree may imply that longer paths have to be taken to reach destinations and this increases the overall energy consumption in the network. Figure 15.5 shows a comparison of the average node degree levels.

The LM-SPT(Opt) algorithm records a great reduction in the average node degree for all the networks in 1 to 10 with nodes ranging from 1 to 100, while maintaining node connectivity.

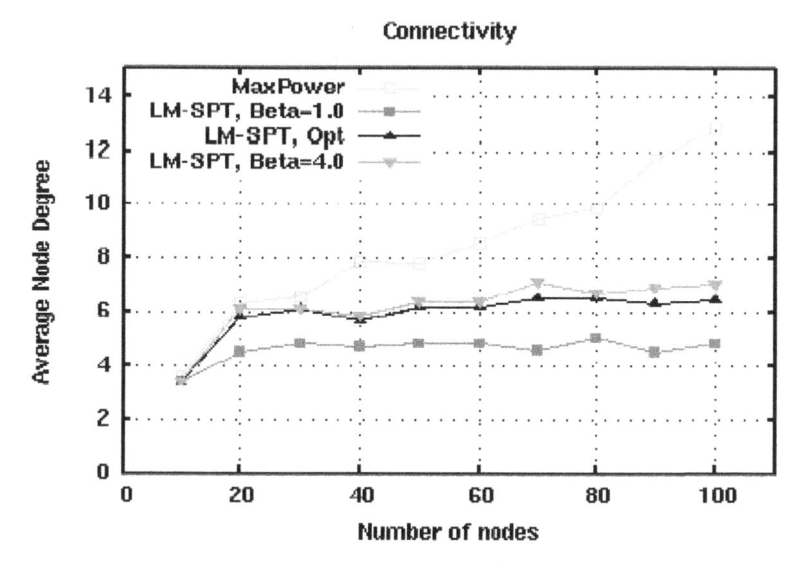

Fig. 15.5 Comparison of the average node degree for each of the networks.

In order to verify the maintenance in the network connectivity, topology graphs under the different schemes are generated as shown in Figure 15.6. It is noted that, under LM-SPT, $\beta = 1.0$, the topology is connected, however crucial minimum power paths are removed. At optimized transmission power level, the network is fully connected and the minimum power paths are maintained. When, $\beta = 4.0$, the topology is connected, however, there is not much reduction in transmission powers leading to shorter lifetimes as shown in section 15.6.2.

15.6.2 Network Lifetime

The lifetime of each network instance is considered. This is measured in terms of the number of nodes that remain alive after a period of time as defined in definition 15.7 of section 15.4. Figure 15.7 depicts performance in regard to the network lifetime. It is observable that the lifetime under LM-SPT, $\beta = 4.0$ ends at around 150 s which is just slightly better than MaxPower. This is because the transmit powers are highly increased leading to more power consumption. LM-SPT, $\beta = 1.0$ and LM-SPT(Opt) are almost equal although the former is slightly better. However, because of the reduction in the shortest power paths, spatial reuse is reduced thus reduced capacity in LM-SPT, $\beta = 1.0$.

15.6.3 Throughput Capacity

The performance is further evaluated with respect to the throughput defined as the fraction of packets that are sent from source and received successfully at the intended destination. Figure 15.8 depicts the rate of throughput per node in the 90 nodes network. It is observed that LM-SPT(Opt) performs better as compared to LM-SPT, $\beta = 1.0$ and Max-Power. This is due to the fact that minimum power paths are maintained and hence increase in spatial reuse. Under the LM-SPT (Opt) a node a can communicate to another node b as node c simultaneously communicates to node d. LM-SPT, $\beta = 4.0$ leads to high throughputs because at high transmission powers minimum delays are incurred, however, this is cancelled by the poor performance in the network lifetime as was shown in Figure 15.7.

It should be noted that, the maximum calculated transmission power has to be made optimal enough to achieve an optimal average node degree that enables the maintenance of shorter paths to destinations.

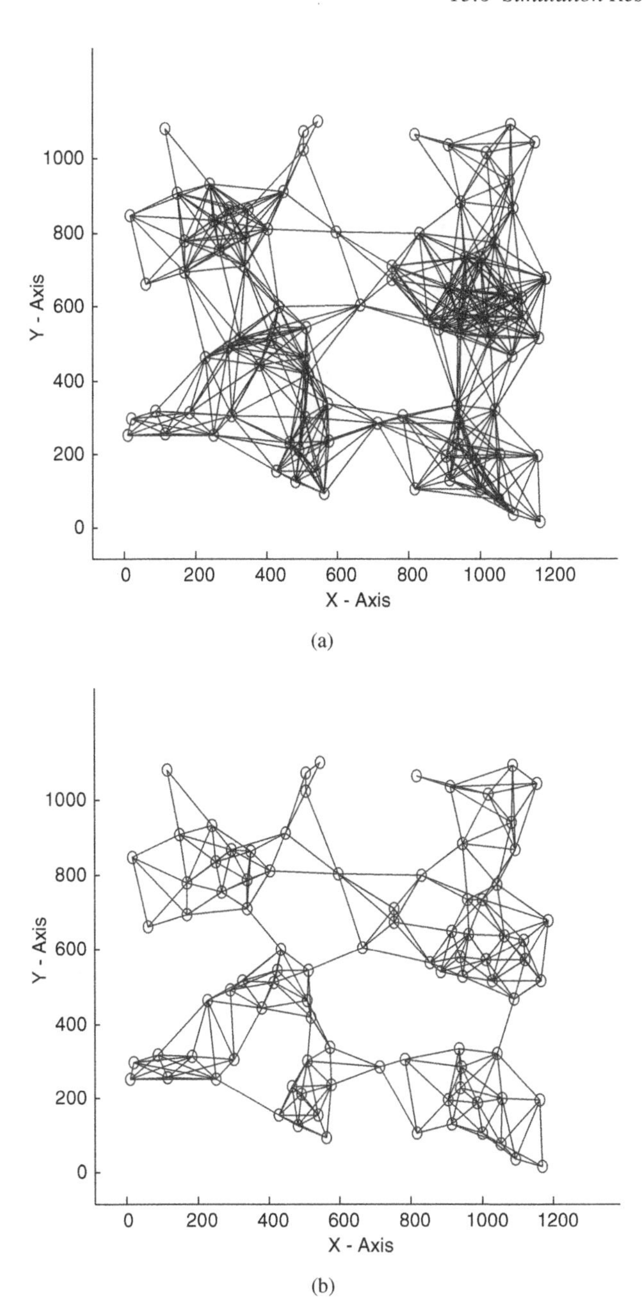

Fig. 15.6 Connectivity Graphs Generated Under the Different Schemes. (a) Topology graph generated from the 90 nodes network with the IEEE 802.11b at maximum transmission power. (b) Topology graph generated from the 90 nodes network with the LM-SPT (Opt) scheme. (c) Topology graph generated from the 90 nodes network with the LM-SPT scheme, beta=1.0. (d) Topology graph generated from the 90 nodes network with the LM-SPT scheme, beta=4.0

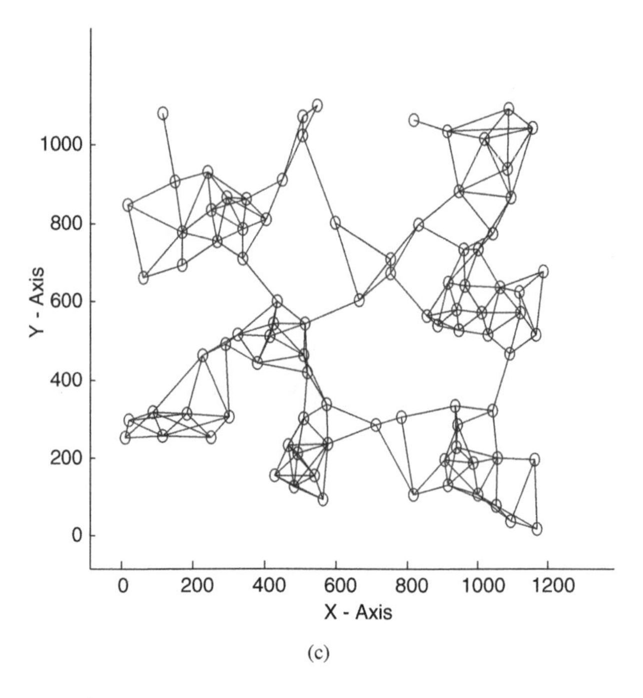

(c)

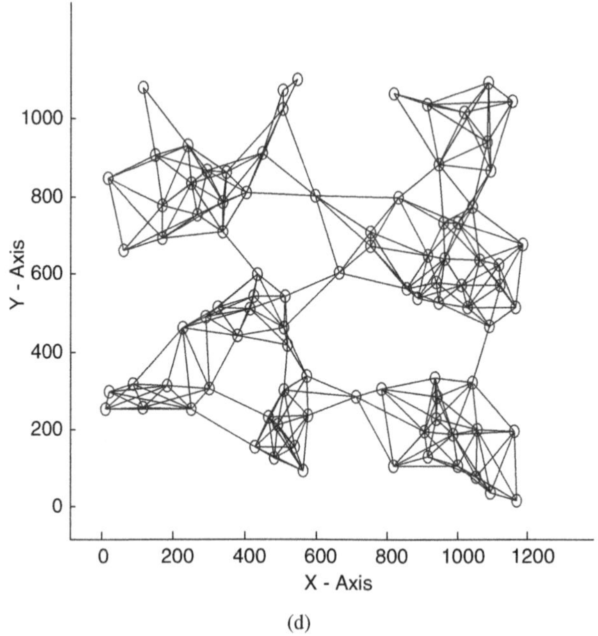

(d)

Fig. 15.6 *Continued*

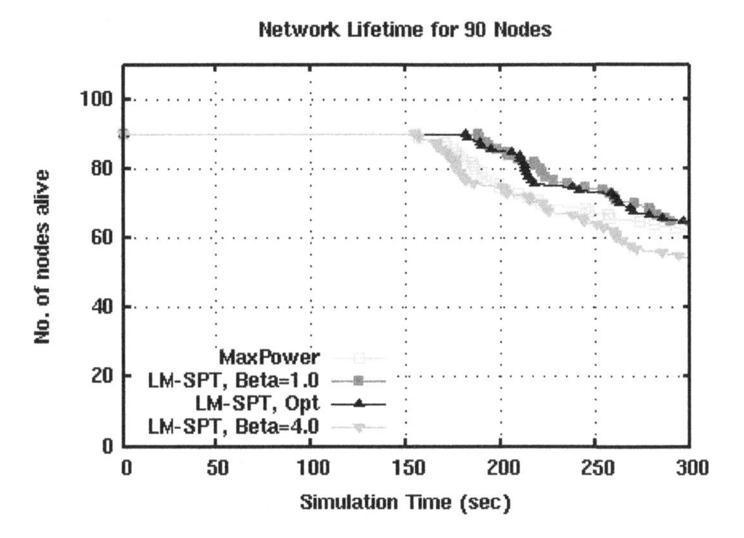

Fig. 15.7 The network lifetime over time, for the various schemes.

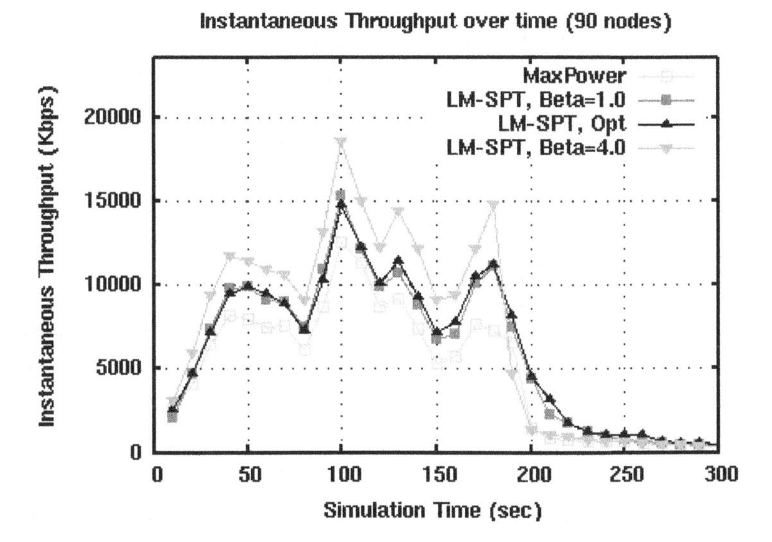

Fig. 15.8 Performance comparison of the total packets that reach their destinations successfully in the network.

This in effect ensures reduced energy consumption in the network which leads to increased network lifetime. Further, it leads to increased spatial channel re-use and hence an overall increase in the networks average throughput. Figure 15.9 depicts the performance comparisons in regard to the average

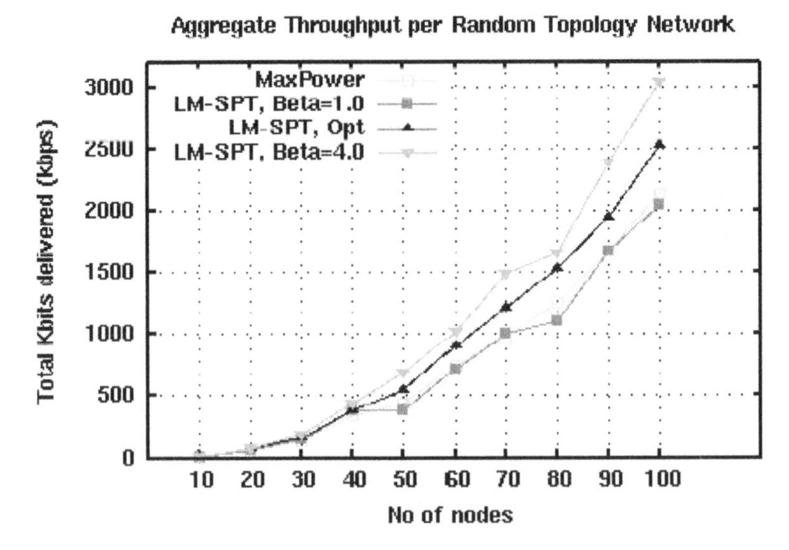

Fig. 15.9 Performance comparison in terms of the average throughput for each of the networks.

throughputs for the various schemes. Again it is notable that LM-SPT(Opt) performs better than the MaxPower scheme and the LM-SPT, $\beta = 1.0$. The performance of LM-SPT, $\beta = 4.0$ is slightly higher, but as was explained before, this is cancelled by a poor performance in the network lifetime as was shown in Figure 15.7.

15.7 Conclusion

In this chapter, a simple localized distributed topology control algorithm is proposed for applications in WMNs. Through simulations, the LM-SPT(opt) algorithm is shown to lead to an optimal reduction in the average physical node degree. This optimality leads to an efficient use of the power in the nodes thereby having longer network lifetimes. It is also shown that throughput is increased. Future work is suggested to cover a heterogeneous network where the Mesh backbone routers have varied degrees of maximum transmission range abilities.

References

[1] R. Bruno, M. Conti, and E. Gregori, "Mesh networks: commodity multihop ad hoc networks," IEEE Communications Magazine, 43(3), pp. 123–131, 2005.

[2] I. F. Akyildiz, X. Wang , and W. Wang, "Wireless mesh networks: a survey," In Computer Networks, vol. 47, pp. 445–487, 2005.
[3] P. N. Thai, H. Won-Joo,"Hierarchical Routing in Wireless Mesh Network", In the proceedings of Advanced Communication Technology (ICACT), pp. 1275–1280, February 2007.
[4] Nortel Networks Wireless Mesh Networks Solution. http://www.nortelnetworks.com/solutions/wrlsmesh/architecture.html. [Accessed Nov 2007].
[5] T. Rappaport, Wireless Communications: Principles and Practices (2nd Edition). Upper Saddle River: Prentice Hall, 2002.
[6] L. Li, J. Y. Halpern, P. Bahl, Y. Wang, and R. Wattenhofer. "Analysis of Cone-Based Distributed Topology Control Algorithm for Wireless Multi-hop Networks", In Proc. of ACM Principles of Distributed Computing Conference (PODC'01), pp. 264–273, 2001.
[7] J. Avonts, N. van den Wijngaert, C. Blondia, "Distributed channel allocation in multiradio wireless mesh networks", In proceedings of Computer Communications and Networks (ICCN), pp. 939–944, August 2007.
[8] V. Rodoplu and T. H. Meng, "Minimum energy mobile wireless networks". IEEE Journal on Selected Areas in Communications, 17(8), pp. 1333–1344, 1999.
[9] L. Li and J. Halpern, "Minimum energy mobile wireless networks revised", in: Proceedings IEEE ICC, June 2001.
[10] Y. Shen, Y. Cai, and X. Xu, "A shortest-path-based topology control algorithm in wireless multihop networks", in Computer Communication Review, 37(5), pp. 29–38, 2007.
[11] N. Li, J. Hou, and L. Sha, "Design and analysis of an MST-based distributed topology control algorithm", in Proceedings IEEE INFOCOM, June 2003.
[12] S. C. Wang, D. S. L. Wei, and S. Y. Kuo, "A topology control algorithm for constructing power efficient wireless ad hoc networks", in Proceedings IEEE GLOBECOM, December 2003.
[13] X.-Y. Li, "Approximate MST for UDG locally", in Proceedings International Computing and Combinatorics Conference (COCOON), July 2003.
[14] N. Li and J. Hou, "Topology Control in Heterogeneous Wireless Networks: Problems and Solutions", in proceedings of IEEE Infocom'04, 2004.
[15] Xiang-Yang Li, Wen-Zhan Song, and Yu Wang, "Localized topology control for heterogeneous wireless sensor networks," ACM Trans. Sen. Networks., 2(1), 2006.
[16] S. Waharte, R. Boutaba, Y. Iraqi, and B. Ishibashi, "Routing protocols in wireless mesh networks: challenges and design considerations," to appear in the Multimedia Tools and Applications (MTAP) Journal, Special Issue on Advances in Consumer Communications and Networking, 29(3), pp. 285–303, 2006.
[17] N. Bulusu, J. Heidemann, and D. Estrin, "Gps-less low-cost outdoor localization for very small devices," IEEE PersonalCommunication, October 2000.
[18] L. Li and J. Y. Halpern, "A minimum-energy path-preserving topology-control algorithm". IEEE Transactions on Wireless Communications, 3(3), pp. 910–921, 2004.
[19] F. O. Aron, T. O. Olwal, A. Kurien, and M. O. Odhiambo, "Energy efficient topology control algorithm for wireless mesh networks", in the proceedings of the IWCMC08, 2008.
[20] Paolo Santi. "Topology control in wireless Ad Hoc and sensor networks." ACM Computing Survey, 37(2), pp. 164–194, 2005.

16

Cross-Layer Performance Modeling and Control of Wireless Channels

Dmitri Moltchanov

Tampere University of Technology, Finland

To optimize performance of applications state-of-the-art wireless access technologies incorporate a number of advanced channel adaptation mechanisms at different layers of the protocol stack. Having the same aim to improve performance of applications they affect the way communication is done differently and their joint effect is often difficult to predict. Recently, to evaluate joint operation of these mechanisms cross-layer performance models started to appear. These models abstract functionality of layers providing channel adaptation and characterize performance of information transmission at the data-link or higher layers, where it is usually measured.

This chapter is organized as follows. We briefly state the need and motivation for cross-layer performance modeling and control in modern wireless access networks in Section 16.1 Then, in Section 16.2 we review channel adaptation mechanisms and demonstrate their effect on performance metrics provided to higher layers. Basic principles and modular structure of cross-layer modeling is discussed in Section 16.3. Design implications revealed by these studies are summarized in Section 16.4. Example of centralized performance control system is discussed in Section 16.5. Summary is given in the last section.

16.1 Introduction

To optimize performance of applications in wired networks it is often sufficient to control performance degradation caused by packet forwarding in network routers. Even though this is not trivial task, dealing with wireless networks we also have to take into account performance degradation caused by incorrect reception of channel symbols at the air interface. These errors propagate to higher layers leading to corruption of protocol data units (PDU). Even if higher layer protocols are able to detect and retransmit corrupted PDUs it leads to unnecessary increase of delay experienced by these PDUs.

Being inherently prone to transmission errors, modern wireless access technologies incorporate a number of advanced channel adaptation mechanisms including multiple-in multiple-out (MIMO) antenna design, adaptive modulation and coding (AMC), automatic repeat request (ARQ), forward error correction (FEC), power control, etc. Implemented at different layers of the protocol stack all these functionalities try to improve performance of information transmission over wireless channels. Moreover, various channel adaptation mechanisms affect performance provided to applications differently and their joint effect is often difficult to predict. To optimize functionality of these mechanisms a cross-layer control performance control system is needed.

To enable cross-layer optimization capabilities, different layers of the protocol stack should be allowed to interact with each other exchanging some control information. This information is to be used by a certain performance control entity to determine the set of protocol parameters providing optimized performance at any given instant of time. Up to date there were a number of proposals for cross-layers design of the protocol stack at the air interface. Depending on the state of the wireless channel they dynamically change protocol parameters at different layers such that the performance metric of interest is optimized. However, most of those frameworks are mainly based on intuition and not supported by detailed cross-layer performance modeling studies. Recently, such studies started to appear. It is expected that they provide new insights into joint behavior of channel adaptation mechanisms and reveal which cross-layer interactions are necessary and sufficient for their coordinated and optimized operation.

Unfortunately, there are no cross-layer performance modeling approaches that are versatile enough to apply to any wireless technology. Instead, recent studies consider a certain combination of channel adaptation mechanisms. Given already existing database of models proposed so far, it is extremely difficult for a newcomer to the field to choose a point to start from. In this chapter we review cross-layer performance modeling frameworks proposed to date for various organizations of the protocol stack at the air interface. We highlight basic features, review applicability and discuss conclusions of these frameworks. We also summarize design implications revealed by these studies. We also believe that this chapter will be helpful for experienced analysts as it provides detailed overview of the work has been done to date.

16.2 Adaptation Mechanisms

There are a number of factors that influence performance of wireless channels. The most important are traffic characteristics of applications, error-prone nature of wireless channels, and protocols with a set of their parameters. Each application is characterized by its own traffic characteristics and they may affect the way traffic is treated in servicing systems in the network. Environmental characteristics of landscapes and movement of a user are stochastic factors determining propagation characteristics of a wireless channel. Protocols and their parameters determine how a given traffic is treated on a wireless channel. Thus, evaluating performance that a given application receives running over wireless channels is complicated task involving a number of interdependent stochastic and deterministic factors.

To provide optimized performance, state-of-the-art wireless access technologies implement a number of channel adaptation mechanisms. These are FEC techniques, ARQ mechanisms, adaptive size of PDUs at different layers, AMC schemes, MIMO antenna design, etc. At the application layer adaptive compression and coding (ACC) can be enforced to reduce the rate required from the network. All these mechanisms are implemented at different layers of the protocol stack and affect performance provided to applications differently. To provide their unified and optimized operation, wireless access technologies call for novel design of the protocol stack that should now include cross-layer performance optimization and control functionalities. Various channel adaptation mechanisms and their places in the protocol stack are marked by grey color in Figure 16.1, where nRT stands for non-real-time applications, RT refers to real-time applications. To understand what performance level can be

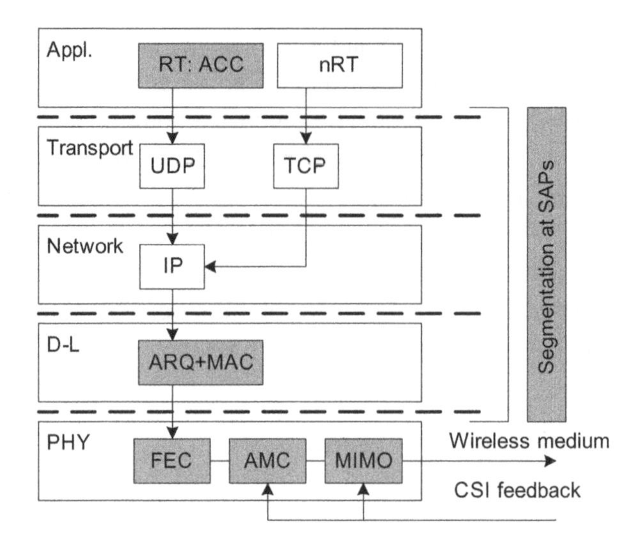

Fig. 16.1 Adaptation mechanisms affecting performance of wireless channels.

provided to applications under different wireless channel conditions, studies of joint operation of these mechanisms are required.

Performance of wireless channels is conventionally estimated at the data-link layer. There are a number of reasons behind that. First of all, data-link layer incorporates medium access procedures that are expected to significantly affect performance provided to higher layers. Secondly, channel adaptation mechanisms are usually defined for data-link or physical layer protocols. From this point of view, performance models at the data-link layer abstracts functionality of underlying protocols describing performance provided to higher layers. Recently, models describing performance of applications at layers higher than data-link started to appear. The reason is that any modern wireless access technology is expected to be an integral part of IP-based Internet. Performance parameters in in IP networks is usually measured and standardized at IP layer. As a result, those models take a new so-called cross-layer approach and incorporate additional features including explicit modeling of error control mechanisms, segmentation and reassembly between different layers of the protocol stack, source rate adaptation, multiple buffers at different layers, etc. Such models are versatile enough allowing to quantify the effect of various parameters of the protocol stack as well as characteristics of propagation environment.

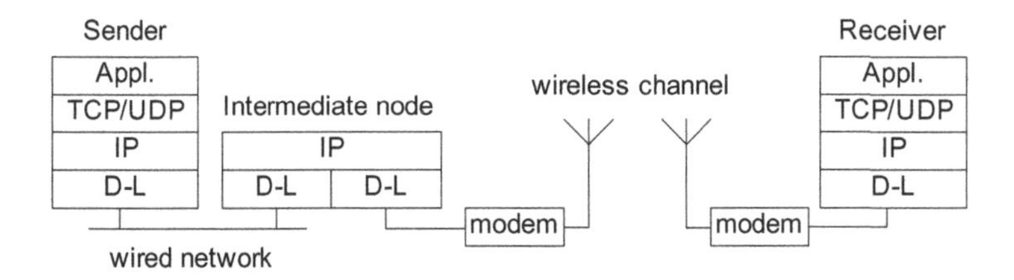

Fig. 16.2 The system model.

Further in this chapter we consider numerical examples demonstrating the effect of various channel adaptation schemes on delay and loss performance provided to IP layer. They are based on the cross-layer performance evaluation model introduced in [1]. This model allows to evaluate the effect of different channel adaptation mechanisms on performance provided to applications in terms of packet loss probability and mean packet delay. Supported channel adaptation mechanisms include FEC, ARQ, buffering space at the IP layer, segmentation procedures between different layers, adaptive applications, etc. The system model used in [1] is shown in Figure 16.2. This system best corresponds (but not limited to) to the downlink of a cellular wireless system, where constant bit rate (CBR) wireless channel is exclusively assigned to a single user. Depending on the type of application either TCP or RTP/UDP are assumed at the transport layer. The wireless channel is considered as a bottleneck meaning that IP packet losses may occur due to both insufficient buffer space and error-prone nature of wireless channel. This is not restrictive assumption as long as rates of wireless technologies are still far below than those of wired links. Since such scenario is common in today's networks, most studies do not resort to a particular wireless technology. Instead, a number of channel adaptation mechanisms are assumed to be implemented and their effect on performance experienced at higher layers is evaluated. This system model is often used in cross-layer modeling and control studies.

Forward Error Correction

Conventional error correction techniques such as forward error correction (FEC) and automatic repeat requests (ARQ) are crucial for satisfactory performance of wireless channels. FEC procedures use proactive approach eliminating the influence of bit errors in advance introducing error correction redundancy. This redundancy is exploited at the receiver to recover from bit

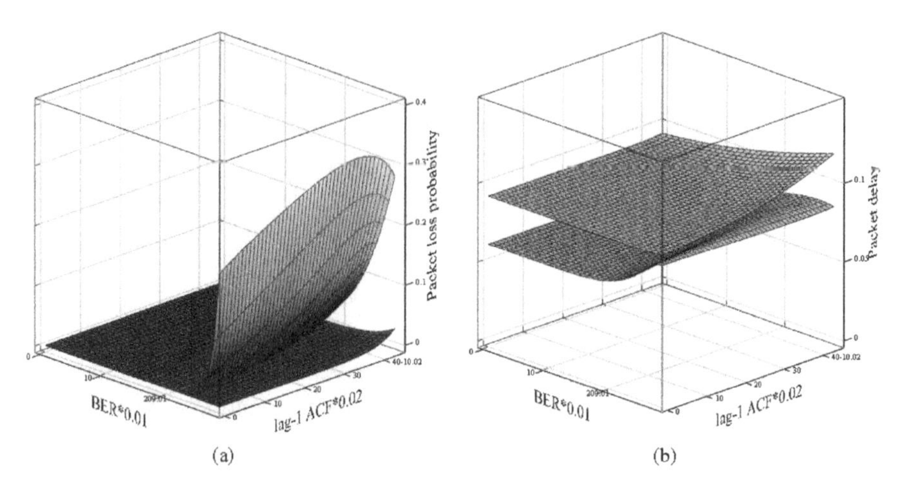

Fig. 16.3 The effect of FEC on delay and loss performance. (a) Packet loss probability, (b) Packet delay.

errors. The advantage of FEC schemes is that they do not introduce retransmission delays allowing some information to be lost. Due to sufficient complexity of implementation and high processing requirements FEC is mainly used in wide and metropolitan area wireless networks.

The effect of FEC on performance provided to higher layers can be quite complicated as illustrated in Figure 16.3, where delay and loss metrics of a wireless channel as a function of bit error rate (BER) and lag-1 autocorrelation coefficient of the bit error process are shown. The upper and lower planes in Figure 16.3a correspond to (255,131,18) and (255,87,26) Bose-Chaudhuri- Hocquenghem (BCH) codes, respectively. In Figure 16.3b the upper plane is for (255,87,26) FEC code while the lower one corresponds to (255,131,18) FEC code. Considering loss performance in isolation it is easy to notice that (255,87,26) FEC code performs better for all values of BER and lag-1 ACF. However, taking into account mean delay experienced by a packet (255,131,18) FEC code shows better results for small values of BER and lag-1 ACF. However, for bigger values of BER and lag-1 ACF this code becomes unacceptable due to drastic increase in packet loss probability. It is also interesting to observe that further increase in the BER results in decrease of the mean delay for (255,131,18) FEC code. However, this is achieved at the expense of increased packet loss probability that approaches 0.4 and becomes unacceptable for both real-time and non-real-time applications. Note that further increase in BER would eventually result in loss becoming closer to 1. Still for small values of BER and lag-1 ACF (255,131,18) FEC code is preferable. When these parameters increase (255,87,26) FEC code demonstrates better results.

Automatic Repeat Request

ARQ eliminates the influence of bit errors allowing to retransmit incorrectly received frames. To notify the sender about the erroneously received frame, ARQ protocols require a feedback channel. We distinguish between Stop-and-Wait (SW), Go-Back-N (GBN), and Selective Repeat (SR) ARQ schemes. According to the former approach the source transmits a frame and then waits for acknowledgement frame from the receiver. Go-Back-N ARQ is a scheme where frames are consecutively transmitted. When a frame is incorrectly received the receiver asks to retransmit all frames starting from incorrectly received one. According to SR-ARQ scheme only incorrectly received frames are retransmitted. When the channel conditions are relatively "bad" ARQ may introduce significant delays.

Figure 16.4 illustrates the effect of FEC. The upper plane in Figure 16.4a corresponds to 3 retransmission attempts allowed for ARQ, while the lower one is for 6 retransmission attempts. It is easy to notice that with 3 retransmission attempts the delay and loss performance provided by wireless channel becomes unacceptable for big values of BER. A bit different effect is observed in Figure 16.4b where mean delay is shown for 3 (lower plane)

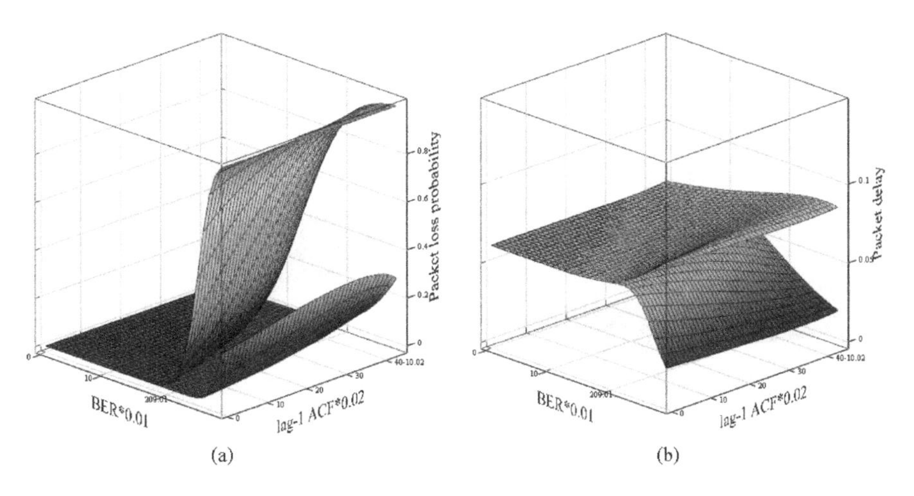

Fig. 16.4 The effect of ARQ on delay and loss performance. (a) Packet loss probability, (b) Packet delay.

and 6 (upper plane) retransmission attempts. When both BER and lag-1 ACF increase mean delay experienced by IP packets decreases. However, it mainly comes at the expense of serous losses and obviously caused by unsuccessful attempts to transmit first few frames in a packet. Further increase in the number of allowed retransmission attempts will result in corresponding increase of mean delay. Considering the effect of the maximum number of retransmissions it is easy to observe that for moderate values of BER and lag-1 ACF better performance can be obtained increasing it.

Hybrid ARQ/FEC schemes

Due to complementary advantages, FEC and ARQ are often used in combination. We distinguish between type I and type II hybrid ARQ (HARQ). In the former case the FEC code applied to a codeword remains constant for all retransmissions performed by ARQ. Type II HARQ allows to incrementally increase the correction capability of FEC codes applied to codewords in case of unsuccessfully transmission attempt. An important advantage of type II HARQ is that it does not need additional feedback from the opposite site of the link as the channel status is implicitly returned using negative or positive acknowledgements. Type II HARQ is also known as incremental redundancy ARQ.

In Figure 16.5 we illustrate the effect of type I hybrid ARQ scheme. Figure 16.5a compares the loss performance of (255,131,18) FEC code

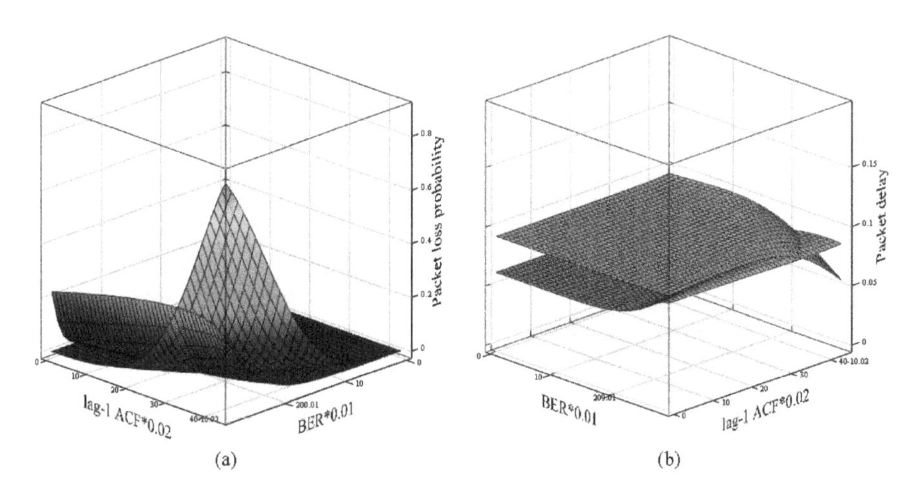

Fig. 16.5 The effect of hybrid ARQ/FEC on delay and loss performance. (a) Packet loss probability, (b) Packet delay.

combined with 6 allowed retransmission attempts (a plane with higher peak) and (255,87,26) FEC code with 3 retransmission attempts. As one can observe the scheme with (255,87,26) FEC code and 3 retransmission attempts provides better performance for most values of BER and lag-1 ACF except for extremely large values where (255,131,18) FEC code combined with 6 retransmission attempts demonstrates better results. This effect is mainly due to larger number of retransmission attempts. However, adding mean delay to the mix (Figure 16.5b) we see that for small values of BER and lag-1 ACF (255,131,18) FEC code with 6 allowed retransmission attempts (lower plane) is preferable for both real-time and non-real-time applications. Indeed, the probability of packet loss is roughly the same for both schemes while the mean delay is lower for (255,131,18) scheme. These results reveal that providing the best possible quality to applications running over wireless channel is complex task that should involve real-time monitoring of wireless channel statistics and subsequent control of parameters of channel adaptation mechanisms.

Buffering

Buffering is one of the major elements in the protocol stack that may significantly affect performance provided to higher layers. Intuitively, large buffer space may lead to significant delays and/or variation of delay introduced to the stream of packets, especially, when extensive error correction is used making channel almost fully reliable. From the other hand, maintaining small buffer size may lead to unnecessary losses prior to packets injection into the network. As a result, one can expect a complex trade-off between performance parameters provided to applications and the amount of buffer space required for reliable communication. However, note that nowadays buffer memory is not a scarce resource even for mobile terminals.

IP packet loss probability and mean delay experienced by IP packets are shown in Figure 16.6. Note that these metrics include delays and losses induced by both buffering at the IP layer and operation of the data-link layers including both FEC and ARQ. For all figures presented here both FEC code and number of retransmission attempts were kept constant. The upper plane in Figure 16.6a corresponds to the buffer size of 60 packets, while the lower one is for 30 packets. As we expected, increase in the buffer size leads to better quality in terms of packet loss probability. Indeed, for large values of BER packet loss probability reaches 1 for buffer size set to 30 packets. Increasing it two

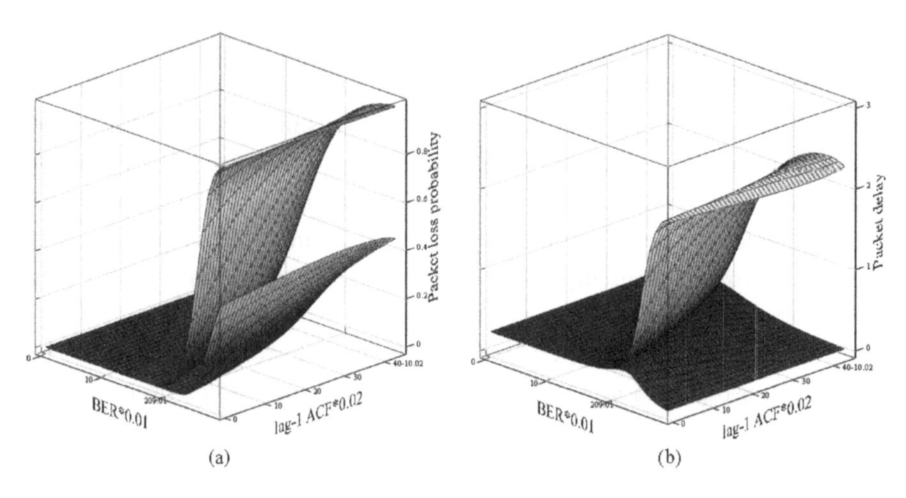

Fig. 16.6 The effect of buffering on delay and loss performance. (a) Packet loss probability, (b) Packet delay.

times significantly decreases packet loss probability. Mean delay is illustrated in Figure 16.6b, where upper plane corresponds to the buffer size set to 60 packets, and lower plane is for 30 packets. It is easy to observe that decrease in the packet loss probability for buffer size of 60 packets comes at the expense of increased delay. While non-real-time application may adapt to this delay without serious disruptive effects, real-time applications may not work well. Note that small delays corresponding to buffer size of 30 packets is mainly due to loss of most packets at the data-link layer.

Considering buffer space as a part of channel adaptation procedures it is important to note that increase in the buffer space always leads to better performance in terms of packet loss probability. However, delay caused by extensive buffering may not be appropriate for real-time applications. In this case performance of wireless channel should better be improved increasing the correction strength of FEC code or increasing the maximum number of retransmission attempts allowed for a single frame at the data-link layer. As a result, the choice of particular channel adaptation scheme also depends on the type of application that uses resources of a wireless channel.

Adaptive Compression and Coding

Adaptive compression and coding is another approach to improve wireless channel performance. Although it does not actually affect wireless channel characteristics it tries to adapt application requirements to current wireless channel conditions providing a user with truly best possible performance at any instant of time. Indeed, sometimes wireless channel conditions are such that there are no channel adaptation schemes suitable to provide a rate required by application with appropriate performance in terms of delays and losses. In this situation the rate at which application injects packets into the network can be decreased employing a different compression and coding scheme with lower output rate. When wireless channel conditions return to normal the compression and coding scheme can be changed back. Note that this capability is conditioned on ability of applications to change the way data are encoded on-the-fly and communicate this decision to their peer applications. This requirement is realistic for modern real-time applications such as IP telephony. This adaptation scheme is only feasible for real-time applications only.

The effect of changing application rate is illustrated in Figure 16.7, where delay and loss performance is shown for different arrival rates. Other parameters (buffer space, FEC code, number of retransmission attempt allowed for a single frame) were kept fixed. The raw rate of the wireless channel was set to 384 Kbps. The upper plane in Figure 16.7a corresponds to the application rate of 260Kbps, the one in the middle to the rate of 210 Kbps, and the latter one to the case of 160 Kbps. It is easy to see that the IP packet loss probability is high for applications rates of 260 Kbps and 210 Kbps and getting unacceptable

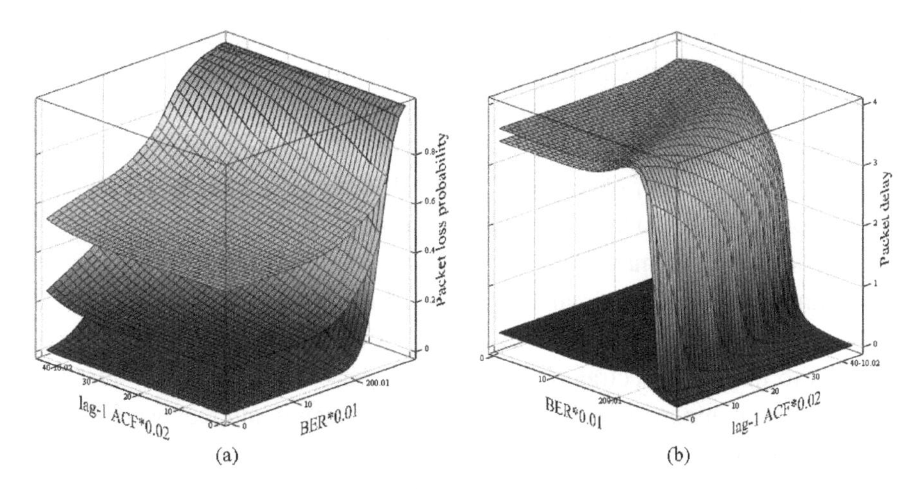

Fig. 16.7 The effect of application rate on delay and loss performance. (a) Packet loss probability, (b) Packet delay.

for big values of BER and lag-1 ACF. One the other hand, for 160 Kbps the packet loss probability remains constant and close to zero for low and moderate values of BER and lag-1 ACF. In this context it is interesting to observe delay performance under different arrival rates illustrated in Figure 16.7b. In this figure the lower plane corresponds to 160 Kbps, the one in the middle to 210Kbps, and the upper one to the case of 270 Kbps. It is interesting to observe that for 160 Kbps mean delay remains almost the same for all values of BER and lag-1 ACF. For small values of BER this is due to the fact that all incorrectly received channel symbols are successfully corrected by FEC code. For high values of BER this is mainly due to extensive losses experienced at the data-link layer. For other application rates the delay is high and decreases only when most of the packets start to get lost due to high BER. Such performance is unacceptable for both real-time and non-real-time applications.

Multiple-in Multiple-out System

Wireless channels behavior is complex and dynamic in nature as the received signal strength varies in time, frequency, and space. To effectively deal with channel fading problems, time and frequency diversity techniques are already used. The spatial diversity is another type of diversity that currently receives a lot of attention from research community. This type of diversity is implemented by maintaining arrays of antennas at receiver and/or transmitters. Depending on the application MIMO techniques are classified into two categories. First type of MIMO systems tries to improve throughput using spatial multiplexing. Second type is aimed at improving reliability of transmission using special codes. Nowadays, MIMO schemes are widely used in wireless access technologies including UMTS, IEEE 802.16, WiBRO, etc.

There are a number of transmit/receive diversity schemes. When channel state information (CSI) in terms of the signal-to-noise (SNR) ratio is available at both sides of a link maximum ratio transmission (MRT) and maximum ratio combining (MRC) are optimal transmit and receive diversity schemes. When no CSI is available at the transmitter, space-time block coding (STBC) can be used to improve spectral efficiency of wireless transmission. In this case selection combining (SC) can also be used for transmit and receive diversity. For in-depth look at MIMO systems see, for example, [2, 3].

The effect of MIMO systems on performance experienced by applications is rather simple, especially, compared to the complexity of implementation.

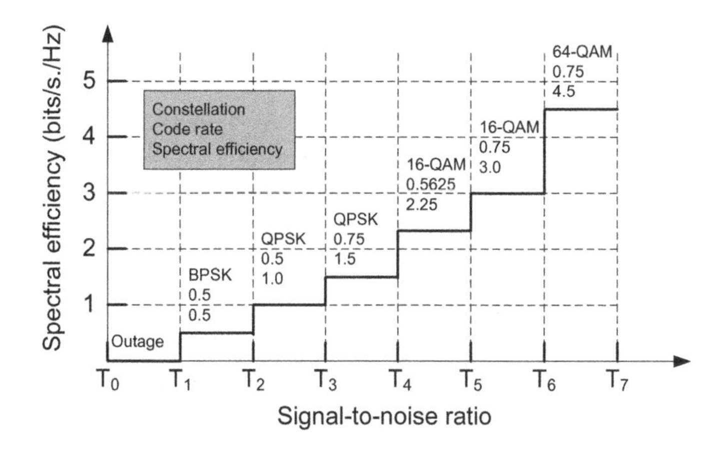

Fig. 16.8 An example of constellation and coding scheme used in AMC system.

Essentially, MIMO systems increase SNR that eventually results in lower bit error rate experienced at the physical layer. Performance gains using MIMO system can be substantial.

Adaptive Modulation and Coding

When CSI is constantly fed back to the transmitter, AMC system can be used to improve performance of information transmission. AMC is responsible for dynamic changes of constellation and coding rate such that the spectral efficiency of wireless channels is maximized. Generic AMC system works as follows. The range of the received SNR is divided into a number of intervals. These intervals are chosen such that a certain combination of constellation and coding rate provide the best possible spectral efficiency in each range of SNR. CSI in terms of the received SNR is constantly fed back to the transmitter on a frame-by-frame basis. When SNR changes, new constellation and coding rate are chosen and further used at the wireless channel. Note, that AMC itself provides significant gains in terms of optimal channel usage. However, its performance heavily depends on accuracy of SNR estimation at the receiver. Additionally, to effectively use AMC system the channel fading process must be slower than the SNR feedback sent from the receiver. Nowadays, AMC system is used in many modern wireless access technologies including UMTS, IEEE 802.16, etc. Combined with MIMO system, AMC may provide good results even in very complicated propagation environments.

AMC substantially increase performance parameters provided to applications running over wireless channels. An example of constellation and coding schemes is show in Figure 16.8, where performance of wireless channel is given using application independent metric called spectral efficiency. Spectral efficiency is defined as the maximum throughput divided by the bandwidth in hertz of a communication channel or a data link and measured in bits per second per herz of the bandwidth. For example, a transmission technique using one kilohertz of bandwidth to transmit 1000 bits per second has a spectral efficiency or modulation efficiency of 1 (bit/s)/Hz. Note that the best spectral efficiency using AMC system is usually achieved close to the transmitter when the received SNR is high. It is easy to observe from Figure 16.8 that the constellation and coding scheme used when the signal is between T_6 and T_7 results in exactly eight times better throughput compared to the case when the SNR is between T_1 and T_2.

The wireless channel implementing AMC incorporates a complex feedback-based system that receives SNR measurements from the other side of a communicating link and decides which modulation and FEC code should be used to transmit the next codeword. The main aim of the AMC system is to increase SNR using less comprehensive modulation scheme and further use the best possible FEC code such that codeword error rate is minimized. Essentially, AMC is FEC and modulation optimization scheme. Note that incorrectly received codewords are still allowed when AMC system is used. It is up to higher layer protocols to recover from them using data-link and transport layer retransmissions.

Power Control

To ensure acceptable quality of the wireless channel, some wireless access technologies incorporate features allowing to control power of the transmitter such that the target bit error rate (BER) is maintained. An example of these technologies is WCDMA air interface of UMTS. In UMTS power control consists of three loops: open loop power control, inner loop power control, and outer loop power control. Open loop power control is the ability of the transmitter to set its output power to a specific value. The inner loop power control is the ability of the transmitter to adjust its output power in accordance with one or more transmit power control commands received from the remote end. Practically, the inner loop power control eliminates influence of fast changes of propagation environment. The decision regarding whether to increase or decrease transmission power is taken by opposite site of the wireless channel. This decision is based on measurements of the received SNR and then immediately communicated to remote side of a link. Outer loop power control is used to maintain the quality of communication at the level of service requirement using as low power as possible. It is responsible for setting a target SNR for each individual inner loop power control. The target SNR is updated for each transmitter/receiver pair according to the estimated BER for each connection.

The power control is mainly used in WCDMA networks, where the whole bandwidth is simultaneously used by a number of transmitter/receiver pairs. Providing rather simple way to improve performance of a connection, power control mechanism still suffers from a number of inherent shortcomings with "cell breathing" being the most important. Cell breathing is a consequence of the power control resulting in dynamic adjustments of capacity of the system. Indeed, the lower the interference level in the network, the higher the capacity that can be achieved. It is also important to note that usage of the power control does not imply that other channel adaptation mechanisms are not used at the wireless channel. Indeed, cell breathing effect restrains us from unlimited increase of the transmission power. To ensure that interference between cells is minimized careful choice of other channel adaptation schemes' parameters should be done. However, usage of power control makes provision of acceptable performance parameters an easier task. For example, a constant modulation and FEC coding can be used at the physical layer.

16.3 Cross-layer Modeling Principles

The basic approach to cross-layer performance modeling consists of three basic steps (i) wireless channel modeling (ii) cross-layer extension of the wireless channel model to the layer of interest (iii) performance evaluation model at the layer of interest. Most cross-layer models proposed so far differs in a way they approach these problems.

Wireless Channel Modeling

Accurate modeling of wireless channel characteristics provides a starting point in any cross-layer performance modeling framework. Wireless channel characteristics are often represented in terms of the stochastic process as a function of propagation environment and transmission characteristics such as emitted power, modulation, noise, length of the codeword, length of the frame, etc. Depending on how many of these properties are implicitly or explicitly taken into account we distinguish between (i) models representing the received signal strength or related process at the physical layer (ii) PDU error models at a certain layer of interest. There are a number of models proposed so far in both categories.

To estimate performance of wireless channels, propagation models are often used. In general, we distinguish between two types of propagation models [4]. These are large-scale propagation models and small-scale propagation models. The former models focus on predicting the received local average signal strength (RLASS) over large distances between the transmitter and a receiver. Propagation models characterizing fluctuations of the received signal strength over short time durations or short travel distances are called small-scale propagation models.

When a mobile user moves away from the transmitter over large distances RLASS gradually decreases. The RLASS value can be predicted using large-scale propagation models. They usually compute it by averaging the received signal strength over movements of 1 to 10 meters. There are a number of large-scale propagation models available in literature. Free space propagation and two-ray ground models are one of the frequently used examples [4]. Most analytical and empirical large-scale propagation models assume that the RLASS decays as the power law function of distance between the transmitter and a receiver. However, neither outdoor nor indoor models take into account movement of a user between areas with different RLASS. It is easy to observe that during an active session a mobile user may move between these areas and experience different RLASS. Additionally, large-scale propagation models do not take into account rapid fluctuations of the received signal strength. Indeed, it may vary rapidly causing bit errors even when the "average channel conditions" are good. Therefore, to effectively use large-scale propagation models in performance evaluation studies they must be properly extended to capture mobility behavior of users between areas with different RLASS and local fluctuations of the signal strength in each area. The main application area of these models is at the planning stage of a mobile system where they are used to provide rough estimates of the coverage area served by a single transmitter.

When a mobile user moves over short distances the instantaneous signal strength varies rapidly. The reason is that the received signal is a sum of many components coming from different directions due to reflection, diffraction and scattering. Since the phases, amplitudes and arriving times of components are random, the resulting signal fluctuates. To capture small-scale propagation of wireless channels we have to distinguish between two cases: there is a LOS between the transmitter and the receiver and there is no LOS between the transmitter and the receiver. When there are an infinite number of scatterers in the channels that contribute to the signal at the receiver, the envelop of the channel response at any time instant has a Rayleigh probability density function (pdf)

$$p(r) = \frac{r}{\sigma^2} \exp\left(-\frac{r^2}{2\sigma^2}\right), \quad r \geq 0, \tag{16.1}$$

where σ is the scale parameters [4].

When there is a dominant LOS propagation path, the small-scale fading envelop distribution is Rician. In this case, random multipath components arriving at different angles are superimposed on a dominant signal. The effect of the dominant signal arriving with many weaker multipath signals results in Rician pdf

$$p(r) = \frac{r}{\sigma^2} \exp\left(-\frac{r^2 + A^2}{2\sigma^2}\right) I_0\left(\frac{Ar}{\sigma^2}\right), \quad r \geq 0, \quad A \geq 0, \tag{16.2}$$

where A denotes the peak amplitude of the dominant signal and $I_0(.)$ is the modified Bessel function of the first kind and zero-order. When $A \rightarrow 0$, the dominant component fades away and Rician distribution degenerates to Rayleigh one.

Small-scale and large-scale propagation models can be classified to theoretical and empirical models. Theoretical models capture fundamental principles of radio propagation such as diffraction, reflection and scattering. The abovementioned Rician and Railegh models are examples of theoretical small-scale propagation models. Unfortunately, most of those are relatively simple and cannot provide the required accuracy (see e.g., [4]). To provide better description of the received signal strength measurement-based empirical models started to appear. In such models, all physical propagation phenomena are taken into account regardless of whether they can be isolated and separately recognized. Their accuracy depends on similarities between the environment under consideration and the environment where measurements have been carried out. For this reason empirical models are not the best choice for cross-layer performance modeling studies.

In [17] authors analyzed SNR measurements collected over IEEE 802.11b wireless channel. They found them to follow normal distribution with geometrically decaying autocorrelation function (ACF). They proposed to capture these properties using autoregressive process of order 1, AR (1), in the form $X(n) = \varphi_0 + \varphi_1 X(n-1) + \varepsilon(n)$, $n = 0, 1, \ldots$, where φ_0 and φ_1 are some constants, $\{\varepsilon(n), n = 0, 1, \ldots\}$ are independently and identically distributed (iid) random variables having the same normal distribution with zero mean and variance $\sigma^2[\varepsilon]$. AR(1) process is fully characterized by a triplet $(\varphi_0, \varphi_1, \sigma^2[\varepsilon])$. Authors demonstrated that these parameters of AR(1) models can be found as follows

$$\varphi_1 = K_Y(1), \quad \varphi_0 = \mu_Y(1 - \varphi_1), \quad \sigma^2[\varepsilon] = \sigma^2[Y](1 - \varphi_1^2), \tag{16.3}$$

where $K_Y(1)$, $\mu_Y(1)$, and $\sigma^2[Y]$ are the lag-1 autocorrelation coefficient, mean and variance of SNR observations, respectively. It is important to note that the proposed model is valid when SNR is sampled over relatively long time intervals, says, 0.1–1s.

Recent measurement studies of the received signal strength and related processes demonstrated that they are characterized by strong degree of autocorrelation. To model these processes hidden Markov models (HMM) are often used. HMMs, also known as Markov modulated processes, are probabilistic functions of Markov chains [6]. If the stochastic variable of interest is the state of the Markov process, we are given a conventional Markov model whose outputs are directly observable. A straightforward way is to associate each state of the "observable" Markov model with a certain random variable. Simulating this model we do not directly observe the state of the Markov chain, but stochastic process defined "over" the Markov chain. It has been shown that HMMs can be effectively used to model a wide-variety of fading characteristics, including frequency-selective [7], flat [8], slow [5], and fast fading channels [9]. See [5] for review.

Propagation or SNR models, while were shown to be appropriate for design of transceivers, cannot be directly used in performance evaluation studies. To successfully use propagation models in performance evaluation studies, they must be properly extended to higher layers, providing convenient characterization of the dynamic nature of the wireless channel at the layer of interest. For example, based on the specific propagation model and modulation and detection techniques bit error probability can be obtained. From this point of view, bit error models are seen as an extension of the small-scale propagation models. To provide this extension we have to take into account modulation scheme at the physical layer. Given an advanced modulation scheme, analytical expressions for bit error probability are available for simple modulation schemes only [5].

PDU error models including bit and frame error models are often represented by binary stochastic processes where 0 stands for correct PDU reception, 1 refers to incorrectly received PDU. Both bit and frame error processes are conventionally modeled using Markov chain starting from simple two-state Gilbert model [10] to quite complicated large-states high-order Markov models [11]. For comprehensive review see [12]. In general, HMMs of PDU error processes can be divided into two broad categories. These are models, based on partitioning of the received signal strength or SNR processes and models, based on direct fitting of parameters of HMMs.

Markov model, based on partitioning of the SNR was firstly proposed by Wang and Moayeri [8]. The authors assumed a Rayleigh fading channel and used a Markov process with finite number of states as follows. Assume a discrete-time Markov chain $\{S(n), n = 0, 1, \ldots\}$, $S(n) \in \{1, 2, \ldots, M\}$. Each state of this model is associated with the so-called binary symmetric channel (BSC). BSC associated with state i, $i = 1, 2, \ldots, M$, determines how the symbol is transmitted while the channel is in the state i. The term "symmetric" stems from the fact that both 0 and 1 are incorrectly received with the same probability. Very often this property is implicitly assumed. To parameterize such a model we have to determine the transition probability matrix D, and the so-called crossover probability vector $\vec{p} = (p_1, p_2, \ldots, p_M)$ each element of which determines the probability of having a symbol in error in appropriate state of $\{S(n), n = 0, 1, \ldots\}$. State transition diagram of such model and BSCs associated with states of this model are shown in Figure 16.9.

Elements of $\vec{p}$ must be related to the received SNR. For this relation to be possible we have to take into account the modulation scheme used at the physical layer. For example, given a specific modulation scheme it is sometimes possible to compute a probability of having a symbol in error as a function

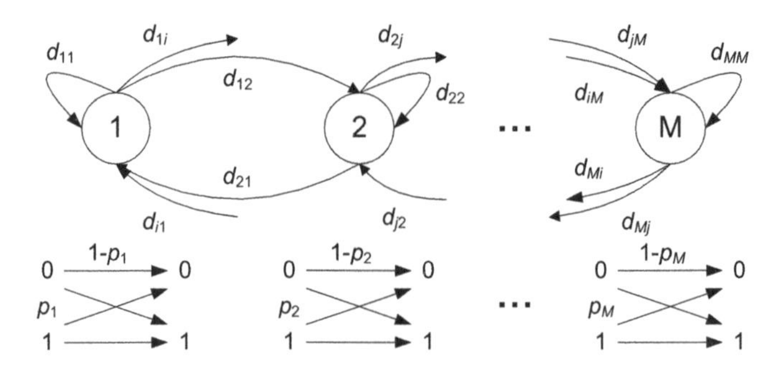

Fig. 16.9 State transition diagram of the HMM and associated BSCs.

of the received SNR. Therefore, we have to partition the received SNR into a finite number of intervals, M, where each interval represents the state of the Markov chain. Elements of the transition probability matrix, D, can be calculated using those intervals to which SNR is partitioned as shown in [8]. There are a number of issues that are still open in parameterizations of such models, i.e., how to best choose the number of intervals to which the SNR is partitioned, for which modulation techniques construction of this model is possible, etc. The authors in [5] provided comprehensive review of such models. To date usage of this model with binary phase shift keying (BPSK) [8] and differential quadrature (DQPSK) have been reported [13].

To represent bit error observations a number of models based on direct fitting to measurements data have been proposed. The first work that mathematically described the bit error pattern observed on a wireless channel is due to Gilbert [10]. The model has only two states, one of which is error-free while the other one is associated with a certain bit error probability, p. Elliott [14] extended the model allowing both states to have non-zero bit error probabilities. This model captures first-order statistics of the bit error process more accurately. The next extension is due to Fritchman [15], who allowed a Markov chain to have more than one error-free state. It was shown that this model captures error-free intervals precisely. In [16] authors reviewed previous studies and compared performance of different models.

Assume that the bit error process is covariance stationary in nature. Our aim is to model its first- and second-order properties including BER and lag-1 ACF. In most studies it is assumed that ACF is geometrically distributed, i.e., $y(m) = \lambda^m$, where λ is some constant coefficient. In order to capture properties of this process it is sufficient to use two-state Markov process. Let $\{W_E(n), n = 0, 1, \ldots\}$, $W_E(n) \in \{0, 1\}$, denote the bit error model with modulating Markov chain $\{S_E(n), n = 0, 1, \ldots\}$, $S_E(n) \in \{0, 1\}$. The model is completely defined using the set of matrices $D_E(k)$, $k = 0, 1$, containing transition probabilities from state i to state j with correct ($k = 0$) and incorrect ($k = 1$) reception of a channel symbol. To parameterize a covariance stationary binary process, only the mean and the lag-1 ACF coefficient have to be captured. Authors in [18] demonstrated that there is a unique switched Bernoulli process (SBP), matching the mean and lag-1 autocorrelation of covariance stationary bit error observations. This model is given by

$$\begin{cases} \alpha_E = (1 - K_E(1))E[W_E], \\ \beta_E = (1 - K_E(1))(1 - E[W_E]), \end{cases} \quad \begin{cases} f_{1,E}(1) = 0, \\ f_{2,E}(1) = 1, \end{cases} \tag{16.4}$$

where α_E and β_E are transition probabilities from state 1 to state 2 and from state 2 to state 1, respectively, $K_E(1)$ is the lag-1 autocorrelation of bit error observations, $E[W_E]$ is the mean of bit error observations, $f_{1,E}(1)$ and $f_{2,E}(1)$ are probabilities of error in states 1 and 2, respectively. Parameters $E[W_E]$ and $K_E(1)$ have to be estimated from empirical data. Note that this model is suitable to represent any binary process with geometrically decaying ACF. Although extension to the case of a FSMC is rather straightforward more advanced fitting procedures have to be used [26].

In [19] Lui *et al.* developed a model for wireless channel with AMC functionality at the physical layer. The model itself a finite-state Markov chain (FSMC) where each state corresponds to a certain SNR range. Authors did not allow SNR dynamics within the ranges as the middle of a range is always chosen when the current SNR value falls within this range. The number of ranges was chosen to be 7 resulting in 8 boundary points denoted by $\{T_k, k = 0, 1, \ldots, K\}$. These points are such that $T_0 < T_1 < \cdots < T_K$, $T_0 = 0$, $T_K = \infty$. Code rates in these modes were chosen to be 0, 0.5, 0.5, 0.75, 0.5625, 0.75 and 0.75, respectively. Corresponding constellations are binary phase shift keying (BPSK), quaternary PSK (QPSK), 16-quadrature amplitude modulation (QAM), 16-QAM, and 64-QAM. With codes rates ranging from 0 to 0.75 and modulation ranging from BPSK to 64-QAM the spectral efficiency of each mode is 0, 0.5, 1.0, 1.5, 2.25, 3.0 and 4.5 b/s/Hz. When SNR gets better, a faster mode is selected. When SNR gets worse, the lower spectral efficiency is chosen. For any given frame error rate (FER) and performance requirements of a service, the boundaries $\{T_k, k = 0, 1, \ldots, K\}$ can be derived numerically using probability density function of the SNR. When CSI is enabled and assumed to fed back timely, this model captures changes in modulation schemes and code rates. As a result, each state of the FSMC can be associated with a certain bit error probability. The channel was assumed to remain constant during the frame transmission time. As a result, modulation and coding are adjusted on a frame-by-frame basis.

It is important to note that the considered AMC model cannot be used in cross-layer performance control of wireless channels. Recall that the choice of a particular modulation and coding scheme at any given time depends on many factors including the distance between the transmitter and the receiver, direction and velocity of the user, configuration of the landscape, etc. As a result, it is not possible to know in advance which scheme will be invoked and for how long. Indeed, it is only possible when the user's trajectory on a landscape as well as movement of other objects in a channel is known in advance which is, however, impossible. While this questions appropriateness of such model to study performance of AMC-enabled wireless channels in real operational environment, the model still allows to numerically evaluate performance of AMC system.

Note that when used in cross-layer performance evaluation studies bit error models do not allow to evaluate the effect of modulation and reception techniques at the physical layer. This also includes operation of MIMO system. Finally, note that there are a number of PDU error models for higher layers developed so far. As these models are mainly based on direct fitting to measurements they also abstracts operation of FEC codes at the physical layer. Keeping in mind that the ultimate goal of cross-layer studies is to represent the effect of channel adaptation mechanisms on performance of higher layer protocols frame error models are not best suited for these kind of studies.

Extension to the Layer of Interest

The received signal strength, SNR or PDU error models cannot be directly used in cross-layer studies at the IP or higher layers and must be previously extended to the layer at which performance of applications is to be evaluated. For such extension to be accurate, we have to take into account specific peculiarities of underlying layers including physical layer channel adaptation mechanisms, data-link error correction techniques, segmentation between layers, etc.

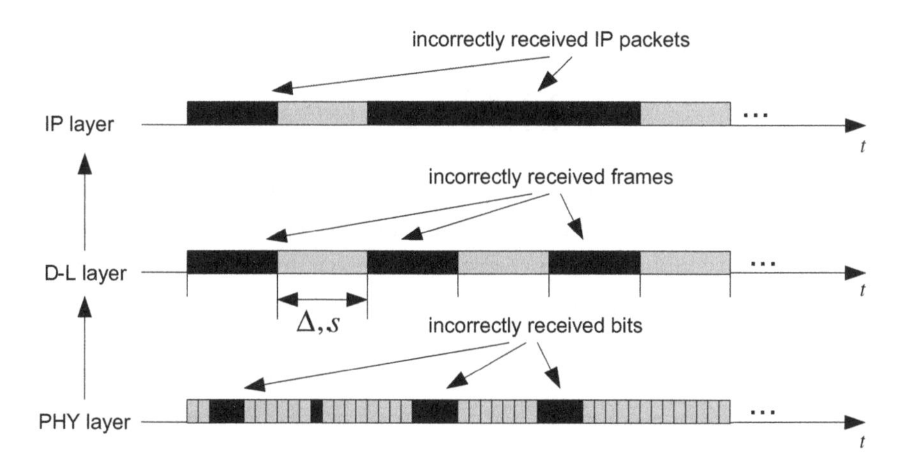

Fig. 16.10 Cross-layer performance modeling of wireless channels.

Basic principles of cross-layer extension are illustrated in Figure 16.10, where black rectangles denote incorrectly received PDUs, grey rectangles stand for correctly received PDUs. In this example, we assume that FEC is capable to correct at most one incorrectly received bit and ARQ is not used at the data-link layer. Since no error correction procedures are defined for IP layer, even a single lost frame within a packet leads to loss of a whole packet. Following those approaches described later in this section, performance metrics at the IP of higher layers can now be obtained as a function of underlying layers parameters. The approaches may differ in detail and complexity, especially, when ARQ, AMC, and MIMO functionalities are also incorporated.

Let us assume that a single codeword at the physical layer consists of exactly m bits. Let now p_B be the bit error probability. If the bit error process is uncorrelated the probability of incorrect reception of a codeword is readily found as

$$p_C = \sum_{i=1}^{m} (1 - p_B)^i p_B. \tag{16.5}$$

In (16.5) we assumed that a single bit error results in incorrect reception of the whole codeword. In practice, however, channel coding is used to deal with incorrect reception of bits. If we assume that FEC code can correct up to l incorrectly received bits we can approximate the codeword error probability using probabilities of binomial distribution

$$p_C = \sum_{i=l}^{m} \binom{m}{i} p_B^i (1 - p_B)^{m-i}. \tag{16.6}$$

It is important to note that that (16.5) and (16.6) are approximate results resembling basic operation of channel coding. However, they were found to provide fairly accurate results in most cases. Also note that in real systems the length of a frame is much larger than the length of the codeword at the physical layer. As a result, we have additional segmentation of data between data-link and physical layer. Frame error probability can be found similarly to (16.5).

Previously, we assumed that the bit error process is completely uncorrelated. In practice wireless channel statistics is often characterized by strong memory properties represented using Markov models. Let $\{W_E(l), l = 0, 1, \dots\}$ be the bit error process with underlying Markov chain $\{S_E(l), l = 0, 1, \dots\}$. Let $D_B(k)$, $k = 0, 1$ denote transition probabilities between states of the Markov chain accompanied with correct ($k = 0$) or incorrect ($k = 1$) reception of a bit. To parameterize the codeword error process we have to determine m-step transition probabilities of the modulating Markov chain $\{S_E(n), n = 0, 1, \dots\}$, $n = lm$, with exactly k, $k = 0, 1, \dots, m$, incorrectly received bits. Denote the probability of transition from state i to state j for the Markov chain $\{S_N(n), n = 0, 1, \dots\}$ with exactly k, $k = 0, 1, \dots, m$, incorrectly received bits in a pattern of length m by $d_{N,ij}(k) = \Pr\{W_N(n) = k, S_N(n) = j | S_N(n-1) = i\}$ and let the set of matrices $D_N(k)$, $k = 0, 1, \dots, m$ to contain these transition probabilities. These matrices can be found using $D_B(k)$, $k = 0, 1$, as follows

$$\begin{aligned}
&D_N(0) = D_E^m(0), \\
&D_N(1) = \sum_{k=m-1}^{0} D_E^{m-k-1}(0) D_E(1) D_E^k(0), \\
&D_N(2) = \sum_{k=0}^{m-2} D_E^k(0) D_E(1) \sum_{i=m-k-2}^{0} D_E^{m-i-k-2}(0) D_E(1) D_E^i(0), \\
&\dots \\
&D_N(m-1) = \sum_{k=m-1}^{0} D_E^{m-k-1}(1) D_E(0) D_E^k(1), \\
&D_N(m) = D_E^m(1).
\end{aligned} \tag{16.7}$$

Note that computation according to (2) is a challenging task and becomes impossible when m is large. Instead, one may use the recursive method as outlined below. Let us extend the definition of $D_N(k)$, $k = 0, 1, \ldots, m$, as follows. We denote the probability of transition from state i to state j for the Markov chain $\{S_N(n), n = 0, 1, \ldots\}$ with exactly k, $k = 0, 1, \ldots, m$, incorrectly received bits in a bit pattern of length m by $d_{N,ij}(k, m)$. Let the set of matrices $D_N(k, m)$, $k = 0, 1, \ldots, m$, contain these transition probabilities. Since at most two errors may occur in two consecutive slots, we have the following expression for $D_N(i, 2)$, $i = 0, 1, 2$

$$D_N(i,2) = \sum_{k=0}^{i} D_N(k) D_E(i-k), \quad i = 0, 1, 2, \tag{16.8}$$

where $D_E(2)$ is the matrix of zeros.

Recursively, we get

$$\begin{aligned} D_N(i,3) &= \sum_{k=0}^{i} D_N(k,2) D_E(i-k), \quad i = 0, 1, \ldots 3, \\ D_N(i,4) &= \sum_{k=0}^{i} D_N(k,3) D_E(i-k), \quad i = 0, 1, \ldots 4, \\ \ldots & \qquad\qquad\qquad\qquad \ldots \\ D_N(i,3) &= \sum_{k=0}^{i} D_N(k,2) D_E(i-k), \quad i = 0, 1, \ldots, m, \end{aligned} \tag{16.9}$$

where $D_E(k)$, $k \geq 2$, and $D_N(i, m)$, $i \geq m + 1$, are all zero matrices.

Once $D_N(k, m)$, $k = 0, 1, \ldots, m$, are obtained we can estimate probability of incorrect reception of a codeword as

$$p_C = \vec{\pi}_C \sum_{i=l}^{m} D(i,m) \vec{e} \tag{16.10}$$

where $\vec{\pi}_C$ is steady-state vector of Markov process $\{S_N(n), n = 0, 1, \ldots\}$, l is the number of errors that can be corrected by a FEC code, and $\vec{e}$ is the vector of ones of appropriate size. The steady-state vector of $\{S_N(n), n = 0, 1, \ldots\}$ can be computed as the solution of the following system

$$\begin{cases} \vec{\pi} D_N = \vec{\pi} \\ \vec{\pi} \vec{e} = 1 \end{cases}, \tag{16.11}$$

where $D_N = \sum_{i=0}^{m} D(i, m)$ is the transition probability matrix of $\{S_N(n), n = 0, 1, \ldots\}$.

Using similar technique the codeword error process can be extended to the data-link layer to obtain parameters of the frame error process. When ARQ is not used at the wireless channel this technique can be also applied to derive IP packet error process. When ARQ is used we may proceed as follows. Assuming that the wireless channel statistics are covariance stationary or can be represented as piecewise covariance stationary process, we observe that the frame error process becomes memoryless, i.e., can be completely defined using the frame error probability. We assume that the wireless channel in the reverse direction is reliable and feedback frames are delivered instantaneously. Indeed, feedback frames are usually small in size and well protected by a FEC code. Thus, it implies that packet losses happen only in the direction from the sender to the receiver due to buffer overflow at the intermediate system. These assumptions were used in many studies and found to be appropriate for wireless links with high rates and short propagation time. Moreover, these assumptions allow to use a single model to capture different variants of ARQ, including SW, GBN, and SR regimes.

In what follows, we are interested in performance metrics provided to IP layer, namely, IP packet loss probability and delay of an arbitrary packet. Since persistent ARQ is the special case of non-persistent one with the number of retransmission attempts allowed for a single frame set to infinity we consider the latter. Let the number of frames to which a single IP packet is segmented be v and the number of retransmission attempts that can be made for a single frame is limited to r. We assume that whenever r successive times a frame failed to be correctly transmitted, the whole IP packet is dropped irrespective of the number of frames that have already been successfully received.

Let $d(k)$, $k = \min(r, v), \min(r, v) + 1, \dots, (r - 1)(v - 1) + r$, be the probability function of the delay of a single IP packet. Note that we define the delay irrespective of whether a given packet is successfully transmitted or not. Thus, the delay of a packet is either the amount of time to successfully transmit a packet or the amount of time till the packet is lost due to excessive number of retransmissions. The minimum delay is, therefore, $\min(r, v)$, while the maximum delay is bounded by $(r - 1)(v - 1) + r$. We obtain IP packet transmission delay using

$$d(k) = c(k) + b(k), \tag{16.12}$$

where $c(k)$ is the delay component corresponding to the case when the packet is successfully transmitted; $b(k)$ is the delay component induced by unsuccessful transmission of the packet.

Let g_L be the probability that a frame is lost due to excessive number of retransmission attempts. We can obtain g_L using the frame error probability $f(1)$ as follows

$$g_L = \sum_{i=r+1}^{\infty} f^{i-1}(1) f(0) = 1 - \sum_{i=1}^{r} f^{i-1}(1) f(0). \tag{16.13}$$

Let $g_{L,i}, i = 1, 2, \dots, v$, be the probability that the IP packet is dropped due to excessive number of retransmissions made for the i-th frame. Since successive frame transmission times are independent, $g_{L,i}, i = 1, 2, \dots, v$, can be found using g_L as follows

$$g_{L,i} = (1 - g_L)^{i-1} g_L, \tag{16.14}$$

Using (16.14), we can also obtain the loss probability of an IP packet due to excessive number of retransmission attempts made for a frame at the data link layer as

$$p_L = 1 - (1 - g_L)^v. \tag{16.15}$$

Let now A be the event of successful frame transmission and I_A be its indicator, i.e., $I_A = 1$ when the frame is successfully received and $I_A = 0$ otherwise. Also let U, $U = 1, 2, \dots$ be the random variable denoting the number of transmission attempts required to transmit a frame given that it is successfully transmitted and let $g_U(k, 1)$ be its probability distribution, i.e., $g_U(k) = \Pr\{U = i | I_A = 1\}$. Since probabilities of incorrect frame reception in successive slots are independent from each other, time to successfully transmit a single frame is given by geometrical distribution

$$g_U(k) = \Pr\{U = k | I_A = 1\} = \frac{(1 - f(0))^{k-1} f(0)}{\sum_{i=1}^{r} f^{i-1}(1) f(0)}, \quad k = 1, 2, \dots, r - 1. \tag{16.16}$$

Now consider that a packet consists of exactly i, $i = 1, 2, \dots, v$, frames. Let $a(i, j), i = 1, 2, \dots, v, j = i, i + 1, \dots, i(r - 1)$, be probability distribution of delay of a single packet on j time slots given that it consists of exactly

i frames and all of them are successfully transmitted. Obviously, $g_U(k) = a(1,k)$.We find $a(2,j)$ using convolution of (10) as follows

$$a(2,j) = \sum_{k=0}^{j} a(1,k)a(1,j-k). \tag{16.17}$$

Extending (16.17) to the case $i = 2, 3, \ldots, v$, we get

$$a(i,j) = \sum_{k=0}^{j} a(i-1,k)a(1,j-k), \quad j = 1, 2, \ldots, r-1. \tag{16.18}$$

Note that in (16.18) the maximum delay for each i is limited by $i(r-1)$ time slots. To compute the delay induced by successful packet transmission, we need only $a(v,j)$, $j = r, r+1, \ldots, v(r-1)$. Removing conditioning on successful packet transmission, we obtain $c(k)$ component of (16.12) as

$$c(k) = a(v,k)(1-p_L), \quad k = r, r+1, \ldots, v(r-1). \tag{16.19}$$

where p_L is the probability of packet loss obtained in (16.15).

Let $b(i,j)$, $i = 1, 2, \ldots, v$, $j = i-1+r, \ldots, (i-1)(r-1)+r$, be delay induced by a single packet given that the i-th frame failed to be transmitted due to excessive number of retransmissions. In order to determine this delay component we need $a(i,j)$, $i = 1, 2, \ldots, v-1$, $j = r, r+1, \ldots, i(r-1)$ obtained in (16.18). Observing that unsuccessful transmission of the i-th frame in a packet increases the delay component $a(i-1,j)$ by exactly r time slots we have the following relation

$$b(i, j+r) = a(i-1,j), \quad j = r, r+1, \ldots, i(r-1)+r. \tag{16.20}$$

Removing conditioning on unsuccessful packet transmission and summing over all possible numbers of unsuccessful packet transmissions, we get

$$b(k) = \sum_{i=1}^{r} b(i,k)(1-g_L)^{i-1} g_L, \tag{16.21}$$

where g_L is the probability of unsuccessful frame transmission obtained in (8). Substituting (16.19) and (16.21) into (16.12) we get a final expression for $d(k)$.

Note that the cross-layer extension affects autocorrelation properties that may exist in received signal, SNR, or bit error processes. To illustrate it let

$\{X(n), n = 0, 1, \ldots\}, X(n) \in \{0, 1\}$, be the bit error process. Assume that codewords are consecutively transmitted over wireless channel and each codeword consists of exactly m bits. It is easy to show that the process $\{Y(i), i = 0, 1, \ldots\}$, $Y(i) \in \{0, 1, \ldots, m\}$ defined as $Y(i) = \sum_{i=ni}^{n(i+1)} X(j), m = 0, 1, \ldots$ describing the number of incorrectly received bits in a codeword quickly loses its correlational properties as m increases. Typically, the length of codeword is large enough ($m > 100$) such that $\{Y(i), i = 0, 1, \ldots\}$ is completely uncorrelated. This implies that the process describing correct reception of codewords is also uncorrelated. This property is used in many cross-layer modeling studies to relax computational complexity of the performance model at the layer of interest. Practically, it means that instead of the queuing system with autocorrelated service process, one with iid service times can be used. We also note that this property is even stronger when there are guard intervals between transmission of codewords or when distributed channel access is used.

It is important to note that PDU error processes at upper layers become iid if the initial process is both covariance stationary and short-range dependent. However, there were indications that these processes may not retain these properties. Particularly, in [20], authors studied statistical characteristics of frame error traces collected at IEEE 802.11b wireless channel. Among other conclusions they revealed that the frame error process still has strong memory. A number of other studies reported similar properties of PDU error processes at layers higher than physical (see e.g., [11]). This clearly indicates that either the underlying bit error process has exceptionally strong memory or it is not stationary at all. Authors in [20] tried to capture properties of frame error observations using a number of conventional models including HMM, firsrt- and high-order Markov models and Markov models with exceptionally large state spaces. Obtained results show some inconsistency in the choice of the candidate model. Particularly, no models were found to be satisfactory for different traces captured under the same environmental conditions. In [11] authors also noticed uncertainty in accuracy of Markov models with large state spaces. Particularly, the number of states required for accurate modeling is exceptionally large (more than 100) and differs from trace to trace. Although authors did not suggest any particular reason for this phenomenon, observations made in those studies serve as an indicator of possible non-stationary behavior of PDU error processes. For example, in [21] authors firstly isolated covariance stationary parts of their traces and then demonstrated that first- and second-order statistics of the stationary frame error process in terms of the error rate and ACF can be well represented using two-states first-order Markov model similar to the Gilbert's one. Note that another plausible explanation of complex behavior of PDU error statistics is indeed long-term memory caused by shadowing of the line-of-sight (LOS) by objects in the channel. However, to date no studies reported long-range dependence of PDU error observations.

Most wireless channel models proposed so far either implicitly or explicitly assumed covariance stationarity for their observations providing no particular reason for that. The first who questioned this assumption were Konrad *et al.* [22]. Authors considered GSM bit error traces and highlighted their possible non-stationary behavior. They also proposed an algorithm to extract covariance stationary parts and demonstrated that isolated parts can be sufficiently well described using conventional Markov modeling techniques. Unfortunately, no clear theoretical basis for their segmentation algorithm was proposed. Their work was continued in [21] where authors observed similar piecewise covariance stationary behavior of SNR observations measured on IEEE 802.11b channel. They further used exponentially-weighted moving average (EWMA) change-point statistical test to distinguish between covariance stationary parts. Similar to [22] these parts in isolation are accurately modeled using AR(1) process. An example of change-point detection is shown in Figure 16.11, where the initial SNR process is shown in the left figure, smoothed EWMA statistics and control limits are shown in the right figure. In [23] this approach was extended to bit and frame error statistics observed on IEEE 802.11b wireless link.

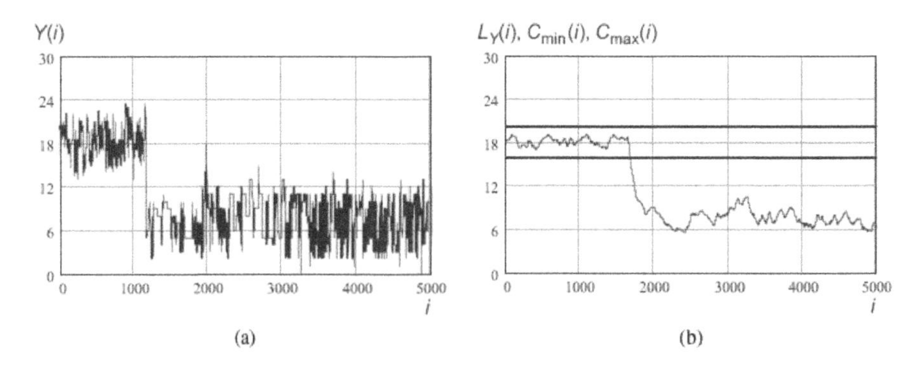

Fig. 16.11 An example of change-detection for SNR process. (a) Initial SNR process, (b) EWMA statistics and control limits.

Unfortunately, there are no simple yet effective methods to be statistically strict concluding whether a limited set of observations is covariance stationary or not. The reason for that is the gap between theoretical understanding of this phenomenon and practical limitations. Dealing with a just few realizations of a process, non-stationarity can be attributed to local changes in the observed process. However, even if the wireless channel characteristics are indeed covariance stationary with long-range dependence property it still affects the way how cross-layer modeling studies need to be used in performance optimization of wireless channels. Indeed, in order to provide the best possible performance of information transmission *at any given instant of time* we are only interested in the local behavior of wireless channel statistics. This duration is given by the minimum time granularity of channel adaptation mechanisms implemented at the wireless channel. The question whether models proposed so far are still valid for such short time duration needs to be further investigated.

Performance Model at the Layer of Interest

Performance evaluation model at the layer of interest is the final step in cross-layer modeling of wireless channels. The input parameters to such models are delay and loss properties of the PDU service process obtained at the previous step. There are a large number of models proposed so far to evaluate performance of real-time and non-real-time applications at different layers of the protocol stack. Usually, they are suitable for use in cross-layer modeling studies.

Nowadays, to evaluate performance of real-time applications in wireless environment queuing models are often used. Early studies did not consider queuing theory as an appropriate tool in wireless domain. The reason is that PDU service times are often observed to be correlated rather than independent. While the theory of queues with correlated service processes is relatively well-studied, solution of those systems involves finding steady-state probabilities of high-order Markov chains of large dimensions leading to computationally demanding models. Nowadays, the most popular performance model is based on G/GI/1/K framework. For example, in [24] finite capacity D-BMAP/G/1/K queuing model was used at the IP layer, where the service times were assumed to be iid. Authors in [25] used D-BMAP/PH/1 queue at the data-link layer with persistent ARQ, where phase-type distribution is used to model time till successful reception of a frame. In [18], to represent service process of frames at the data-link layer with autocorrelated arrival and service processes authors considered non-preemptive D-BMAP+D-MAP/D/1/K queuing system, where the second arrival process is of absolute priority and represents one-to-one mapping from the frame error process. All these systems can be analyzed using well-known matrix analytic techniques. We will briefly consider one of these models in what follows.

Consider D-BMAP/G/1/K queuing system. According to such a system frames/packets arrive in batches, batches arrive just before the end of slots. Arrivals are not allowed to seize the server immediately and the service of any arrival starts at the beginning of a slot. Arrivals depart from the system at the slot boundaries, just after batch arrivals (if any). The state of the system is observed just after the departure (if any) and these points are imbedded Markov points. This system is known as "late arrival model with delayed access" (LAS-DA). The sojourn (service) time is counted as the number of slots spent by a customer in the system. The system can accommodate at most K customers. We assume partial batch acceptance strategy. According to this strategy, if a batch of R customers arrives when k customers are already in the system and $R > K - k$, only $K - k$ of them are accommodated and $R - K + k$ are discarded.

Complete description of the queuing system requires two-dimensional Markov chain $\{S_Q(n), S_A(n), n = 0, 1, \ldots\}$ imbedded at the moments of frame departures from the system, where $S_A(n)$ is the state space of the arrival process, and $S_Q(n) \in \{0, 1, \ldots, K - 1\}$ is the number of frames in the system just after frame departures. To parameterize this process we have to determine transition probabilities between imbedded Markov points.

Let $D_A(k, m), k = 0, 1, \ldots, m = 0, 1, \ldots$, be the set of matrices describing transitions from state i to state $j, i, j \in \{0, 1, \ldots, M\}$ of the modulating Markov chain of the arrival process in m steps with exactly k packet arrivals. Setting $D_A(1, m) = D_A(k), k = 0, 1, \ldots$ we can find them as

$$D_A(k, m) = \sum_{l=0}^{m} D_A(k - 1, l) D_A(m - l). \tag{16.22}$$

Let $D_Q(k)$, $k = 0, 1, \ldots$, be the set of matrices describing transitions from state i to state $j, i, j \in \{0, 1, \ldots, M\}$ of the modulating Markov chain of the arrival process with exactly k arrivals in a service time of a single packet. We determine them using service time distribution $d(i)$ as

$$D_Q(k) = \sum_{i=0}^{\infty} D_A(k,i)d(i), \quad (16.23)$$

Let $C(i,j)$, $i,j \in \{0,1,\ldots,K\}$ be transition probability matrices describing time evolution of the $\{S_Q(n), S_A(n), n = 0,1,\ldots\}$. These matrices can be defined using $D_Q(k)$, $k = 0,1,\ldots$ as follows

$$C(0,j) = \begin{cases} D_Q(k) & j \neq K \\ \sum_{m=K}^{\infty} D_Q(m) & j = K \end{cases},$$

$$C(i,j) = \begin{cases} D_Q(j-i+1) & j \neq K \quad j \leq i-2 \\ \sum_{m=K-i}^{\infty} D_Q(m) & j = K \quad j > i-1 \end{cases}. \quad (16.24)$$

Let now T be transition probability matrix containing $C(i,j)$, $i,j \in \{0,1,\ldots,K\}$ as its (i,j) elements and let $\vec{x}_D = (x_{D,01}, x_{D,11}, \ldots, x_{D,K-1M})$ be its steady-state distribution. Solving $\vec{x}_D T = \vec{x}_D$, $\vec{x}_D \vec{e} = 1$, we get $x_{kj} = \lim_{n\to\infty} \Pr\{S_Q(n) = k, S(n) = j\}$. There are a number of algorithms to compute these probabilities. For example, one may notice that $\vec{x}_D$ is the eigenvector corresponding to unit eigenvalue of T.

In order to obtain expressions for probability functions of the number of lost packets in a slot and delay of a packet, in addition to $\vec{x}_D$ we also need steady-state probabilities of the number of packets in the system in the arbitrary slot and just before batch arrival to the system. Using flow balance principle it was shown in [26] that for D-BMAP/G/1/K late arrival systems with delayed access there are explicit relations between steady-state distributions at the departure, arrival and arbitrary time instants. Thus, using $\vec{x}_D$ we can further obtain delay and loss performance of the system. For example, the probability function of the number of lost packets in a slot is

$$f_L(k) = \sum_{i=0}^{K} \vec{x}_{A,i} D_A(K-i+k)\vec{e}, \quad (16.25)$$

where $\vec{x}_{A,i}$ is the steady-state distribution as seen by arrival batch of IP packets.

The effective capacity is another approach to estimate performance of a wireless channel. The effective capacity is the similar problem to well-known effective bandwidth. The latter has been extensively studied for wired links in the beginning of 90s. The effective capacity and effective bandwidth allow to analyze the statistical delay bound violation probability, which is of paramount importance for wireless networks. Observing that the service process of wireless channels is variable in nature Wu and Negi in [28] formulated and solved the problem of finding a maximum arrival traffic that can be supported over a given wireless channels such that performance parameters of interest (e.g., delay bound) are satisfied. However, the concept of effective capacity as was proposed in [28] assumes a constant arrival traffic, which is not realistic for most practical applications. In [27] authors extended the model allowing arbitrary stationary service and arrival processes. Particularly, it was shown that the effective capacity of a wireless channel modeled by FSMC is given by

$$E(\theta) = -\frac{1}{\theta}\log(p\{P\Phi(\theta)\}), \tag{16.26}$$

where θ is the QoS exponent, P is the one-step transition probability matrix governing FSMC of a wireless channel, $\rho(.)$ is the spectral radius of P and $\Phi(\theta)$ is given by

$$\Phi(\theta) = diag(e^{-\mu_1\theta}, e^{-\mu_2\theta}, \dots, e^{-\mu_K\theta}), \tag{16.27}$$

where μ_i, $i = 1, 2, \dots, K$ is the number of bits transmitted in the state i of the FSMC, K is the number of states. The effective capacity may provide an alternative approach to classic queuing analysis. Moreover, this approach is numerically simpler compared to queuing analysis.

Completely different models are required to evaluate performance of non-real-time applications that use TCP at the transport layer. The number of models proposed to evaluate throughput of different TCP versions is abundant. Up-to-date review with extensive list of references can be found in [30]. While most models have been developed having wired networks in mind, it was shown that they can also be adapted to wireless environment. Basically, we distinguish between models based on renewal theory and those based on fixed-point approximation. The former ones represent throughput as a ratio between the number of segments sent in a renewal cycle to the duration of a cycle. In the simple case when losses are caused by duplicate acknowledgements only the renewal cycle is defined as time between two losses caused by buffer overflow. When packets are lost in bursts PFTK model [31] was found to provide rather accurate results. When losses are iid Mathis model [32] is more appropriate.

In [33] authors extended their cross-layer framework originally proposed in [18] to the case of real-time applications using TCP with SACK option at the transport layer. The proposed framework includes FEC at the physical layer, persistent ARQ at the data-link layer, and also captures segmentation between different layers in the protocol stacks. The buffer at the IP layer was of finite length and droptail queuing discipline was assumed. Also note that usage of persistent ARQ implies that data-link frames are always correctly received irrespective of the amount of retransmission attempts it may take. As a result, packet losses can only occur due to buffer overflow. To represent throughput of TCP SACK a modified PFTK model was used. The proposed cross-layer model allows to estimate TCP throughput as a function of rate and lag-1 ACF of the bit error process, FEC code, and length of PDUs at different layers. Authors demonstrated that the choice of FEC codes as well as PDU length at different layers significantly affects performance experienced by TCP. Particularly, failure to identify an optimal FEC code may lead to significant loss in throughput obtained by TCP. Another important conclusion revealed by the authors is that timeouts almost never happens for a given configuration of a wireless channel. As a result, the evolution of the congestion window is mainly driven by duplicate acknowledgements.

In [34] authors extended their work to the case of non-persistent (truncated) ARQ at the data-link layer. Since now a packet can be lost due to both excessive amount of retransmission attempts made for a single frame and buffer overflow at the IP layer the TCP SACK model used in [33] has been modified to distinguish between these two cases. According to the first case losses are mainly driven by excessive retransmission attempts made for a single frame in a packet. Thus, packet losses are completely random as there is no residual correlation left at the IP layer and TCP model similar to Mathis one can be used to estimate throughput of a connection. When losses caused by buffer overflow dominate they are considered to be correlated due to droptail queuing and the modified PFTK model is used instead. Authors demonstrated that the presence of non-persistent ARQ does not qualitatively affect throughput obtained by TCP meaning that the evolution of the congestion window is again mainly driven by duplicate acknowledgements. On the other hand, the choice of the maximum number of retransmission attempts has a great quantitative impact on TCP throughput. However this effect is not trivial as there is a trade-off between longer delays and lower losses caused by high values of this parameter and smaller delays and high losses otherwise.

Fixed point approach involves solution of the equation in the form $f(p) = p$, where p is the unknown variable (see e.g., [35]). To obtain solution a system of equations in the form $W = f(p, T_R)$, $(p, T_R) = g(W)$ is to be defined, where W is the average size of the congestion window, p is the probability of segment loss, T_R is the RTT, $f(.)$ and $g(.)$ denote some functional relationships. The first equation provides the average TCP window size as a function of constant packet loss probability and RTT. Examples of these models include the Mathis and the PFTK models. Second equation describes the relationship between p, T_R, and average window size W. This relationship can be obtained by solving an appropriate queuing system.

In [36] authors applied fixed-point approximation to obtain TCP Reno throughput. The fixed-point approximation was firstly used to obtain throughput of a number of TCP connections sharing the same bottleneck in wired networks. Following [34] they distinguished between two modes of operation: (i) segment losses are primarily caused by wireless channels and (ii) loss process is dominated by buffer overflows. To obtain $W = f(p, T_R)$ case Mathis model is used in the former case while PFTK is used otherwise. After extending the wireless channel model to IP layer authors use queuing system of D-BMAP/G/1/K type to obtain $(p, T_R) = g(W)$. Although the obtained results are qualitatively similar to those presented in [34] this approach can also be used to estimate TCP throughput when active queue management such as random early detection is used. While the fixed-point approach is approximate in nature and can only be solved numerically it still intuitively clear compared to renewal analysis. However, there are still many open questions concerning accuracy of the approach. The general belief is that it is affected by the choice of queuing model to approximate packet arrival process from a single or multiple TCP sources. Note that the queuing model used by authors can be changed whenever new insights into its choice appear.

To the best of our knowledge the work of Le et al. [37] is the only another attempt so far to use fixed-point approximation in a cross-layer study to estimate TCP performance running over wireless channels. To capture properties of wireless channel with adaptive modulation and coding (AMC), the authors used the Markov model originally proposed in [19]. Then, they introduced the queuing model that implicitly captures non-persistent ARQ at the data-link layer. The second independent equation to form a system of equations is obtained applying the original PFTK model. The choice of PFTK model was dictated by droptail queuing assumed in the buffering system of a mobile node. Although the proposed framework is quite general, it does not incorporate some important features of wireless channels. First of all, packet losses are only assumed to occur due to excessive amount of retransmission attempts meaning that, effectively, the amount of the buffering space is unlimited. This assumption does not hold in some special cases. For example, given ideal wireless channel conditions due to inherent properties of TCP to achieve the whole path capacity even a single TCP source sharing the buffer will experience losses due to buffer overflow. Additionally, only a single TCP connection is assumed to use resources of the wireless channel. This assumption is inappropriate for modern wireless access technologies where a number of TCP connections may compete for link's resources. However, this assumptions is easy to relax using those approaches developed to date for fixed-point approximation in wired networks. Finally, the effective channel rate was assumed to vary according to homogenous Markov chain modeling the AMC. As we already discussed, this assumption only holds when we know trajectory of user movement over the landscape. Finally, note that [37] considered the case of droptail queuing. This assumption is however easy to relax.

Liu *et al.* also considered performance of TCP transmission over wireless channels with AMC implemented at the physical layer, finite queue length and truncated ARQ at the data-link layer [38]. Assumptions regarding physical and data-link layers were similar to those taken in [19]. The considered TCP Reno model was adopted from [31]. The only modification made to the PFTK model was that the delay bound at the wireless part was added to the final expression. Note that later, the model proposed in [31] was shown to overestimate the actual TCP performance [39]. However, qualitatively results of [38] remain valid. Observe, however, that PFTK model was initially developed for droptail queuing procedure and implicitly assumes severely correlated loss process meaning that if a packet in a sending window of a certain length is lost all subsequent packets in this window are also lost. It implies that obtained results are valid when losses caused buffer overflow dominate those caused by excessive amount of retransmission attempts.

16.4 Cross-layer Design of Wireless Channels

The Need for Cross-layer Interactions

Both ITU-T OSI and TCP/IP protocol models separate and isolate functionalities of each layer of the protocol stack. In these models each layer is responsible for a certain set of functions, communicates directly with the same layer of a peer communication entity and is unaware of specific functions of other layers. Both architectures do not allow direct communication of any kind between non-adjacent layers. Communications within the protocol stack are only allowed between adjacent layers using the so-called request-response primitives defined for service access points (SAP). Higher layers use functions provided by adjacent lower layers.

As was demonstrated using cross-layer performance modeling, to optimize performance of applications wireless access technologies require novel organization of the protocol stack at the air interface. Although interfaces between adjacent layers are still preferable, there is the need for direct interactions between non-adjacent layers. Particularly, transmission parameters including transmission model, channel coding, modulation scheme, ARQ retransmissions must be related to application characteristics (e.g., type of information, source coding, etc.), network characteristics, user preferences and context of use. In order to take decisions on traffic management, data-link layer protocols should be aware of higher layers including network and transport layers' parameters and vice versa. In future wireless access technologies we can refer to the air interface protocol architecture with interactions among different layers.

Following [40] we define the cross-layer design of the protocol stack as a design that violates the layered structure of communication protocols. Up to date there were a number of proposals for cross-layers communication that can be classified into three categories: creating new interfaces between layers, design coupling, and merging of adjacent layers as shown in Figure 16.12. The first approach creates a way to exchange information between non-adjacent layers at the runtime. According to the design coupling no information is exchanged between non-adjacent layers at the runtime. Instead, two protocols are just made aware of operational parameters of each other at the design phase. Merging of adjacent layers refers to joint implementation of two or more adjacent protocols in the protocol stack. This technique allows to avoid new

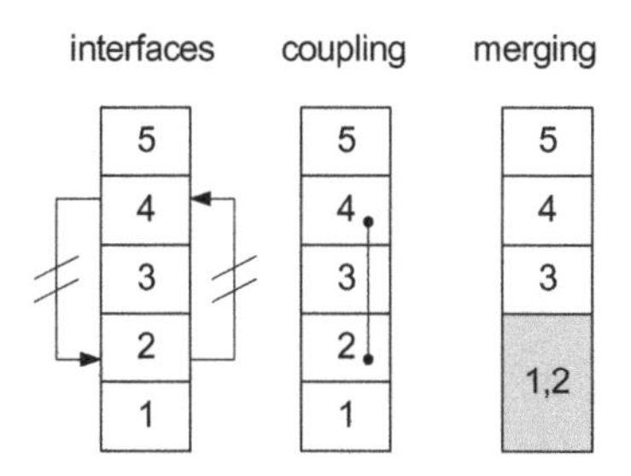

Fig. 16.12 The cross-layer design proposals.

interfaces at the expense of more complicated implementation. This approach is not inherently cross-layer but still violates the layered architecture of the protocol stack. Moreover, this is the only approach used in modern wireless access technologies. Detailed review of these approaches can be found in [40]. Several examples of the cross-layer design methodologies are discussed in [41].

The common aim of abovementioned cross-layer communication schemes is to explicitly or implicitly exchange information between layers of the protocol stack whether at the runtime or at the design phase. Note that the latter approach actually provides no tools to optimize performance. Indeed, as was demonstrated by cross-layer modeling studies even small deviation in channel characteristics may lead to drastic changes in performance experienced by applications and require new parameters of channel adaptation mechanism. The former approach is also referred to as vertical calibration of parameters across the layers. It refers to the case when parameters of protocols at different layers are adjusted at the runtime such that a certain performance metric of interest is controlled and optimized. This approach potentially allows to respond to arbitrarily small changes in wireless channel behavior.

The Cross-layer Signaling

The cross-layer design schemes proposed to date do not specify how information is actually exchanged between different layers. Indeed, for non-adjacent layers to communicate between each other some sort of cross-layer signaling scheme is needed. Up to date a number of such schemes were proposed. We distinguish between in-band and out-of-band signaling as shown in Figure 16.13. In order to communicate between TCP and radio link protocol (RLP) in wireless IP-enabled networks authors in [42] proposed to use the wireless extension

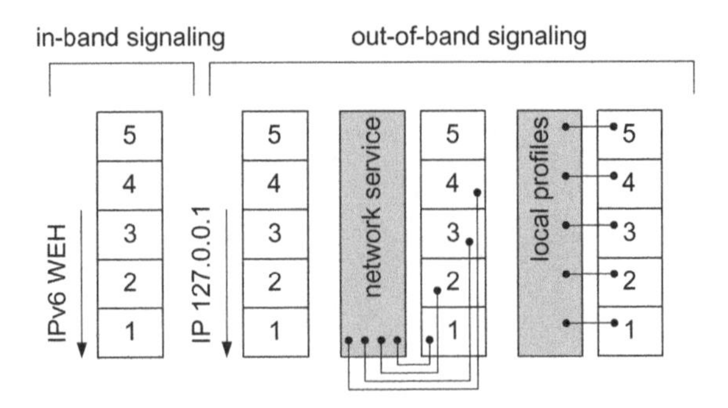

Fig. 16.13 The cross-layer signaling in the protocol stack.

header (WEH) of IPv6 protocol. The advantage of this method is that it makes use of IP data packets as in-band signaling messages. Another method was proposed in [43]. According to it ICMP message is generated whenever a certain parameter changes. The common shortcoming of two abovementioned approaches is that only few layers can actually exchange information. A different "network" approach was proposed in [44]. According to the proposal a special network service that gathers, stores, manages, and distributes information about current parameters used at mobile hosts is introduced. Those protocols that are interested in a certain parameter can access this network service. This approach provides the cross-layer functionality via a "third party" service. Usage of local profiles instead of remote network profiles was proposed in [45]. The concept is similar to that one proposed in [44]. The only difference is that information is stored locally and there is no need to access it via the network. This leads to low overhead and low delay associated with information exchange. In [46] authors proposed a dedicated cross-layer signaling protocol for communication between layers in the protocol stack. The major advantage of this protocol is that non-adjacent layers can exchange control information directly without processing at intermediate layers. This approach, however, requires additional complexity to be introduced directly to the protocol stack.

Authors in [46] also provided a comparison between abovementioned signaling schemes and advocated out-of-band signaling. They argue that the signaling propagation path across the protocol stack is not efficient due to unnecessary processing of messages at intermediate layers. Additionally, the signaling message formats provided by in-band signaling schemes are not flexible enough for signaling in both upward and downward directions or not optimized for wireless environment where the need for a new parameter to be exchanged between non-adjacent layer may occur. Finally, we note that the signaling scheme itself does not provide any advantages for a communicating entity. Unfortunately, those approaches, cited above, do not explicitly take into account the ultimate goal of the cross-layer signaling. One of the most promising applications of the cross-layer signaling is the optimization of parameters of protocols at different layers in real-time to provide the best possible performance at any instant of time for any channel and traffic conditions. As a result, information carried out by the cross-layer signaling scheme should be efficiently exploited to optimize operational parameters of protocols at different layers at the runtime.

Centralized vs. Distributed Optimization

Considering the cross-layer signaling and the optimization of application performance jointly, one has to choose whether to use distributed or centralized control of wireless channel performance. The in-band cross-layer signaling proposals imply the distributed performance control strategy. According to those proposals, layers exchange their information and this information should then be used by the performance control entities implemented at each layer that participates in information exchange and allows its parameters to be dynamically controlled. Such approach is referred to as distributed performance control and requires significant modifications to be introduced *to each layer* of the protocol stack. It was pointed out in [47] that there are a number of problems associated with distributed control strategy. Indeed, when the decision regarding changes of parameters is taken independently at each layer the resulting effect may not be straightforward. According to out-of-band signaling proposals [44, 45], layers export their current operational parameters to a certain external performance control entity via a predefined set of interfaces. This external entity not only saves information of all layers but optimize performance of wireless channel distributing information regarding what kind of protocol parameters should be used at the air interface. Thus, it makes an external intelligent cross-layer performance optimization system that incorporates features of the out-of-band cross-layer signaling system. This method is centralized in nature and generally in accordance with the signaling scheme proposed in [45]. Note that distributed control of parameters is also possible with out-of-band signaling scheme.

The Layered Architecture and the Cross-layer Design

Any cross-layer design of the protocol stack violates the well-known layered design concept. Dealing with cross-layer design we need to remember the problems it may bring. As it was pointed out in [47] the strict layered structure of communication protocols has already proven itself to be simple and easily manageable in wired networks. Isolated layered architecture provides a modular system design that is often important to understand the operation of the whole system. At the development phase system designers must specify a number of layers, functionalities each layer should provide and interfaces to adjacent layers. The isolated layered design of the system also simplifies implementation and further manufacturing allowing, for example, reuse of components (e.g., protocols, interfaces). Protocols of the system can be developed in isolation assuming a certain set of services that a given protocol receives and provides from/to adjacent layers.

Cross-layer protocol design may result in significant increase of complexity at the development and implementation phases. For example, functionality of the whole system may not be clearly understood due to a number of multi-layer loops. Additionally, since the modular design is no longer feasible, implementation and manufacturing costs can be high. Modification to any component of the system does not only change its own behavior but may also affect the performance of the whole system. Results of this influence are often difficult to predict. Thus, additional efforts are required to ensure stability of the system.

Summarizing, one may conclude that there should be a rational trade-off between the layered architecture and the performance optimization of wireless channel performance using cross-layer design. The cross-layer performance optimization tries to fulfill the short-term gains in terms of better performance for a given wireless access technology. Clear and easily understandable layered architecture allows a number of long-term benefits. Among others, the low per-unit cost for a certain performance is one of the most important driving factors. As a result, dealing with cross-layer optimization of wireless channel performance we also have to take into account architectural considerations, that is, making as less cross-layer interactions as possible and implementing the performance control system as well isolated from the protocol stack as feasible.

16.5 The Performance Control System: An Example

In this section we consider an example of the performance control system for wireless channels. The system is responsible for dynamic adaptation of protocol parameters at different layers of the protocol stack at the runtime and based on centralized control.

The Layered Architecture and the Cross-layer Design

The structure of the proposed performance control system is shown in Figure 16.14, where CPOS stands for cross-layer performance optimization subsystem. The protocol stack is divided into three groups of protocols. The first group consists of an application itself that falls into a certain traffic class. We assume that a network is intended to deliver four traffic classes. These are conversational, streaming, interactive, and background classes. Applications that fall in conversational class require to preserve time relation between information entities of the traffic stream and require strict guarantees of end-to-end delivery. Examples of these applications include real-time two-way voice communications, audio multicasting, etc. Applications that are classified to the streaming class require the network to preserve time relation between information entities of the traffic stream and do not require strict guarantees

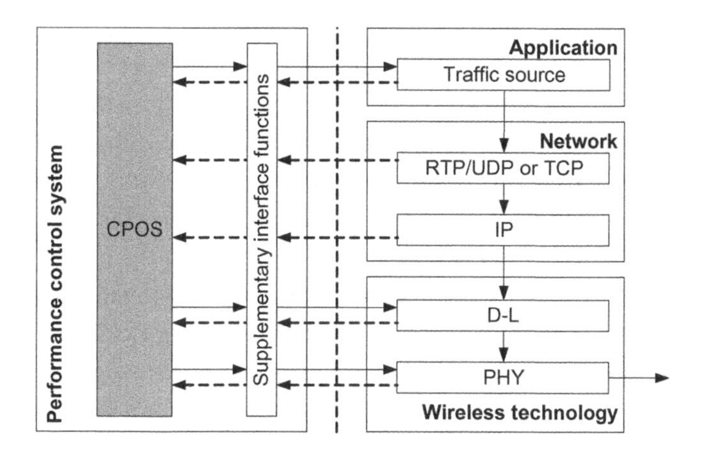

Fig. 16.14 The cross-layer performance optimization system.

of end-to-end delivery. The most common example of these applications is the streaming video. Both interactive and background classes expect a network to guarantee the reliable delivery of information units. The difference between these two classes is that interactive class includes applications operating in request-response mode, thus, posing additional requirements on end-to-end delay. Applications that fall in background class are characterized by the so-called bulk transfers and do not require bounded delay in the network. Although the functionality of the performance control system for conversational and steaming applications is only considered here, the system can also be used for interactive and background applications.

Considering the defined traffic classes, one may observe that there is strict correspondence between the traffic class and protocols at the transport and network layers. Those applications that require strict delay requirements usually use (RTP)UDP/IP as the combination of the transport and network layer protocols. Those applications that require a network to preserve the content of the transmission use TCP and IP at the transport and network layers, respectively. Nowadays, TCP, UDP, and IP protocols are well standardized. For the sake of interoperability with existing implementations, no modifications should be made to these protocols. Indeed, changes introduced to any of these protocols may require network-wide modifications. For this reason, we require that protocols of the transport and network layers and their parameters should not be controlled.

Wireless access technology determines how the traffic is treated at the wireless channel. It defines protocols of the data-link and physical layers. These protocols are usually specific for a given wireless access technology and may incorporate advanced features such as dynamic choice of the parameters to achieve the best possible performance for given wireless channel conditions. Since the main performance degradation in wireless networks stems from the stochastic nature of wireless channel characteristics, these features provide a feasible option for performance control.

According to operation of the performance control system, the application firstly determines the network protocol suit (TCP/IP or (RTP)UDP/IP) to be used during the active session. This decision is taken independently of the performance control system and the mapping is strict for a given application. The application may implicitly notify the CPOS about protocols that are used at the transport and network layers providing the traffic class on informational output. It should also provide information concerning the expected performance level that should be provided at the local wireless channel. Alternatively, this information can be stored in CPOS. During the whole duration of a session CPOS monitors states of wireless channel and application in terms of covariance-stationary stochastic process. The current state of the application (traffic model), the current state of the wireless channel (wireless channel model) and protocols parameters at the data-link and physical layers are used to determine performance parameters that are important for a given application. These parameters may include frame loss rate, frame delay, delay variation, etc. Then, CPOS should determine which actions should be taken to provide the best possible performance for a given application at the current instant of time, that is, whether current protocol parameters should be changed and, if yes, what changes are required. The list of actions should include change of protocol parameters at the data-link and physical layers. This capability is already available in most state-of-the-art wireless access technologies. Additionally, it may also include change of the applications parameters (e.g., rate of the codec for video and audio applications), change of the buffer space at the data-link layer, change of PDU size at different layers. The former capability is usually available for real-time applications such as streaming video or two-way voice communications. Note that when a certain application does not allow to change the rate at which the traffic is fed to the network, the feedback regarding the current rate should still exist. However, when this capability is available, controlling inputs should be provided to the source.

At the beginning of the session, the CPOS should also be made aware of controllable protocols in the protocol stack. It can be made statically at the development phase. It is important to note that protocols can be initialized with default parameters. In this case, these parameters are immediately communicated to the CPOS. Another approach is to setup a predefined set of initial parameters for each class of applications and allow CPOS to initialize protocol parameters. During an active session the CPOS controls the performance perceived by an application by setting protocol parameters in response to changing traffic and channel conditions.

The Cross-layer Performance Optimization Subsystem

The core of the proposed performance control system is the CPOS. The structure of the CPOS is shown in Figure 16.15. Three major components of this

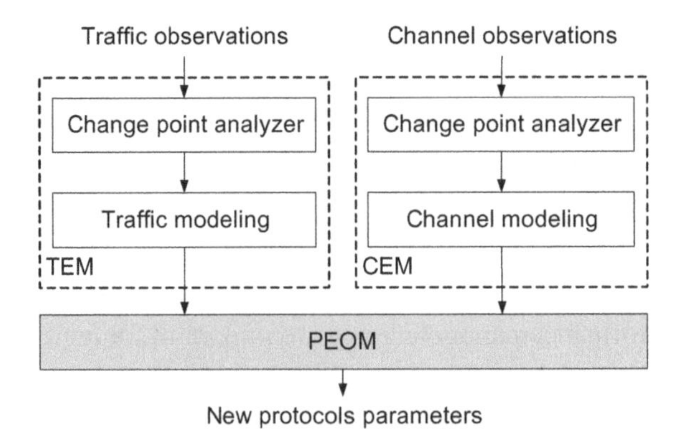

Fig. 16.15 The structure of the performance optimization subsystem.

system are the real-time channel estimation module (rt-CEM), the real-time traffic estimation module (rt-TEM) and the performance evaluation and optimization module (PEOM). The rt-CEM is responsible for detecting changes in wireless channel statistics and estimation of the channel state in terms of the mathematical model. The rt-TEM performs the same functions for traffic observations. To enable these capabilities, wireless channel and traffic statistics are observed in real-time, pre-processed and then fed to the input of the respective change-point analyzer. Note that usage of rt-TEM is only mandatory for real-time applications that have unexpectable traffic patterns. The most common example of these applications is variable bit rate (VBR) streaming video. When the traffic pattern of an application is known in advance this block should be omitted and the predefined model should be used. Example of these applications includes voice communications.

Change-point analyzers test incoming observations for changes in parameters that affect performance of applications. Emerged methods of measurement-based traffic modeling allowed to recognize major statistical characteristics of the traffic affecting its service performance in a network [48,49]. According to Li and Hwang [48] the major impact on performance parameters of the service process is produced by the empirical distribution of the arrival process and the structure of its autocorrelation function (ACF). Hayek and He [49] highlighted importance of empirical distributions of the number of arrivals showing that the queuing response may vary for inputs with the same mean and ACF. It was also shown [48] that accurate approximation of empirical data can be achieved when both marginal distribution and ACF of the model match their empirical counterparts well. Recently, it was also shown that the wireless channel statistics including BER and the lag-1 autocorrelation of the bit error process significantly affect performance parameters of applications running over wireless channels at the data-link layer with hybrid ARQ/FEC [18]. Similar conclusions have been made in [50], where authors considered the effect of bit errors on performance of applications at the IP layer.

To monitor wireless channel statistics, SNR, bit error, or frame error processes can be used. The reason to use the bit error process is twofold. Firstly, it allows to abstract functionality of the physical layer of different wireless access technologies. As a result, a single cross-layer performance control system can be potentially applied to different wireless channels. Secondly, the bit error process is binary in nature. It allows to significantly decrease the complexity of the modeling algorithm. It is also allowed to use the SNR process instead of the bit error process if the relationship between the bit error probability and the SNR value is known. Finally, the frame error process can also be used. Note, that frame error statistics can be directly obtained observing operation of ARQ protocols at the data-link layer.

Monitoring the frame error process introduces significant delays in detection of the channel state. In this case, the system may not react timely to changes in wireless channel conditions. The advantage of monitoring SNR or bit error processes is that the reaction time decreases significantly. When the relationship between the SNR value and bit error probability is already available (e.g., obtained via filed measurements) the proposed scheme can be used for SNR observations too. Direct monitoring of the bit error process of the wireless channel may provide a feasible alternative to this approach. However, in order to estimate statistics of the bit error process in real-time the source should periodically transmit predefined information at the wireless channel such that the receiver is aware of the content of this transmission and its exact placement. This feature can be implemented using either channel equalization bits or synchronization information. However, it is still unclear how much information should be transmitted to provide a satisfactory estimator of the channel state.

The change-point analyzers must signal those points when a change in either traffic or wireless channel statistics is detected. When a change is detected, the current wireless channel and traffic models are parameterized in the respective modeling blocks and then immediately fed to the input of PEOM. The current traffic and channel models are also stored in the respective modeling blocks for further usage. Note that it is allowed for PEOM to be activated in response to the change in either channel or traffic statistics only. Otherwise, no actions are taken except for continuous monitoring of the channel and traffic statistics.

The structure of PEOM is shown in Figure 16.16. According to the system design the current traffic and channel models are fed to the input of the decision module. Taking the reference performance of a given application at appropriate layer (e.g., data-link or IP) as another input, this module decides whether the current performance is satisfactory. In order to take this decision the module containing the performance evaluation framework (PEOF) is activated. If the performance is satisfactory, no changes are required and current protocol parameters are further used. Otherwise, the current wireless channel and traffic models are used to decide whether performance can be improved and, if so, which parameters have to be changed and how. Depending on particular protocols of the protocol stack and type of the application, new protocol parameters resulting in best possible performance for a given wireless channel and traffic statistics are computed in the PEOF and then fed back to the decision module. These parameters are used till the next change in input wireless channel or traffic statistics. The PEOF may implement the performance evaluation framework or just contain a set of pre-computed performance curves

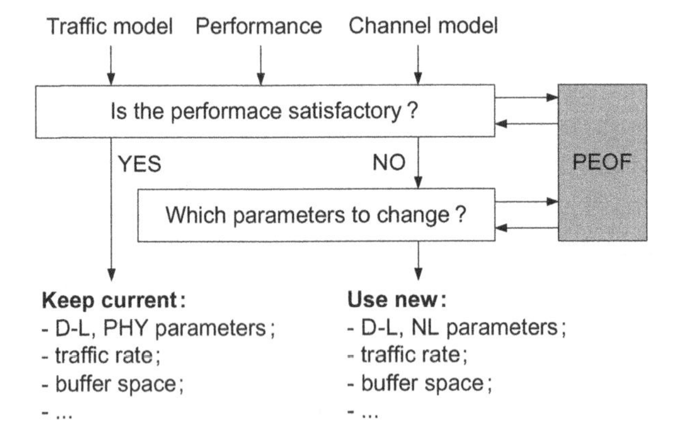

Fig. 16.16 The performance evaluation and optimization module.

corresponding to a wide range of wireless channel and traffic statistics and different configurations of the protocol stack. Due to the real-time nature of the system the latter approach is preferable.

To implement the system, the following has to be developed:

- test for detecting changes in channel and traffic statistics;
- model for channel and traffic observations;
- cross-layer extension for wireless channel model;
- performance evaluation model at IP layer.

16.6 Summary of the Chapter

Cross-layer performance modeling and control is relatively new topic in wireless communications that recently receives a lot of attention from research community. In this chapter we discussed why conventional organization of the protocol stack is no longer feasible for wireless access technologies and needs to be supplemented with cross-layer performance control capabilities. To determine which parameters of channel adaption schemes lead to best possible performance provided to applications performance evaluation studies are needed. We considered the cross-layer modeling approach used in recent literature and highlighted basic features of each step. Finally, we discussed possibilities for performance control of applications using both centralized and decentralized approaches.

Questions for self-control

1. What are the reasons for performance degradation in wireless networks?
2. Which channel adaptation mechanisms are used in modern wireless systems?
3. Does AMC incorporate FEC mechanism?
4. What is the major negative side effect of power control in WCDMA?
5. What are the basic steps in cross-layer performance modeling?
6. Are frame error models suitable for cross-layer modeling?
7. Why propagation models cannot be directly used in performance evaluation studies?

8. Is there evidence that wireless channel statistics is stationary?
9. Does autocorrelation of wireless channel statistics affect performance at higher layers?
10. Is it possible to use Markov models to model AMC? Why?
11. At which layer wireless channel performance is conventionally assessed?
12. Does channel memory propagate to layers higher than data-link?
13. Which channel model is easier to extent to IP layer, uncorrelated or Markov one?
14. What is the difference between performance evaluation of RT and nRT applications?
15. What kind of queuing systems are used to estimate performance of RT applications?
16. What is the difference between PFTK and Mathis TCP models?
17. Is it possible to use unaltered PFTK model in wireless environment? Why?
18. What is the basic idea of fixed-point approximation?
19. Why is FPA approach approximate in nature?
20. Why is cross-layer design of wireless channels beneficial?
21. What are shortcomings of cross-layer design?
22. What are advantages of strictly layered design of protocols?
23. What kind of approaches exists to exchange control information in the protocol stack?
24. What are advantages and shortcomings of centralized control of wireless channels?
25. Can we modify TCP to include wireless channel adaption algorithms? Why?

Problems

1. Assume that the bit error process observed on a wireless channel is completely uncorrelated and each codeword consists of exactly 255 bits. Estimate codeword error probability for the following FEC codes: (255,131,18), (255,87,26) for BER in the range $\{0.01, 0.02, \ldots, 0.10\}$. Which of these codes perform better in terms of codeword error probability for 0.01 BER. For 0.10 BER?

Assume that the channel rate is 3.84E5 and estimate throughput for all values of BER.

2. Assume now that the wireless channel is characterized by BER 0.03 and lag-1 ACF 0.5 and described using Markov based SBP model. Estimate parameters of the bit error process. Estimate parameters of the codeword error process including codeword error probability assuming than no FEC is used and the length of a codeword is 511 bits. Is codeword error probability different compared to the previous problem for the same BER?
3. Let BER be 0.05 and lag-1 ACF be 0.4. The length of IP packet is 100 bytes. Each packet is divided into codewords. ARQ is applied directly to codewords and the maximum number of retransmission attempts allowed for a single codewords is 3. Whenever a codeword is not successfully transmitted in 3 attempts whole IP packet to which it belongs to is dropped. Estimate codeword error rate. Estimate packet loss probability due to error-prone nature of wireless channel.
4. Assuming parameters of previous problem apply M/M/1/K, ($K = 10$, $\lambda = 50$ packets per second), queuing system to estimate packet loss due to queuing and mean packet delay. Note: you need to estimate mean transmission time of a packet over wireless channel. Packet loss probability, p_K, and packet delay, $E[W]$, in M/M/1/K system are given by

$$p_K = \frac{(1-\rho)\rho^K}{1-\rho^{K+1}}, \quad E[W] = \frac{1/\mu}{1-\rho}, \quad \rho = \frac{\lambda}{\mu},$$

where λ and μ are arrival and service rates, measured in packet per time unit.
5. Let the rate of the wireless channel be 3.84E5. Bit error process is uncorrelated and BER is 0.03. (255,131,18) FEC code is used and the length of MTU is 1500 bytes. Each TCP segment is segmented to a number of codewords directly. Assuming M/M/1 queuing system with infinite buffer space and a single TCP connection sharing the wireless channel estimate the TCP throughput using Mathis formula. Is it different from error-free conditions? Why? Mathis

expression is

$$r(p, T_R) = \frac{\sqrt{3/4}L}{T_R\sqrt{p}},$$

where L is the length of MTU, p is the packet loss probability, T_R is the RTT. Note that RTT consists of time require to transmit a packet in wired network, τ, and over wireless channel. Let $\tau = 3E - 2$s. Is it possible to use PFTK expression in this environment?

References

[1] D. Moltchanov, "Cross-layer performance control of wireless channels using active local profiles," J. on Communications Software and Systems, 3(3), pp. 148–164, November 2007.

[2] A. Paulraj et al. "An overview of MIMO communications - a key to gigabit wireless," Proceeding of IEEE, 92(2), pp. 198–218, February 2004.

[3] A. Gesbert et al. "From theory to practice: an overview of MIMO space-time coded wireless systems," IEEE JSAC, 21(3), pp. 281–302, April 2003.

[4] T. Rappaport, Wireless communications: principles and practice, 2nd ed., Prentice Hall, 2002.

[5] J. Arauz and P. Krishnamurthy "Discrete Rayleigh Fading Channel Modeling," J. Wireless Comm. and Mobile Computing, 5(1), pp. 413-425, July 2004.

[6] L. Rabiner "A tutorial on hidden {M}arkov models and selected applications in speech recognition," Proceeding of the IEEE, 77(2), pp. 257–286, February 1989.

[7] H. Kong and E. Shwedyk "Markov characterization of frequency selective Rayleigh fading channels," IEEE Pacific Rim Conf. on Communications, Computing and Signal Processing, pp. 359–362, 1995.

[8] H.-S. Wang and N. Moayeri "Finite-state Markov channel - a useful model for wireless communications channels," IEEE Trans. On Vehicular Tech., 44(1), pp. 163–171, February 1995.

[9] Y.-Y. Kim and S.-Q. Li "Modeling fast fading channel dynamics for packet data performance analysis," IEEE INFOCOM, pp. 1292–1300, March/April 1998.

[10] E. Gilbert, "Capacity of a burst-noise channel," Bell Systems Technical Journal, vol. 39, pp. 1253–1265, 1960.

[11] S. Khayam and H. Radha, "Markov-based modelling of wireless local area networks," ACM Conference on Modelling, Analysis and Simulation of Wireless and Mobile Systems, pp. 100–107, September 2003.

[12] H. Bai and M. Atiquzzaman "Error modeling schemes for fading channel in wireless communications: a survey," IEEE Comm. Surveys, 4th quarter 2003, on-line.

[13] H.-S. Wang and P.-C. Chang "On verifying the first-order Markovian assumption for a Rayleigh fading channel model," IEEE Trans. On Vehicular Tech., 45(2), pp. 353–357, May 1996.

[14] E. Elliott, "Estimates of error rates for codes on burst-noise channel," Bell Systems Technical Journal, pp. 1977–1997, 1963.

[15] B. Fritchman, "A binary channel characterization using partitioned Markov chain," {IEEE} Trans. Inform. Theory, vol. 13, pp. 221–227, 1967.
[16] J. Swarts and H. Ferreira "On the evaluation and application of Markov channel models in wireless communications," IEEE Vehicular Tech. Conference, pp. 117–121, 1999.
[17] D. Moltchanov et al. "Simple, accurate and computationally efficient wireless channel modelling algorithm," Conference on Wireless/wired Internet Communications, pp. 234–245, 2005.
[18] D. Moltchanov et al. "Loss performance model for wireless channels with autocorrelated arrivals and losses," Computer Communications, 29(13/14), pp. 2646–2660, June 2006.
[19] Q. Liu et al. "Cross-layer combining of adaptive modulation and coding with truncated ARQ over wireless links," IEEE Trans. Wireless Comm., 3(5), pp. 1746–1755, September 2004.
[20] G. Nguyen et al. "A trace-based approach for modelling wireless channel behaviour," Winter simulation conference, pp. 597–604, 1996.
[21] D. Moltchanov "State description of wireless channels using change-point statistical tests," Conference on Wireless/Wired Internet Communications, pp. 275–286, May 2006.
[22] A. Konrad et al. "Markov-based channel model algorithm for wireless networks," Wireless Networks, 9(3), pp. 189–199, June 2003.
[23] D. Moltchanov "Monitoring the state of wireless channels in terms of the covariance stationary PDU error process," International Conference on Telecommunications, pp. 456–462, May 2006.
[24] D. Moltchanov "The effect of data-link layer reliability on performance of wireless channels," IEEE PIMRC, pp. 1–6, September 2008.
[25] J.-A. Zhao "MPEG-4 video transmission over wireless networks: a link level performance study," Wireless Networks, 10(2), pp. 133–146, March 2004.
[26] N. Kim et al. "On the relationships among queue length at arrival, departure, and random epochs in the discrete-time queue with D-BMAP arrivals," Operation Research Letters, vol. 30, pp. 25–32, 2002.
[27] X. Zhang et al. "Cross-layer-based modeling for quality of service guarantees in mobile wireless networks," IEEE Comm. Mag., 44(1), pp. 100–106, January 2006.
[28] D. Wu and R. Negi "Effective capacity: a wireless link model for support of quality of service," IEEE Trans. Vehicular Tech., 2(4), pp. 630–643, July 2003.
[29] J. Tang and X. Zhang "Cross-layer modeling for quality of service guarantees over mobile wireless networks," IEEE Trans. Wireless Comm., 6(12), pp. 4504–4512, December 2007.
[30] J. Olsen, Stochastic modeling and simulation of the TCP protocol, PhD thesis, Uppsala University, 2003, available at: http://publications.uu.se/theses/abstract.xsql?dbid=3534, accessed on 18.02.2009.
[31] J. Padhye et al. "Modeling TCP Reno performance: a simple model and its empirical validation," IEEE Trans. Netw., 8(2), pp. 133–145, April 2000.
[32] M. Mathis "The macroscopic behavior of the TCP congestion avoidance algorithm," Computer Communications Review, 27(3), pp. 67–82, 1997.
[33] D. Moltchanov et al. "Cross-layer modeling of TCP SACK performance over wireless channels with completely reliable ARQ/FEC," Conference on Wireless/Wired Internet Communications, pp. 13–26, May 2008.

[34] D. Moltchanov and R. Dunaytsev "Modeling TCP performance over Wireless Channels with a Semi-reliable Data-Link Layer," 11th IEEE Conference on Communication Systems, pp. 912–918, November 2008.

[35] A. Misra "The window distribution of multiple TCPs with random loss queues," IEEE GLOBECOM, pp. 1714–1726, March 1999.

[36] D. Moltchanov and R. Dunaytsev "Modeling TCP performance over wireless channels using fixed-point approximation," International Conference on Telecommunications, pp. 1–10, June 2008.

[37] L. Le et al, "Interaction between radio link level truncated ARQ and TCP in multirate wireless networks: a cross-layer performance analysis," IET Communications, 1(5), pp. 821–830, December 2007.

[38] Q. Liu "TCP performance in wireless access with adaptive modulation and coding," IEEE International Conference on Communications, pp. 3989–3993, June 2004.

[39] R. Dunaytsev et al. "The PFTK-model revised," Computer Communications, 29(13/14), pp. 2671–2679, August 2006.

[40] V. Srivastava and M. Montani "Cross-layer design: a survey and the road ahead," IEEE Comm. Mag., 43(12), pp. 112–119, December 2005.

[41] V. Raisinghani and S. Lyer "Cross-layer design optimizations in wireless protocol stacks," Computer Communications, 27(8), pp. 720–724, April 2004.

[42] G. Wu et al. "Interactions between TCP and RLP in wireless Internet," IEEE GLOBECOM, pp. 661–666, December 1999.

[43] P. Sudame and B. Badrinath "On providing support for protocol adaptation in mobile wireless networks," Mobile Networks and Applications, 6(1), pp. 43–55, January/February 2001.

[44] B.-J. Kim "A network service providing wireless channel information for adaptive mobile applications," IEEE ICC, pp. 1345–1251, June 2001.

[45] K. Chen "Cross-layer design for accessibility in mobile ad hoc networks," IEEE Wireless Pers. Comm., 21(1), pp. 49–76, April 2002.

[46] Q. Wang and M. Abu-Rgheff "Cross-layer signaling for next-generation wireless systems," IEEE Wireless Communications and Networking Conference, pp. 1084–1089, March 2003.

[47] V. Kawadia and P.R. Kumar "A cautionary perspective on cross-layer design," IEEE Pers. Comm., 12(1), pp. 3–11, February 2005.

[48] S.-Q. Li and C.-L. Hwang "Queue response to input correlation functions: discrete spectral analysis," IEEE Trans. Netw., 7(1), pp. 522–533, October 1997.

[49] B. Hajek and L. He "On variations of queue response for inputs with the same mean and autocorrelation function," IEEE Trans. Netw., 6(5), pp. 588–598, October 1998.

[50] Y.-Y. Kim and S.-Q. Li "Capturing important statistics of a fading/shadowing channel for network performance analysis," IEEE JSAC, 17(5), pp. 888–901, May 1999.

17

Open Access to Resource Management in Multimedia Networks

Evelina Pencheva and Ivaylo Atanasov

Technical University of Sofia, Bulgaria

The chapter is dedicated to mechanisms for open access to resource management in the Internet Protocol (IP) multimedia networks. First we present the concept of IP Multimedia Subsystem (IMS) and explain the IMS functional architecture, principles of quality of service management and service control in IMS. Then we describe the idea behind the opening of network interfaces for third parties so that others besides the network operator can create and deploy services. Open Service Access (OSA) and Parlay appear to be the technologies for value-added service delivery in multimedia networks. In the chapter we take a closer look to the Parlay/OSA interfaces that allow third party applications to access the resource management functions in IMS. OSA "Connectivity Manager" interfaces and OSA "Policy Management" interfaces are considered. Parlay X Web Services interfaces provide a higher level of abstraction than Parlay/OSA interfaces and gain an amazing amount of support among service developers. We address "Application-driven Quality of Service" Parlay X Web Service and "Policy" Parlay X Web Service also.

17.1 Internet Protocol Multimedia Subsystem

Internet Protocol Multimedia Subsystem concept

Internet Protocol Multimedia Subsystem (IMS) was born as an architectural framework for service delivering in third generation mobile networks. Now it

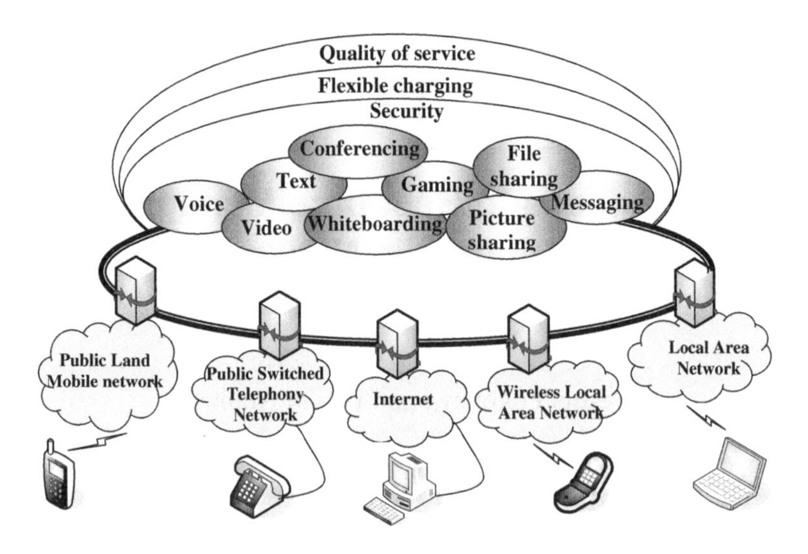

Fig. 17.1 The IMS — All IP core network multimedia domain.

is adopted by next generation networks as a key technology for fixed-mobile convergence.

The IMS is all about services. It enables operators to offer multimedia services based on and built upon Internet applications, services and protocols. IMS facilitates convergence of and access to voice, video, messaging, data and web-based technologies [1, 2]. One of the main principles behind the IMS shown in Figure 17.1 is access independence. It allows services to be provided over any network that supports IP.

IMS facilitates efficient introduction of new multimedia services. The services themselves are not standardizes but tailored to customer needs. Service customization is achieved by the use of service capabilities in both networks and terminals. The service capabilities are service component that can be used to create IP multimedia applications.

IMS functional architecture

The layering approach is used to define the IMS architecture [3]. The IMS consists of three separate planes: User plane, Control plane and Application plane. Figure 17.2 shows IMS functions defined at the three layers.

The User plane is composed of traffic caring network elements like switches, routers, media gateways and access elements at the borders of the

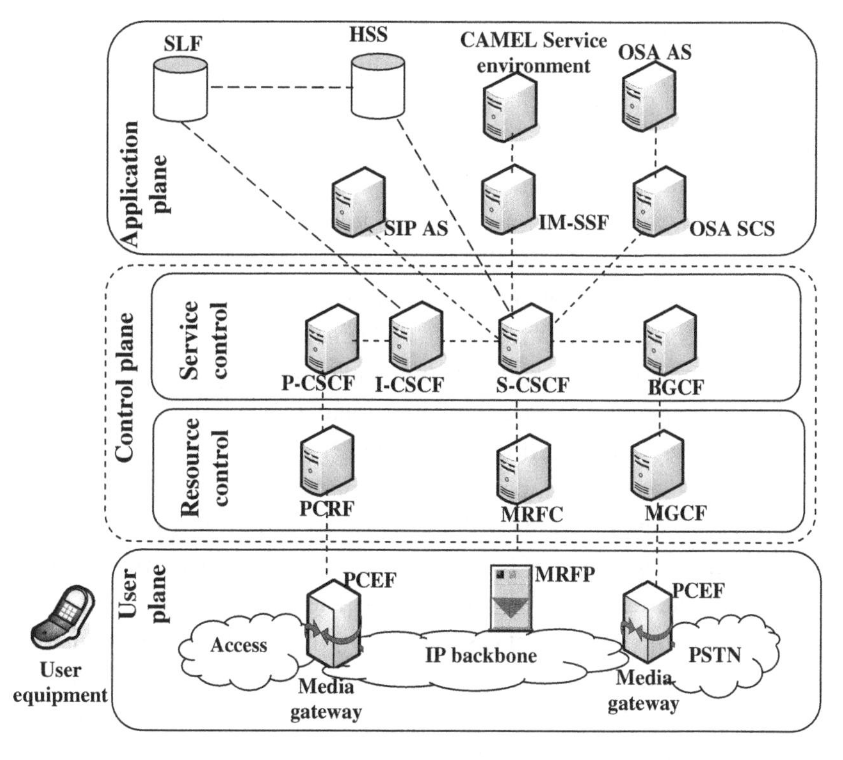

Fig. 17.2 The IMS functional architecture.

core network. The Policy and Charging Enforcement Function (PCEF) encompasses service data flow detection, policy enforcement and flow based charging functionalities. The PCEF is located at the media gateway. The media gateway provides service data flow detection, user traffic handling, triggering control plane session management, quality of service handling, and service data flow measurement as well as online and offline charging interactions. The User plane also contains Media Resource Function Processor (MRFP) providing media resources for playing announcements and conferencing.

The Control plane comprises control servers. The control functions are further dived into service control functions and resource control functions.

Resource control functions provide connectivity control. The Policy and Charging Rule Function (PCRF) encompasses policy control decision and flow based charging control functionalities [4]. It is responsible for finding routes in the network that meet requirements for quality of service. The PCRF provides network control regarding the service data flow detection, gating, quality of service and flow based charging (except credit management) towards the PCEF. Media Resource Function Controller (MRFC) controls media resources. The Media Gateway Control Function (MGCF) controls the media gateway interfacing to circuited switched networks.

Service control functions are common functions for a number of services. The central role is played by Call Session Control Functions (CSCFs) used for multimedia session control and address translation function. In addition, the CSCFs manage service control, voice codec negotiation for audio communication, and authentication, authorization and accounting. The session control including management of dynamic inclusion/exclusion of elements in a session relies on Session Initiation Protocol (SIP) signalling [5]. The CSCFs are responsible for secure routing of the SIP messages, monitoring of the SIP sessions and communications with the PCRF to support media authorization. There are three kinds of CSCFs: Serving CSCF (S-CSCF), Proxy CSCF (P-CSCF) and Interrogating CSCF (I-CSCF). The P-CSCF is the first point of contact for the user equipment (user terminal) in the IMS. It is responsible for the security of SIP signalling between the user and the IMS, and SIP compression. The P-CSCF communicates with the PCRF to allocate resources for media flows. The I-CSCF plays a role of a SIP signalling gateway to external networks. It is responsible for assignment of a S-CSCF to the user during registration and for routing the incoming requests to an assigned S-CSCF or application server. The S-CSCF is the brain of the IMS and is always located at the home network. It performs session control and registration of user terminals. The Border Gateway Control Function (BGCF) is used in breakout scenarios to circuit switched networks.

The Application plane contains applications that extend network services using call control, messaging, user interaction, user location and flexible charging. The Application Server (AS) delivers value-added services. If the S-CSCF determines that the AS must be involved, it delegates the session control to that AS. The interface IMS Service Control (ISC) between the S-CSCF and the AS is based on SIP.

There are several types of application servers [6]. The SIP AS hosts applications that are able to influence the session control by receiving and emitting SIP signalling. The IP Multimedia Service Switching Function (IM-SSF) is an AS that allows services based on Customized Application for Mobile Enhanced service Logic (CAMEL) to be involved in session control. The Open Service Access (OSA) defines service architecture for third party application control on multimedia sessions through open interfaces. The OSA Application Programming Interfaces (APIs) define a standardized way for access to network functions (like call and session control, messaging, user interaction, location etc.) and hide underlying network technology and protocol complexity from application developers. Network functions that can be accessed by external application through OSA APIs are defined as service capabilities. The OSA Service Capability Server (OSA SCS) interfaces between OSA applications and IMS service control. The OSA AS hosts third party applications that use network functions exposed through OSA APIs.

Home Subscriber Server (HSS) is an authentication server that stores authentication parameters applied for the users and user profiles that contain information about the media types that the users are authorized to use, and about the services that are to be applied to the users. The Subscription Locator Function (SLF) locates the HSS responsible for holding the user-related data for given user.

17.2 Quality of Service in IMS

Quality of Service Mechanisms

Resource management is the efficient and effective deployment for network resources when they are needed. The resource management comprises different mechanisms to support the quality of service (QoS).

As to [7] Quality of Service is "the collective effect of service performance which determines the degree of satisfaction of a user of the service". The QoS is aimed to support the characteristics and properties of specific applications. Different applications may have quite different needs. For example, for e-commerce, the accuracy of the delivery is more important than overall delay or packet delay variation (i.e., jitter), while for IP telephony, jitter and delay are key and must be minimized.

To deliver service performance that determines the degree of user satisfaction of the service, an architectural framework for QoS support is defined. The QoS architectural framework is a set of generic network mechanisms for controlling the network service response to a service request, which can be specific to a network element, or for signalling between network elements, or

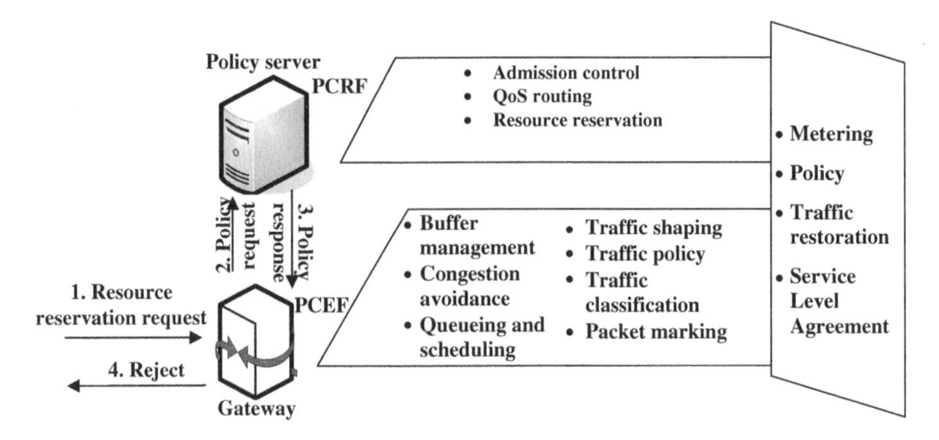

Fig. 17.3 QoS mechanisms.

for controlling and administering traffic across a network. As it is shown in Figure 17.3, the mechanisms are classified in three groups:

- Control mechanisms deal with the pathways through which user traffic travels. These mechanisms include admission control, QoS routing, and resource reservation. These mechanisms are realized by the PCRF.
- Data mechanisms deal with the user traffic directly. These mechanisms include buffer management, congestion avoidance, packet marking, queueing and scheduling, traffic classification, traffic policing and traffic shaping. These mechanisms are realized by the PCEF.
- Management mechanisms deal with the operation, administration and management aspects of the network. These mechanisms include Service Level Agreement (SLA), traffic restoration, metering and recording, and policy. These mechanisms are implemented in both PCRF and PCEF.

In IMS, service requirements are signalled at the Control plane and reflected on the underlying IP access and transport networks. Without interaction between User plane and Control plane the operator will not be able to provide the required QoS. The gateway contains a PCEF that has the capability of policing packet flow into the IP network, and restricting the set of IP destinations that may be reached according to a packet classifier. This service-based policy "gate" function has an external control interface that allows the gate to be selectively "opened" or "closed" on the basis of IP destination address and port. When open, the gate allows packets to pass through (to the destination specified in the classifier) and when closed, no packets are allowed to pass through. The control is performed by the PCRF which interacts with the P-CSCF. When the PCRF is implemented in a separate physical node, the interface between the PCRF and the P-CSCF is based on Diameter protocol [8].

The overall interaction between the network elements at the user plane and IMS control functions is called Service-Based Local Policy (SBLP) control [9]. There are eight interactions defined for Service-Based Local Policy:

- Authorization of QoS resources
- Resource reservation with Service-Based Local Policy
- Enabling media flows
- Disabling media flows
- Revoke authorization for QoS resources
- Indication of bearer release from the PCEF to the PCRF
- Authorization of QoS resource modification
- Indication of QoS resource modification from the PCEF to the PCRF.

Authorization of QoS Resources

The Session Description Protocol (SDP) is typically used to describe the desired session characteristics and the QoS requirements that must be met in order to successfully set up a session [10]. During SIP session establishment P-CSCF (PCRF) uses the SDP contained in the SIP signalling to calculate the proper authorization token. The token contains all the information needed to perform resource reservation. The P-CSCF includes the token in the response back to the user. The authorization is expressed in terms of the IP resources to be authorized and includes limits on IP packet flows, and may include restrictions on IP destination address and port. The principle of QoS resource authorization is illustrated in Figure 17.4.

Resource Reservation with Service-based Local Policy

Resource reservation is always initiated by the user equipment after successful authorization. The protocol used for resource reservation is referred to

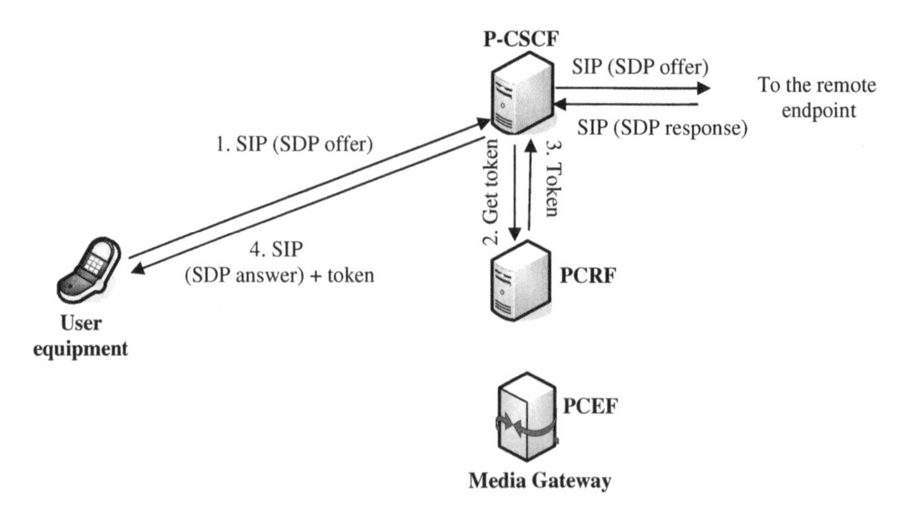

Fig. 17.4 Authorization of QoS resources.

Resource Reservation Protocol (RSVP) [11]. The user equipment includes in the RSVP message for resource reservation the authentication token granted. With request for QoS resource reservation, the PCEF in the gateway needs to assure that the requested resources match to the authorized resources. The PCEF forwards the token, together with the requested QoS parameters, to the PCRF. The PCRF checks if the corresponding requested QoS resources are within the limit of what was negotiated in the SDP exchange. The PCRF uses the token as the key to find the stored negotiated SDP. The principle of QoS resource reservation is illustrated in Figure 17.5.

Enabling Media Flows

The PCRF makes policy decisions and provides an indication to the PCEF that the user is allowed to use the allocated QoS resources. The PCEF enforces the policy decisions and 'opens the door' for the user traffic.

Disabling Media Flows

The PCRF makes policy decisions and provides an indication to the PCEF about revoking the user's capacity to use the allocated QoS resources for per-session authorizations. The PCEF enforces the policy decisions and 'closes the door' blocking the user's media flows.

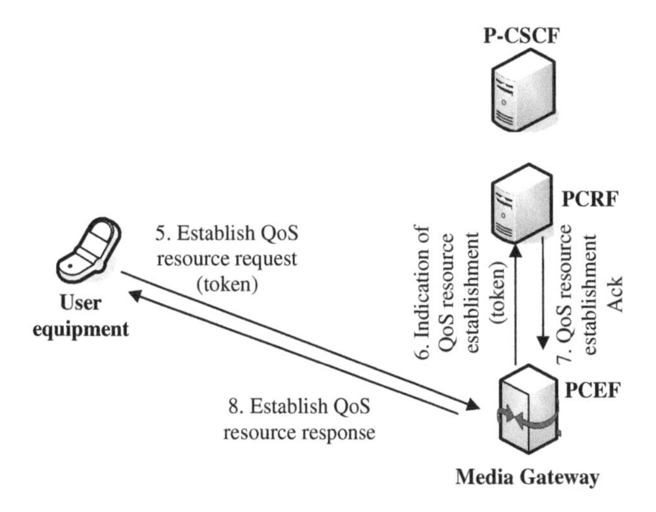

Fig. 17.5 Resource reservation of QoS resources.

Revoke Authorization for QoS Resources

At IP multimedia session release, the user equipment should deactivate the QoS resources used for the IP multimedia session. In various cases, such as loss of signal from the mobile, the user equipment is unable to perform this release itself. The PCRF provides indication to the PCEF when the resources previously authorized, and possibly allocated by the user equipment are to be released. The QoS resources are deactivated.

Indication of QoS Resource Release

Any release of QoS resources that were established based on authorization from the PCRF are reported to the PCRF by the PCEF. This indication is forwarded to the P-CSCF and may be used by the P-CSCF to initiate a session release towards the remote part. The principle of QoS resource release is illustrated in Figure 17.6.

Authorization of QoS Resource Modification

When a QoS resource is modified by the user equipment, such that the requested QoS falls outside of the limits which were authorized, then the PCEF needs to verify the authorization of this QoS resource modification. If the PCEF does not have sufficient information to authorize the QoS resource modification request, the PCEF sends an authorization request to the PCRF.

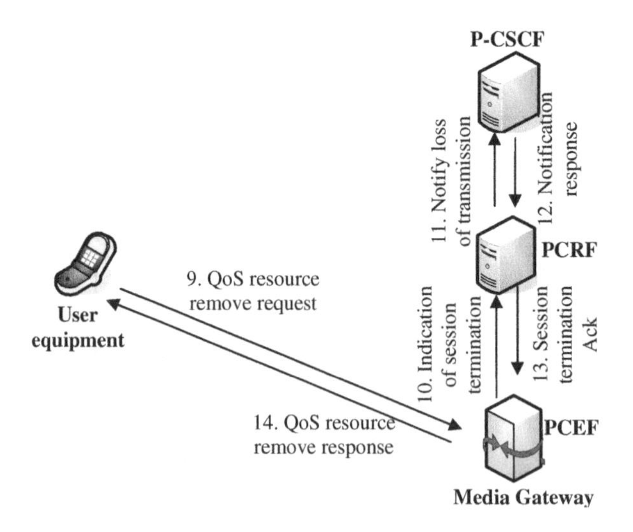

Fig. 17.6 Release of QoS resources.

The PCRF authorizes the modified QoS resources based on the current session information.

It is also possible for the P-CSCF to send an update of the session information in case of a modification of a SIP session which results in an update of the authorization. The principle of QoS resource modification is illustrated in Figure 17.7.

Indication of QoS Resource Modification

When a QoS resource is modified such that the maximum bit rate (downlink and uplink) is downgraded to 0 bit/s or changed from 0 kbps to a value that falls within the limits that were authorized, then the PCEF has to report this to the PCRF. This indication is forwarded to the P-CSCF. The P-CSCF uses this information to initiate a session release towards the remote endpoint.

Event and Information Distribution

The S-CSCF and Application Servers (SIP-AS, IM-SSF, OSA-SCS) must be able to send service information messages to endpoints. This is done using a SIP Request/Response information exchange containing the service information and/or a list of addresses pointing to the location of information represented in other media formats. The stimulus for initiating the service event

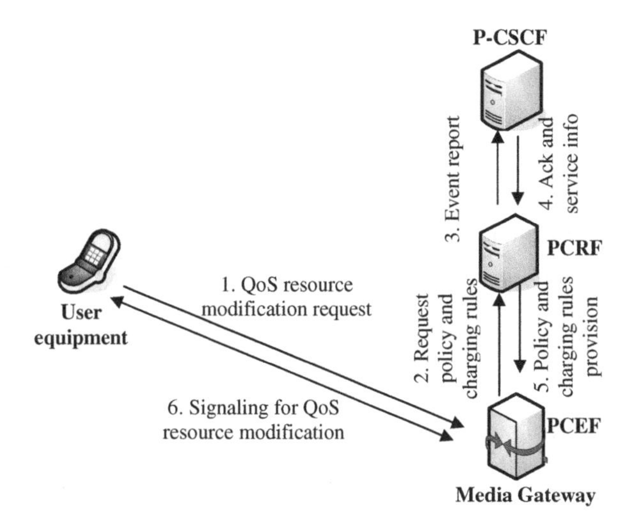

Fig. 17.7 Modification of QoS resources.

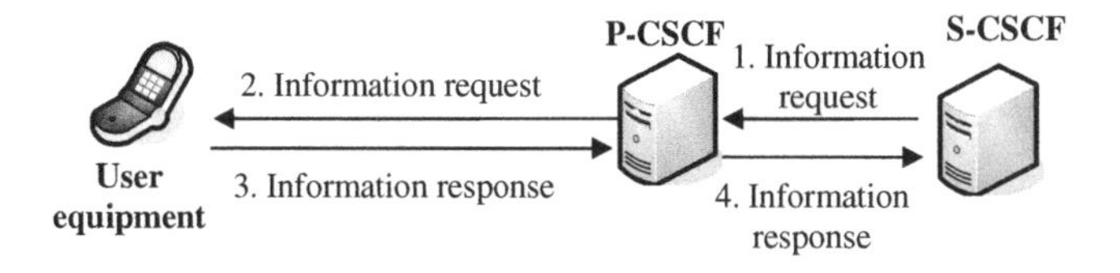

Fig. 17.8 Distribution of QoS events and information.

related information message may come from e.g., a service logic residing in an AS.

In addition, the endpoints are also able to send information to each other. This information is delivered using SIP based messages. The corresponding SIP messages are forwarded along the IMS SIP signalling path. This includes the S-CSCF but may also include SIP Application Servers.

The service event related information exchange may either take place in the context of a session, or independently outside the context of any existing session.

The principle of QoS event and information distribution is shown in Figure 17.8.

An Application Server offering value-added services including QoS management applications resides either in the user's home network or at a third party location. The third party could be a network or simply a stand-alone AS.

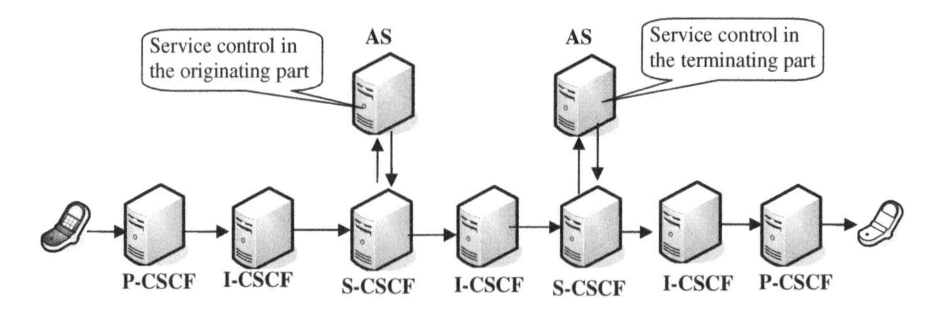

Fig. 17.9 Applying application control in the originating and terminating parts.

For resource control purposes, some applications may interact with the PCRF, while others rely on the S-CSCF providing the basic session over which the value-added application is built.

To present the way in which applications can manage the QoS, it is necessary to have a notion of service control.

17.3 Service Control

Service Triggering

Applications that are hosted by and execute on an application server may be invoked by a request from the S-CSCF or may be initiated by another mechanism. Applications may issue session control signalling via the S-CSCF and may therefore influence the multimedia session. Information about the applications that are to be triggered on behalf of the user (i.e., about the application servers that will need to be contacted whenever the user issues a request) is stored in HSS. The application might be invoked for the originating part and for the terminating part as shown in Figure 17.9.

Initial filter criteria (iFC) describe the way in which the user can access the multimedia services. The iFC contain information about service triggering and describe when an incoming SIP message is further routed to a specific AS. They are downloaded to the S-CSCF upon user registration as shown in Figure 17.10.

On receiving a SIP request, the S-CSCF checks the conditions for service triggering, i.e., determines if some of the iFC are met. If it is so, the S-CSCF delegates the service control to the appropriate AS. The principle of service triggering is shown in Figure 17.11.

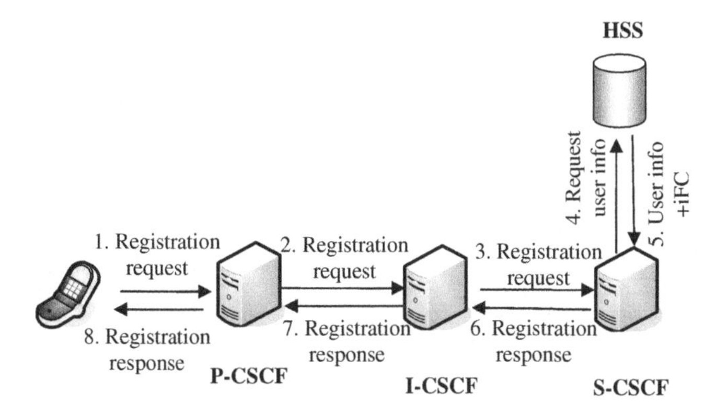

Fig. 17.10 Downloading iFC into the S-CSCF on user registration.

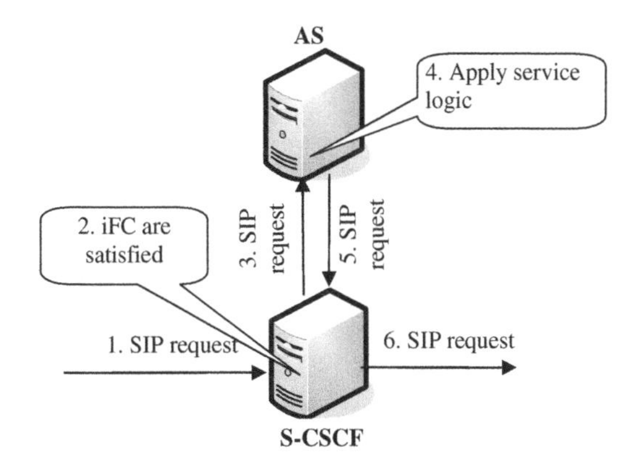

Fig. 17.11 Service triggering.

In processing the SIP request the AS can play different roles. The AS roles are illustrated in Figure 17.12.

In case (a) of acting as a redirect server, the AS receives the incoming SIP request and uses the service information to send a redirection response. In case (b) of acting as a terminating user agent, the AS can establish a session with the originating user agent. In case (c) of acting as an originating user agent, the AS generates a SIP request and sends it to the S-CSCF which then proxies it towards destination. Case (d) occurs when the AS acting as a SIP proxy receives incoming SIP request and proxies it back to the S-CSCF which then proxies it towards the destination. Case (e) shows an AS acting as a routing Back-to-Back User Agent (B2BUA) which performs third party call

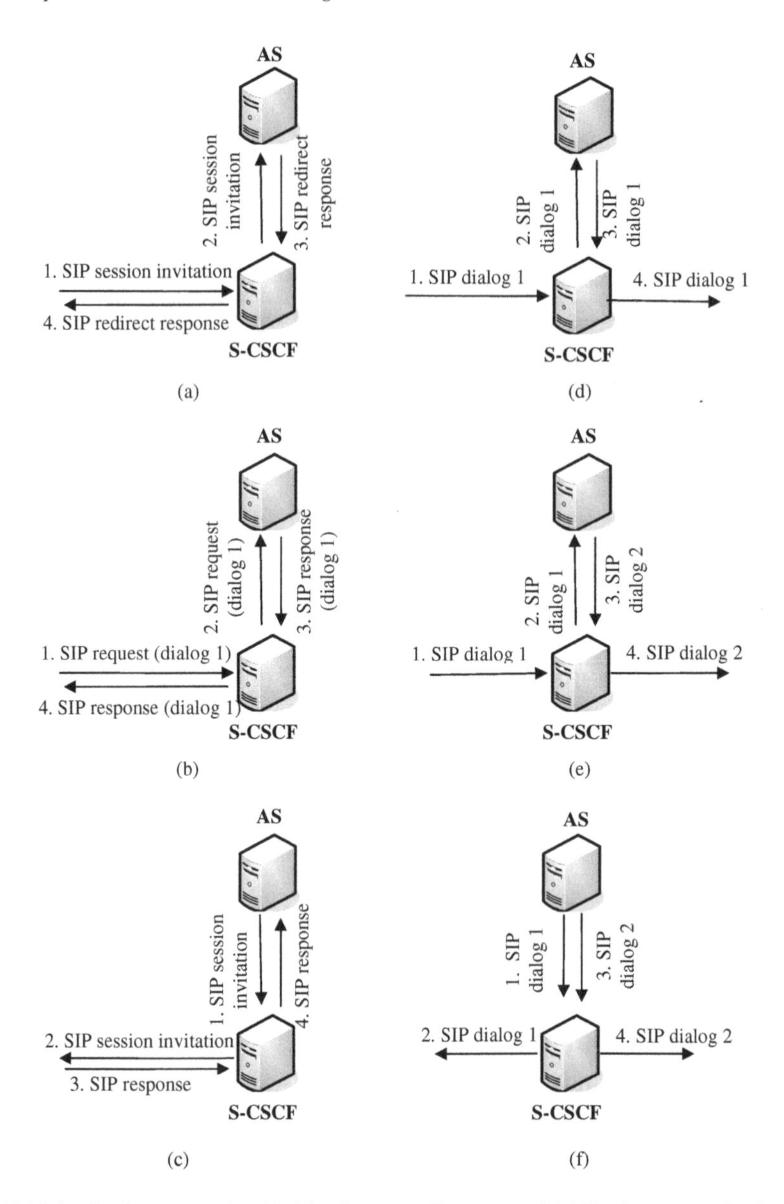

Fig. 17.12 Application server roles. (a) AS acting as a redirect server; (b) AS acting as a terminating user; (c) AS acting as an originating user; (d) AS acting as a proxy server; (e) AS acting as a routing B2BUA; (f) AS acting as an initiating B2BUA.

control. In this case the AS receives the incoming SIP request, generates a new SIP request and sends it to the S-CSCF which in turn proxies it towards the destination. Case (f) shows another type of third party application control by an AS. In this case the AS acting as an initiating B2BUA initiate two new requests within different SIP dialogs. These requests are proxied through the S-CSCF which in turn forwards them to their destinations.

Usually a native SIP AS offers purely SIP-based services. An operator can also offer access to services based on the Customized Applications for Mobile network Enhanced Logic (CAMEL) Service Environment and the Open Service Access (OSA) for its IMS subscribers. The IM-SSF AS provides access for IMS users to existing Intelligent Network services. The OSA service capability servers offer the OSA interface to the OSA Application Server running third party applications. In comparison to CAMEL-based services which are restricted to the operator domain, OSA opens network interfaces for external applications and allows secure access to functions such as call and session control, location management, charging, messaging and QoS management.

In the next section we explain in brief basic concepts of the open service access and then focus on OSA service capabilities for resource management.

17.4 Open Access to Network Functionality

Open Service Access and Parlay

Open Service Access (OSA) is defined as a service architecture for delivering value-added services in third generation mobile networks [12]. It adopts the approach of the classical Intelligent Network for service assembling out of components. But while the Intelligent Network concept fits well to circuit-switched networks, the OSA provides underlying network independence.

In OSA, the functions provided by the network are defined as service capabilities servers (SCS). The elementary SCS functions are grouped into Service Capability Features (SCF). A SCF is a unit of functionality that can be used as a reusable building block for services. Examples of features are call and session control, user interaction, messaging, accessing user status and user location and so on. Figure 17.13 gives an overview of OSA.

To make easier service creation, OSA defines object-oriented application programming interfaces (APIs) for each SCF. The APIs abstract the details of the underlying network service capabilities and data. They are specified in an implementation-independent way in a form of Interface Definition Language (IDL) definitions and Unified Modelling Language (UML) descriptions [12]. The APIs are implementable in a distributed computing environment. The

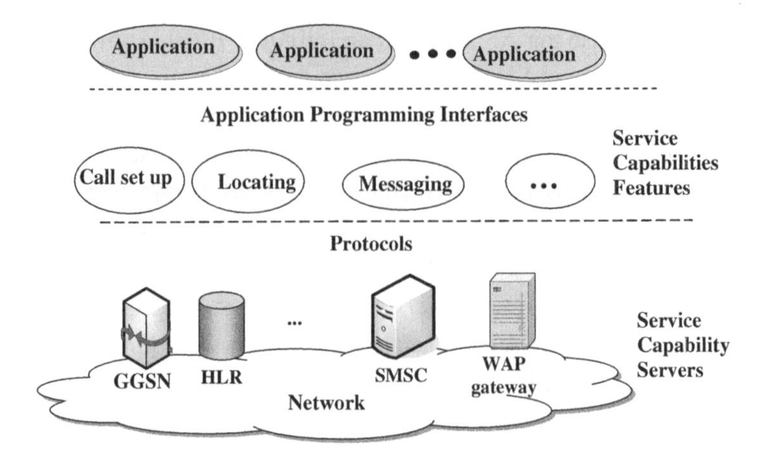

Fig. 17.13 The concept of Open Service Access.

usage of API for service creation hides for developers the complexity of the network control protocols and encapsulates network implementation.

The OSA API definition is strongly influenced by Parlay programming interfaces. The concept of Parlay is to open network interfaces for third party software developers (other than network operator and service provider). The Parlay interfaces also allow access to network functions, such as call and session control, messaging, charging, QoS management. Network access to third party applications is secured. Before using network functions third party applications are subject to authentication and authorization. To avoid repudiation applications are required to digitally sign an on-line agreement for the use of certain features.

Because of their similarity the Parlay and OSA interface specifications have been aligned.

OSA and Parlay consists of 13 interface groups, listed in Table 17.1. The Framework is a special interface with common functions for authentication, discovery and manageability required to enable services to work together in a coherent manner. The other interfaces are service interfaces that allow applications to control network resources.

The OSA APIs are split into three types of interface classes as shown in Figure 17.14. Interface classes between the applications and the framework provide applications with basic mechanisms enabling applications to make use of the service capabilities in the network. Interface classes between

Table 17.1. Parlay/OSA interfaces.

Interface	Short description
Framework	Authentication, authorization, service discovery, service subscription, integrity management
Call control	Setup and control of multiparty multimedia calls and conferences, notification of call and connection-related events
User interaction	Playing announcements, retrieving user input, sending short messages
Mobility	Provisioning information for user status and user location
Terminal capabilities	Getting capabilities of end user terminals
Data session control	Setup and control of packet data sessions
Connectivity management	Negotiation and management of QoS in IP networks
Account management	Creating, modifying and deleting subscriber accounts
Charging	Reservation and charging of units of volume or money against subscriber accounts, split charging
Policy management	Defining policy information including service level agreements, evaluating policies and subscription for and notification of policy events
Presence and availability management	Managing, retrieving and publishing user-related information including identities, communication capabilities, content delivery capabilities, state information, presence and availability of entities for different contexts and communication methods
Multimedia messaging	Sending, storing and receiving multimedia messages, manipulating mailboxes and mail directories
Service brokering	Registering the application interest in particular traffic as part of service interactions

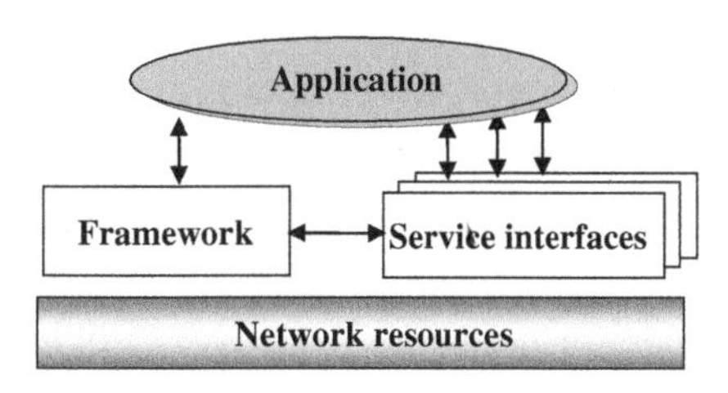

Fig. 17.14 OSA interface types.

applications and SCFs allow application to access to network functions. Interface classes between the framework and the SCFs provide the mechanisms necessary for a multi-vendor environment. Network side interfaces' names are prefixed with "Ip" and application side interfaces' names are prefixed with "IpApp".

Each OSA interface is defined as a set of object types (classes). A class defines the methods that can be called on the object and their parameters. The method definition includes also the method type and exceptions. The method type defines the type of the result returned. The exceptions define errors that may arise during the method execution.

Two OSA SCFs are defined for the purpose of resource management in IMS:

- OSA Connectivity Manager API
- OSA Policy Management API.

OSA Connectivity Management

The Connectivity Management is a set of functions that provide configuration and control of both the attributes of IP connectivity and policies governing IP connectivity, within and between IP domains. Such attributes include QoS, security, and routing policy.

The "OSA Connectivity Manager" SCF is defined to establish QoS parameters for an enterprise network traffic travelling through a provider network. Assuming that the underlying packet network can be configured as a virtual private network (VPN), the Connectivity Manager interfaces provide methods that allow management applications to configure inter-site virtual connections.

The "Connectivity Manager SCF" can be used by a VPN client (enterprise operator subscribed for VPN services) that has entered a relationship with a VPN provider (network operator) to set up a provisioned QoS. Connectivity Manager includes API between VPN client and VPN provider to establish QoS parameters for VPN packets passing through the provider network.

The API requires any specific QoS method to be used neither in the VPN network nor in the operator network. To deliver QoS between networks the differentiated services approach is used which is based on giving preferential treatment to some packets over others in the edge routers. Each packet arriving from the VPN client network into the VPN provider network is marked with a tag called DSCP (differentiated services code point). Only marked packets can enjoy the QoS service provisioned in the VPN provider network.

The VPN client may be an enterprise operator that owns a number of enterprise sites connected via virtual private network provided by a VPN provider. The VPN provider is available at a number of sites by service access points

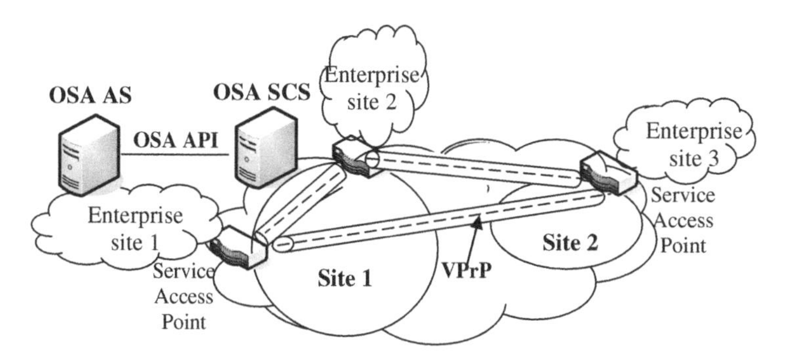

Fig. 17.15 An imaginary Virtual Private Network of an enterprise operator.

for the VPN client. An application using "Connectivity Manager" SCF is hosted at the VPN client domain. The telecom operator gives the VPN provider access to the "Connectivity Manager" SCF. The access through an OSA SCS is subject to the safeguards provided by the Framework. An imaginary VPN configuration is shown in Figure 17.15.

The VPN provider offers configuration service to the VPN client. Using the "Connectivity Manager" API the VPN client can create virtual provisioned pipes (VPrP) in the VPN provider network to carry the enterprise traffic and support it with pre-specified QoS. The VPrP defines QoS parameters for traffic flowing through the provider network between two specified enterprise endpoints. The VPN provider offers a set of templates that are used by the VPN client to specify a VPrP. For instance, the provider may offer templates for video conferencing, audio conferencing, Gold Service, Silver Service, etc. Using these templates the VPN client can select and provision a VPrP with specific QoS attributes. Elements that can be specified for a VPrP include attributes such as packet delay and packet loss, and traffic characteristics such as maximum rate and burst rate. The collection of all the VPrPs, provisioned within the enterprise VPN, constitutes the Virtual Provisioned Network (VPrN).

Figure 17.16 shows the UML class diagram of the "Connectivity Manager" interfaces. The figure shows the aggregation association between classes with the multiplicity notations near the end of the associations. The Connectivity Manager interface is the entry point to this service. From this interface the client application can get reference to the VPN client network interface and to the QoS menu interface. The QoS menu interface provides a list of templates, each of which specifies the QoS service parameters that are offered by the

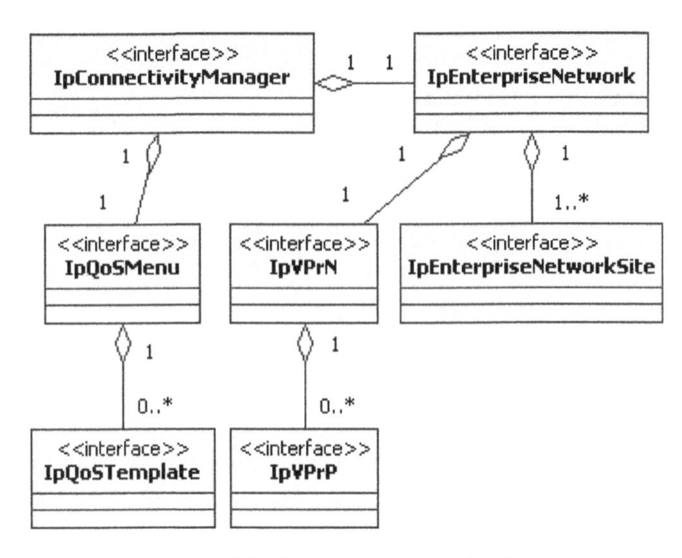

Fig. 17.16 OSA Connectivity Manager interfaces.

VPN provider. The service is composed of components that are associated with a provisioned QoS. The client application can use the template interface to specify the service parameters that are offered by the VPN provider, and also, to store the parameters that the VPN client selects temporarily. The VPN client network interface is associated with two components: enterprise sites, and the VPrN that has been already provisioned in the provider network. The Virtual Provisioned Network interface contains references to all the VPrPs already established. The client application can use the QoS Menu to get references to all the QoS templates offered by the provider. Once the VPN client selects the QoS parameters provided in the QoS template, and submits the request to create a new VPrP, the VPN provider validates the information submitted and if the request is approved, the new VPrP is activated.

After successful authentication and authorization the enterprise OSA application may use the methods supported by the IpConnectivityManager interface to get the handle to the menu of QoS services offered by the VPN provider, and to get a handle to the enterprise network interface that holds information about current services that the provider network delivers to the enterprise network.

The IpQoSMenu interface holds the QoS menu offered by the provider. Each QoS service offered (e.g., Gold, Silver) is specified in a separate template. When the VPN client asks for a specific template from the list of templates, a temporary template interface is created. This temporary template interface holds all the parameters (e.g., all the Gold parameters) and their default values offered by the provider for this template.

The IpQoSTemplate interface provides access to a specific QoS template, such as Gold, offered by the provider. This interface provides getters to discover the QoS service details, and setters to set the requested values for a new VPrP.

The VPN client can create a new Virtual Provisioned Pipe (VPrP) in an existing VPN by the IpVPrN interface. Each such pipe is associated with QoS parameters identified by a specific DiffServ Codepoint. A packet that arrives at a service access point with a specific Codepoint, is "directed" to the VPrP that supports the QoS parameters provisioned for this pipe. The VPN client can create new VPrPs and delete existing VPrP using the IpVPrN interface. This interface provides also methods to get the list of already provisioned VPrPs, and a handle to a specific VPrP interface that holds information for this VPrP.

Figure 17.17 illustrates the sequence in which VPN client selects service components and creates a new VPrP. In the figure, the client application collects the information required to select a service, then selects service parameters, and finally submits it to the Connectivity manager.

The IpEnterpriseNetwork interface stores enterprise network information maintained by the provider as it relates to the VPN service and the virtual provisioned network service that the VPN client had already established with the provider network. The VPN client can only retrieve information regarding an existing VPrN, list the sites connected to the VPN, and get the handle to a specific site interface that stores information about the site.

Figure 17.18 illustrates the way in which a VPN client browses a VPrP. Using the IpVPrP interface the client application browses and collects information regarding existing VPrP, including all the QoS parameters that have been set for this pipe.

The IpVPrP interface provides information on a VPrP whose status can be in one of the following states:

- *Active*: a previously established VPrP, which indicates that a previous request to create the VPrP was granted by the provider. Packets that belong to this VPrP and meet the validity time requirements are admitted to the VPrN.

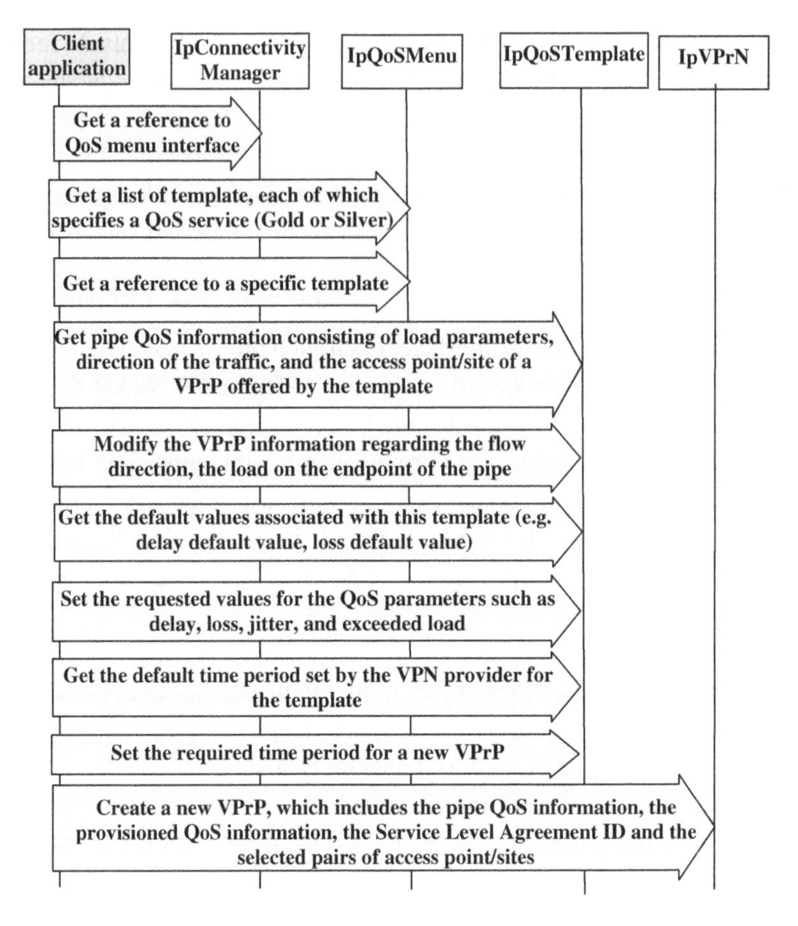

Fig. 17.17 The application creates a new virtual provisioned pipe.

- *Pending*: a request to create a new VPrP is still pending response from the provider, indicating that the provider is still processing the request to create a new VPrP. Packets that belong to this VPrP are not admitted to the VPrN.
- *Disallowed*: a request to create a new VPrP was denied. A description parameter may include the reason for the denial. This is a disallowed VPrP and packets that belong to this VPrP are not admitted to the VPrN.

The IpEnterpriseNetworkSite stores site information of the VPN client network. This information is maintained by the VPN provider.

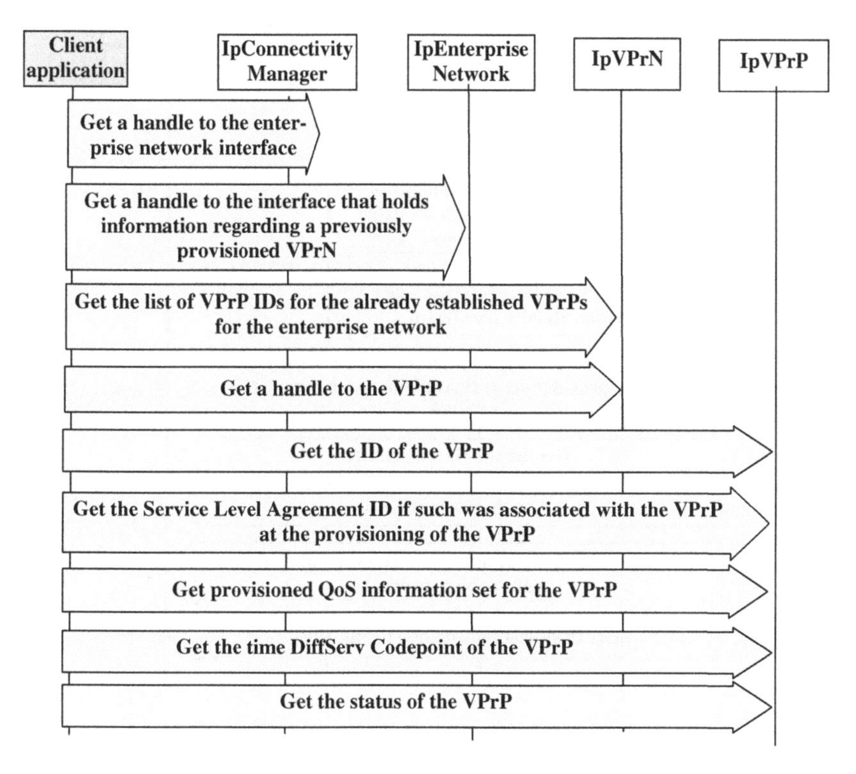

Fig. 17.18 The application browses a virtual provisioned pipe.

Figure 17.19 shows the way in which a VPN client browses service access points and sites. The client application browses service access points and sites to retrieve information for a site and its service access points.

The OSA "Connectivity Manager" SCF is defined in [13].

OSA Policy Management

The "Policy Management" SCF is defined to support policy-enabled services.

Policy is an ordered combination of policy rules that define how to administer, manage, and control access to resources. Policy rule is a combination of conditions and actions to be performed if the condition is true. Policy evaluation is the process of evaluating the policy conditions and executing the associated policy actions up to the point that the end of the policy is reached. The repository is meant to hold unattached conditions and actions. The network operator can populate the repository with the conditions and actions that it can support.

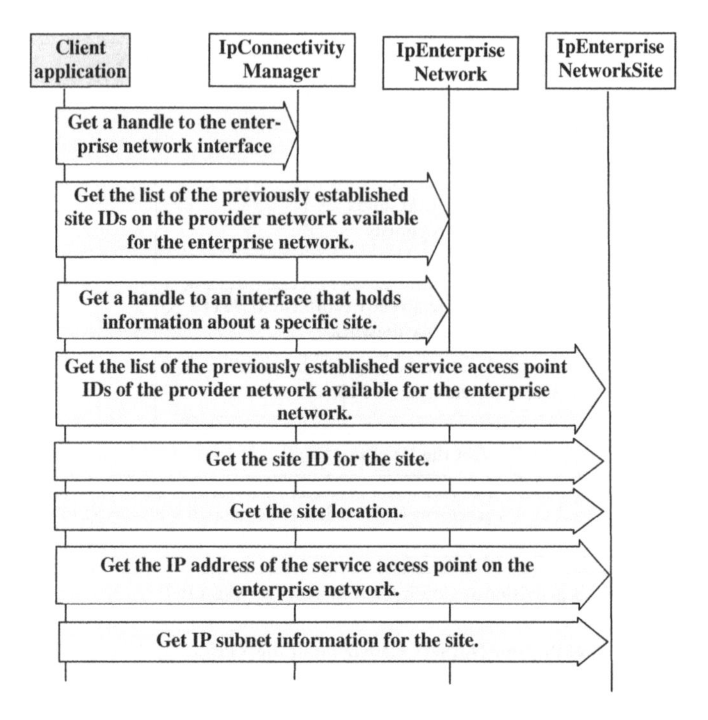

Fig. 17.19 The application browses service access points and sites.

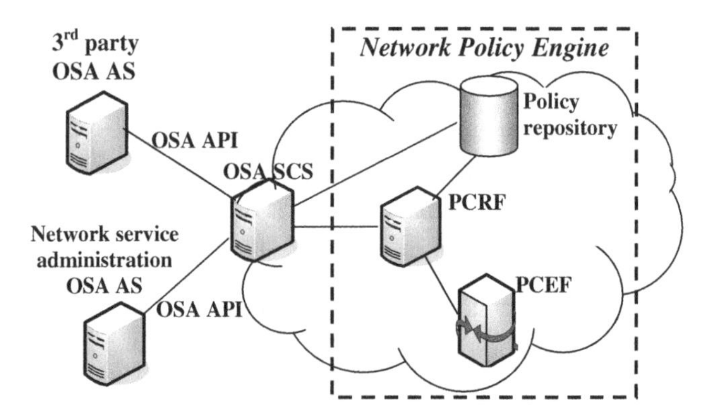

Fig. 17.20 Policy management architecture.

Figure 17.20 shows the OSA Policy Management architecture in IMS environment. The PCRF, PCEF and Policy repository form network policy engine.

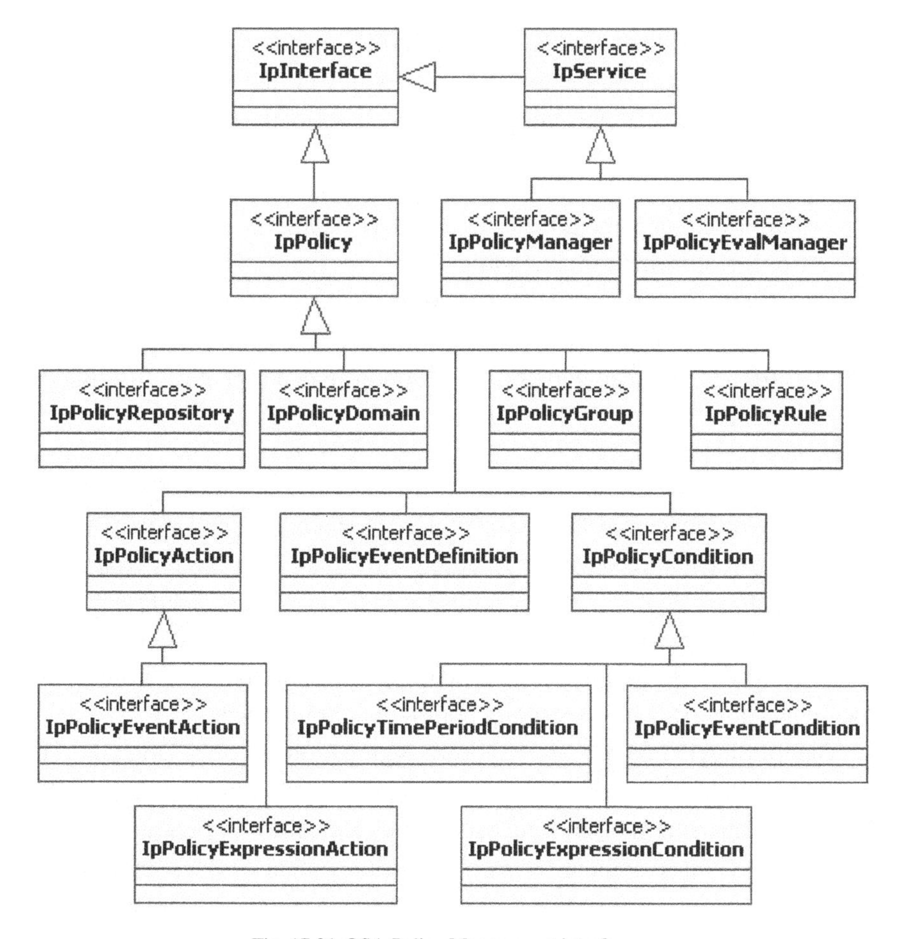

Fig. 17.21 OSA Policy Management interfaces.

The UML class diagram of "Policy Management" interfaces is shown in Figure 17.21. The figure presents the interface inheritance hierarchy. Some of the interfaces are omitted for simplicity.

The OSA "Policy Management" interfaces allow policies to be provisioned and compliance of service usage with policies to be evaluated. Policy-based management provides a way to allocate network resources, primarily network bandwidth, quality of service (QoS), and security (such as firewalls), according to defined business policies.

Client (3rd party) applications can use the Policy Management API to create, update or view policy information for any policy enabled service. It is possible for an application to subscribe to policy events, to request evaluation of policies and to request the generation of policy events.

The OSA "Policy Management" SCF comprises of the following interfaces classes:

- Policy Management Provisioning Service interfaces used to define policy information, including policy rules, policy events, etc., and to update and view this information.
- Policy Management Evaluation interfaces used to request evaluation of policies and for subscription and notification to policy events.

The client applications participating in Policy Management use the IpPolicy-Manager to reference a policy domain of interest, create a new policy domain or remove an existing one. Client applications also can reference a policy repository, to create a new policy repository or remove an existing one.

The IpPolicy is the base interface from which are derived all of the Policy interfaces (except IpPolicyManager and IpPolicyEvalManager). This interface documents attributes for describing policy-related instances.

The IpPolicyDomain interface allows aggregation of Policy Domains, Policy Groups, Policy Rules, or Policy Event Definitions in a single container. The Policy Domains and their nesting capabilities are shown in Figure 17.22. For example, the Policy domain for resource management can nest other Policy domains that provide specialized rules regarding to QoS management.

The IpPolicyGroup interface allows aggregation of Policy Rules or other Policy Groups in a single container. The Policy Groups and their nesting

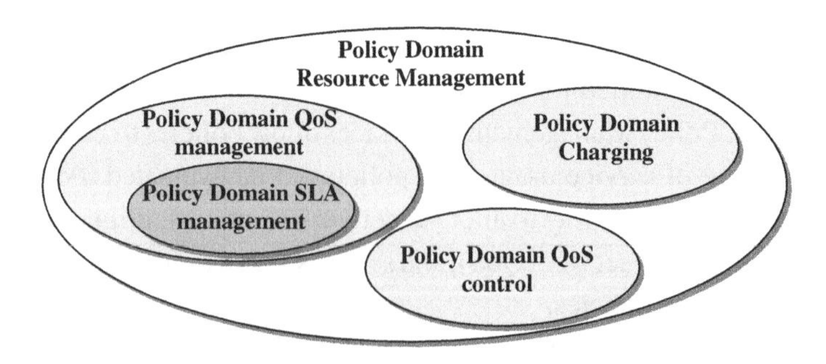

Fig. 17.22 Policy domains.

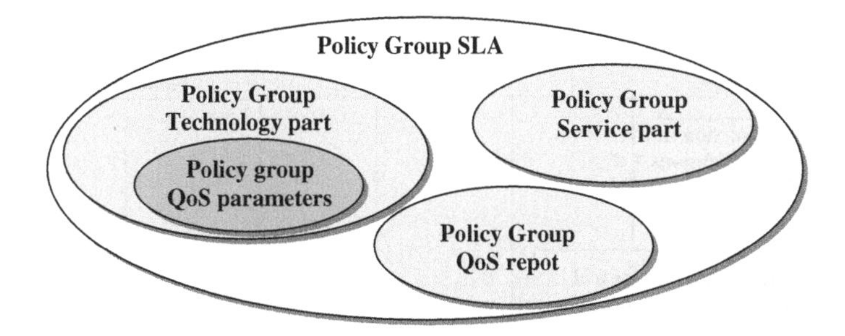

Fig. 17.23 Policy groups.

capabilities are shown in Figure 17.23. For example, the Policy Group SLA is used to define the policy based on the content of the Service Level Agreement (SLA) templates. It can nest Policy Groups for QoS parameters regarding the technology part of the SLA template.

The IpPolicyRepository interface represents a container for reusable policy-related information. The Policy Repository containing Policy Conditions and Policy Actions can be used in definition of one or more Policy Rules.

The IpPolicyRule represents the semantics of conditions and actions associated with a policy. A policy rule is define in a form of "if Condition then Action". The conditions and actions associated with a policy rule are modelled respectively with IpPolicyCondition and IpPolicyAction interfaces. A policy rule may also be associated with one or more policy time periods indicating when the policy rule is active or inactive. The policy time periods are modelled by the IpPolicyTimePeriodCondition interface. For example, a policy rule may define actions for traffic shaping in case the user has exceeded the negotiated in SLA maximum rate for more than an hour. IpPolicyExpressionCondition interface is used to generate a Policy condition regarded to a specified event. The IpPolicyExpressionAction interface is used to evaluate an expression.

The example shown in Figure 17.24 may be used to implement Service-based local policy control, which is a way of managing the access network through policies. It allows policy control over IP bearer resources.

As described in Section 17.2, it is important for an operator to correlate the QoS requested at the session layer (through session control signalling, such as SIP) with the actual QoS provided at the bearer level (initiation of QoS resources). The binding between the media components specified at the session layer and the corresponding QoS resources maintained at the edge

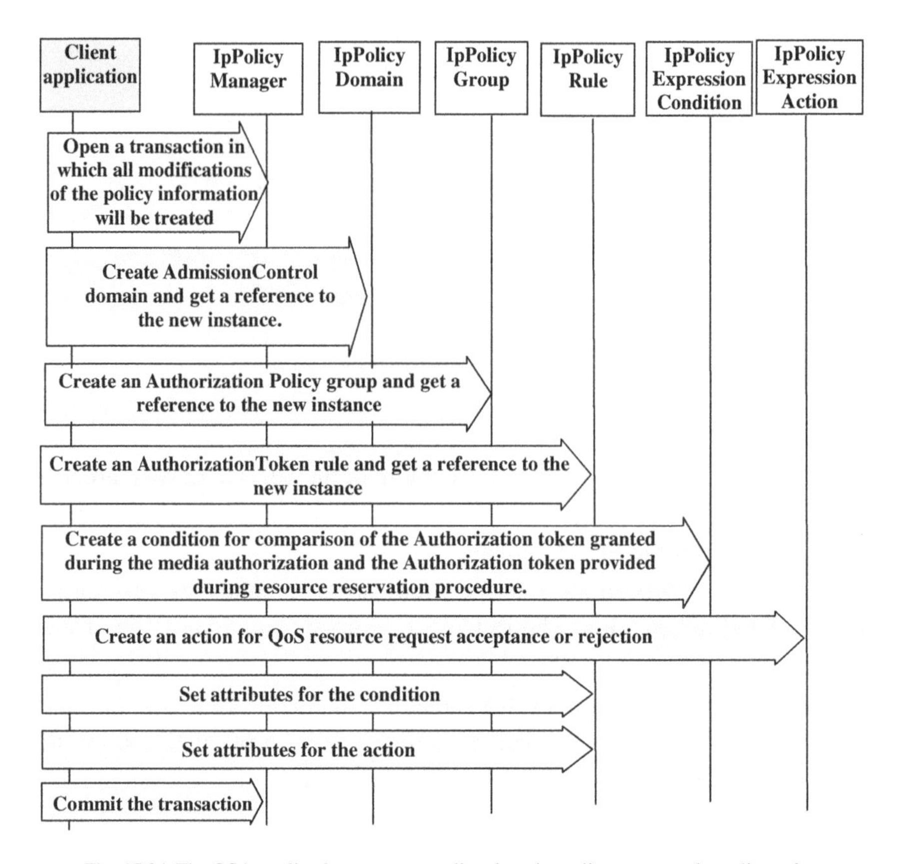

Fig. 17.24 The OSA application creates a policy domain, policy group and a policy rule.

router, is ensured by using an authorization token. One authorization token is assigned per IMS (SIP) session; each media component (e.g., video or audio) in a SIP session is identified by a sequence number. The token provided during the authorization of the media session and sent in the QoS resource reservation request is used as the mechanism to enable the edge router to contact the PCRF that generated the token. First, the PCRF verifies that the QoS resource activation request corresponds to an ongoing session. Second, it verifies that the requested bearer QoS corresponds to media resource information authorized by the P-CSCF.

Figure 17.24 shows the sequence in which the operator can define a policy domain regarded to Admission Control procedures, a policy group with policy rule for authorization of QoS resources.

The IpPolicyEventDefinition interface specifies the required and optional attributes of events that can be subscribed to, specified as conditions, and generated by clients or actions. A Policy Condition that is satisfied when the specified event with the matching attributes is generated, is modelled by the IpPolicyEventCondition interface. The IpPolicyEventAction interface represents specified events.

A client application can use the IpPolicyEvalManager interface to evaluate policy rules, to subscribe to and receive notifications of policy events and to generate events.

The IpAppPolicyDomain interface is supported by the client application and returns values.

Figure 17.25 shows how policy events are used. For example, a client application may define an event concerning unsuccessful resource reservation based on comparison of requested QoS resources and the authorized session QoS resources.

Figure 17.26 illustrates how a client subscribes to a policy event and receives notification when the event is triggered. Assuming that the policy

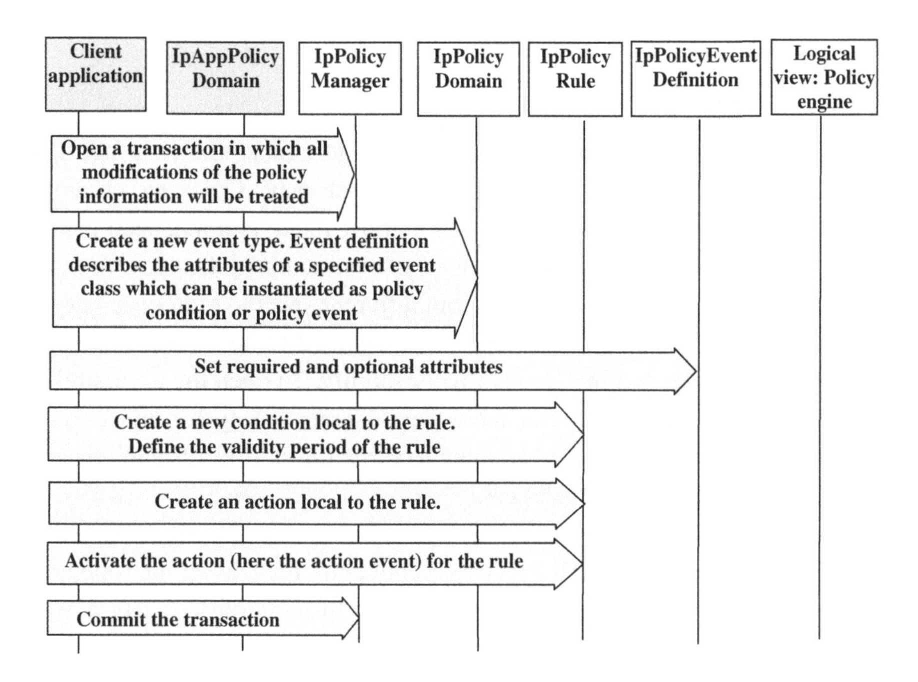

Fig. 17.25 The OSA application uses a template to define allowable events.

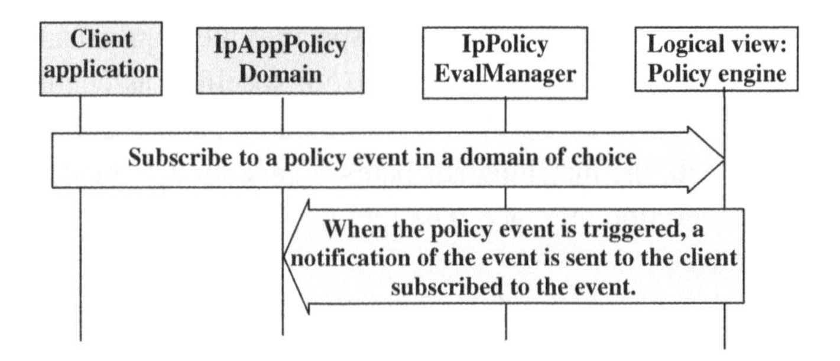

Fig. 17.26 The OSA application registers for and receives notification of a policy event.

event has been defined, the event is triggered when the action part of a rule fires. This may happen when for example the Authorization token provided during the QoS resource reservation procedure does not match to the Authorization token granted during the media authorization of session QoS resources causing the condition of a policy rule to be satisfied thus resulting in the action part to be executed.

The OSA "Policy Management" SCF is defined in [14].

17.5 Parlay X Web Services

Parlay X Overview

The Parlay/OSA APIs expose telecommunication functions in a network technology and programming language neutral way. Covering common programmability aspects of (converged) mobile, fixed, and managed packet networks, the APIs provide a medium level of abstraction of the network capabilities. They provide an abstraction from different specific protocols, but the abstraction level of Parlay/OSA APIs is not judged to be oriented to traditional IT-developers and this could affect usability. To open the accessibility of the network capabilities to a much wider audience Parlay X provides a set of high level interfaces that are oriented towards the skill levels and telecom knowledge levels of web developers.

The Parlay X is the name of the interface for accessing Parlay/OSA APIs using Web Services. The Web Services architecture realizes an interoperable network of services focused on service reuse. In addition to being a Web Services interface, the Parlay X interface is much simplified presentation of Parlay/OSA APIs.

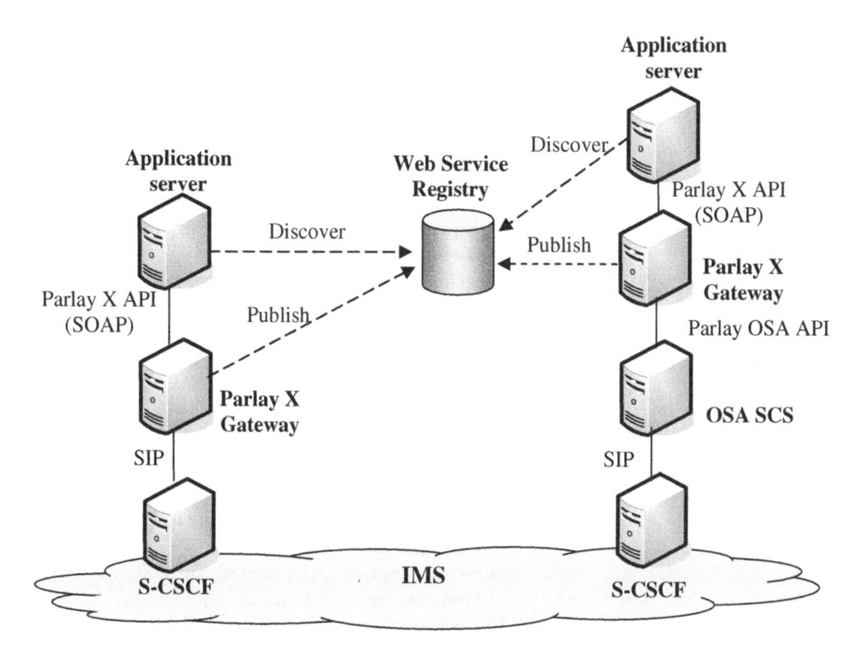

Fig. 17.27 Parlay X Web Services deployment in IMS.

Web Services operations and access are described by the Web Services Definition Language (WSDL). The Web Service provider publishes Web Services through a registry, making those Web Services available for discovery. Web Services applications use the Simple Object Access Protocol (SOAP) to exchange information.

Figure 17.27 shows how Parlay X Web Services may be deployed in IMS environment.

The Application server (AS) hosts Parlay X applications. The information published to the Web Services Registry provides applications with the connection information required to connect with the Parlay X Web Services Gateway. The applications access Web services through the Parlay X Gateway and the implementation behind the gateway is not visible for them. The Parlay X Gateway may be attached directly to the network element (S-CSCF) or to the OSA SCS. In the first case, the Parlay X Gateway plays a role of SIP AS and 'talks' SIP at the side of the network. In the other case, the Parlay X Gateway attaches to the OSA SCS through OSA interfaces.

The list of presently available Parlay X Web services is shown in Table 17.2.

Table 17.2. Parlay X Web Services.

Parlay X Web Service	Short description
Third party call	Call initiation from one caller to another by third party
Network initiated third party call	Call control on calls in the network by third party
Short message service	Submitting and receiving text messages
Multimedia message	Submitting and receiving multimedia messages
Payment	Initiation of payment sessions according to different charging methods
Account management	Checking account status, such as account balance, account credit expiration, etc.
Terminal status	Getting terminal status information
Terminal location	Getting location information
Call handling	Call control by third party
Audio call	Supporting a call with associated audio content
Multimedia conferencing	Creation of a multimedia conference and dynamic management of the participants involved
Address list management	Management of address groups and management of members within a group, supporting add, delete and query operations.
Presence	Getting presence information about one or more users and registering presence for the same
Message broadcast	Message sending to all the fixed or mobile terminals in a specified geographical area
Geocoding	Getting geographical coordinates at which a terminal is located
Application-driven quality of service (QoS)	Dynamically change the quality of service available on end user network connections by third party applications
Device capabilities and configuration	Getting information about device capabilities and pushing device configuration to a device
Multimedia streaming control	Support of streaming multimedia
Multimedia multicast session management	Third party control on multicast sessions, their members and multimedia stream, and obtaining channel presence information
Content management	Uploading content into the network (or a third party content provider) and consuming content from the network (or a third party content provider)
Policy	Policy provisioning and evaluation

Up to now, two Parlay X Web services that can be used for resource management purposes in IMS are defined:

- "Application-driven Quality of Service" Parlay X Web Service
- "Policy" Parlay X Web Service.

Application-driven Quality of Service Parlay X Web Service

Using the "Application Driven QoS" Parlay X Web service applications can dynamically change the QoS available on end user network connections.

Configurable service attributes are upstream bandwidth rate, downstream bandwidth rate and other QoS properties specified by the service provider. Changes in QoS may be applied on either for a defined period of time, or each time a user connects to the network. Application-driven QoS Web Service enables applications to register with the service for notifications about network events that affect QoS temporary configured on the end-user's connection. On such event occurrence the service notifies the applications.

The Application Driven QoS Web Service provides three interfaces.

The ApplicationQoS interface supports methods that allow the Application to apply a new QoS feature on end user connections, to modify active QoS features on end user connections or to release temporary QoS currently active on end used connection. Using the ApplicationQoS interface the application can retrieve the status of end user connections. The ApplicationQoSNotificationManager interface is used by the Application to manage the registration for notifications. The ApplicationQoSNotification interface provides method for notifying the Application about the impact of certain events on QoS features that were active on the end user connection where these events occurred.

Figure 17.28 shows an example of where the Application applies a temporary QoS feature to an end user connection, then modifies the active temporary QoS feature and at last removes the active QoS feature. This scenario might be applied for an end user with contracted 1 Mbps DSL service wishing to stream

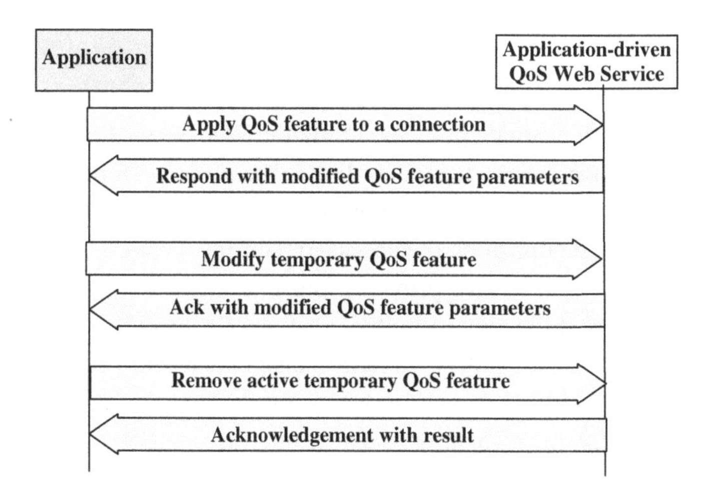

Fig. 17.28 The Parlay X application sets a temporary QoS feature to an end user connection.

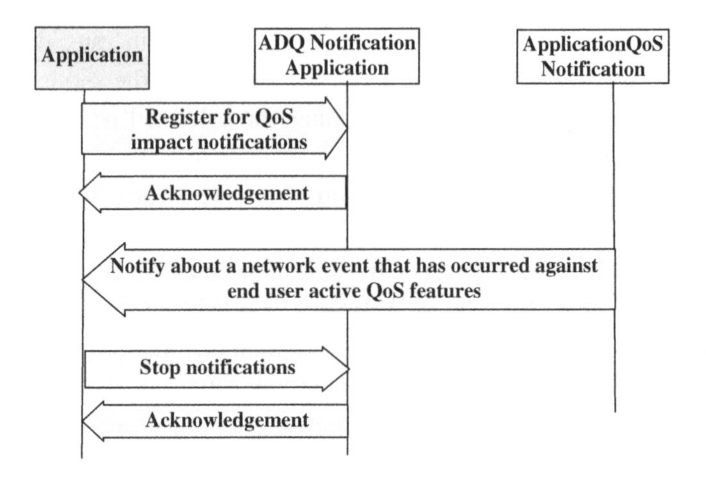

Fig. 17.29 The Parlay X application registers for QoS events and receives notifications.

a piece of video content. The video stream lasts two hours and a temporary bandwidth upgrade is required to support streaming.

Figure 17.29 shows an example of where the Application registers its interest in receiving notifications of specific event types in the context of given end user. The notifications report network events that have occurred against end user active QoS feature(s).

The “Application-driven Quality of Service” Parlay X Web Service is defined in [15].

Policy Parlay X Web Service

The “Policy” Web Service allows third party applications to manage policy information and to evaluate policies. Using the Web Service interfaces applications can personalize services according to their own preferences expressed as policies. On the other side, network operators and service providers can apply policy-based control on the access to their resources.

The “Policy” Web service provides four interfaces. The PolicyProvisioning interface is used for requesting the creation of a specified policy domain. It supports methods for policy domain management and policy rules management. The PolicyEvaluation interface is used to request an evaluation of a rule. The PolicyEventNotificationManager interface supports methods to subscribe for notifications about events and to request end of notifications. The PolicyEventNotification is used to deliver to the Applications with the event information when the monitored policy rule changes, e.g., when packet rates exceed.

Figure 17.30 shows an example, where an enterprise operator being a VPN client decides to cut down the expenses for the traffic generated by the employees over the weekends. To accomplish that the VPN client defines a policy allowing weekend packet rates twice lower than the ones over business days.

Figure 17.31 shows how to request the evaluation of a policy rule. The application creates first a signature in the domain before requesting the evaluation. The request for evaluation contains the name of signature.

The Policy Parlay X Web Service is defined in [16].

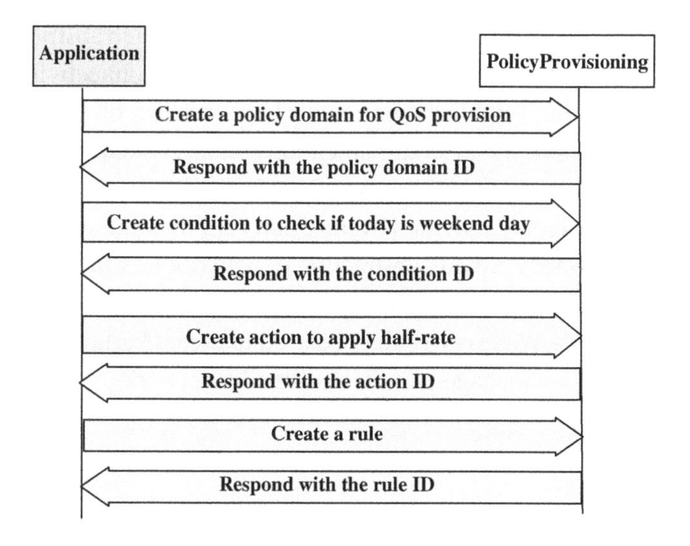

Fig. 17.30 The Parlay X application creates a policy domain and policy rule.

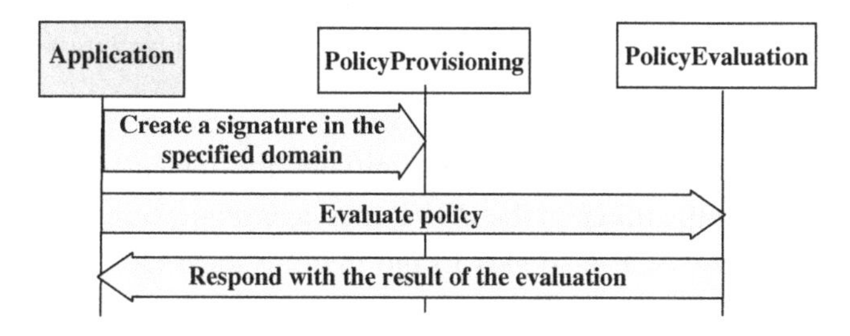

Fig. 17.31 The Parlay X application requests the evaluation of a policy rule.

17.6 Summary of Chapter

The purpose of resource management is to enable the telecommunication network to provide customers with the services they demand in a way that creates the greatest possible satisfaction and to enable operators to have these services provided at the lowest possible cost. Different mechanisms for resource management in IP-multimedia networks are defined. The access network and the transport network implement packet classification, scheduling and admission control. The operator uses mechanisms to assure that the negotiated QoS at SIP/SDP level is enforced at the access and transport network level. Resource reservation is realized in a coordinated manner with session establishment.

The opening network interfaces for third party software developers creates opportunities for provisioning of customer-oriented value-added services. The Open Service Access defined service architecture which abstracts applications from underlying network technology. The OSA Connectivity Manager interfaces enable the establishment, control and release of semi-permanent connections requested by telecommunications services such as virtual private networks. The OSA Policy Management interfaces allow network operators to host policy enabled service written by third party application service providers.

Parlay X interfaces provide abstraction of network functions by the use of Web Services. Application-driven Quality of Service Parlay X Web service and Policy Parlay X Web service may be used to create Parlay X applications for quality of service management in multimedia networks.

Problems

Problem 17.1

Let us assume that Bob wants to set up a multimedia session with Mary. Both Bob and Mary have contracted QoS with their service provider. Enlist the following high level steps in the procedure of media authorization in order of occurance.

> *A*: The Bob's user equipment initiates the resource reservation, including the token in the RSVP message requesting QoS.
> *B*: The P-CSCF responds to the inviting message including the authorization token. The token contains all the information needed to the Bob's user equipment to perform resource reservation.

C: The PCRF checks if the corresponding requested QoS parameters are within the limits of what was negotiated in the SDP exchange. The PCRF uses the token to find the stored negotiated media description.
D: The P-CSCF receives the inviting message, authenticates Bob and verifies that he is authorized to obtain QoS.
E: If the check is passed, the PRCF allows reservation and sends back a positive response to the PERF.
F: The PERF reserves the QoS resources and forwards the RSVP message.
G: The P-CSCF contacts the PCRF and obtains an authorization token. The authorization token is stored in the PCRF together with the negotiated session description.
H: Bob's user equipment sends an inviting SIP message with SDP offer to his P-CSCF. The SDP contains the description of the media that Bob desires to use for the communication the bandwidth requested.
I: On receiving the message, the PERF forwards the token together with the requested QoS parameters to the PCRF.

Problem 17.2

Find out the roles which the Application Server (AS) plays controlling the following services:

- The AS stores a black list containing addresses the user is allowed to contact. On receiving an inviting message the S-CSCF forwards the message to the AS which in turn allows or denies the session establishment.
- The AS hosts a "Wake up" application sending a reminding voice message. The AS sends an inviting message to MRFC via the S-CSCF requesting playing announcement to the user.
- The AS provides a call forwarding service. On receiving an inviting message the S-CSCF directs the message to the AS which in turn answers with a redirection response containing the currently valid address.

- The AS contains a conference application. The end user initiates a prearranged conference and communicates with the application via a web interface. By dragging and dropping names from the address book, the end-user adds parties to the conference.
- The AS provides an application monitoring a call between party A and party B in order to collect call information at the end of the call for charging and statistic purposes. The service includes number translation of the dialled number and special charging (e.g., a premium rate service).

Problem 17.3

Mark as true or false the following:

1. The Open Service Access (OSA) allows third parties others than the network operator and service provider to deploy services.
2. The interface between OSA Application Server and OSA Service Capability Server is SIP-based.
3. The third party applications need to deal with control protocol details in order to access network functions.
4. The OSA Application Programming Interfaces are underlying network technology agnostic.
5. The Connectivity Management is a set of functions that provide configuration and control of attributes such as QoS, security, and routing policy.
6. Policy Management includes policy specification, deployment, reasoning over policies, updating and maintaining policies, and enforcement.
7. OSA Connectivity Manager interfaces are more of real time control than an operation support nature.
8. OSA Policy Management interfaces can not be used for evaluation of the negotiated QoS.
9. OSA Policy Management interfaces can be used to add load balancing policy.

Problem 17.4

The OSA Connectivity Manager interfaces provide functions that are used to:

- retrieve QoS services offered by provider, stored in QoS templates

- set up a new VPrP
- retrieve information on VPrN and its virtual provisioned pipes

These functions are listed in the table below.

Retrieve QoS services offered by provider, stored in QoS templates	Set up a new VPrP	Retrieve information on VPrN and its virtual provisioned pipes
getQoSMenu (IpConnectivityManager)	createVPrP (IpVPrN)	getVPrN (IpEterpriseNetwork)
getTemplateList (IpQoSMenu)	deleteVPrP (IpVPrN)	getVPrPList(IpVPrN)
getTemplate (IpQoSMenu)	setSlaID (IpQoSTemplate)	getVPrP (IpVPrN)
getTemplateType (IpQoSTemplate)	setPipeQoSInfo (IpQoSTemplate)	getVPrPID (IpVPrP)
getDescription (IpQoSTemplate)	setValidityInfo (IpQoSTemplate)	getSlaID (IpVPrP)
getPipeQoSInfo (IpQoSTemplate)	setProvisionedQoSInfo (IpQoSTemplate)	getStatus (IpVPrP)
getValidityInfo (IpQoSTemplate)		getProvisionedQoSInfo (IpVPrP)
getProvisionedQoSInfo (IpQoSTemplate)		getPipeQoSInfo (IpVPrP)
getDsCodepoint (IpQoSTemplate)		getDsCodepoint (IpVPrP)

Go back to the example shown in Figure 17.17 where the VPN client selects service components and creates a new VPrP. Determine the functions used to perform the steps in the way illustrated for the first 3 steps.

1. Get a reference to QoS menu interface — getQoSMenu().
2. Get a list of template, each of which specifies a QoS service (Gold or Silver) — getTemplateList().
3. Get a reference to a specific template — getTemplate().
4. Get pipe QoS information consisting of load parameters, direction of the traffic, and the access point/site of a VPrP offered by the template.
5. Modify VPrP information regarding the flow direction, the load on the endpoint of the pipe.
6. Get the default values associated with this template (e.g., delay, default value, loss default value).
7. Set the requested values for the QoS parameters.
8. Get the default time period set by the VPN provider for the template.

9. Set the required time period for a new VPrP.
10. Create a new VPrP, which includes the pipe QoS information, the provisioned QoS information, the Service Level Agreement ID and the selected pairs of access point/sites.

Problem 17.5

The procedure described below gives an example of introduction of a condition and action into a rule using OSA Policy Management interfaces.

1. Open a transaction in which all modifications of the policy information will be treated.
2. Create a group object in the group by passing the name as parameter.
3. Commit the transaction.
4. Open a transaction in which all modifications of the policy information will be treated.
5. Create a condition as a part of the rule.
6. Commit the transaction.
7. Get a reference to the repository from the Policy manager. The repository will be used to retrieve an action that has to be reused in the rule.
8. Get a reference to the specified action.
9. Open a transaction in which all modifications of the policy information will be treated.
10. Set the action list for this rule.
11. Set the condition list for this rule.
12. Commit the transaction.

Sketch a sequence diagram including the steps described. Fill the interfaces involved in each step.

Problem 17.6

The "Application-driven Quality of Service" Parlay X Web Service can be used for example by an end user of WiMAX service requesting permanently to upgrade from the existing service offering (e.g., 1 Mbps) to a higher bandwidth service (e.g., 5 Mbps). The applyQoSFeature method of the ApplicationQoS interfaces is used by the Application to request either permanent or a

Table 17.3. Message applyQoSFeatureRequest parameters.

Part name	Part type	Description
endUserIdentifier	anyURI	Identifies the end user in the network
qosFeatureIdentifier	string	Identifies the QoS feature that is applied to the end user connection — e.g., "Gold724"
defaultQoSFeature	boolean	Specifies whether the requested QoS Feature is to be applied as the default QoS Feature on an end users connection.
modifyExistingSession	boolean	Only required where defaultQoSFeature = TRUE. Specifies whether the default QoS feature should be applied to the ongoing connection of the end user or to the subsequent connection when the end user next comes online.
qosFeatureProperties	QoSFeatureProperties	Not required for a default QoS Feature request. Optional for a temporary QoS Feature request. Specifics values for the configurable service attributes that govern the temporary QoS feature (e.g., Duration).

temporary QoS feature to be set on the end user connection. Table 17.3 shows the parameters of the applyQoSFeatureRequest.

How the applyQoSFeatureRequest message parameters have to be used for requesting permanent upgrade?

Problem 17.7

Across

1. The opening of network interfaces allows third party .... to access network functions through application programming interfaces.
2. The application can register its interest in receiving of notifications for specific ....types.
3. .... is an ordered combination of policy rules that define how to administer, manage, and control access to resources.
4. The service level, quality of service, priorities and duties negotiated between the service provider and the client are defined in ....
5. After successful QoS resource authorization the user equipment initiates ....reservation.

6. .... is service architecture for delivering value-added services in third generation mobile networks.
7. Quality of .... is "the collective effect of service performance which determines the degree of satisfaction of a user of the service".

Down

8. The OSA APIs are underlying .... independent.
9. The OSA APIs are agnostic from network control ....
10. Using the OSA Connectivity management interfaces the Application can set up a new virtual provisioned ....
11. The concept of .... is to open network interfaces for third party software developers.
12. An Application Programming .... is defined as a set of object types (classes) with methods that can be called on the object and their parameters.
13. Session Description Protocol is used to describe the desired multimedia .... characteristics.

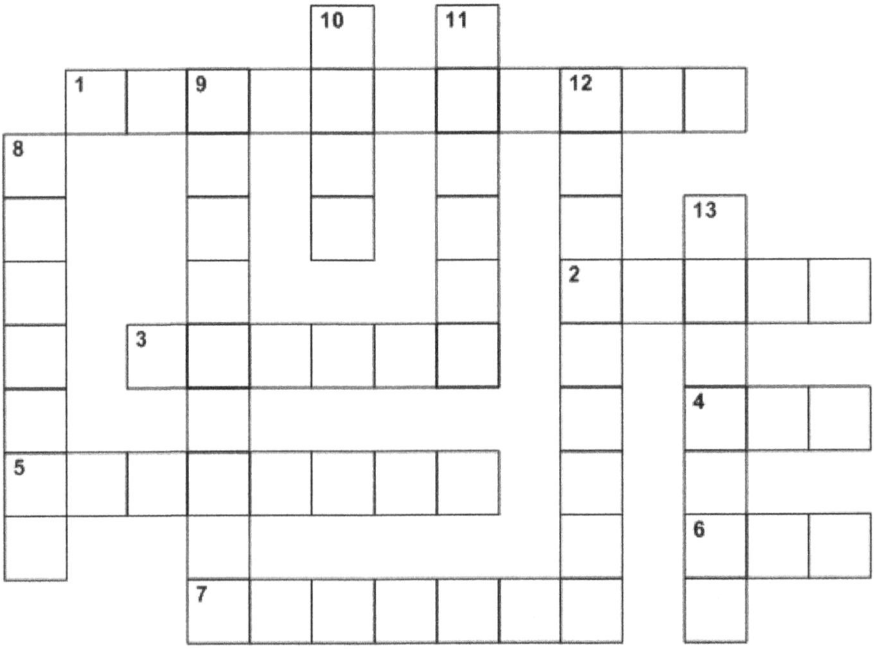

References

[1] Miikka Poikselka, et al., The IMS Multimedia Concepts and Services, Wiley, 2008.
[2] Rogelio Perea, Internet Multimedia Communications Using SIP, A modern approach including Java practice, Elsevier, 2008.
[3] 3GPP TS 23.002 Network architecture.
[4] 3GPP TS 23.203 v8.4.0, Policy and charging control architecture.
[5] Alan Johnson, SIP: Understanding the session Initiation Protocol, Artech House, 2004.
[6] 3GPP TS 23.218 V8.2.0, IP Multimedia (IM) session handling, IM call model.
[7] ITU-T Y.1291, An architectural framework for support of Quality of Service in packet networks.
[8] ETSI TS 183 017 v1.1.1, Resource and Admission Control: DIAMETER protocol for session based policy set-up information exchange between the Application Function (AF) and the Service Policy Decision Function (SPDF); Protocol specification.
[9] 3GPP TS 23.107 v8.0.0, Quality of Service (QoS) concept and Architecture.
[10] 3GPP TS 24.229 v5.21.0 IP Multimedia Call Control Protocol based on Session Initiation Protocol (SIP) and Session Description Protocol (SDP).
[11] IETF RFC 2205 Braden (R.), *et al.*: Resource Reservation Protocol (RSVP) — Version 1, Functional Specification.
[12] Hu Hanrahan, Network Convergence, Services, Applications, Transport and Operations Support, Wiley, 2008.
[13] 3GPP TS 23.198-10 v8.0.0, Open Service Access (OSA); Application Programming Interface (API); Part 10: Connectivity Manager Service Capability Feature (SCF).
[14] 3GPP TS 29.198-13 v7.0.0, Open Service Access (OSA); Application Programming Interface (API); Part 13: Policy Management Service Capability Feature (SCF).
[15] 3GPP TS 29.199-17 v8.0.0, Open Service Access (OSA); Parlay X Web Services; Part 17: Application-driven Quality of Service (QoS)
[16] 3GPP TS 29.199-22 v8.0.0, Open Service Access (OSA); Parlay X Web Services; Part 22: Policy
[17] Muslim Elkotob (2008), Autonomic Resource Management in IEEE 802.11 Open Access networks, http://epubl.ltu.se/1402-1757/ 2008/38/LTU-LIC-0838-SE.pdf
[18] Research Project DO02-135/2008, Research on Cross Layer Optimization of Telecommunication Resource Allocation, Bulgarian Ministry of Education and Science.

Index

802.16, 41
802.16e, 39

Access Service Network (ASN), 36
Access service network gateway, 45
Access service network gateway (ASN-GW), 45
Acquisition, 448
acquisition, 360
ad-hoc on demand distance vector protocol (AODV), 216
adaptive modulation, 47
adaptive modulation and coding (AMC), 47
additive white Gaussian noise (AWGN), 110
AF class (Assured Forwarding), 101
AMC, 11
application for cell phones, 148
Application for Portable Systems, 148
application programming interface, 559
application programming interfaces, 592
application server, 395
Application-driven Quality of Service Parlay X Web Service, 555
architecture, 33
art gallery problem, 310
ATM, 35
Authentication, Authorization and Accounting (AAA), 38
auto-correlation, 456
Automated meter reading, 158
automatic repeat request (ARQ), 502

backhaul, 99
bandwidth, 35
BE class (Best Effort), 101
beamforming, 44
Beyond 3G (B3G), 1
Bit Error Rate, 59
blocking probability, 84
body loss, 107
Bose-Chaudhuri-Hocquenghem (BCH) codes, 506
box muller, 460
BPSK, 41

cable replacement, 139
call session control functions, 558
camera placement, 311
carrier frequency, 48
CDMA, 3
cell, 2
cell breathing, 2
cell range, 3
cell splitting, 62
channel, 9
channel state information (CSI), 44
channelization, 42
chaotic sequences, 447
Characteristics of Frequency Bands, 139
circular convolution, 54
clock drift, 447
cluster, 50
coexistence, 165

cognitive radios, 173
coherence bandwidth, 69
connectivity, 37
connectivity management, 571
control quality of service mechanisms, 559
convolutional coding, 413
cooperative game, 277
cooperative relaying, 265
cooperative sensing, 252
cooperative services, 267
coordination, 230
Core Service Network (CSN), 36
correlation, 64
COST 231 model, 121
Coverage, 1
cross over distance, 489
cross-layer optimization, 271
cyclic prefix, 40

Data Regions, 47
DC carrier, 48
delay spread, 41
Digital Video Broadcast-Handheld (DVB-H), 46
dimensioning, 1
dipole, 107
directional sensors, 306
Discrete Fourier Transform (DFT), 53
DL Fully Used Sub-channelization (FUSC), 50
DL optional FUSC (DL OFUSC), 51
DL PUSC, 50
DL-MAP, 42
Doppler, 48
Downlink, 2
downlink frame prefix (DLFP), 42
dynamic frequency selection, 225
Dynamic source routing protocol (DSR), 216

EF class (Expedited Forwarding), 101
Effective isotropic radiated power (EIRP), 107
effective radiated power (ERP), 107
Erceg Model, 126
error function, 9

fade margin, 107
fading, 3
false alarm, 251
fast Fourier transform (FFT), 53
Field of View, 308
forward error correction (FEC), 42
frame, 39
free space propagation, 116
frequency agility, 225
frequency bands, 139
Frequency Division Duplex (FDD), 35
frequency division multiple access (FDMA), 419
frequency planning, 85
frequency reuse, 39
Friis formula, 118

game theory, 265
Gateway, 37
Gaussian noise, 42
Gaussian random variable, 127
Global network interface, 138
global positioning system (GPS), 40
GPRS, 46
grade of service (GoS), 86
graph-constrained — shortest route, 344
guard time, 40

handoff, 38
Handoff Technique, 141
Hata-Okamura model, 120
heterogeneous, 141
HSPA, 1

ICMP, 540
IEEE 802.16h, 240
IEEE 802.19, 241
IEEE802.11 y/k, 243
implementation margin, 110
IMS architecture, 556
IMS control plane, 561
IMS user plane, 556
integer linear programming, 313
inter-modulation noise, 110
interference, 3
interleaving, 42

Internet Protocol Multimedia Subsystem (IMS), 555
intersymbol interference, 73
intersymbol interference (ISI), 41
inverse discrete Fourier transform (IDFT), 55
inverse fast fourier transform (IFFT), 414
IP address assignment, 38
isotropic sensors, 312
ITU-T, 86

license and unlicensed, 229
line of sight (LOS), 112
linear optimization, 276
link budget, 2
local minimum shortest path, 488
location tracking system, 151
low cost handheld messaging system, 152
LTE, 33

MAC, 18
macro cooperation, 267
MAN-OFDM, 34
MAN-SC, 34
MAN-Sca, 34
mapper0, 337
Markov chain, 518
Markov model, 518
MCSP, 307
mesh access point, 481
mesh router, 208
message passing system for PAN, 154
micro cooperation, 267
minimum cost sensor placement, 307
minimum sensor placement, 307
mobile stations (MS), 45
mobility, 33
mobility management, 37
mobility management in PAN, 140
mobility models, 331
mobility prediction, 265
MSP, 307
multi-protocol networks, 171
MultiBand Orthogonal Frequency Division Multiplexing (MB-OFDM) UWB, 157
multicarrier modulation, 69
multipath, 47
multiple access interference (MAI), 110
multiple input multiple output, 260
multiple input, single output (MISO), 61
multisphere reference model, 162
multiuser diversity, 58

network lifetime, 314
network service provider (NSP), 37
next generation networks, 169
noise figure, 109
noise rise, 3
noise-like sequences, 447
non-line of sight (NLOS), 130

open service access (OSA), 555
open spectrum, 174
optimal sensor configuration, 305
orthogonal frequency division multiple access (OFDMA), 242
orthogonal frequency division multiplexing (OFDM), 409
orthogonality, 6
orthogonality interval, 74
OSA connectivity manager API, 572
OSA policy management API, 572
OSC, 307
over subscription ratio (OSR), 92
oversampling, 49

packet analyzers, 145
PAN standards, 139
parlay, 555
Parlay X, 555
path loss, 3
PDU, 193
performance, 15
permutation zones, 49
phase, 2
PHY, 39
Physical slot, 41

pilot carrier, 50
policy and charging enforcement function (PCEF), 557
policy and charging rule function (PCRF), 557
policy evaluation, 577
policy management, 46
policy parlay X Web service, 588
policy, policy domain, policy rule, 582
power control, 2
preamble, 41
probability of detection, 250
probability of failure, 452
proxy call session control function (P-CSCF), 558
Pseudo random binary sequences (PRBS), 447
PUSC, 44

QAM, 43
QoS, 3
QPSK, 42
quality of service (QoS), 35

radio resource management, 37
randomization, 42
Rayleigh probability density function (pdf), 517
receiver operating characteristics (ROC), 454
regulatory issues, 156
resource authorization, 561
resource modification, 561
resource release, 563
resource reservation, 265
Rician distribution, 517
RTP, 505

sampling factor, 68
Sampling frequency, 41
sectoring, 82
sectors, 47
security issues in PAN, 143
security & privacy issues, 163
self healing, 204
self-symbol-interference (SSI), 74
sensing range, 305
sensing region, 305
sensor based pollution monitoring system, 159
sensor network monitor, 149
sensor networks, 158
sensor placement, 305
service capability features (SCF), 569
service capability server (SCS), 559
service triggering, 566
serving call session control function (S-CSCF), 558
session initiation protocol, 558
shadowing, 1
single input single output (SISO), 50
slot, 41
slot allocation, 47
smart antennas, 34
snoop, 146
software define radio, 224
spatial division multiplexing, 43
spectrum efficiency, 58
spectrum management, 174
spectrum mobility, 174
spectrum sensing, 173
spectrum sharing, 174
spreading sequences, 447
Stanford University Interim (USI) model, 116
state transition, 519
STC, 44
sub-carrier, 47
sub-channelization, 42
subcarrier spacing, 40
symbol, 40
symbol time, 40
synchronization, 39

TCP/IP, 171
tcpdump, 146
test tools and network management applications, 147
throughput, 1
time division duplex (TDD), 35
timing, 41
timing jitter, 466
topology control, 481

topology dissemination based on reverse-path forward (TPRPF), 216
total usage of sub-channels (TUSC1 and TUSC2), 51
tracking, 11
traffic engineering, 85
transmit time interval (TTI), 9

UDP, 177
UL-MAP, 42
UMTS, 1
uplink, 2
UWB, 130

Virtual Provisioned Network (VPrN), 573
virtual provisioned pipes (VPrP), 573
voice over IP (VOIP), 33

Walfisch-Ikegami models, 120
Web Services, 555
WiBRO, 40
WiMAX, 33
wireless sensor networks, 166
wireless mesh networks, 201
WPAN standards, 164

zero order hold, 470
zone avoiding — border route, 343
zone avoiding — shortest route, 343

RIVER PUBLISHERS SERIES IN COMMUNICATIONS

Other books in this series:

Volume 1
4G Mobile & Wireless Communications Technologies
Sofoklis Kyriazakos, Ioannis Soldatos, and George Karetsos
September 2008
ISBN: 978-87-92329-02-8

Volume 2
Advances in Broadband Communication and Networks
Johnson I. Agbinya, Oya Sevimli, Sara All, Selvakennedy Selvadurai, Adel Al-Jumaily, Yonghui Li, and Sam Reisenfeld
October 2008
ISBN: 978-87-92329-00-4

Volume 3
Aerospace Technologies and Applications for Dual Use A New World of Defense and Commercial in 21st Century Security
General Pietro Finocchio, Ramjee Prasad, and Marina Ruggieri
November 2008
ISBN: 978-87-92329-04-2

Volume 4
Ultra Wideband Demystified Technologies, Applications, and System Design Considerations
Sunil Jogi and Manoj Choudhary
January 2009
ISBN: 978-87-92329-14-1

Volume 5
Single- and Multi-Carrier MIMO Transmission for Broadband Wireless Systems
Ramjee Prasad, Muhammad Imadur Rahman, Suvra Sekhar Das, and Nicola Marchetti
April 2009
ISBN: 978-87-92329-06-6

Volume 6
Principles of Communications: A First Course in Communications
Kwang-Cheng Chen
June 2009
ISBN: 978-87-92329-10-3

Volume 7
Link Adaptation for Relay-Based Cellular Networks
Başak Can
November 2009
ISBN: 978-87-92329-30-1

Lightning Source UK Ltd.
Milton Keynes UK
UKOW06f1836120415

249521UK00001B/1/P

9 788792 329240